COMPANION TO PUBLIC SPACE

The *Companion to Public Space* draws together an outstanding multidisciplinary collection of specially commissioned chapters that offer the most current and relevant intellectual discourse, scholarship, and research regarding public space.

Thematically, the volume crosses disciplinary boundaries and traverses territories to address the philosophical, political, legal, planning, design, and management issues in the social construction of public space. The *Companion* uniquely assembles important voices from diverse fields of philosophy, political science, geography, anthropology, sociology, urban design and planning, architecture, art, and many more, under one cover. It addresses the complete ecology of the topic to expose the interrelated issues, challenges, and opportunities of public space in the twenty-first century.

The book is primarily intended for scholars and graduate students for whom it will provide an invaluable and up-to-date guide to current thinking across the range of disciplines that converge in the study of public space. The *Companion* will also be of use to practitioners and public officials who deal with the planning, design, and management of public spaces.

Vikas Mehta is an Associate Professor of urbanism at the School of Planning at the University of Cincinnati. His research explores the various dimensions of urbanity through the exploration of place as a social and ecological setting and as a sensorial art. Mehta is the editor of *Public Space*, an anthology of 98 chapters; author of *The Street: A Quintessential Social Public Space*, which received the 2014 Book Award from the Environmental Design Research Association; and with Matthew Frederick, *101 Things I Learned in Urban Design School*.

Danilo Palazzo is a Professor and Director of the School of Planning at the University of Cincinnati. Prior to moving to Cincinnati, Palazzo was on the faculty at the Polytechnic University of Milan. His major fields of interest are urbanism, urban design, and planning. Palazzo, with Frederick Steiner, has authored *Urban Ecological Design: A Process for Regenerative Places* (translated in Chinese in 2018), as well as contributing a chapter on "Pedagogical Tradition" to the *Companion to Urban Design*.

COMPANION TO PUBLIC SPACE

Edited by Vikas Mehta and Danilo Palazzo

Routledge
Taylor & Francis Group

LONDON AND NEW YORK

First published 2020
by Routledge
4 Park Square, Milton Park, Abingdon, Oxon OX14 4RN
605 Third Avenue, New York, NY 10017

First issued in paperback 2023

Routledge is an imprint of the Taylor & Francis Group, an informa business

British Library Cataloguing-in-Publication Data
A catalogue record for this book is available from the British Library

Library of Congress Cataloging-in-Publication Data
Names: Mehta, Vikas, 1966- editor. | Palazzo, Danilo, 1962- editor.
Title: Companion to public space / Vikas Mehta and Danilo Palazzo.
Description: New York : Routledge, 2020. | Includes bibliographical references and index.
Identifiers: LCCN 2019059292 (print) | LCCN 2019059293 (ebook)
Subjects: LCSH: Public spaces–Social aspects. | City planning. | Historic districts.
Classification: LCC HT185 .C66 2020 (print) | LCC HT185 (ebook) | DDC 307.1/216–dc23
LC record available at https://lccn.loc.gov/2019059292
LC ebook record available at https://lccn.loc.gov/2019059293

ISBN: 978-1-03-257031-0 (pbk)
ISBN: 978-1-138-54972-2 (hbk)
ISBN: 978-1-351-00218-9 (ebk)

DOI: 10.4324/9781351002189

Typeset in Bembo
by Swales & Willis, Exeter, Devon, UK

Publisher's Note
The publisher has gone to great lengths to ensure the quality of this reprint but points out that some
imperfections in the original copies may be apparent.

CONTENTS

List of figures   *x*
List of tables   *xvi*
List of contributors   *xvii*
Acknowledgments   *xxv*

Introduction   1
Vikas Mehta and Danilo Palazzo

PART 1
Perspectives   **5**

  1 A critique of public space: between interaction and attraction   7
     Ali Madanipour

  2 Researching public space: from place-based to process-oriented approaches
     and methods   16
     Rianne van Melik and Bas Spierings

  3 The collective outdoors: memories, desires and becoming local in an era
     of mobility   27
     Clare Rishbeth

  4 Appropriation of public space: a dialectical approach in designing publicness   35
     Elahe Karimnia and Tigran Haas

  5 Landshape urbanism: the topography of public space   46
     Karl Kullmann

6 Social justice as a framework for evaluating public space   59
   Setha Low

PART 2
Influences   **71**

7 Planetary public space: scale, context, and politics   73
   Jason Luger and Loretta Lees

8 Public space and the New Urban Agenda: fostering a human-centered
   approach for the future of our cities   85
   Luisa Bravo

9 Safety in public spaces from a place-oriented public health perspective 94
   Vania Ceccato

10 Inclusiveness and exclusion in public open spaces for visually
   impaired persons   106
   Kin Wai Michael Siu, Yi Lin Wong and Jiao Xin Xiao

11 Public space, austerity, and innovation (the London case)   121
   Matthew Carmona

12 Visibility in public space and socially inclusive cities: a new conceptual
   tool for urban design and planning   137
   Ceren Sezer

13 (Post?)-colonial parks: urban public space in Trinidad and Zanzibar 152
   Garth Myers

14 Global homogenization of public space? A comparison of "Western" and
   "Eastern" contexts   165
   Tigor W. S. Panjaitan, Dorina Pojani, and Sébastien Darchen

15 Right to the city (at night): spectacle and surveillance in public space 182
   Su-Jan Yeo

16 The strange idea of the public: no, *hiroba* (広場) is not public space;
   so, what?!   191
   Darko Radović

PART 3
Types 205

17 Types: descriptive and analytic tools in public space research 209
    Karen A. Franck and Te-Sheng Huang

18 Public space use: a classification 221
    Sverre Bjerkeset and Jonny Aspen

19 Mapping the publicness of public space: an access/control typology 234
    Kim Dovey and Elek Pafka

20 Private, hybrid, and public spaces: urban design assessment, comparisons,
    and recommendations 249
    Els Leclercq and Dorina Pojani

21 Throwntogether spaces: disassembling 'urban beaches' 267
    Quentin Stevens

22 Designing parks for older adults 282
    Anastasia Loukaitou-Sideris

23 Expanding common ground 296
    Ken Greenberg

24 The skyscraper and public space: an uneasy history and the capacity
    for radical reinvention 309
    Vuk Radović

25 In pursuit of inclusive spaces? Memory, monuments, and the politics
    of public story telling 320
    Renia Ehrenfeucht

PART 4
Actions 333

26 Public space as a space of resistance and democratic resilience 335
    Jeffrey Hou

27 Public space and the political: reconnecting urban resistance and
    urban emancipation 346
    Sabine Knierbein

28 Alternating narratives: the dynamic between public spaces, protests, and
   meanings   358
   Tali Hatuka

29 Publics, pavements, and public-facing art in post-colonial urban Africa   366
   Rike Sitas

30 Bridging and bonding: public space and immigrant integration in
   Barcelona's *el Raval*   378
   Jeremy Németh

31 Public space challenges and possibilities in Latin America: the city's
   socio-political dimensions through the lens of everyday life   390
   Andrei M. Z. Crestani and Clara Irazábal

32 Typologies of the temporary: constructing public space   399
   Mona El Khafif

33 Bringing public spaces to life: the animation of public space   414
   Troy D. Glover

PART 5
Futures   **427**

34 Public space and the terrorism of time   429
   Mark Kingwell

35 Renewing the public trust doctrine: a solidarist account of public space   438
   Margaret Kohn

36 Events on urban parkland: scrutinizing public–private partnerships in parks
   governance regimes   447
   Susanna F. Schaller and Elizabeth Nisbet

37 The private lives of public spaces   457
   Michael W. Mehaffy and Peter Elmlund

38 Using big data to support public space research   467
   Avigail Vantu and Kristen Day

39 Digital technologies and public space in contemporary China   478
   Tim Jachna

40 What if? Forecasting and composing public spaces   486
    Mark C. Childs

41 The idea of the urban commons: challenges of enclosure, encroachment,
    and exclusion   499
    Tridib Banerjee

Epilogue   *513*
Index   *515*

FIGURES

2.1	Here and there	18
2.2	Then and now	20
2.3a	Quiet shortcut in Utrecht	23
2.3b	Official cycle lane in Utrecht	23
5.1	Contour map contrasting the street grid and underlying topography of San Francisco's Russian Hill and Telegraph Hill districts	47
5.2	Constructed postglacial wilderness creates an explorative retreat from the city in Central Park, New York	49
5.3	Variable playing surface encourages a site-specific version of basketball at Volkspark, Potsdam	51
5.4	Shallow concaved form enhances the square's potency as a central political gathering space on Federation Square, Melbourne	52
5.5	Convex and concaved shaped artificial topography at Gasworks Park, Seattle	53
5.6	Horseshoe shaped mound creates a space framed by a topographic threshold on the Esplanade, Fremantle	55
5.7	Deployment of diminutive level changes on Marlene-Dietrich-Platz, Berlin	56
5.8	Cobblestones amplifying a previously worn desire path at Park Rabet, East Leipzig	57
7.1	Speakers' Corner/Hong Lim Park: Singapore hybrid public space?	78
9.1	Term maps of the research field center around, "public place" while, with center in "crime," publications linking crime and fear to public health and built environment, 1968–2018, density visualization map of keywords in literature from Scopus only with at least five repetitions of key words	97
9.2	Hotspots of falls among older adult pedestrians in Stockholm mapped by street segment	101
10.1	Piled rental bikes blocking the street	111
10.2	Environment in a sitting-out area in Beijing	112

10.3 Seating area in a small public open space surrounded by buildings in Taipei 112

10.4 Entrance to a seating area in Taipei 113

10.5 Tactile guided path in front of staircases in a public open space in Hong Kong 114

10.6 Environment in a public open space in Hong Kong 114

10.7 Policy-implementation-management model 117

10.8 One of the possible adjustments in the policy-implementation-management model for VIPs 118

11.1 Windrush Square, Brixton: Boris Johnson also benefitted hugely from the legacy of work on public spaces conducted during the Livingstone era, as, like the gift that keeps on giving, public space projects continued to mature throughout the austerity years 123

11.2 For example, the new 'public space' – Sky Garden – on top of the Walkie Talkie tower negotiated as part of the planning permission 124

11.3 Exploratory cycleway layouts in London's Bloomsbury, implemented at minimal cost prior to more permanent investments being made 127

11.4 The Queen Elizabeth Olympic Park, Stratford – post-games 128

11.5 The rise in rough sleepers represents a direct and highly visible impact of austerity 130

11.6 The Line art walk, from the O2 to the Queen Elizabeth Olympic Park, Damien Hirst's *Sensation* (2003) as unintended plaything 131

11.7 Barriers installed following the London terror attacks of 2017 132

12.1 The streets with the clusters of Turkish amenities in Amsterdam and the location of the case streets, Burgemeester de Vlugtlaan and Javastraat 143

12.2 Neighborhood level analysis of Turkish amenities in terms of their personalization and legibility at street level 144

12.3 Turkish amenities in Javastraat in 2007 145

12.4 User intensity and user activities in Javastraat, 2007 146

12.5 Turkish amenities in Burgemeester de Vlugtlaan in 2007 147

12.6 User intensity and user activities in Burgemeester de Vlugtlaan, 2007 148

13.1 Royal Botanical Garden, Kapok Valley, Port of Spain 153

13.2 Queen's Park Savannah, Port of Spain 154

13.3 Migombani Botanical Garden, Zanzibar 156

13.4 Mnazimmoja Park, Eid celebrations, Zanzibar 156

13.5 Laventille Hills, Port of Spain 159

13.6 Fondes Amandes Community Re-Forestation Project, St. Anns 160

13.7 Mnazimmoja Park looking out toward Peace Memorial, Vuga, Zanzibar 161

13.8 Kapok tree, Migombani Botanical Garden, Zanzibar 162

14.1 Street barber in Bangkok 170

14.2 Street food in Kuala Lumpur 171

14.3a Surabaya street is crowded by vehicles during the daytime 172

14.3b The street becomes a culinary center at night 173

14.4 Patuxai Park in Vientiane 174

14.5 Traffic congestion and restricted pedestrian space in Manila 175

14.6   Neglected informal settlement in Yangon   176
14.7   Open concept mall in Bandung   177
14.8   Singapore's Chinatown   178
16.1   Japanese roji in Nezu, Tokyo: spatial expression of Japanese residential culture   193
16.2   Jiyūgaoka *Hiroba*: on normal day, when occupied by vehicles, and pedestrianized, with preparations for strictly choreographed and controlled events   198
16.3   Central Kuhonbutsugawa Ryokudō: during the cherry blossom season, at its aesthetic best, and full of people, even during a rare snowy day in Tokyo   200
16.4   Central Kuhonbutsugawa Ryokudō: the view from former café LaManda, one of early incubators of authentic, yet cosmopolitan urban flavour   201
16.5   Central Jiyūgaoka and its fake gondola, in an appropriately fake canal, next to the fake piazza, in a block-sized, scaled-down "Venice"   202
17.1   Analysis of 24 Interior Privately Owned Public Spaces in New York City   217
18.1   Aerial photo of the Tjuvholmen and Aker Brygge neighborhoods, 2014   227
18.2   Bryggetorget ('Harbour Square'), Aker Brygge neighborhood   228
18.3   Holmens gate ('Holmen's Street'), Aker Brygge neighborhood   229
19.1   Maps of Rome by Giambattista Nolli (1748) and Naples by Giovanni Carafa (1775) Scale: 400 × 400 meters   238
19.2   Six Types of Publicness – an Access/Control Typology   239
19.3   Three Melbourne Sites   241
19.4   Central Melbourne   243
19.5   Victoria Harbour   244
19.6   Waterfront City/The District   245
19.7   Accessible versus Restricted Space   246
20.1   Case study locations in Liverpool   251
20.2a Case studies after redevelopment: Liverpool ONE   252
20.2b Case studies after redevelopment: the Ropewalks   253
20.2c Case studies after redevelopment: Granby4Streets   254
20.3   Assessed street sections in each case study area   257
20.4   Elevated walkways at South John St (Liverpool ONE) provide enclosure   259
20.5   Boarded up residential properties on Cairns St (Granby4Streets)   261
20.6   Wood St is too narrow to accommodate any trees (the Ropewalks)   261
20.7   Private security guards ('red coats') at Liverpool ONE   263
20.8   Flower pots and picnic tables have been contributed by residents along a sidewalk at Granby4Streets   264
21.1   Typical city beaches in Germany, with portable pools, decking, deck chairs, potted palm trees, and surfboard props. Strandsalon, Lübeck and La Playa, Leipzig   268
21.2   Skybeach, Stuttgart, on the roof of the parking garage of the Galeria Kaufhof department store, adjacent to the city's main pedestrian axis, Königstraße   270

21.3   Temporary and relocatable buildings, play equipment, awnings, south sea umbrellas, palm trees, and a ship create a beach atmosphere. Bundespressestrand, Berlin   271

21.4   Demonstration by local gymnastics club. Strandleben, Vaihingen an der Enz   272

21.5   Public Viewing of World Cup football matches. Strandbar, Magdeburg   275

21.6   South Pacific atmosphere through extensive use of thatching. Strand Pauli, Hamburg   276

21.7   The six sets of actors that are thrown together to create city beaches   279

22.1   Age-Friendly City Elements, according to the World Health Organization   283

22.2   Elements for a senior-friendly park   285

22.3   A distinctive gate and sign makes the presence of this park visible from afar   286

22.4   Multiple benches along pathways make it easier for older adults to rest while walking   289

22.5   A small water pond brings the natural element of water into the park   291

22.6   Seating without a backrest and surrounding traffic make this park setting uncomfortable for older adults   292

23.1   Derives—A Site Specific Performance—June 2019   297

23.2   Singing Out choir performs on The Bentway Skate Trail opening day   298

23.3   Skaters skate past Janine Miedzik's installation Pro Tem   300

23.4   A Fall Day at the Bentway   300

23.5   Visitors watch a streetdance battle at The Bentway Block Party   303

23.6   Sunday Social Yoga with Muse Movement   305

23.7   Bocce at the opening of CITE   305

23.8   Playing in The Bentway water feature   307

24.1   The open-space skyscraper concept   311

24.2   Leadenhall Cheesgrater building by Roger Stirk Harbour + Partners (2013)   313

24.3   20 Fenchurch Street (Walkie-Talkie) building by Rafael Viñoly (2014)   314

25.1   Stumbling blocks in Chemnitz, Germany, artist Gunter Demnig   324

25.2   The Memorial to the Murdered Jews of Europe   325

25.3   Ghost bike in Albuquerque, 2018   326

25.4   Jefferson Davis, New Orleans, 2016   328

25.5   The site where Jefferson Davis stood   329

26.1   During Occupy Wall Street, Zuccotti Park was transformed from an ordinary plaza in a financial district into a site of occupation and protest   336

26.2   Spontaneous conversation between protestors and passersby during Occupy Wall Street   340

26.3   Women's March 2017, USA: The battle was won not on the street but in the ballot box during the 2018 midterm election in which minority and woman candidates made significant gains in the Congress   343

27.1   Volunteers at the Hauptbahnhof (Vienna New Main Station) organ-
       ized the care of thousands of refugees   351
27.2   Protest against the WKR–Ball in Vienna in 2014; the building of the
       Austrian Parliament is visible in the background   354
28.1a Public spaces are public when they are not only "mapped" by sover-
       eign powers but also "used" and "instituted" by civic practices,
       debates, forms of representations, and social conflicts. Washington
       Mall, Washington, DC   359
28.1b The discussion of the role and power of narration of public space is
       associated with the debate surrounding historical narration, the ways
       in which it is constructed, and those who construct it. Tiananmen
       Square, Beijing   360
28.2   Reconstruction is an approach aims to reconstruct the meaning of
       space with what has been lost, adding missing components or empha-
       sizing neglected parts. Plaza De Mayo, Buenos Aires   362
28.3   Alternating narratives of public spaces   363
29.1   Chale Wote Street Art Festival, Accra, Ghana   370
29.2   *pumflet*, Cape Town, South Africa   373
29.3   CityWalk, Durban, South Africa   374
30.1   Figure ground showing changes to urban fabric between 1956
       and 2012   382
30.2   Map showing key public spaces in *el Raval*   383
30.3   Rambla del Raval   384
30.4   Plaça dels Àngels   385
30.5   Carrer de Joaquín Costa   386
32.1   Site map of precedent projects   402
32.2   Programming strategy Museums Quarter Vienna   403
32.3   Before and after: Herald Square   405
32.4   Process oriented planning Berlin Airfield Tempelhof   407
32.5   Urban Prototyping Event 2012   410
32.6   Urban Prototyping Event 2015   410
32.7   The Urban Script   412
37.1   Segregation of urban elements can resolve potential conflicts between
       them, like these two homes   460
37.2   Another method to resolve conflicts is to create mediating structures
       that both separate and connect, usually through adjoining public
       spaces   461
37.3   An ordinary "high street" in London yet featuring a very complex set
       of public and private spaces – a "place network"   463
37.4   Transformation of plots of land as subsequent owners made subdivisions
       and agreements to create new streets, shown in Ben Hamouche (2009)   464
40.1   Hénard's vision of the future street including a Parisian boulevard,
       Da Vinci-like flying machines, and the latest conceptions of
       infrastructure   487
40.2   "Parked Bench" in London, U.K. designed by WMBstudio, 2015   492
40.3   An example of the artistic reframing of everyday built forms in Marfa,
       Texas; the shop and exhibition space Pure Joy is housed in a reused
       fuel-storage facility (Pure Joy 2019)   493

40.4	The light filled atrium of the Bradbury Building, Los Angeles, CA	494
40.5	Cycle of built forms as ideas and structures	495
41.1	Political protest, Naples	501
41.2	Millennium Park, Chicago	502
41.3	Gated community, Rajarhat, Kolkata	505
41.4	Ephemeral urbanism – encroachment of the urban commons	506
41.5	Informing video surveillance	507
41.6	Competing demands on the urban commons – hawker stalls occupying sidewalk space, Kolkata	509
41.7	The new pedestrian Times Square, New York	510
41.8	High Line linear park, New York	510

TABLES

9.1	Results of the searches in PubMed and Scopus	95
10.1	Districts, population densities, and number of public open spaces visited in each city	109
10.2	Inclusive facilities for VIPs in public open space in Beijing, Taipei, and Hong Kong	115
10.3	VIP interview comments	115
14.1	Public space issues in Western settings	174
17.1	Twenty cases of interior privately owned public spaces in 19 buildings in Manhattan	215
18.1	A classification of public space use	225
20.1	Analytical framework. Adapted from Ewing and Clemente (2013)	255
20.2	Quantitative assessment of five urban design aspects	258
20.3	Qualitative assessment of six urban design aspects	259

CONTRIBUTORS

Jonny Aspen is a Professor of Urban Theory at the Institute of Urbanism and Landscape, Oslo School of Architecture and Design. He does research on issues of urban planning history, urban cultures, contemporary urban development, and digital cities. His latest book, co-written with John Pløger, is *Den vitale byen* [*The Vitalist City*] (2015).

Tridib Banerjee holds the James Irvine Chair in Urban and Regional Planning at USC's Price School of Public Policy. His research and writings focus on the design and planning of the built environment and their human consequences. *The New Companion to Urban Design* (2019), co-edited with Anastasia Loukaitou-Sideris is his most recent publication.

Sverre Bjerkeset is a Ph.D. Fellow in Urbanism at the Institute of Urbanism and Landscape, Oslo School of Architecture and Design. Based on field work in central Oslo, he investigates uses of urban public space, with a particular focus on circumstances that condition interaction among strangers.

Luisa Bravo is Adjunct Professor in Urban Design at the University of Florence in Italy. She is the Founder of City Space Architecture, a non-profit organization promoting public space culture, and the Founder and Editor-in-Chief of *The Journal of Public Space*, established in partnership with UN-Habitat (journalpublicspace.org).

Matthew Carmona is Professor of Planning and Urban Design at The Bartlett, UCL. He is an architect/planner and researches design governance, public spaces, and the value of urban design. He chairs the Place Alliance and edits www.place-value-wiki.net. His research can be found at https://matthew-carmona.com.

Vania Ceccato is a Professor in the Department of Urban Planning and Environment, School of Architecture and the Built Environment, KTH Royal Institute of Technology, Stockholm, Sweden. Ceccato is the (co-)author of seven books, the most recent are, *Transit Crime and Sexual Violence in Cities* and *Crime and Fear in Public Places*. She coordinates the national network Safeplaces (Säkraplatser) funded by The Swedish National Crime Prevention Council (BRÅ).

Mark C. Childs' award-winning books include *Imagine a City that Remembers* (with A. Anella, UNM Press 2018); *The Zeon Files: Art and Design of Historic Route 66 Signs* (with E. Babcock, UNM Press 2016); *Urban Composition* (Princeton Architectural Press 2012); *Squares: A Public Space Design Guide* (UNM Press 2004); and *Parking Spaces* (McGraw-Hill 1999). Mark serves as Interim Dean of the School of Architecture and Planning at the University of New Mexico.

Andrei M. Z. Crestani, Ph.D., is a Professor in the undergraduate program in architecture and urbanism and the graduate specialization program in urban planning and landscape architecture at the University Positivo (UP). His research focuses on the following topics: the production and configuration of the public space, urban territories and places, and urban landscape.

Sébastien Darchen is Senior Lecturer in urban planning at the University of Queensland, Australia. He studies the political economy of the built environment with a focus on regeneration processes.

Kristen Day is Vice Provost of New York University. She oversees Faculty Affairs on behalf of the university. She is also Professor in NYU's Department of Technology, Culture and Society. Kristen is an expert in urban planning strategies that promote equity and well-being. Her current research focuses on the impacts of air pollution on people's everyday lives in China. Dr. Day's research has been supported by the Council on Tall Buildings and Urban Habitat, the National Institutes of Mental Health, the California State Department of Transportation, and others.

Kim Dovey is Professor of Architecture and Urban Design and Director of InfUr- (Informal Urbanism Research Hub) at the University of Melbourne where he leads research projects on urban morphology and informal settlements. Authored books include *Framing Places, Becoming Places*, and *Urban Design Thinking*.

Renia Ehrenfeucht is a Professor in the Community and Regional Planning Department at the University of New Mexico. Her research investigates public spaces and the politics of everyday life as well as urban change in shrinking cities and communities. She has written two books and numerous articles.

Mona El Khafif, Dr. techn., is an Associate Professor at UVA School of Architecture, co-author of *URBANbuild: Local/Global* (with Ila Berman) and author of *Staged Urbanism* (published in German). Her design research operates at multiple scales, examining the interdisciplinary aspects of urban regeneration, temporary urbanism, urban prototyping, and strategies for the smart city. At UVA, El Khafif co-directs UVA Smart Environments and MainStreet21, engaging underserved rural communities with digital technologies.

Peter Elmlund is Director of the Urban City Research Programme at the Ax:son Johnson Foundation, and Research Affiliate at KTH Royal Institute of Technology, both in Stockholm, Sweden. He led the partnership with UN-Habitat to develop language on public space for Habitat III and its outcome document, the New Urban Agenda.

Karen A. Franck, Ph.D., is a professor in the College of Architecture and Design at the New Jersey Institute of Technology in Newark, NJ where she also serves as Director of the

Joint Ph.D. Program in Urban Systems. Over the years, her research interests have spanned several topics including alternative housing (*New Households, New Housing*, 1989), building and place types (*Ordering Space*, 1994), the design process (*Architecture from the Inside Out*, 2007 and *Design through Dialogue*, 2010). The interest she has pursued continuously is public space (*Loose Space*, 2007 and *Memorials as Spaces of Engagement*, 2015).

Troy D. Glover is Professor and Chair of the Department of Recreation and Leisure Studies and director of the Healthy Communities Research Network at the University of Waterloo. His research explores the role(s) of leisure in advancing or deterring community, primarily through the development of social capital and transformative placemaking.

Ken Greenberg is an urban designer, teacher, writer, former Director of Urban Design and Architecture for the City of Toronto and Principal of Greenberg Consultants. He was selected as a Member of the Order of Canada in 2020. He was also awarded an Honorary Doctorate by the University of Toronto in 2020 for his outstanding service for the public good as a tireless advocate for restoring the vitality, relevance and sustainability of the public realm in urban life. He is the author of *Walking Home: The Life and Lessons of a City Builder* and *Toronto Reborn: Design Successes and Challenges*.

Tigran Haas, Ph.D., is the Tenured Associate Professor of Urban Planning + Urban Design. He is currently the Director of the Centre for the Future of Places at KTH. His current research focuses on contemporary trends and paradigms in urban planning and design, new urbanism, sustainable urbanism, social housing and urban transformations, city development, ageing society, design and medialization of urban form. His latest book for Routledge is *In the Post-Urban World—Emergent Transformations of Cities and Regions in The Innovative Global Economy* (with Hans Westlund).

Tali Hatuka (B.Arch., MSc., Ph.D.) is Associate Professor of Urban Planning and Head (and founder) of the Laboratory of contemporary Urban Design, in Tel Aviv University (lcud.tau.ac.il). Her work focuses on: (1) the urban realm and society (i.e., public space, conflicts, and dissent); and (2) urban development and city design (i.e., housing and industrial areas and technology). Her most recent book is *The Design of Protest: Choreographing Political Demonstrations in Public Space* (University of Texas Press, 2018).

Jeffrey Hou, Ph.D., is Professor of Landscape Architecture at the University of Washington, Seattle. His work focuses on democracy and public space. Hou is recognized for his pioneering work on guerrilla urbanism and bottom-up placemaking through publications including *Insurgent Public Space: Guerrilla Urbanism and the Remaking of Contemporary Cities* (2010).

Te-Sheng Huang has a master's degree in Architecture from Cheng Kung University and a Ph.D. in Urban Systems from the New Jersey Institute of Technology. He has conducted research on interior privately owned public spaces in New York. After completing his Ph.D., he taught in the School of Architecture at Feng Chia University and opened an architectural firm that focuses on public space, urban design and housing design.

Clara Irazábal is Professor of Urban Planning in the Department of Architecture, Urban Planning and Design (AUPD), and Director of the Latinx and Latin American Studies Program

(LLAS) at the University of Missouri—Kansas City. Her research deals with community development and justice in Latin America and the US.

Timothy Jachna, Ph.D., has worked as an architect, urban designer and academic in Chicago, Berlin and Hong Kong. He is currently the Dean of the College of Design, Architecture, Art, and Planning (DAAP) at the University of Cincinnati. His research deals with urban processes and systems, with a particular focus on digital technologies.

Elahe Karimnia, Ph.D., is a practicing architect and urbanist. She is an Associate at Theatrum Mundi (London), leading the Urban Research and Spatial Practice. Previously, Elahe was a doctoral fellow and design tutor in Urban and Regional Studies at KTH Royal Institute of Technology (Stockholm), and a Visiting Scholar at University of Toronto. Her work is at the intersection of urban design and critical theory, and currently she is involved in two research projects: Making Cultural Infrastructure and Choreographing the City.

Mark Kingwell is Professor of Philosophy at the University of Toronto and a contributing editor of Harper's Magazine. Recent works include a study of sports and error, *Fail Better* (2017); a monograph on the future of work, *Nach der Arbeit* (2018); a philosophical study of boredom, *Wish I Were Here* (2019); and an analysis of the political dimensions of risk (forthcoming 2020).

Sabine Knierbein, Ph.D., is Associate Professor and the Director of the Interdisciplinary Centre for Urban Culture and Public Space, Faculty of Architecture and Planning, TU Wien. She has co-edited *Public Space and the Challenges of Urban Transformation in Europe* (2014), *Public Space and Relational Perspectives* (2015), *City Unsilenced: Urban Resistance and Public Space in the Age of Shrinking Democracy* (2017), and *Public Space Unbound: Urban Emancipation and the Post-Political Condition* (2018), all with Routledge. Research foci include: theory of urbanization, lived space and critique of everyday life, planning theory, and civic and open innovation.

Margaret Kohn is a Professor of political theory at the University of Toronto. Her primary research interests are in the areas of the history of political thought, critical theory, social justice, and urbanism. Her most recent book, *The Death and Life of the Urban Commonwealth* (Oxford University Press 2016) won the David Easton Award for Best Book in Political Theory and the Judd Award for Best Book in Urban and Local Politics. She is the author of *Radical Space: Building the House of the People* (Cornell University Press 2003), *Brave New Neighborhoods: The Privatization of Public Space* (Routledge 2004), and *Political Theories of Decolonization* (with K. McBride, Oxford University Press 2011).

Karl Kullmann is Associate Professor of Landscape Architecture and Urban Design at the University of California, Berkeley, where he teaches design studios and courses in landscape theory and digital representation. Karl's research currently explores the cultural agency of topographically complex landscapes.

Els Leclercq is a Postdoctoral Fellow at Delft University of Technology and Amsterdam Institute for Advanced Metropolitan Solutions, The Netherlands. She conducts research on topics related to the transition towards a sustainable urban environment including circular economy, digitalization ("smart city"), new democracy and citizen engagement. Her Ph.D. dissertation focused on the privatization of public space in European cities. She is also Principal of StudioAitken, an urban design consultancy based in the Netherlands and the UK.

Loretta Lees (FAcSS, FRSA, FHEA) is Professor of Human Geography at the University of Leicester, UK. She is an urban geographer who is internationally known for her research on gentrification/urban regeneration, global urbanism, urban policy, urban public space, architecture, and urban social theory. Since 2009 she has co-organized The Urban Salon: A London forum for architecture, cities and international urbanism (see www.theurbansalon. org) and since 2016 the Leicester Urban Observatory (www.leicesterurbanobservatoryword press.com). She has been identified as the 17th most referenced author in urban geography worldwide (Urban Studies, 2017) and the only woman in the Top 20.

Anastasia Loukaitou-Sideris is Professor of Urban Planning at UCLA and Associate Dean of the UCLA Luskin School of Public Affairs. She is the co-author of *Urban Design Downtown* (1998), *Sidewalks* (2009), *Transit-Oriented Displacement or Community Dividends?* (2019), and the co-editor of *Jobs and Economic Development in Minority Communities* (2006), *Companion to Urban Design* (2011), *Informal American City* (2014), *and The New Companion to Urban Design* (2019).

Setha Low is Distinguished Professor of Environmental Psychology, Geography, Anthropology, and Women's Studies, and Director of the Public Space Research Group at The Graduate Center, City University of New York. She has been awarded a Getty Fellowship, a NEH fellowship, a Fulbright Senior Fellowship, a Future of Places Fellowship, and a Guggenheim for her ethnographic research on public space in Latin America and the United States. Her most recent books are *Spatializing Culture: The Ethnography of Space and Place* (2017), *Anthropology and the City* (2019), and *Spaces of Security* (with M. Maguire, 2019).

Jason Luger is an urban geographer at the University of California, Berkeley, College of Environmental Design, with research focusing on the production of urban space and urban spatial politics. He published a co-edited volume *Art and the City* (with J. Ren, Routledge 2017) and is an Assistant Editor of the *Journal of Urban Cultural Studies*.

Ali Madanipour is Professor of Urban Design and a founding member of the Global Urban Research Unit (GURU) at the School of Architecture, Planning and Landscape, Newcastle University, UK. His latest books include the *Handbook of Planning Theory* (Routledge 2018) and *Cities in Time: Temporary Urbanism and the Future of the City* (Bloomsbury 2017).

Michael W. Mehaffy is Senior Researcher with the Ax:son Johnson Foundation and KTH Royal Institute of Technology in Stockholm. He consults internationally to governments, NGOs, and businesses and has held teaching and/or research appointments at seven graduate institutions in six countries. His Ph.D. is in architecture from Delft University of Technology.

Garth Myers is Director of the Center for Urban & Global Studies and Professor of Urban International Studies at Trinity College. He focuses on African urban geography and planning, comparative urbanism, and urban political ecology. He has published six books and more than 70 book chapters and articles on these themes.

Jeremy Németh is Associate Professor of Urban and Regional Planning at the University of Colorado Denver, where he also directs the Ph.D. program in Geography, Planning, and Design. He was a Fulbright Scholar in Rome and served as Chair of the Department of Urban and Regional Planning and Director of the Master of Urban Design program. His research looks at how planners, designers, and city dwellers can help create more socially and environmentally just places, and he is particularly interested in the relationship between urban design and social equity.

Elizabeth Nisbet is Assistant Professor of Public Management at John Jay College of Criminal Justice in New York. Her research focuses on how public policy shapes inequality and inequities in public services and labor markets and establishes boundaries between the public and private sectors.

Elek Pafka is Lecturer in Urban Planning and Urban Design at the University of Melbourne. His research focuses on the relationship between material density, urban form and the intensity of urban life, as well as methods of mapping the "pulse" of the city. He has co-edited the book *Mapping Urbanities*.

Tigor W. S. Panjaitan is a Ph.D. student at the University of Queensland, Australia. His research focuses on public space issues in Southeast Asia. He teaches urban design at the University of 17 Agustus 1945, Surabaya, Indonesia, and works as an urban design consultant.

Dorina Pojani is Senior Lecturer in urban planning at The University of Queensland, Australia. Her research interests encompass built environment topics (urban design, urban transport, and urban housing) in both the Global North and South. She completed her graduate studies in the US, Albania, and Belgium, and her postdoctoral residency was in the Netherlands. She has held guest teaching and/or research positions in Austria, Chile, and Italy. Prior to joining academia, she worked for several years in urban design and planning in California.

Darko Radović is Professor of Architecture and Urban Design at Keio University, Tokyo. He has taught, researched, and practiced architecture and urbanism in Europe, Australia, and Asia in the fields of architecture, urbanism, and strategic thinking. At Keio, he leads the research laboratory co+labo radović. He is published in English, Serbo-Croatian, Japanese, Korean, Italian, and Thai languages.

Vuk Radović is an Australian architect and urbanist of Serbian background, with a professional career spanning a decade working in Australia, United Kingdom, and Japan. His specialization has been on residential and commercial high-rise in Asia and Oceania. He is currently completing his Ph.D. (2019) at Keio University (Japan) at the School of Media and Governance, Faculty of Environmental Governance—on the topic relating to the integration of supertall residential skyscrapers into their cultural contexts.

Clare Rishbeth is a Senior Lecturer in the Department of Landscape Architecture, University of Sheffield, UK. Her research focuses on migration histories and the experiential qualities of place, developing a landscape specific contribution within a broad field of literature encompassing belonging and isolation, conviviality, racism, and transnational connections. Her approach and social values are focused on profiles of marginalization (shaped by intersections of ethnicity, class, and gender) and how these inform the civic ambition of public space. She has led a range of research projects on these themes, most recently *The Bench Project* (2015) and #refugeeswelcome in parks (2017).

Susanna F. Schaller is Assistant Professor in Urban Studies and Planning at The City College of New York. Her 2019 book *Business Improvement Districts and the Contradictions of Placemaking: BID Urbanism in Washington, D.C.* (University of Georgia Press) examines how private-partnership regimes have restructured Washington's landscape and consolidated gentrification processes.

Ceren Sezer is the DAAD Research Fellow at the Institute for Urban Design and European Urbanism of RWTH Aachen University in Germany. She is joint editor of special issues entitled *Marketplaces as an Urban Development Strategy (2013), Public Space and Urban Justice (2017)* and *The Politics of Visibility in Public Space (2020)*. Her research interests include livability and sustainability of public spaces; urban form and social life in the city, and urban regeneration and renewal processes. Ceren is co-founder and coordinator of an international research group Public Spaces and Urban Cultures established under the Association of European Schools of Planning (AESOP)

Rike Sitas joined the African Centre for Cities in 2011 and drives the cultural justice and urban humanities work at the Centre, straddling the academic world of urban studies and creative practice. She is fascinated by the intersection of culture and cities, and more specifically on the role of art, culture and heritage in urban life.

Kin Wai Michael Siu is Associate Dean (Research), Eric C. Yim Professor in Inclusive Design and Chair Professor of Public Design, School of Design, The Hong Kong Polytechnic University. He is Founder and Leader of Public Design Lab. His research areas are in public design, user reception, inclusive and universal design, and social design.

Bas Spierings is Assistant Professor in Urban Geography at the Department of Human Geography and Spatial Planning at Utrecht University in the Netherlands. His research focuses on the nexus between urban consumption and public space with specific interests in urban competition, touristification, retailing, (cross-border) shopping, walking mobilities, and encounters with difference.

Quentin Stevens is Associate Professor of Urban Design at RMIT University, Melbourne. His books include *The Ludic City* and *Loose Space*. He studied city beaches through an Alexander von Humboldt Foundation Senior Research Fellowship at Humboldt University Berlin, and currently leads an Australian Research Council funded project exploring temporary and tactical urbanism through ANT and assemblage thinking.

Rianne van Melik is Assistant Professor in Urban Geography at the Institute of Management Research at Radboud University Nijmegen in the Netherlands. Her research focuses on contemporary cities and their public spaces with specific interests in the design, management, use, and perception of different kinds of public spaces.

Avigail Vantu currently holds a position as a Research Scientist at the New York University College of Global Public Health. She is interested in the use of data analytics and visualizations to understand people and cities. In 2016, Avigail received her Urban Informatics MS from New York University. Prior to that, Avigail worked on urban design and spatial analyses projects in the public sector, in tech, as well as in academia. Avigail also spent time as an independent journalist in which she wrote for several leading Israeli publications.

Yi Lin Wong is a Lecturer at the Hong Kong Baptist University. She has participated in various qualitative and quantitative design research related to inclusive design and participatory design. She has been actively involved in social design projects to promote social and cultural sustainability through creative and innovative design methods.

Jiao Xin Xiao is a Teaching Fellow of School of Art and Design, Guangdong University of Technology. Her research focuses on participatory design, sustainable research, and inclusive design. She has been involved in several funded research projects related to participatory action research and sustainable design.

Su-Jan Yeo is a Lecturer in the School of Community and Regional Planning at the University of British Columbia. Her research on cities and urban transformation brings into focus the interplay between public space use, design/planning, and public policy vis-à-vis global processes.

ACKNOWLEDGMENTS

First, we want to thank the 55 contributors for their original contributions and the patience in answering, replying, and sometimes challenging our comments and editing requests. It has been a long journey and we are glad that you have decided to travel with us.

We would like to thank Garrett Stone, a doctoral student at the School of Planning, University of Cincinnati, who diligently edited most of the chapters and associated documents and organized the whole manuscript.

We thank the common academic home of the editors, the School of Planning, College of Design, Architecture, Art, and Planning, at the University of Cincinnati.

Finally, we would like to thank Andrew Mould, our acquisition editor, who came up with the idea for the *Companion to Public Space*, and Egle Zigaite, editorial assistant at Routledge.

INTRODUCTION

Vikas Mehta and Danilo Palazzo

The past decade has definitely erased any doubts of the relevance and resurgence of public space. Citizens are taking to the streets to protest and demand justice. Cities and regions are using public space to reclaim and revitalize derelict areas to create walkable humane places, adapting public spaces to create specialized places to accommodate new needs, or to employ ecological practices and engage with nature in urban settings. Local communities are applying bottom-up actions and interventions to build and strengthen identity, challenging monuments and symbols of contested pasts, or using public spaces to nurture an ethos of diversity and inclusiveness. On the contrary, the significance, power, and imperative of public space is visible in its annexation by the private sector or in public–private partnerships for economic gain, in the control of access to public space by the state to limit dissent, or in the surveillance of public space for monitoring behaviors, bringing some societies at the brink of a panoptic state.

In the increasing heterogeneous and urban twenty-first century, many more groups are voicing their views in the public sphere and claiming their right to public space. In cities around the world, a growing number of contemporary public spaces are being produced by a much more dynamic, open, and grassroots process that goes beyond the primarily visual "placemaking" paradigm for the production of space. This new landscape of urban public space engages the social, political, and economic forces where the diversity of use, lifestyle, and aesthetic preference are celebrated in the making and appropriation of space. In the contemporary city, the boundaries of public space have been blurred in more than one way. On one hand, public life now commonly exists in public as well as quasi-public and even in privatized space. On the other, more and more publicly owned yet non-traditional space within the city is being appropriated for public use. These changes call for a new interpretation of public space as a place of diversity, but they should equally inspire us to critically defend the role of public space as a place for assembly. As many more groups claim the city, the need for a neutral field for expression, discussion, and display has become even greater for the survival of an engaged civil society.

The vast literature on public space demonstrates a wide spectrum of possibilities, significance, and potential approaches as well as political, social, economic, and psychological value. We believe that there is a heightened awareness of these possibilities in contemporary times. And, by closely examining its public spaces, and the bottom-up and top-down actions that have been taken by citizens, communities, public institutions, and private organizations, we can decipher the social, cultural, and political life of a city. The access and availability to public spaces can show how public spaces are, or not, an arena for public life: a place for individual and group expression; a forum for dialogue, debate, and contestation; a space for conviviality, leisure, performance, and display; a place for economic survival and refuge; a site for exchange of information and ideas; and a setting for nature to exist in the city and to support the well-being of its inhabitants.

To understand these dimensions of public life, the study and understanding of public space must cross disciplinary boundaries, and public space—as a physical manifestation of the public realm and public sphere—must be viewed as a reflection of the values of its citizenry and be at the center of the debate about the contemporary ethical and moral issues of society. Through the selection of authors and themes, the *Companion to Public Space* emphasizes that it is inappropriate to pigeonhole public space to a discipline or even a field. The inquiry of public space in myriad disciplines brings a rich and wide range of viewpoints in examining it through the lens of varying perspectives and theories. Scholars, researchers, and educators should approach the exploration of public space without boundaries, even if this may challenge their own disciplinary alliances and belief systems. This is the concept that the *Companion to Public Space* advances. Thematically, the *Companion* crosses disciplinary boundaries and territories to address the philosophical, political, social, legal, planning, design, management, and other issues in the social construction of public space. The main objective of the *Companion* is to address the complete ecology of public space under one cover and expose the student, scholar, or practitioner in the public or private sector to interrelated issues of public space in the twenty-first century. Chapters in the *Companion* address theories on and related to public space in reference to the current state of thinking about them. The trajectory of public spaces and the narratives on them are discussed, along with chapters on meanings, politics, collective and collectivistic culture, and changing roles of public spaces. Interpretations of public space are also provided by authors representing different design-oriented or artistic disciplines that participate in the production and activation of public space. Other chapters present examples of invention or reinvention.

The *Companion to Public Space* is proposed principally as an academic text for a broad audience. Faculty teaching graduate and undergraduate courses on urbanism and public space typically emphasize and reflect upon the multidisciplinarity of the subject including, as reading materials, book chapters and journal articles from the social sciences, legal, arts, planning, and design disciplines. Graduate and undergraduate students in these classes come from a wide range of schools and departments including city planning, urban design, architecture, landscape architecture, art, design, geography, environmental policy, sociology, anthropology, law, engineering, and more. Although the academe has for long advocated and experimented with multidisciplinary learning, there is a clear lag in public space literature. Such a multi-disciplinary approach has been considered essential to this volume contributing to the discussion regarding public space. The politics, planning, design, and management of public space have been ubiquitous in the history and development of human settlements and a common and shared concern across cultures and countries. Accordingly, the text in the *Companion* refers to, engages, and emerges from political, cultural, and social issues related to public space across the globe. Although the nature of the volume is academic, it can also be of much use to practitioners in the private sector as well as planners and managers of public space working in the public sector.

The essays that comprise this *Companion* have been organized in five parts. This structure offers a map for the reader to explore the most current reflections on public space.

Contributions that provide readings and interpretations of public space deriving from different disciplines' points of view are combined in *Perspectives*. This part sets the tone of the Companion anticipating the multiplicity of potential entry points to investigating public space in relation to its physical, social, and political dimension. This part also takes into account the interpretations that derive from the cultural and educational settings of the authors, as well as from the actual meanings of public space in the different geographic context and by the practices determined by different users.

Public space production and practice are constantly in the process of being reframed in relation to changes in state and city régimes, addition of new populations and communities in urban settings, fluctuations in city financing capabilities, impacts of global agendas or issues, or reactions to episodes of violent protests. *Influences* explores, through the authors' original contributions, the transformations and variations in the roles and significance of public space due to several reasons elaborated in the chapters.

Classification is a component of scientific or cultural processes. Starting from Carl Linnaeus in the eighteenth century, taxonomy has become a branch of science concerned with systematic reading of living species. Taxonomy has evolved and is embraced by disciplines that do not necessarily deal with living species, such as architecture and urban design. Space typologies or types can be described, as Rafael Moneo (1978) puts in an elegant and synthetic way, "the act of thinking in group" (23). *Types* assembles various attempts to classify public spaces as a means to offer a common—and shared—ground to the understanding of this vital component of our cities.

A public space, without being actioned and activated by people, is just another component of the physical part of the city. People and communities, through the act of playing or protesting, occupying or abandoning, ornamenting or de-facing, and altering or protecting, activate and provide new meanings to public spaces. In *Actions* the authors' contributions look at different ways animations of public space occur in various cultural and geographical contexts.

Cities are places of change. They drive change, are subject to and are the testbeds of change. Public spaces are the workrooms of these changes. Interactions between virtual spaces and physical spaces have become common practices in the use of public space. Changes in public space are also determined by their adaptation in response to new challenges and threats. The last part of the *Companion to Public Space* is, inevitably, about its *futures*. The authors of the chapters, composing this part, reflect on current practices in public space and how they might be altered in the coming decades due to the inevitable and already occurring phenomena that are pressing our cities.

Reference

Moneo, R. (1978) On Typology. *Opposition*, 13(Summer): 23–45.

PART 1

Perspectives

In these challenging times of diminishing democracy, there is a heightened awareness of the value of public space. At the global scale, the social and political uprisings and resistance movements in Istanbul, Caracas, Madrid, Hong Kong, Taipei, Paris, St. Louis, the Arab Spring, Occupy Wall Street, and the #MeToo and Black Lives Matter movements have drawn attention to the civic and political significance of public space. Public spaces provide a physical platform for active democracy and a locus of the struggle between the citizenry and the power of the state. At the local scale, more groups have been identifying, claiming, and making space public in locations and ways unexplored in the past. Social media has enabled and strengthened cyberspace as the virtual public space for effortless and instant contacts and exchange that transcends the dichotomy of local–global space and its material components. Once thought of as the nemesis of physical public space, cyberspace, through its ability of instant and perpetual connectivity and information sharing is, in numerous ways, aiding the activation, use, and, appropriation of public space. In these new ways, public space has gained legitimacy, appropriately, not only as space but also as an act and event. In fact, we now experience new ways in which physical space and cyberspace seamlessly connect to create a new typology of the public sphere that delivers a unique experience of publicness.

The inquiry of public space in myriad disciplines brings a wide range of perspectives that examine public space through numerous and diverse lenses/concepts. Simultaneously, the rapidly evolving changes in global and local political landscapes become remarkably visible in public space. These transformations present new challenges, opportunities and perspectives to critique and understand public space. In this context, established and new scholars have a newfound interest reflected in contemporary views and voices. The opening part of the *Companion* builds on the trajectory of this discourse and narrative and also presents evolving and new perspectives on public space that are relevant today. The opening six chapters highlight important ideas and paradigms that shape the understanding of public space in contemporary times ranging from a critique of the rhetoric of civic and democratic space versus the practice of economic gain; publicness as a field of study; public space as dynamic, fluid and relational space; benefits and challenges of being in public space; topography as an influence on experience and performance of public space; and an evaluation criteria for public space based on social justice.

Much contemporary literature on public space, directly or indirectly, points to the clear divide between the heightened awareness and dialogue on public space as a democratic right and its adoption by neoliberal capitalist regimes. The opening chapter by Ali Madanipour focuses on this and demonstrates how the rhetoric of public space as a space of civic and democratic interaction continues in contemporary times, while simultaneously being used increasingly as an instrument to attract and generate economic benefits.

Moving from the political to the experiential aspects, Rianne van Melik and Bas Spierings' chapter emphasizes the relational nature of public space in both time and space, and makes the case for its dynamic and fluid interpretation. The authors explore alternative epistemologies that give users an active role in research, which in turn helps researchers better understand how people combine and relate to public spaces across time and space.

The ever-heterogenizing city means new interpretations and meanings of public space for newcomers. Examining the mobile demographics of the twenty-first-century city, Clare Rishbeth presents the challenges and benefits of being collectively in public. For people new to a place, the outdoors in public space provides an opportunity of belonging, but being outdoors also exposes them to be visible as the "other," resulting in an un-belonging. Exploring the idea and practice of "curated sociability," the chapter highlights the potential for public spaces to be places of everyday and ordinary inclusion, where the process of "becoming local" is grounded in shared collective joys.

Building on this dichotomous outcome, Elahe Karimnia and Tigran Haas' chapter brings to fore the challenges embedded in the process of spatial production of public space, particularly as it relates to achieving publicness. The authors call for a field of study where the design of publicness is an explicit focus. They argue that the design of publicness can be a locus of criticism in the way public space is used or misused as a justification for urban development, and as an active undertaking to understand how the right to appropriate space and produce publicness can be stimulated by urban design.

As tangible and physical, the experience of public space depends on the shape of the land that gives it form and influences the experience and performance of public space. Karl Kullmann's chapter examines the topography of public space at numerous scales, from the city to the specific site. The author presents strategies for re-grounding public spaces and re-amplifying engagement with the topography of the city.

Emerging from the quest to define and understand public space, particularly in an evolving political landscape, an important question has emerged recently. How do we measure and evaluate public space? Since public space is multidimensional, the publicness of public space must be examined in ways that address issues of politics and democracy, sociability, leisure and recreation, economic exchange, symbolic value and social justice. Founded on 20 years of ethnographic research in numerous cities and a careful reading of the literature on the just city, right to the city and social justice planning, Setha Low's chapter proposes an expanded framework for evaluating public space based on five dimensions of social justice designed to enhance diversity and equity.

1

A CRITIQUE OF PUBLIC SPACE

Between interaction and attraction

Ali Madanipour

Introduction

Public space has been widely discussed in the literature, and its development and provision have become a widespread policy embraced by many public, private and voluntary agencies (e.g. Carmona et al., 2008; De Souza et al., 2012; Hou, 2010; Low and Smith, 2006; Orum and Zachary, 2010; Parkinson, 2012; Sadeh, 2010; Watson, 2006). Public spaces have always been an integral part of the city, a key component in the vocabulary of urbanism (Benevolo, 1980; Morris, 1994), and their social and political significance has long been recognized (Arendt, 1998; Habermas, 1989); so what are the reasons for the renewed interest in something as old as the city itself?

The growing emphasis on the significance of the public space emerged as a critique of social fragmentation and privatization of urban development processes, as the urban space was increasingly being produced and controlled by private interests, undermining the democratic potentials of urban public spaces. The campaigners for public spaces saw them as a means of turning fragmented cities into integrated places, improving the quality of urban life and offering an alternative to the suburban sprawl. In this process, however, the idea of public space has been transformed from a critique to an orthodoxy, taken for granted by most stakeholders as an important ingredient of urban life.

The rising attention to the public space is a welcome development, as few people would doubt its value, but we may also wonder whether all the different actors who are involved in urban development have the same approach towards the public space. Do the property developers, advertising agencies, architects, city planners, urban designers, municipal organizations and local communities have the same understanding of the public space? More broadly, do the producers, regulators, and users of the built environment have the same approaches to and expectations from the public space (Madanipour, 2006)?

This chapter argues that in the journey of the idea from a critique to an orthodoxy, and in its pervasive acceptance in the urban development process, the approach towards public spaces has metamorphosed and a gap has emerged between the rhetoric and reality. The chapter develops a critique of this process of metamorphosis, showing how the widely used rhetoric of the public space as a multidimensional space of social and civic interaction may be at odds with the practice of creating of a space of attraction, an instrument of commercialization and gentrification, luring investment and maximizing rewards. The chapter examines this transition in four arguments within the broad processes of political, economic, and cultural transformation.[1]

The changing relationship between the public and private spheres

An important reason for the rising concern for public spaces lies in the changing relationship between the public and private spheres, as urban spaces are increasingly produced and managed by private agents for private use. Historically, public authorities have been responsible for the development and management of public spaces (Madanipour, 2003). However, as the public authorities have become more entrepreneurial and market-oriented, the question is whether their approach towards public space has remained the same. *How public are the spaces produced by these public authorities?*

For a generation after the Second World War, a model of development emerged that was based on a stronger presence of the state in the economy; a welfare state involved in the provision of public services and the production of the built environment. This model led to large-scale urban-development projects, public-housing schemes, and comprehensive planning. The actual extent of the state's role and intervention varied widely in different countries, but the Keynesian paradigm of state-market relations prevailed in the Western countries and spread around the world (Madanipour, 2011). Following the economic crises of the 1970s, this model was replaced by a neoliberal paradigm, in which the state withdrew from many fields, while the market was given a more prevalent role, transferring the production and management of the built environment to the private sector (Aglietta, 2008; Lipietz, 1987). This shift radically changed the balance between the public and private agencies and spheres.

In the first model, modernist designs were used to replace the overcrowded and badly built cities with mass-produced buildings in car-dominated environments (Le Corbusier, [1929] 1987). In the modernist design philosophy, public spaces were breathing spaces at the service of buildings, enveloping, and supporting them for health and recreation (Sert, 1944). With the end of this model, which came with deindustrialization and globalization, these buildings and neighborhoods declined, and their public spaces became a huge problem (Castell, 2010). The close connection between housing and public spaces was broken, as public authorities started to abandon their role in housing provision, and emphasis on the public space was a rather convenient substitute for this shortcoming.

In the neoliberal model that followed, the resources of the private firms were mobilized, which had access to productive capacities that could transform large parts of cities and regenerate declining areas. However, these firms had a limited remit, responsible towards their shareholders, rather than delivering services and spaces for the public. Urban development projects still needed common spaces, but these new spaces were functional intermediate spaces rather than publicly accessible ones. In an increasingly unequal society, these intermediate spaces were privately controlled, sometimes with the help of guards, walls, gates, and cameras, setting boundaries that would limit access to these spaces. This reduction in supply and access opened up a crisis of confidence and a rising sense of anxiety about public services and spaces, and by extension a crisis for the city as a whole (London Assembly, 2011). So much of the debate about the public space reflected anxiety about this changing relationship, which is a mirror of the broader relationships between the market and the state, and between the individual and society (Carmona et al., 2008; Low and Smith, 2006). The campaign for the public space, in this sense, was a campaign for the integrity of the city and society.

The early phase of criticizing the privatization of public spaces was based on the idea that the lines between the public and private agencies should be sharply drawn. In political theory, the public sphere is often the sphere of the state, distinct from the private sphere of individuals and households. The two spheres are kept apart, as the intervention of the public sphere into the private sphere would result in the loss of privacy and individual freedom, while the encroachment of the private sphere into the public sphere may create individual gain and collective loss (Nolan, 1995; Wacks, 1993). These lines, however, are increasingly blurred, as the public authorities adopt private sector approaches and enter partnerships with the private developers. Publicness, even when produced and managed by the public authorities, becomes a relative concept.

The early concerns about the privatization of the public space, therefore, have been compounded by concerns about the character of the public institutions, which has direct implications for the public spaces that they produce and manage. The rhetoric of the public space has been widely adopted by public authorities. However, these authorities now operate on a basis that is far closer to the way private companies function with their motives of risk and reward. This is evident in the growth of markets and quasi markets within public services, which is known as the New Public Management (Ferlie, 2017). The issue has changed from the relationship between the public and private institutions to a metamorphosis of the public institutions. The outcome would therefore be a transfiguration of the public space that is produced under these conditions. Particularly after the global financial crisis of 2008, and the dwindling budgets of public authorities, their attitude towards public spaces has become far more entrepreneurial, using these spaces as a source of much needed income (Cheshmehzangi, 2012).

Therefore, first there has been a shift in the production and management of space from public to private agencies, and then a transformation in the character of the public authorities, who behave as if they were private firms. The character of the public spaces they produce, therefore, is a continuing concern, which is closely related to the primacy of economic considerations.

The prevalence of economic considerations

A related cause for concern is the instrumental use of public spaces for economic outcomes, as a means of attracting attention and investment. As economic considerations become a primary motive for public authorities, the question becomes: how far does this emphasis on economics shape the content and character of public spaces?

Facilitated by technological change, the major economic shift in recent decades has been the globalization of industrial production, relocating the manufacturing industries from their old centers to new ones (Madanipour, 2011). This was initiated by the companies that looked for cheaper factors of production, and for being free from labor disputes and environmental regulations (Bell, 1973; Esping-Andersen, 1999; Touraine, 1995). This fundamental economic change has had considerable impact on the social and spatial organization of the city. It has fueled urbanization in industrializing countries, like in China, which has experienced what may be the largest wave of urbanization in human history. It has also fueled transition to services in deindustrialized cities, which have been looking for a knowledge-based economy, an alternative economic rationale to fill the gap (OECD, 1996; Stiglitz, 1999; UNESCO, 2005). In almost all cases, public spaces play a mediating and facilitating role in these economic transformations, in the forms of attraction and interaction that would stimulate innovation, investment, and consumption.

In globalized economies, cities are engines of economic development, where the production, exchange, and consumption of goods and services take place. A key driver of economic development is innovation, which is often thought to be enabled through the encounters between different perspectives, where the minds meet and are able to develop new ideas and products (Schumpeter, 2003). Such a meeting of minds, it is thought, would be partly facilitated by the composition of the urban environment and support from a vibrant public sphere (Madanipour, 2011). Clustering the new companies in science parks and cultural quarters has become the holy grail of local economic development, thought to generate the critical mass and the space of interaction that is needed for such innovation. International organizations such as the UNESCO (2010) advocate the development of science parks, while many municipalities promote the development of cultural quarters (e.g. Creative Sheffield, 2010). Stimulating innovation that would trigger economic development is therefore expected to benefit from the possibility of interaction that such scientific and cultural districts and their common infrastructure can provide.

Economic development also draws on attracting private investment, which is partly leveraged through public investment in public spaces, making cities attractive and competitive (EC, 2006). Public spaces become "soft locational factors," which are "important for attracting knowledge industry businesses, a qualified and creative workforce and for tourism" (EC, 2007, 3). In the context of globalization, where cities compete with each other for investment, high-quality public spaces, tall buildings, and expressive architecture are all seen as symbolic assets, enhancing the image and quality of a city on the global stage. These prominent urban features become devices for distinguishing a city in the crowded global marketplace, much in the same way that advertising is meant to differentiate goods on supermarket shelves.

More specifically, public spaces have a direct role in the real estate market, using the public infrastructure to encourage private investment and to increase land and property values. The economic roles of public space at the local level include building market confidence, creating attractive conditions for private developers to invest in an area, making and enhancing the land and the property market. Research has shown the positive impact of public space in the demand for residential space and higher values in such properties. In some cities, proximity to a green space could add up to half the price of some types of dwelling (McCord et al., 2014). Some public authorities look for economic justification for investment in public spaces, and they find this justification in confidence building for the market, laying the foundations of a property market in declining areas, where none existed, attracting private investors to an area, and seeing the rise in the land value as the ultimate justification for investment in the public space. For private developers, good public spaces provide a clear competitive advantage for the quality and market value of their development, especially if public authorities cover the cost of providing these public spaces.

The economic role of the public space is also evident in its support for leisure and retail activities, which drive the urban economy in many cities. The consumption of goods and services, so thriving through globalization, is a major driver for the global economy; the more we consume goods and services the faster the wheels of the global economy spin. Consumption becomes a goal in itself, whether or not we need those products, to the extent that consumerism has become a primary identifier of the rich urban societies. Investment in public space is an essential ingredient of boosting this consumerism and experience economy. Public spaces provide the atmosphere of glitz and spectacle that would draw people to particular places, where we can enjoy the pleasure of apparent abundance and being with others. In response to the economic crisis of 2008, which threatened the retail and leisure spaces of British city centers (Local Data Company, 2013), a solution by a government advisor was to think of entire city centers as a commercial space: "getting our town centers running like businesses" (Portas, 2011, 18). This approach has partially transformed the character of public spaces in city centers, bringing them in line with the commercial logic of shops and restaurants.

The rhetoric of the public space has been adopted at the macroscale level of urban development in globalizing economies, and at the microscale of property development and commercial support. In many of these promotions of the idea, however, the public space is used as a vehicle of attracting investment from companies, builders, buyers, and visitors. The promotional rhetoric therefore tends to see the public space as an instrument at the service of economic aims, which may not necessarily serve the social and environmental needs and expectations of large parts of the urban society.

Technological change and dispersing cities

The third important reason for giving prominence to public spaces is the problem of urban spatial fragmentation and dispersion, which has had social and environmental consequences (Madanipour et al., 2014). Transport technologies have long allowed the growing cities to disperse in all directions, a trend that continues to this day, with major social and environmental implications. Suburbanization has been an ongoing trend since the nineteenth century, facilitated through the invention of trains and cars, and in the twentieth century supported and encouraged through government subsidies, planning policies and cultural preferences (Abercrombie, 1945; Briggs, 1968; Cullingworth and Caves, 2013; Keating and Krumholz, 1999). It is a trend that continues in most forms of urban expansion around the world. The dispersion of the urban population into low-density suburbs enabled the middle-class households to live a private life, but it undermined the possibility of creating common spaces for shared experiences (Fishman, 1987). It reflected a fragmentation of society into atomized units without sufficient spatial links to one another. The cry for public space was partly a cry for the reintegration of this fragmented fabric through the introduction of connective tissues. The urban sprawl was also associated with the ecological crisis (EC, 1994). The response included advocating for high-density, compact urban environments made liveable by high-quality public open space. The public open space, therefore, became a central theme in the policy documents that advocate sustainable development (European Commission (EC), 2007; 2010; EEA, 2009).

Linked with suburbanization, the car had hollowed out the city itself. Despite early criticisms by Camillo Sitte (1986) in the late nineteenth century, the reorganization of urban space for vehicular access fascinated the early modernists and was written into the core agenda of urban development (Le Corbusier, 1987; Sert, 1944; Buchanan, 1963). The street was losing its social value and turned into a functional tool for rapid travel. The campaign for public space was partly an endeavor to turn this tide and reclaim the streets and squares for sociability. Pedestrian movement would allow the urban population to linger and repose, and as such to be able to develop spaces of interaction and sociability, rather than mere functionality.

With the arrival of the information and communication technologies, it was thought that cities would disappear altogether (Martindale, 1966). Time and space were thought to have been annihilated and life was going to take place in a space of flows (Castells, 1996). It was no longer important where you were, as you could have access to resources and services from any location. The possibility of connecting to anyone anywhere, and the creation of online communities, would herald a new type of non-spatial public space. These technical possibilities, however, have not removed the need for cities, but in fact cities have become more vibrant. The economies of scale, the changing nature of economic relations, the need for mutual social support and the cultural texture of social life have all stimulated the growth of cities, with inevitable need for the public space provision and improvement (O'Sullivan, 2012). While the dramatic emergence of the information and communication technologies has stimulated the growth of a digital public sphere, it has enhanced, rather than impeded, the need for face-to-face interaction and communication associated with physical co-presence.

The problem, however, lies in the gap between the need and the availability of resources, between the rhetoric and practice in the provision and distribution of public spaces. Suburbanization continues and, under the conditions of dwindling public budgets, the provision and maintenance of public spaces is under threat. An example is the situation of public parks in the UK, which are considered to be increasingly unaffordable by the public authorities (Heritage Lottery Fund, 2016). The provision and maintenance of public spaces, which are so essential for the social integration of fragmented societies and the spatial reintegration of wasteful urban sprawl, are not supported by the practices of urban development and management.

Social diversification and inequality

The spatial dispersion of cities unfolds alongside their social diversification and inequality, together creating a mosaic of difference and segregation. The fourth dimension of a social critique of the current wave of rhetoric about public spaces, therefore, is whether the provision of public spaces takes into account and responds to the problems of diversity, inequality, vulnerability, and exclusion, or it exacerbates them by becoming a vehicle of gentrification and a barrier to access (Madanipour, 2010).

In an urban world, where larger cities are growing rapidly, public spaces are particularly significant on many levels (UN Habitat, 2012; RWI et al., 2010). Alongside the growth of urban populations, social diversity and inequality has increased. As various reports by the EU, OECD, and the UK government show, social inequality has grown in the last three decades, alongside the neoliberal model of economic development and the shifting boundaries between public and private spheres (EC, 2010; OECD, 2008). With globalization and international migration, smaller households and increasing variety of lifestyles, the urban populations are more diverse than ever before. As more people migrate to cities, they need the essential spaces that facilitate social life, a common infrastructure of institutions and spaces that is a vital prerequisite for making collective life possible. It is in the DNA of urban life, as evident in informal settlements around the world, where we can witness the birth of an urban area, where consolidation of housing is followed by the development of local public spaces (Hernàndez Bonilla, 2010).

In the transition from manufacturing to services, the organization of social groups and urban spaces has been changing. Blue-collar workers had once shaped the industrial cities, with their rigid routines of life and mass patterns of consumption and socialization. In service cities, the norm is flexible routines of work and diversified patterns of consumption, served by an international army of casual and underpaid workers. Public space improvements, whether by public authorities, civil society activists, or private companies, adjust the urban space for its new activities and inhabitants, but in doing so, they might knowingly or unknowingly facilitate displacement and gentrification (Atkinson and Bridge, 2005). On the receiving end, ghettoization, homelessness, and sudden bursts of anger in the form of protests and riots, are some of the ways that these changes find expression in public spaces.

Meanwhile, a series of social movements have pushed for broadening the meaning of the public. The word public refers to people as a whole and theoretically includes everyone. But in practice, it has tended towards a narrow definition, without taking the diversity of society and the different positions and needs of its members into account. Women have argued that cities have historically been built and run by men, undermining women's roles and needs. In the distinction between the public and the private, men have dominated the public sphere of work and politics, pushing women to a domestic sphere in which they could be controlled and suppressed. City design clearly reflected this unequal arrangement, whereby industrialization separated the world of work from home, suburbanization trapped women in isolated peripheries, socialization became limited to the spaces of consumption, and the design and management of urban spaces remained insensitive to women's needs. Alongside women's movements, ethnic and cultural minorities have also argued for their right to the city, overcoming the actual and symbolic barriers that deny them access to particular places and activities.

In the design of the urban environment, the standards were set by the able-bodied and mobile populations, while the elderly and the disabled were often ignored, and their reduced mobility was seen as a regrettable but inevitable fact of life. In ageing societies, however, addressing their needs becomes a pressing concern. For a person with reduced mobility, moving in most public spaces is a struggle, continually negotiating impassable barriers. Many cities have started adopting measures for widening access to those with reduced mobility, either in a wheelchair, pushchair, or just having difficulty in negotiating the steps and steep slopes. Children's presence in the public space has been managed through a combination of ordering and protection, limiting their presence to specialist and monitored places. Young people in public places, meanwhile, are closely watched for any misbehavior that would unsettle the calm order of the city. When fear of crime has risen, all vulnerable groups have withdrawn from public spaces.

The key feature of public spaces is their accessibility. The more accessible a place, the more public it becomes. Access is not abstract and universal: it is the expression of relationships between people, an expression of power and control over territory, an interplay of inclusion and exclusion. So it always takes different forms and levels, and that is why a city is full of shades of public–private relations, from the most public to the most private places. The problem lies in the provision of access in highly diverse and unequal urban societies.

Conclusion: between rhetoric and reality

The concept of public space is in need of a critical evaluation: emerging as a critique of neoliberal urban development, but widely becoming co-opted in that development process. This co-option comes face to face with the need for inclusive and accessible public spaces, at a time of dwindling public budgets and growing social inequality. As public authorities have become more entrepreneurial, their approaches to the public space have also changed. The rhetoric of the public space as a space of civic interaction continues to be used, but increasingly as an instrument of attracting resources. The preceding four arguments revolved around the changing nature of and relationship between different agencies and spheres, the predominance of economic considerations, the impact of technological change, and the challenges of social diversification and inequality, showing how the rhetoric of the public space may have grown stronger, but its character being metamorphosed in practice.

Note

1 An earlier, longer version of this paper has been published in *Urban Design International* (Madanipour, 2019).

References

Abercrombie, P. (1945) *Greater London Plan 1944*. London: His Majesty's Stationery Office.

Aglietta, M. (2008) Into a New Growth Regime. *New Left Review*, 54(Nov–Dec): 61–74.

Arendt, H. (1998) *The Human Condition*, 2nd edition. Chicago: The University of Chicago Press.

Atkinson, R. and Bridge, G. (eds). (2005) *Gentrification in a Global Context: The New Urban Colonialism*. London: Routledge.

Bell, D. (1973) *The Coming of Post-Industrial Society: A Venture in Social Forecasting*. New York: Basic Books.

Benevolo, L. (1980) *The History of the City*. London: Scholar Press.

Briggs, A. (1968) *Victorian Cities*. Middlesex: Penguin.

Buchanan, P. (1963) *Traffic in Towns*. London: HMSO.

Carmona, M., de Magalhães, C. and Hammond, L. (2008) *Public Space: The Management Dimension*. London: Routledge.

Castell, P. (2010) *Managing Yards and Togetherness*. Gothenburg: Chalmers University of Technology.

Castells, M. (1996) *The Rise of the Network Society*. Oxford: Blackwell.

Cheshmehzangi, A. (2012) "Reviving Urban Identities: Temporary Use of Public Realms and Its Influences on Socio-Environmental Values and Spatial Interrelations." Unpublished PhD diss., University of Nottingham.

Corbusier, L. (1987 [1929]). *The City of Tomorrow and Its Planning*. London: The Architectural Press.

Creative Sheffield. (2010) "Creative Sheffield: Transforming Sheffield's Economy." Accessed 21 April 2010. www.creativesheffield.co.uk/.

Cullingworth, J. B. and Caves, R. (2013) *Planning in the USA: Policies, Issues and Processes*, 4th edition. London: Routledge.

De Souza, E., Silva, A. and Frith, J. (2012) *Mobile Interfaces in Public Spaces: Locational Privacy, Control and Urban Sociability*. London: Routledge.

Esping-Andersen, G. (1999) *Social Foundations of Postindustrial Economies*. Oxford: Oxford University Press.

European Commission (EC). (1994) *The Aalborg Charter*. Brussels: European Commission.

European Commission (EC). (2010) *Why Socio-Economic Inequalities Increase?* Brussels: European Commission.

European Commission (EC). (2006) *Thematic Strategy on the Urban Environment*. Brussels: European Commission.

European Commission (EC). (2007) *The Leipzig Charter on Sustainable European Cities*. Brussels: European Commission.

European Environment Agency (EEA). (2009) *Ensuring Quality of Life in Europe's Cities and Towns*. Copenhagen: European Environment Agency.

Ferlie, E. (2017) The New Public Management and Public Management Studies. *Oxford Research Encyclopedias: Business and Management*. Oxford: Oxford University Press.

Fishman, R. (1987) *Bourgeois Utopias: The Rise and Fall of Suburbia*. New York: Basic Books.

Habermas, J. (1989) *The Structural Transformation of the Public Sphere: An Inquiry into a Category of Bourgeois Society*. Cambridge: Polity Press.

Heritage Lottery Fund. (2016) *State of UK Public Parks 2016*. London: Heritage Lottery Fund.

Hernàndez, B. M. (2010) Making Public Space in Low-Income Neighbourhoods in Mexico. In *Whose Public Space?* Madanipour, A. (ed). London: Routledge, pp. 191–211.

Hou, J. (2010) *Insurgent Public Space: Guerrilla Urbanism and the Remaking of Contemporary Cities*. London: Routledge.

Keating, D. and Krumholz, N. (eds). (1999) *Rebuilding Urban Neighborhoods: Achievements, Opportunities and Limits*. Thousand Oaks: Sage.

Lipietz, A. (1987) *Mirages and Miracles: The Crises of Global Fordism*. London: Verso.

Local Data Company (2013) "The Knowledge Centre." 23 May. www.localdatacompany.com/knowledge.

London Assembly. (2011) *Public Life in Private Hands: Managing London's Public Space.* London: Greater London Authority.

Low, S. and Smith, N. (2006) *Politics of Public Space.* London: Routledge.

Madanipour, A. (2003) *Public and Private Spaces of the City.* London: Routledge.

Madanipour, A. (2006) Roles and Challenges of Urban Design. *Journal of Urban Design,* 11(2): 173–193.

Madanipour, A. (ed). (2010) *Whose Public Space? International Case Studies in Urban Design and Development.* London: Routledge.

Madanipour, A. (2011) *Knowledge Economy and the City: Spaces of Knowledge.* London: Routledge.

Madanipour, A. (2019) Rethinking Public Space: Between Rhetoric and Reality. *Urban Design International,* 24(1): 38–46.

Madanipour, A., Knierbein, S. and Degros, A. (eds). (2014) *Public Space and the Challenges of Urban Transformation in Europe.* London: Routledge.

Martindale, D. (1966) Prefatory Remarks: The Theory of the City. In: *The City,* Weber, M. (ed.) New York: The Free Press, pp. 9–62.

McCord, J., McCord, M., McCluskey, W., Davis, P. T., McIlhatton, D. and Haran, M. (2014) Effect of Public Green Space on Residential Property Values in Belfast Metropolitan Area. *Journal of Financial Management of Property and Construction,* 19(2): 117–137.

Morris, A. E. J. (1994) *History of Urban Form: Before the Industrial Revolution,* 3rd edition. Harlow: Longman.

Nolan, L. (1995) *Standards in Public Life, Volume 1: Report.* London: HMSO.

O'Sullivan, A. (2012) *Urban Economics,* 8th edition. New York: McGraw-Hill.

Organization for Economic Cooperation and Development (OECD). (1996) *The Knowledge-Based Economy.* Paris: Organization for Economic Cooperation and Development.

Organization for Economic Cooperation and Development (OECD). (2008) *Growing Unequal? Income Distribution and Poverty in OECD Countries.* Paris: Organization for Economic Cooperation and Development.

Orum, A. M. and Zachary, N. (2010) *Common Ground? Readings and Reflections on Public Space.* London: Routledge.

Parkinson, J. (2012) *Democracy and Public Space: The Physical Sites of Democratic Performance.* Oxford: Oxford University Press.

Portas, M. (2011) *The Portas Review: An Independent Review into the Future of Our High Streets.* London: Department for Business, Innovation and Skills.

RWI, DIFU, NEA and PRAC. (2010) *Second State of European Cities Report.* Brussels: European Commission.

Sadeh, E. (2010) *Politics of Public Space: A Survey.* London: Routledge.

Schumpeter, J. (2003) *Capitalism, Socialism and Democracy.* London: Taylor and Francis.

Sert, J. L. (1944) *Can Our Cities Survive? An ABC of Urban Problems, Their Analysis, Their Solution.* Cambridge: Harvard.

Sitte, C. (1986 [1889]) City Planning According to Artistic Principles. In: *Camillo Sitte: The Birth of Modern City Planning,* Collins, G. and Collins, C. (eds) New York: Rizzoli, pp. 129–332.

Stiglitz, J. (1999) "Public Policy for a Knowledge Economy." Accessed 6 July 2009. www.worldbank.org/html/extdr/extme/jssp012799a.htm.

Touraine, A. (1995) *Critique of Modernity.* Oxford: Blackwell.

UNESCO. (2005) *Towards Knowledge Societies: UNESCO World Report.* Paris: United Nations Educational, Scientific and Cultural Organization.

UNESCO. (2010) "Science, Technology and Innovation Policy." Accessed 8 April 2010. www.unesco.org/science/psd/thm_innov/unispar/sc_parks/parks.shtml.

UN-Habitat. (2012) *Join the World Urban Campaign: Better City, Better Life.* Nairobi: UN-Habitat.

Wacks, R. (ed.). (1993) *The International Library of Essays in Law and Legal Theory, Volume 1: The Concept of Privacy.* New York: New York University Press.

Watson, S. (2006) *City Publics: The (Dis)enchantments of Urban Encounters.* New York: Routledge.

2

RESEARCHING PUBLIC SPACE

From place-based to process-oriented approaches and methods

Rianne van Melik and Bas Spierings

Introduction

Public space has been researched from a myriad of disciplinary positions, including philosophy, psychology, sociology, geography, planning, and urban design (Varna, 2014). Depending on the vantage point, studied themes have amongst others been everyday interactions between public-space users (Goffman, 1963; Lofland, 1973; 1998; Sennett, 1978; Watson, 2006), the democratic, political and liberating function of public space (Habermas, 1989; Hénaff and Strong, 2001; Low and Smith, 2006; Parkinson, 2012; Shiffman et al., 2012), or its use, design and ownership (Gehl, 2001; Hou, 2010; Whyte, 1980). According to Mitchell (2017), public-space research was relatively scarce until the late 1980s and largely understood to be a design issue. However, two key publications – Davis' (1990) *City of Quartz* and Sorkin's (1992) *Variations on a Theme Park* – 'catalyzed new, sharply critical, and eventually wide-ranging research on the role of public space in making more or less just cities' (Mitchell, 2017, 504). Both publications sparked an exponential growth of published work on public space; first mainly addressing issues of spatial justice, but gradually touching upon a wider variety of issues including – but not limited to – urban encounters, privatization, commercialization, and embodied experiences (Blomley, 2011). Harding and Blokland (2014) ascribe this growth in the number and variety of public-space studies to two trends: the process of 'privatization' in the neo-liberal city and the 'cultural turn' in urban studies – generating more attention to the ways in which social diversity is expressed, creates conflict, is governed, controlled and negotiated in public space.

Most studies on public space tend to concentrate on particular places, often as geographically bounded areas – ranging from streets (Blomley, 2011; Mehta, 2013; Zukin et al., 2016) and sidewalks (Kim, 2015) to squares (Childs, 2004; Low, 2000; Webb, 1990), parks (Low et al., 2005; Mitchell, 2017) and waterfronts (Varna, 2014). Such a 'place-based approach' greatly enhances our understanding of particular places. It acknowledges the fact that public spaces should be conceived as different and diverse in terms of uses, perceptions, experiences, and meanings. Some spaces may be more mono-functional in design and use, for instance playgrounds, while others host many different activities that may also conflict with each other, such as in the case of city squares (Van Melik, 2008).

However, a place-based approach to public space usually provides little information and details regarding, for example, those users and functions *not* (currently) present, relations between different public spaces within and outside the city, and the meaning of these places beyond their physical confines. In this chapter, we aim to discuss and explore possibilities to research public space from a more 'process-oriented' perspective than a place-based approach. In so doing, public spaces are conceived as in flux, open-ended, changing and struggled over, and thus constantly made and remade (Mitchell, 2017; Qian, 2018). A more relational understanding of public space may help us to study public space not a 'as bounded object, specific context or delimited site (…) [but] as an object which is relentlessly being assembled at concrete sites of urban practice' (Farias and Bender, 2010, 2). Public space is not 'static, a container or neutral backdrop in which action unfolds,' but instead 'an outcome of the specific mutual relations between people and places' (Gotham, 2003, 724; Knierbein and Tornaghi, 2015, 5). People can be in one public space 'physically' while simultaneously in many 'mentally' through thoughts and memories – but also interact with people at a distance when using mobile phones, for instance (Kim, 2012; see also the case study below). As such, public spaces do not exist outside the relational process that define them, and this process is also highly dynamic in the sense that the assemblage of relations with other public spaces continuously develops over time (Spierings, 2009). Hence, public spaces should be regarded as dynamic processes constituting and constituted by constant flows of and encounters with people, practices, goods, physical environments, and ideas, or a 'thrown togetherness of bodies, mass and matter, and of many uses and needs' as Amin (2008) would put it.

This chapter provides two sections discussing and exploring more process-oriented approaches for further enriching public-space research: one on studying relations between public spaces in both time and space, and the second on applying mobile methods for process-oriented investigations of public space. The concluding part touches upon some pros and cons of process-oriented investigations. Throughout the chapter, research examples developing a processual understanding of public space will be discussed – drawing on some of our own studies, often dealing with cases from the Netherlands, as well as on studies from other scholars discussing examples from elsewhere.

Case study: the station square adjacent the Hoog Catharijne shopping mall in Utrecht

Figure 2.1 shows a variety of people walking in diverse directions on the station square in Utrecht during a late afternoon. While walking from one public space to another they are 'here,' in one particular space 'physically' – the station square – but may simultaneously be 'there,' in many other public spaces 'mentally.' Some may still be thinking about, for instance, the business park or university campus they were earlier or the shopping mall they just left. Others may already be anticipating the upcoming train station or the neighborhood they will be returning home to soon. In addition, some seem 'virtually' connected with other people and places through the use of their mobile phones, potentially involving public spaces in Utrecht, the Netherlands and beyond.

Figure 2.1 Here and there
Image credit: Bas Spierings

Relational in time and space

Urban public space is often defined not only physically but also socially as open and accessible space where people from all kinds of backgrounds can congregate and interact (e.g. Lofland, 1998; Sennett, 1978; Watson, 2006). These social interactions and encounters can foster mutual relations between people in public space – including brief, fleeting, functional and unfocused encounters as well as more intimate, close, and focused relations (Gehl, 2001; Lofland, 1998; Mehta, 2013; Spierings et al., 2016).

In most of these and other studies on social interactions and encounters, the physical contours of a particular public space define and demarcate the research focus. For example, the rich body of literature on marketplaces has elaborately described specific street markets as meeting grounds supporting inclusive city life, including multicultural exchanges and encounters between and amongst sellers and buyers (e.g. Hou, 2010; Polyák, 2014; Watson, 2009). However, Schappo and Van Melik (2017) illustrate that the integrative potential of street markets stretches beyond the daily selling and buying activities on the actual market square itself. When studying The Hague Market, the biggest street market in the Netherlands, they found that it not only facilitates on-site interactions, but also brings about broader partnerships uniting a large number of stakeholders, including the local government, merchants, inhabitants, entrepreneurs, local schools, and other institutions in the neighborhood. Consequently, they conclude that a focus on marketplaces as bounded research locations leads to limited insights regarding the question as to how these public spaces are (re)produced by the constant incoming and outgoing flows of people, goods, and ideas. This reasoning resonates with the highly influential work *For Space*, in which Massey (2005) pleas for a relational sense of place:

> '[W]hat is special about place is not some romance of a pre-given collective identity ... Rather, what is special about place is precisely that throwntogetherness, the unavoidable challenge of negotiating a here-and-now (itself drawing on a history and a geography of thens and theres)' (140)

This makes the city (and its public spaces) into something much more fluid and flexible rather than fixed; and existing on multiple scales simultaneously, which Hubbard (2006) labels as intransitivity. Take for example Zuccotti Park near Wall Street in New York City, where the first Occupy Wall Street Movement camp emerged in 2011 (Shiffman et al., 2012). This park mainly served as local public space for residents and office workers at one moment, while becoming the site of global protest the next, and vice versa. Besides these more local dynamics, Zuccotti Park also shows that a place is not only connected to its immediate surroundings, but also to places elsewhere, even at the global scale – like other protest camps emerging throughout the US and Europe. This inter-relatedness of cities and their public spaces not only can be found in rather temporary situations but also in more long-term development trajectories. Spierings (2006), for instance, shows how the imaginary of an increasingly mobile and fun-seeking consumer has become dominant and articulated in redevelopment visions and plans for Dutch city centers. The fear of losing and belief in winning consumption capital makes city centers continuously use each other as 'benchmark' in the endeavor to at least keep pace with and ideally leapfrog competitors. In so doing, the imaginary of contemporary consumerism 'travels' from one place to another, is being used to substantiate often rather similar redevelopment projects and gets materialized in cities and their commercial public spaces throughout the Netherlands, and beyond.

Pinpointing the political power of physical places, Kimmelman (in Shiffman et al., 2012, xiii) argues that the earlier-mentioned protest camps illustrate that 'no matter how instrumental new media have become in spreading protest these days, nothing replaces people taking to the streets.' Social media indeed has played an important role in facilitating and fostering protest movements such as the Occupy Wall Street Movement or the urban uprisings in Turkey in 2013 by spreading news, calling for action, getting people together and fostering trust (Haciyakupoglu and Zhang, 2015). As such, physical places are not just tied to 'real' places elsewhere, but also to 'virtual' public spaces. However, these strong relations between real and virtual public space are also not limited to protest settings only, but also emerge in more everyday situations. Van Melik (2018), for instance, describes the practice of stoop-sitting in a gentrifying neighborhood in the Netherlands. A wide variety of residents – both gentrifiers and non-gentrifiers – literally extend their homes into public space, for example by hanging bird cages outside or placing inflatable swimming pools or garden furniture on the sidewalks. Although this at first seems a spontaneous 'insurgent' practice fostering encounters between all residents in the neighborhood, it is in fact often carefully orchestrated via a residential WhatsApp group, which uses 'sidewalk' as code word to call for a brief moment of stoop-sitting for conversation and coffee (Hou, 2010). However, this WhatsApp group mostly consists of social renters; there are hardly any online or real-life conversations with other types of residents in the neighborhood. This example illustrates how focusing on relations occurring within specific bounded contours of public space disregards how, where, when and by whom the space is actually being produced and reproduced.

Also resonating with Massey's plea for a relational sense of place, Degen and Rose (2012) further emphasize the time dimension of a relational understanding of public space by focusing on how perceptual memories mediate present, everyday experiences of town centers and their public spaces. They argue that the present experience of a public space and its design is 'overlaid with memories of how that same environment was encountered in the past' (3279). Moreover, this experience is also mediated in the sense that 'memories of other places induce judgements about their different sensory qualities' – highlighting the importance of relations and comparisons with other town centers that were visited at earlier occasions (3281; see also the case study below).

Case study: the station square adjacent the Hoog Catharijne shopping mall in Utrecht

Figure 2.2 shows a billboard comparing the station square in Utrecht as it looks now ('Nu') (on the right) with how it looked then ('Toen'), a couple of years ago (on the left). Present experiences of this renewed square and shopping mall are intensely mediated through memories of how the same spaces were experienced in the past, especially because the changes happened quite recently and are still on-going (also taking place behind the billboard). These mediating memories are further intensified and triggered by the actual billboard aiming to promote the new shopping mall and square by pinpointing how it has changed or improved – hence judging to a certain extent how it used to look like in the past.

Figure 2.2 Then and now
Image credit: Bas Spierings

Mediating effects of perceptual memories are not necessarily limited to the rather official settings of town centers but also very much emerge in more informal settings of shopping and trading. In their study on German cross-border shoppers in the Polish bazaar in Słubice, Szyntniewski and Spierings (2018), for instance, found that memories of the built environment of the bazaar, provisional market stalls of the old days in particular, are important for understanding present, everyday experiences. For some of the German visitors these good memories were triggered because they could still recognize 'the past' in the renewed bazaar, while for others the memories actually pinpointed a clear distinction between 'now' and 'then' – in the sense that they experience the new bazaar as too commercial and therefore less attractive. Moreover, historical perceptions of social, cultural, and economic differences between Germany and Poland proved of high importance in mediating present experiences of the encounters related with buying and selling in the bazaar. In addition to Degen and Rose's study (2012), Szytniewski and Spierings (2018) found that memories and perceptions can be rather stereotypical and 'sticky'. Despite the economic progress in Poland, memories, and perceptions of, for example, the comparatively low economic development of and living standards in the country still persist – often resulting in visitors being surprised to find out that price differences with Germany were getting smaller. Both studies illustrate the importance of the temporality of everyday experiences in public spaces, requiring an understanding of these spaces as always in the process of becoming and mediated through memories of the same and other spaces – something which is usually paid only little attention to in place-based approaches.

Mobile methods

When our ontological understanding of public space shifts from something that is rather fixed and bounded to more dynamic and fluid, this also requires alternative epistemologies for developing a processual understanding of public space. For inspiration, we draw on the so-called 'new mobilities paradigm' or 'mobility turn' (Sheller and Urry, 2006), which resulted in a large and still growing number of studies focusing on the flows and movement of people, goods, and ideas in social sciences – and in much attention for mobile methods to do so. For a relational understanding and analysis of public space we consider the 'go-along' method to be highly suitable in particular.

Kusenbach (2003) describes the 'go-along' method as a combination of participant observations and in-depth interviews. This implies that 'fieldworkers accompany individual informants on their "natural" outings, and – through asking questions, listening and observing – actively explore their subjects' stream of experiences and practices as they move through, and interact with, their physical and social environment' (463). This method is not only 'participatory' in the sense that fieldworkers accompany their informants on their daily life paths in both time and space, but also because the latter are given a more active and powerful role in the research process – both in pinpointing public spaces and topics for discussion. Van Melik and Pijpers (2017) followed this approach by starting narrative interviews with the elderly by asking what an ordinary day in their lives looks like, while Wildish and Spierings (2019) commenced their in-depth interviews by asking Airbnb users to draw a map of the neighborhood they were staying in. In so doing, public spaces to be discussed in more detail were not pre-selected for analysis by the fieldworkers as 'experts,' but by the informants themselves as the actual experts in using and experiencing these spaces during daily life. Both studies resulted in a discussion of a large variety of public spaces with often unexpected meaningfulness – something which would not have emerged to such a degree when pre-selecting public spaces beforehand.

An essential advantage of 'go-alongs' – or 'mobile interviews' and 'conversations in place' – is that the fieldwork results 'are profoundly informed by the landscapes in which they take place, emphasizing the importance of environmental features in shaping discussions' (Evans and Jones, 2011, 849). Small but often significant and meaningful details of public spaces tend to come much more to the foreground while *in-situ* than during sedentary interviews, simply because the informant would not remember them or perhaps even would think they were too mundane and therefore not relevant enough to be discussed. As such, 'go-alongs' enable a deeper and more comprehensive investigation of people's everyday practices in and experiences of specific public spaces (Finlay and Bowman, 2017). However, more importantly for the purposes of this chapter, they also provide richer access to the relationality of public spaces in both time and space. A telling example of this is provided by Van Duppen and Spierings' (2013) study on the embodied experience of commuter cyclists in the city of Utrecht in the Netherlands. They show how 'riding along' with the informants results in rich descriptions of 'sensescapes' along the trajectories from home to work, and back, through a variety of public spaces. These sensescapes – denoting relationships and interactions between the body and the environment – include a 'twilight zone,' a 'quiet shortcut' and an 'official cycle lane.' The twilight zone is not only discussed as landscape of disorder and openness with many wastelands but also as a zone displaying continuous change, reflecting how present experiences are mediated through perceptual memories of how the same space used to be in the past. Taking the 'quiet shortcut' (see Figure 2.3a) is preferred by most cyclists which they not only describe as quiet but also as green and neighborhood setting, enabling to get 'in the flow' and let thoughts meander. To further substantiate their choice and preference, participants often discuss memories judging the 'official lane' (see Figure 2.3b) – i.e. another cycling path running parallel and designed by the local authorities as an official route – for being busy and noisy, an 'open road' and requiring lots of stopping and waiting for many traffic lights.

Being literally part of the flows and movement of people, goods, and ideas in public space and from one to another enables researchers to develop a better understanding of how people combine and relate diverse public spaces. At the same time, the act of being mobile or mobility as an embodied practice and experience itself becomes a central feature of the research – one of the main starting points and foundations of the 'new mobilities paradigm' (Cresswell, 2010). Walking or cycling, for instance, are not just 'ways of getting around' but essentially also 'modes of experiencing' the public spaces of cities (Wunderlich, 2008). What mobilities of walking and cycling mean for how public spaces are being used and experienced therefore should become involved into the investigation. Analyzing how people move in and across public spaces, while paying close attention to the mobilities involved, also brings to the fore how these public spaces are experienced as more 'fluid' than clearly 'bounded' – especially when higher speed is involved like in the case of cycling (Van Duppen and Spierings, 2013). Moreover, the dynamics with respect to the use and experience of public spaces also become more explicit then. In this context, Wunderlich (2008) talks about 'discursive walking' as a mode of walking involving 'complete awareness of the external environment and participating in it' through which familiarity with the external environment is developed and deepened over time (132). A telling example of this can be found in a study by Wildish and Spierings (2019) on how Airbnb users in Amsterdam as 'first-timers' and 'outsiders' explore the neighborhood they temporarily reside in and the

Figure 2.3a Quiet shortcut in Utrecht
Image credit: Jan van Duppen and Bas Spierings

Figure 2.3b Official cycle lane in Utrecht
Image credit: Jan van Duppen and Bas Spierings

city they are staying in. It is through the act of 'discursive walking,' as performed and experienced in public space, that the Airbnb users over time build a mental map of the local neighborhood and the larger city. In so doing, they not only continuously relate and compare the public spaces they visit with public spaces they are very much familiar with in their hometown. They also gradually become familiar with and develop an understanding of how public spaces are connected and together constitute the Amsterdam neighborhood as well as the city at large.

Conclusion

In this chapter we have discussed the potential of process-oriented approaches and methods to study and develop an understanding of public spaces as unbounded, relational, dynamic, and open-ended spaces. In so doing, our intentions were by no means to discard, deny or downplay the values and merits of place-based research. On the contrary, many of these studies like Whyte's (1980) detailed observations and vivid descriptions of the social life of small urban spaces in New York in the 1970s have hardly lost significance and relevance, and still provide inspiration for public-space researchers today. Instead, the aim of the chapter was to explore how the field of public-space research can be further enriched by loosening the focus and fixity on specific places and applying more process-oriented approaches and methods.

Studying the processual nature of public space by looking at inter-connectedness is nothing new in itself. In previous academic debates and literature on globalization and mobilities, for instance, attention for 'spaces of flows' rather than 'spaces of places' (Castells, 1996) has managed to gain an important position. However, the academic debate on public space has so far been largely preoccupied with analyzing places instead of flows, whereas increasing numbers of scholars exploring and applying process-oriented approaches and methods more recently. At the same time, the academic shift towards thinking public space more relationally has not become commonplace in the policy domain yet. On the contrary, Knierbein and Tornaghi (2015) describe how public space is still being treated as 'an abstract two- or three-dimensional object to be sliced into workable pieces ... tailored to specialized disciplines' (5). In so doing, urban planners and designers working on public space still tend to focus on developing demarcated and bounded places, also through their specific perspective of the respective discipline and vocabulary.

Obviously, process-oriented studies of public space also have their challenges and limitations – perhaps the most important one related with their 'unboundedness' and 'open-endedness' in both time and space. As such, it might be rather difficult to determine where and when to begin the research as well as where and when to end it. In this context, Monteiro (2000) argues – while discussing challenges related with the relational nature and methodological approach of actor network theory – that the notion of an actor network, quite literally, instructs us to map out the sets of elements (the network) that influence, shape, or determine an action. But each of these elements is in turn part of another actor network, and so forth. Hence, if you take this in too literal a sense, unpacking *any* actor network will cause an explosion in terms of complexity (76). Instead of concluding that relational perspectives and methodologies are hence utterly unmanageable, Monteiro continues by arguing that any (relational) investigation requires researchers to make critical judgments about how to delineate the context of their study. Hence, we do not advocate for not making any delineation at all, but rather developing one that emanates from and during the research, instead of pre-selecting boundaries and making a clear delineation of public spaces beforehand.

The application of mobile methods in public space research is also not without its limitations and challenges. Visiting public spaces together with informants often provides more deep and detailed research findings, but the involvement of the fieldworker may have implications for the 'naturalness' of the social situation – its degree depending amongst others on the research design (Kusenbach, 2003). Moreover, being mobile while doing the research assists in developing a better understanding of how people combine and relate diverse public spaces, but the mode of travel may have an impact on the research findings (Finlay and

Bownan 2017). Riding along, for instance, may result in more fragmented interview transcripts than in case of sedentary interviews due to practices such as overtaking other cyclists and avoiding cars when cycling in crowded cities and busy streets (Van Duppen and Spierings, 2013).

Notwithstanding these and other challenges and limitations, we argue that process-oriented approaches and methods seem apt and promising to tap into the processual nature of public space as both relational and evolving. This enables us to develop a dynamic and deeper understanding of how people use and experience the variety of publics spaces they relate with and combine during their daily lives while moving through them, back and forth, in both time and space.

References

Amin, A. (2008) Collective Culture and Urban Public Space. *City*, 12(1): 5–24.

Blomley, N. (2011) *Rights of Passage: Sidewalks and the Regulation of Public Flow*. Abingdon: Routledge.

Castells, M. (1996) *The Rise of the Network Society*. Oxford: Blackwell.

Childs, M. (2004) *Squares: A Public Space Design Guide for Urbanists*. Albuquerque: University of New Mexico Press.

Cresswell, T. (2010) Towards a Politics of Mobility. *Environment and Planning D*, 28(1): 17–31.

Davis, M. (1990) *City of Quartz: Excavating the Future in Los Angeles*. London: Verso.

Degen, M. M. and Rose, G. (2012) The Sensory Experience of Urban Design: The Role of Walking and Perceptual Memory. *Urban Studies*, 49(15): 3271–3287.

Evans, J. and Jones, P. (2011) The Walking Interview: Methodology, Mobility and Place. *Applied Geography*, 31: 849–858.

Farias, I. and Bender, T. (2010) *Urban Assemblages: How Actor-Network Theory Changes Urban Studies*. London: Routledge.

Finlay, J. M. and Bowman, J. A. (2017) Geographies on the Move: A Practical and Theoretical Approach to the Mobile Interview. *The Professional Geographer*, 69(2): 263–274.

Gehl, J. (2001) *Life between Buildings*. Copenhagen: The Danish Architecture Press.

Goffman, E. (1963) *Behavior in Public Places: Notes on the Social Organization of Gatherings*. New York: The Free Press.

Gotham, K. F. (2003) Toward an Understanding of the Spatiality of Urban Poverty: The Urban Poor as Spatial Actors. *International Journal of Urban and Regional Research*, 27(3): 723–7337.

Habermas, J. (1989) *The Structural Transformation of the Public Sphere: An Inquiry into a Category of Bourgeois Society*. Cambridge: Polity Press.

Haciyakupoglu, G. and Zhang, W. (2015) Social Media and Trust during the Gezi Protests in Turkey. *Journal of Computer-Mediated Communication*, 20(4): 450–466.

Harding, A. and Blokland, T. (2014) *Urban Theory: A Critical Introduction to Power, Cities and Urbanism in the 21st Century*. Los Angeles: Sage.

Hénaff, M. and Strong, T. B. (eds.). (2001) *Public Space and Democracy*. Minneapolis: University of Minnesota Press.

Hou, J. (2010) *Insurgent Public Space: Guerrilla Urbanism and the Remaking of Contemporary Cities*. New York: Routledge.

Hubbard, P. (2006) *City*. London: Routledge.

Kim, A. M. (2015) *Sidewalk City: Remapping Public Space in Ho Chi Minh City*. Chicago: The University of Chicago Press.

Kim, E. C. (2012) Nonsocial Transient Behavior: Social Disengagement on the Greyhound Bus. *Symbolic Interaction*, 35(3): 267–283.

Knierbein, S. and Tornaghi, C. (2015) Relational Public Space: New Challenges for Architecture and Planning Education. In: *Public Space and Relational Perspectives: New Challenges for Architecture and Planning*, Tornaghi, C. and Knierbein, S. (eds.). London: Routledge, pp. 1–11.

Kusenbach, M. (2003) Street Phenomenology: The Go-Along as Ethnographic Research Tool. *Ethnography*, 4(3): 455–485.

Lofland, L. (1973) *A World of Strangers: Order and Action in Public Space*. Illinois: Waveland Press.

Lofland, L. (1998) *The Public Realm: Exploring the City's Quintessential Social Territory*. New York: Aldine de Gruyter.

Low, S. (2000) *On the Plaza: The Politics of Public Space and Culture*. Austin: The University of Texas Press.

Low, S. and Smith, N. (eds.). (2006) *The Politics of Public Space*. New York: Routledge.

Low, S., Taplin, D. and Scheld, S. (2005) *Rethinking Urban Parks: Public Space and Cultural Diversity*. Austin: The University of Texas Press.

Massey, D. (2005) *For Space*. London: Sage.

Mehta, V. (2013) *The Street: A Quintessential Social Public Space*. New York: Routledge.

Mitchell, D. (2017) People's Park Again: On the End and Ends of Public Space. *Environment and Planning A*, 49: 503–518.

Monteiro, E. (2000) Actor-Network Theory and Information Infrastructure. In: *From Control to Drift: The Dynamics of Corporate Information Infrastructures*, Ciborra, C. U., et al. (eds.). Oxford: Oxford University Press, pp. 71–83.

Parkinson, J. R. (2012) *Democracy and Public Space: The Physical Sites of Democratic Performance*. Oxford: Oxford University Press.

Polyák, L. (2014) Exchange in the Street: Rethinking Open-Air Markets in Budapest. In: *Public Space and the Challenges of Urban Transformation in Europe*, Madanipour, A., Knierbein, S. and Degros, A. (eds.). New York: Routledge, pp. 48–59.

Qian, J. (2018) Geographies of Public Space: Variegated Publicness, Variegated Epistemologies. *Progress in Human Geography* [online].

Schappo, P. and Van Melik, R. (2017) Meeting on the Marketplace: On the Integrative Potential of the Hague Market. *Journal of Urbanism*, 10(3): 318–332.

Sennett, R. (1978) *The Fall of Public Man*. New York: Vintage Books.

Sheller, M. and Urry, J. (2006) The New Mobilities Paradigm. *Environment and Planning A*, 38(2): 207–226.

Shiffman, R., Bell, R., Brown, L. J. and Elizabeth, L. (eds.). (2012) *Beyond Zuccotti Park: Freedom of Assembly and the Occupation of Public Space*. Oakland: New Village Press.

Sorkin, M. (ed.). (1992) *Variations on a Theme Park: The New American City and the End of Public Space*. New York: Noonday Press.

Spierings, B. (2006) *Cities, Consumption and Competition: The Image of Consumerism and the Making of City Centres*. Nijmegen: Faculteit Managementwetenschappen, Radboud Universiteit Nijmegen.

Spierings, B. (2009) Travelling an Urban Puzzle: The Construction, Experience and Communication of Multi(pli)cities. *Liminalities*, 5(4): 1–8.

Spierings, B., Van Melik, R. and Van Aalst, I. (2016) Parallel Lives on the Plaza: Young Women of Immigrant Descent and Their Feelings of Social Comfort and Control on Rotterdam's Schouwburgplein. *Space and Culture*, 19(2): 150–163.

Szytniewski, B. and Spierings, B. (2018) Place Image Formation and Cross-Border Shopping: German Shoppers in the Polish Bazaar in Słubice. *Journal of Economic and Social Geography*, 109(2): 295–308.

Van Duppen, J. and Spierings, B. (2013) Retracing Trajectories: The Embodied Experience of Cycling, Urban Sensescapes and the Commute between 'Neighbourhood' and 'City' in Utrecht, NL. *Journal of Transport Geography*, 30: 234–243.

Van Melik, R. (2008) *Changing Public Space: The Recent Redevelopment of Dutch City Squares*. Utrecht: KNAG/Faculteit Geowetenschappen, Universiteit Utrecht.

Van Melik, R. (2018) Op de Stoep: Het Straatleven van Klarendal. *Geografie*, 27(5): 8–11.

Van Melik, R. and Pijpers, R. (2017) Older People's Self-Selected Spaces of Encounter in Urban Aging Environments in the Netherlands. *City & Community*, 16(3): 284–303.

Varna, G. (2014) *Measuring Public Space: The Star Model*. London: Routledge.

Watson, S. (2006) *City Publics: The (Dis)enchantment of Urban Encounters*. London: Routledge.

Watson, S. (2009) The Magic of the Marketplace: Sociality in a Neglected Public Space. *Urban Studies*, 46(8): 1577–1597.

Webb, M. (1990) *The City Square: A Historical Evolution*. London: Thames and Hudson.

Whyte, W. H. (1980) *The Social Life of Small Urban Spaces*. Washington: Conservation Foundation.

Wildish, B. and Spierings, B. (2019) Living Like a Local: Amsterdam Airbnb Users and the Blurring of Boundaries between 'Tourists' and 'Residents' in Residential Neighbourhoods. In: *Tourism and Everyday Life in the Contemporary City*, Frisch, T., Sommer, C., Stoltenberg, L. and Stors, N. (eds.). London: Routledge, pp. 139–164.

Wunderlich, F. M. (2008) Walking and Rhythmicity: Sensing Urban Space. *Journal of Urban Design*, 13 (1): 125–139.

Zukin, S., Kasinitz, P. and Chen, X. (eds.). (2016) *Global Cities, Local Streets: Everyday Diversity from New York to Shanghai*. London: Routledge.

3

THE COLLECTIVE OUTDOORS

Memories, desires and becoming local in an era of mobility

Clare Rishbeth

Introduction

It felt very different from my country; there was no sunshine, no desert.

As articulated by Hozan, an asylum seeker, the difference between Iran and England is experienced written in the landscape and reactive on the skin (Rishbeth et al., 2019). Global migrants, whether moving for economic, education or sanctuary seeking reasons, encounter environmental unfamiliarity. This new city, new neighborhood is not known: it holds no memories and is uncertain in its affordances. Belonging and un-belonging is personal and embodied, experienced in the efforts of new learning, in sensory confusion, in the difficulty of anticipating ways to fit in.

Clearly the individual pattern of these is informed by relative levels of resources and personal agency. But even for migrants with privileges of choice and without the impact of poverty, unfamiliarity can be de-skilling. The tacit knowledges of urban legibility – the practicalities of moving around, of reading safe spaces and temporal norms – are culturally and geographically defined. Networks of urban parks, familiar and important resources for recreation in many European and North American cities, may be unfamiliar typologies to migrants from other parts of the world: why do you go there, and what do you do? The experience of uncertainty reflects a 'not-yet-belonging.' A nurse from Jamaica living in Sheffield talked about her initial disquiet at the 'leaves shedding' in her first British autumn (Rishbeth and Powell 2013). A year later this was a past concern, the buds appear in spring, her remembered surprise representative of her arrival story rather than her current situation. Though urban environments are never fully predictable, the measure of 'a local' is partly though a shared understanding of the perimeters of normality; what is to be expected, what is unusual and what is a cause of worry. 'Fitting in' may mean different things in different locations, but an everyday confidence in understanding your local area is an important baseline. The turning of one full year in the life of a migrant is a marker which reflects both person and place, duration measured in part by amassed knowledge – this is what happens, the extents of this can be understood.

Alongside unfamiliarity, migrants also encounter recognition. Compared to most indoor environments, often the qualities of openness and complexity characteristic of most outdoor public spaces can better support nuances of embodied experience and the stationary time travel of recollection. Newcomers, even those from countries that are culturally and eco-logically very different, find points of recognition in oblique ways, from nature connections to opportunities for socializing. In Sheffield, UK, a walk through a wooded valley on the edge of the city reminded Lamin of his home country:

> This place remind me of when I live in Sierra Leone … It remind me about the forest and the mountain and the rock … The bush was so green and the day was foggy you know. It remind me of when I was a little boy and me and my grand-father used to walk in the forest during the rainy season. [This place] surprised me.
>
> *(Rishbeth and Finney 2006, 287)*

Migrant's stories travel with them, past normalities can be related to new ones.

And while for new migrants, much of their larger personal context has changed (proximity to family, occupation, place of living) – especially for those in forced migration who experience this as loss – glimpses of 'being myself' may be found in seemingly inconsequential items and actions. A flower, a bench, a basketball, a barbeque are not merely props but sensory experiences, often relational ones. Amidst all that is strange and new, it can be these small memories and affordances that allow a connection with one's own identity across different parts of the life course, and in time offer a 'way in' to using public spaces and to developing a sense of local belonging.

Visibility and un-belonging

The 'public' and the 'open' in Public Open Space inform a varying visibility of individuals. While taking part in quite specific and straightforward interactions – a member of a kick-about team, a conversation while walking, looking at flowers in a park – there is a parallel public–facing aspect, a legitimate taking up of space and a body present in the public sphere. For new arrivals in the city these times and places provide potential for respite (being 'in the moment,' a distraction, a change in focus) but also potential for representation: I am living here. It is to this collective positioning that we now turn.

Being outdoors is not neutral, and Lefebvre's 'right to the city' is not evenly distributed (Lefebvre 1968). In thinking about the challenges of urban population diversity, we also need to attend to the specifics of the encounter and how we experience 'the being together of strangers' in the city (Young 1990). The public space as pro-social is voiced by many urban theorists, but can position an overly benign viewpoint of spending time outdoors (e.g. Gehl and Gemzoe 1996; Whyte 1980). 'Who you are, [and] who you are seen to be,' has important intersections of gender, class, age, ethnicity, and religion (Rose 2003, 72). For new migrants, and in particular those who are visible as minorities, the realistic option of spending time outside reflects not only the individual confidence of navigating new environments and 'being oneself,' but also is a gauge of exposure to threat. While informal hanging-out can be an important first step to belonging, it can also result in a sharp point of 'un-belonging,' as evidenced by the rise of hate crime and racial harassment (from the 'top down' political discourses from the UK's 'hostile environment' to Trump's 'protect the border' rhetoric (Heuman and González 2018; Liberty 2018). Crime statistics in the UK clearly reflect a greater vulnerability to personal attack for people from Black and Minority Ethnic backgrounds and a rise in racial and religious hate crime in public places (Webster 2017, 208), and are reflective of statistics in other European and

North American contexts. Understanding how these conflate to typologies of place is important for anyone involved in the design or management of the public realm. Zenith, who has lived in the UK for a number of years seeking asylum from Zimbabwe, talks dismissively about Sheffield's suburban parks, categorizing them as 'places for indigenous people and not for people who came from abroad' (Rishbeth et al., 2019). Excluding experiences such as these are contrasted with feeling included and safe in the Peace Gardens in the city center, a busy square with water fountains, ornamental planting, and city wardens.

So notions of 'fitting in' and 'becoming local' can seldom or never be attained simply as a product of individual effort, but require an embedding of diversity as a re-making of public space. In urban areas with highly mixed populations, at best this leads to a commonplace diversity and an acceptance of the 'multiplicities, potentials and practices' of social identities (Wessendorf 2013; Wilson and Darling 2016, 1). But all too often in neighborhoods where migrants are visible as a minority status, stigmatization and fixed narratives of class, socio-economic status and race can become dominant in supporting perceptions of acceptability or un-acceptability. A singular viewpoint exacerbates the potential for misinterpretation, as succinctly highlighted by Clark's account of Roma community in Glasgow, UK.

> What is often socially constructed as morally reprehensible and anti-social for non-Roma is actually seen as being social, hospitable and inclusionary by Roma themselves. For example, 'loitering on street corners' is actually socializing with friends and 'improper' rubbish disposal is actually forms of recycling and income generation.
>
> *(Clark 2014, 8)*

This account is insightful in also reflecting a collective memory by member of the Roma community that shapes ongoing practices and informs ways of performing publicness. In the context of urban parks, Neal et al. (2015) discuss the fluidity of 'park practices,' the ways in which transnational dynamics inform everyday patterns of recreation and connection, but the notion extends to other mundane environments; the square, the bus-stop, the school yard. Ways of being sociable are culturally defined and the use of public space inevitably shape-shifts in response to this.

Thus far we have briefly explored three dimensions of migration and urban public space:

- The potential for connections and memories that can unexpectedly shape pathways to glimpsing belonging.
- An acknowledgement that many migrants experience uncertainty and intimidation in accessing various urban public spaces.
- The need to find ways for collective publicness to be diversely expressed and to inform the ways in which public space is used and is changed.

The call then is both to solidarity and activism: an urbanism that robustly defends the diversity of urban space and users of urban space. This needs to move away from naïve aspirations of 'places for everyone,' and towards design and management practices that actively and collaboratively facilitate equality of engagement and in doing this challenge the fixity of 'normal' ways of being outside. In the concluding part of this chapter, I examine how this might influence practice, presenting findings from a research project that examined the experiences of refugees and asylum seekers living in northern European urban contexts.

#*refugeeswelcome* in parks

Refugees and asylum seekers living in cities have a range of differing histories and experiences, shaped by the contexts of their forced migration and their current legal status (seeking sanctuary, refused sanctuary, or granted fixed term leave to remain) as well as their ethnicity, gender, age and family responsibilities. However, many circumstances are brutally familiar to all but the luckiest: social fragmentation, lack of agency and future uncertainly, poor housing conditions and extremely low living allowances, past and ongoing mental health pressures. In the UK the government has, over many years, implemented a 'hostile environment' policy, which means that in practice the day-to-day lives of asylum seekers are deliberately made difficult and psychologically demoralizing (Liberty 2018). More encouragingly, there is grass-root resistance to this, and a plethora of campaigning and on-the-ground initiatives that strive to demonstrate #refugeeswelcome, both in the UK and across Europe (Karakayali 2019). Many of these were started or gained momentum in 2015 during the height of the exodus from war-ravaged Syria.

Amidst this background of challenges, it could potentially appear facile to seek to understand and develop good professional practice in supporting refugees in the use of parks.[1] The principles previously discussed underpinned the intentions of the '#refugeeswelcome in parks' research project: local environments make a difference to quality of lives, safety concerns are important to address, we may need to be intentional about how we bridge differences and respond to the specific circumstances of those caught up in forced migration. We first deepened our understanding, interviewing sixteen refugee and asylum seekers across three cities and two countries (Sheffield and London, England and Berlin, Germany), and thirty-five different stakeholders in these contexts from both the refugee support sector and the greenspace management sector (Rishbeth et al. 2019). We then drew on our findings and examples of projects from across northern Europe to design resources which could inform good practice (www.refugees welcomeinparks.com).

Throughout the interviews we found that restorative qualities of spending time in parks and other recreational public spaces can offer temporary respite from the combination of waiting (often termed 'the limbo') and anxiety, which is the mundane experience for many seeking sanctuary. A high number of our participants reported benefits to their sense of wellbeing and inclusion from spending time in parks, especially when related to the sensory qualities of nature or human to human connection.

> For letting the stress out and having some relaxation by her own, she goes to the lake near them alone, without the kids or the husband. She walks more than half an hour around the lake ... it is not crowded, full of trees, no one sees what the others are doing, she can even cry lonely when she needs to do this without being embarrassed. The sound of the birds, the sound of the trees and how they look, the water in front of her, altogether makes her feeling relaxed, fresh, and helps her to forget her worries and gives her positive energy that she needs to be able to continue in her life.
>
> *(Fieldworker notes from an interview with Nawaj, female, Berlin/Palestine)*

> When we sit in the park we say hello to people. When we see someone with an Arabic face we talk to them, but we talk to anyone if they can understand our English.
>
> *(Khalid, male, London/Syria)*

> I would see no purpose in going [to parks] on my own, but from a moment where my friend introduced me to running, I take part in park running classes regularly.
>
> *(Firuz, male, Sheffield / Iran)*

However, especially in the initial weeks and months of settling in a new city, many also described feeling lonely, disoriented, and fearful about venturing out. Many have experienced or heard of racial harassment and avoid places where they sense they will stick out. Some will have come from countries in which there are parks and some not, but nearly all will be unfamiliar with the range of types of different types of urban greenspace in a northern European context, from heritage parks to canal-sides, cemeteries, sports grounds, and nature reserves. Local neighborhood parks in particular can be perceived as featureless, and many could not think of reasons why they might visit these places. 'Hanging out' here can feel vulnerable and too reflective of the 'do-nothingness' of their current situation.

'You need a certain boldness to venture out' observed Mercy, a Kenyan woman seeking asylum in London. Boldness is certainly easier to marshal when one has company, and 'going with a friend' has potential to overcome many anxieties. But the social isolation and high incidence of depression experienced by many refugees and asylum seekers means that support might often be needed to provide a sociable context. In the project we defined the need for 'curated sociability' in supporting access to urban greenspace. This reflects a notion of curating, which draws on its etymological roots as 'taking care,' a purposeful sharing of enthusiasms and knowledges that develops meaning and experience. These approaches are relational, providing a human-to-human connection as well as engaging with place. The good practices that we found and promoted included befriending schemes, walking groups, language classes outdoors, gardening projects and easy to access 'pick-up' sports such as basketball, football and table tennis, and were initiated both from the refugee sector and the greenspace management sector (all the following examples cited from Rishbeth et al. 2017).

The START initiative recognized that both students and refugees in Plymouth, UK shared experiences of being newcomers, and as such they organize regular walks together through the city parks and nearby countryside (START n.d.). This simple idea reduces anxiety, supports an ongoing familiarity and confidence with the local area but also bridges social barriers and breaks down stereotypes. In Paris, at 6pm each day, language teachers provide outdoor drop-in French lessons to the steps of the Place de la Bataille de Stalingrad. It is a time for developing friendships as well as learning. Situated in a busy location, but one often associated with rough sleeping and drug dealing, the public visibility of this activity is an important side benefit, providing a counter-narrative to asylum seekers as a social burden. We discovered many examples of gardening projects. Some of these are highly specific to refugees and asylum seekers, such as SLAM in London, initiated by a hospital as part of a holistic approach to addressing post-traumatic stress disorder. Others are local community non-profit organizations, which are open to all and provide the experience and skills to create a welcoming safe space for participants with potential vulnerabilities and specific needs.

> Growing happens, a cup of tea happens, a lot of conversation happens, people are sharing words with each other, recipes are shared on quite a regular basis. People will bring food to share. They might take produce, and then often they will cook it, bring it back.
>
> *(Facilitator at Green City Action, Sheffield, UK)*

Two UK national organizations, City of Sanctuary and Social Farms and Gardens, are working together to extend ideas and support for food growing projects that want to start or extend their inclusion of local refugees and asylum seekers, recognizing both its potential and some of its challenges.

Though the specificity of a neighborhood growing space offers very different affordances and levels of security to spending time in urban parks or squares, understanding the practicalities and relationships that may be embedded in 'curated sociability' approaches can illuminate some of the creativity and community building that can shape inclusion in a wide range of urban greenspaces. Supporting orientation, providing a time of social connection, giving a sense of purpose, and understanding the wellbeing benefits of nature contact can be important to everyone, but inform specific resonances and challenges with regard to refugees and asylum seekers. At the least they provide pause points of respite and distraction amidst extremely difficult life circumstances, at best they provide starting points for local belonging, with the greenspaces themselves re-appropriated, even re-made, through processes of inclusion and shared agency.

Cultures of mundane inclusion

Underpinning all of the above is the assertation that the use of public open space is a cultural practice, one which both reflects and shapes the mundane and diverse ways in which our collective cultures are changing. In my research, I specifically address collective and personal histories of migration. In the 'throwtogetherness' of place Massey (2005) highlighted both integrations and contradictions between the compellingly local and all the 'elsewheres' in our lives. In particular, the sensory qualities of spending time outdoors can represent both familiarity and alienation: smell, touch and visual recognition ground us in the here and now, but also come to us laden with memories, tinged with dimensions of loss and joy.

The presence of individuals is social positioning, a person in public outdoor space is both a viewer and as someone in view. While we all are attuned to the nuances of this, the check in the mirror before we head out, the awareness of these dynamics can be heightened and experienced as vulnerability when an individual is uncertain about their ability to 'fit.' The notion of 'fitting' here reflects both a self-identity (skin color, gender, clothing which affiliates to a particular religion), and the social diversity and complexity of the locality. Ethnographic research seems to indicate that the individual burden of 'fitting' (and the practical limits and multiple losses of this) is partially lifted when the context itself is reflective of difference and fluidity of use (Hou 2013; Hall 2015; Rishbeth and Rogaly 2018). At one level, this is shaped by local demographics: where there are high levels of social mixity and population churn to be 'a newcomer' is inherently a more usual identity, and it therefore it seems more feasible to establish or expand new niches. However, I would argue that the temporality of public open space, how this is reflected and experienced, can also be important. Local place change reflects global connections: social, economic, technological, and ecological. Of course, 'heritage' is never entirely unproblematic, but to be able to trace the layers of history in an environment underpins an often-unarticulated expectation of future change. In a highly publicized strategy to welcome Syrian refugees in Bute, a fairly remote Scottish island, the potential for newcomers to belong was not due to an existing ethnically diverse population, quite the opposite, but reflective of the notion of the island history as fundamentally shaped by both leaving and arriving (Mckenna 2017). In an urban context, the city experienced in a constant shape of re-making is a physical embodiment of top-down and bottom-up successes, failures, and new beginnings. Public open space reflects these many dynamics, the quality of looseness and adaptability allowing a responsiveness to the fluidity of transnational cultures.

I conclude by taking us back to the urban park. At different life stages (and for some more than others), we choose to spend time in this local venue for recreation, fresh air, and socializing. While acknowledging the limits of equality of access, these places still reflect the ethos of 'a public good' and practically offer inclusion within the collective outdoors. In the mundane temporal and spatial patterns of daily use, a diversity of culture is expressed and shaped. Difference is supported and occasionally resisted (no drinking, no skateboarding, do not feed the pigeons). Newcomers to the city chance upon this place and take the opportunity to have a smoke or kick around a football or take a photo of themselves in front of the rhododendron flowers … alongside the other smokers, footballers and selfie-takers who have lived here for decades. These places allow for moments of 'becoming local' through the everydayness and visibility of a shared humanity in this city of many strangers.

Note

1 The '#refugeeswelcome in parks' research project was funded by the Arts and Humanities Research Council, UK specifically to support practical change, 2017. It was a collaboration between the University of Sheffield, the University of Manchester, The Young Foundation (London) and Minor (Berlin).

References

Clark, C. R. (2014) Glasgow's Ellis Island? The Integration and Stigmatisation of Govanhill's Roma Population. *People, Place and Policy*, 8(1): 34–50.

Gehl, J. and Gemzoe, L. (1996) *Public Spaces – Public Life*. Copenhagen: Danish Architectural Press.

Hall, S. M. (2015) Super-Diverse Street: A "Trans-Ethnography" across Migrant Localities. *Ethnic and Racial Studies*, 38(1): 22–37.

Heuman, A. N. and González, A. (2018) Trump's Essentialist Border Rhetoric: Racial Identities and Dangerous Liminalities. *Journal of Intercultural Communication Research*, 47(4): 326–342.

Hou, J. (2013) *Transcultural Cities: Border-Crossing and Placemaking*. New York: Routledge.

Karakayali, S. (2019) The Welcomers: How Volunteers Frame Their Commitment for Refugees. In: *Refugee Protection and Civil Society in Europe*, Feischmidt, M., Pries, L. and Cantat, C. (eds). Cham: Palgrave Macmillan, pp. 221–241.

Lefebvre, H. (1968) *Le Droit À La Ville*. Paris: Anthopos.

Liberty. (2018) *A Guide to the Hostile Environment*. London: Liberty.

Massey, D. (2005) *For Space*. London: Sage.

Mckenna, D. (2017) "The Refugees Who Brought Hope to a Scottish Island." *The Guardian*, 24 December. Accessed 24 November 2018. www.theguardian.com/uk-news/2017/dec/24/bute-scotland-syrian-refugees-asylum

Neal, S., Bennett, K., Jones, H., Cochrane, A. and Mohan, G. (2015) Multiculture and Public Parks: Researching Super-Diversity and Attachment in Public Green Space. *Population, Space and Place*, 21 (5): 463–475.

Rishbeth, C., Blachnicka-Ciacek, D., Bynon, R. and Stapf, T. (2017) *#Refugeeswelcome in Parks: A Resource Book*. Sheffield: University of Sheffield.

Rishbeth, C., Blachnicka-Ciacek, D. and Darling, J. (2019) Participation and Wellbeing in Urban Greenspace: 'Curating Sociability' for Refugees and Asylum Seekers. *Geoforum*, 106: 125–134.

Rishbeth, C. and Finney, N. (2006) Novelty and Nostalgia in Urban Greenspace – Refugee Perspectives. *Tijdschrift Voor Economische En Sociale Geografie*, 7(1): 27–46.

Rishbeth, C. and Powell, M. (2013) Place Attachment and Memory: Landscapes of Belonging as Experienced Post-Migration. *Landscape Research*, 38(2): 160–178.

Rishbeth, C. and Rogaly, B. (2018) Sitting Outside: Conviviality, Self-care and the Design of Benches in Urban Public Space. *Transactions of the Institute of British Geographers*, 43(2): 284–298.

Rose, G. (2003) Landscape, Identity and Estrangement. In: *Deterritorialisations … Revisioning Landscapes and Politics*, Dorrian, M. and Rose, G. (eds). London: Black Dog, p. 72.

START. (n.d.) "Welcome to START." *Students and Refugees Together.* Accessed 27 October 2018. www.studentsandrefugeestogether.com

Webster, C. (2017) Race, Religion, Victims and Crime. In: *Victims, Crime and Society: An Introduction*, 2nd edition, Davies, P., Francis, P. and Greer, C. (eds). London: Sage, pp. 207–228.

Wessendorf, S. (2013) Commonplace Diversity and the "Ethos of Mixing:" Perceptions of Difference in a London Neighbourhood. *Identities*, 20(4): 407–422.

Whyte, W. H. (1980) *The Social Life of Small Urban Spaces.* Washington DC: Conservation Foundation.

Wilson, H. and Darling, J. (2016) The Possibilities of Encounter. In: *Encountering the City: Urban Encounters from Accra to New York*, Wilson, H. and Darling, J. (eds). New York: Routledge, pp. 1–24.

Young, I. M. (1990) *Justice and the Politics of Difference.* Princeton: Princeton University Press.

4

APPROPRIATION OF PUBLIC SPACE

A dialectical approach in designing publicness

Elahe Karimnia and Tigran Haas

Introduction

Publicness is a crucial concept in urban design practice that provides a critical though constructive backdrop to approaches and strategies that go beyond simply designing 'public space.' Planning and building public spaces are among the objectives of urban design projects, while the publicness of spaces is shaped as the aftermath of larger development strategies. Over the last decade, we have witnessed the socio-cultural potential of public spaces through tremendous efforts by activists, practitioners as well as scholars. Scholarly research (mainly in the social sciences) has addressed the multidimensionality and complexity of the publicness of urban spaces, questioning the production of pseudo-public places as an urban design outcome. Addressing the question '*Whose public space?*' casts a negative light on urban design practice for its contribution to systematic exclusions from public spaces (Madanipour 1995; 2010). The publicness of urban spaces is subject to change through the everyday spatial practices that reflect the importance of expanding the notion of 'design' as a process, and embracing the multiplicity of spaces, actors/actions as well as unintended consequences as a driving force in the design process.

In this chapter, we intend to build a deeper understanding of the challenges embedded in spatial production process, and in the required shift in urban design practice when designing publicness. We argue that producing publicness is not a linear process of applying experts' knowledge and approaches conceived in a positivist mode. Publicness is in fact not an outcome of any specific stage of the urban design process; rather, publicness emerges out of a series of purposeful strategies and actions enabled or delimited by public spaces in everyday practice.

The potentials of public space, as infrastructure for social activities, cultural production, and political expression, have been highlighted, specifically over the last decade, in various academic conferences, educational programs, and even policy-making platforms. However, we still lack multi-disciplinary knowledge of these potentials in and for practice (Carmona 2014). A dialectical approach suggests that urban design knowledge can evolve through practice: applying the knowledges of everyday practices, such as the way public space is appropriated, lets urban designers redefine the problems encountered in both practice and design (Inam 2011). Appropriations reveal that public space is a medium and a catalyst where various actors, together with their aspirations, needs, and conflicts, participate in producing publicness.

This chapter concerns the complex processes of producing publicness. The first argument of this chapter concerns how publicness as an abstract 'intention' is translated through urban development processes. We discuss the limitations of spatial production, which is usually driven by political and economic intentions and territorial strategies and is known as the intention–outcome gap in urban design. Urban design is often subject to criticism for the gap (Foroughmand Araabi 2017) arising from the antithesis between a substantive–descriptive understanding of publicness and a normative–prescriptive design of public spaces, with urban design traditionally considered responsible for the latter (Moudon 1992). This argument calls for a deeper understanding of the contribution of practice to capital creation (i.e. neoliberal practices of urban design).

The second argument highlights the appropriations by which space becomes public in action, and the sets of efforts by those whose regular presence, togetherness, and encounters contest the intended publicness. Appropriation is discussed in light of Lefebvre's concept of 'trial by space,' to address everyone's right and intention to practice their being in space, resulting in levels of territorial adaptation and domination that accordingly transform the publicness of space. The third argument of this chapter treats the implications of appropriations as creative conflicts that can identify and challenge the assumptions of urban design practice. Such an approach recommends a multi-scale and consequence-based framework for designing not only the space but also the conditions under which publicness emerges. This represents a shift towards creative urban practices going beyond publicness as intention and public space as outcome.

Intention–outcome gap: materializing publicness

The urban theorists concerned with the larger political effectiveness and social relations of publicness expanded the notion of public space to encompass the 'public realm' or 'public domain.' Although the notion of 'public' as constituting individuals or the collective has been defined in different ways, as has the function of public space, there is no doubt in the potential of materialized forms of the public realm, i.e. public spaces as 'sites of public use and citizen interaction,' having a causal impact on social life (Goodsell 2003, 363; Soja 2003). These concepts suggest going beyond the boundaries of physical space to include other conditions that affect the quality of publicness, for example, privately managed collective spaces that function as the public domain (Hajer and Reijndorp 2001). When public space is discussed at the intersection of urban design and socio-political studies, the type and levels of publicness is critically scrutinized. But in practice, publicness as a highly intended value is limited to physical space to be public, and the various effects and democratic contributions of 'public' spaces, as consequences, are poorly understood and applied.

Whatever the explicit intentions for initiating an urban design project, the driving forces of spatial production define such elaborated and abstract aspirations. Better understanding the actors, knowledges, and uneven power dynamics engaged in spatial production processes helps us understand why social and political theories of cities can be partially and uncritically applied in design and development processes. In this process, urban design actions are merely intended to 'secure a commission, gain support and/or acceptance for a project, obtain funding and/or secure permission for a development or proposal' (Carmona et al.

2010, 331). Designing publicness through conventional urban design processes usually involves narratives of reduction, simplification, and obsession with the production of 'public space.' Publicness is therefore partially produced by the labors of city designers and in different stages of the design process and therefore it requires a 'holistic and strategic understanding, not defined by or limited to the geography of responsibilities' (Karimnia 2018, 191). The publicness of space as a holistic value is identified through larger development strategies and is gradually materialized through long-term processes. Deeper research into urban design practice addresses the challenges of the implementation process whereby spatial strategies are designed and materialized in space (e.g. Carmona 2014; Carmona et al. 2010; Punter 2003).

Since local authorities have lost their power and ability to create and maintain urban spaces, neoliberal approaches define contemporary urbanism as 'megaproject' urbanization and capital influences on the complex process of spatial production, from early stage and within focused territories (Harvey 1975). As a result, urban design is inclined by emancipatory politics and speculative developments, which demands rapid capital turnover. Urban design, under neo-liberalism, takes an entrepreneurial approach, serves as a tool of neo-liberalism, and paves the way for cooperative and regulated competition (Madanipour 2006). Even though space is supposed to serve both private and public interests, in practice the demands of the service economy and the private sector influence the planning, design, and implementation processes (Cuthbert 2010; Harvey 2000; Madanipour 2006). This is now even happening in welfare states where market values define and drive the urban design projects, mainly through public–private collaboration.

Public–private initiatives to design publicness highlight the uneven power dynamics between different actors involved in long-term planning, design, and development process, possessing a wide range of responsibilities and capacities to make decisions and exert influence. These actors also have different periods of engagement and accordingly different interpretations and expectations of project outcomes (Carmona 2017; Carmona et al. 2010; Childs 2010; Madanipour 2006). The process by which designers decide on the types and levels of publicness includes both knowledge and trade-offs, but not in isolation. Urban design, as a social practice engaged in the political economy, is defined as an attempt to 'express an accepted urban meaning in certain urban forms' (Castells 1983, 304).

Urban designers have the ability to discuss the political economy spatially and to visualize the consequences of partial decisions cohesively. However, such knowledge of urban design practice is not powerful enough to challenge and influence decisions on the types and levels of publicness, which are usually negotiated and made jointly by planners, landowners, financers, developers, and their associated architects. Another issue is that urban designers become engaged in the process when decisions about the location (larger relations), social and economic aspects of the project (long-term relations) have already been taken; the territory of publicness or the 'site' has been defined, and the urban form through which public spaces are shaped is the subject of trade-off between the project owners and developers to reduce the cost and the time of implementation. Therefore, publicness of spaces is an aftermath of the design of private spaces and buildings.

This issue is reflected in privatized public spaces where private interests decide on the levels and types of public engagement in urban spaces. The publicness of these spaces is described through abstract, politically correct, and technically savvy intentions and usually referred to as spatial transformation. Later, the publicness is justified by the provision of physical accessibility and high visibility of functions as more 'public' while it is swept under the carpet of privatization and consumption. Public space is exemplified as embodying timeless, static values, though not all publicly accessible space, or public functions are perceived as public. Under capitalist investors, urban space has led to the rise of 'economic space' and spaces with use value and exchange value overlap (Friedmann 1988). Therefore, the boundaries between public and private space in everyday use have become blurry and there is no sharp distinction between economic spaces and life spaces (Zukin 1991). Global empirical research on the levels and types of publicness reveals the complexity and multidimensionality of publicness (Madanipour 1995; Németh and Schmidt 2011; Varna and Tiesdell 2010). To avoid reductiveness, some researchers offer a range of possible criteria for publicness, such as ownership, control, civility, physical configuration, and animation (Varna and Tiesdell 2010). Through these dimensions, Varna and Tisdell described 'more public' and 'less public' spaces while considering the construction of publicness in both design-oriented/physical and managerial/social ways. So, publicness is measured as an interaction between individual's experiences of place (micro-actions) and macro-contextual relations.

In *Publics and the City*, Iveson (2007) argued for the multidimensionality of 'public' space, noting failures in discussing its spatiality using stable 'topographical' approaches. Understanding public space 'procedurally,' as he recommends, is about 'use at a given time for collective action and debate' (Iveson 2007, 3). In this view, public space is not just in a state of being; rather, it becomes a site of power and collective action, while its 'being' constantly changes. Attention has recently been paid to the process rather than the product of urban design practice. Through this process context and power, as Carmona (2014) considers them, are at the core of the ongoing process of shaping and reshaping places. Through a deep analysis of several public spaces in London, Carmona (2014) scrutinized various stages of the design and development (i.e. shaping), use, and management (i.e. reshaping) of space by which it becomes public. However, such a procedural understanding of publicness does not offer any 'new theory' to urban design practice, as it claims, rather expanding its responsibilities to the management level.

Although public space is intended to be accessible for everyone, and to welcome diversity of actors and uses, the urban management aims for 'zero-friction' atmosphere, which supposedly represents collective satisfaction, i.e. 'symbolic' public space that is constantly in use and free from any undesired consequences, such as fear of vandalism and crime. This mistaken ideal results in over-managed spaces with continuous surveillance, certain dominant users, and permanent beneficiaries. They exert their power over these spaces through, for example, CCTV cameras, and community policing, which keep out a large proportion of the public to prevent declining use.

Publicness welcomes high diversity and multiplicity of activities, but urban design can risk reducing it simply by recreating the appearance of these qualities as static objects rather than emerging values. For example, the quality of public gathering is usually understood as use and presence in space, with different material and immaterial conditions that can support it to emerge. But in practice togetherness is about intensified use or increased disposition towards others without considering the conflicts of interest and perceived access to public space by those with less visibility and power. These strong assumptions direct developers' attention towards making spaces that result in increasing consumption and leisure, limited sensitivity to differences, and limited possibilities of appropriation. Therefore, values such as public gathering and togetherness are 'far from predictable when it comes to questions of collective inculcation, mediated as they are by sharp differences in social experience, expectations and conduct' (Amin 2008, 7).

Designing for multi-scale and multidimensional publicness is about the interactions between designed, perceived, and lived space that can enable or restrict publicness. Such interactions are dialectical and go beyond merely applying rational and empirical knowledge. A dialectical approach suggests that urban designers should look beyond patterns of behavior and encounters and critically seek alternative ways as a set of actions able to transform the perception of space and its publicness in everyday life. Regarding appropriations not as *unintended outcomes* but as materialized interactions and tensions between actors invites design practitioners to critically navigate such consequences and exploit them as sources of action. Exploring appropriations as unintended consequences reveal the invisible or unexpected actors in space, or spaces that become public through everyday use, or temporary activities going beyond defined functions. Generally, urban design practice needs to learn from appropriations, that interrupt professional determinism and fundamental thinking are needed to expand the theory of practice. A dialectical approach recommends a body of knowledge extending from effect to cause. Accordingly, designing publicness is about designing conditions based on the relationship and possible tensions between larger strategies and everyday tactics.

Appropriations as driving forces for transforming publicness

We consider appropriations of public space to be multi-scale conscious actions resulting from conceived and perceived publicness. Appropriations are also related to individual and collective users' self-perceptions, rights to space, and interest in space, as well as to the physical environment, all of which affects users' passive or active engagement in space and its publicness. The transformation of publicness needs both time and interruptions. Appropriations not only interrupt the established power, defined meanings, expected uses, spatial order, and what is considered proper, but also become a driving force for transforming publicness. They destabilize and restabilize publicness. In view of recent debates about public space, appropriations can be defined simply as a set of collective actions, such as occupations and protests, to reclaim the space. More than defining types of actions, the focus is on the approach that materializes micro-scale presence in space, possibly blurring the dichotomy between public and private. This approach not only disrupts reductive narratives about 'good public spaces' but illustrates how dialectical understanding is important in navigating the consequences and effects of urban design approaches.

We base our understanding of appropriation on Lefebvre's concept of 'trial by space,' mentioned towards the end of *Production of Space* and emphasizing that 'groups, classes and fractions of classes cannot constitute themselves, or recognize one another, as subjects unless they generate (or produce) a space' (Lefebvre 1991, 416). Through this concept, Lefebvre addresses the power embedded in the space that, metaphorically, is produced by and can produce dialogue between the actors whose interests are materialized. Appropriations are perceived publicness turned into actions based on social or political intentions; primarily, however, they are alternative practices built in relation to and upon the existing setting that can challenge them. Space therefore becomes a condition for action, and public space becomes more than public territory. It becomes a 'spatial actant' that 'brings about a certain effect in a certain situation or place' (Kärrholm 2007, 440).

Perceived publicness can be traced in the remarkable studies conducted in the 1970s by urbanists such as Jane Jacobs, Kevin Lynch, and William H. Whyte, who discussed individuals' contacts with space based on perceived (visual) qualities of the built environment and its surroundings. Environmental perceptions help urban designers figure out the conditions that people (not vulnerable groups though) prefer for adapting themselves to a dominant setting ('adaptation-level theory'), as probabilities deriving from individuals' experience of functioning ('probabilistic functionalism'), or as 'active observers' detecting the environmental structure for functional opportunities ('ecological approach to perception' or affordances) (Nasar 2011, 164–165). The concept of 'affordance,' introduced by Gibson (1979), addresses the possibilities for action based on perceptions of the built environment, which evoke meanings and multiple 'senses of place' (Nasar 1998). Perceived qualities such as safety, legibility, crowding, or comfort affect people's behaviors (e.g. distance estimation for walkability), choices of public space (e.g. historic places or green spaces), and social encounters in public spaces (e.g. inclusiveness, meaningfulness, safety, comfort, and pleasurability) (Mehta 2014). These perceptions based on observing urban micro-actions provide urban designers with a set of assumptions regarding affordances in public spaces (e.g. favorable locations for gathering or sitting) or regarding preferred personal and social distances in public space (see Gehl 1987).

These phenomenological studies suggest that well-being can be cultivated in public space, as a realm for individuals' feelings and satisfaction. Through delayering well-being into the material (i.e. physical space) and immaterial conditions (i.e. urban context, time, relationship with others, and other relations) we argue that publicness emerges from individual and collective actions cannot be either romanticized or simplified and should be understood in its specific context and time (see examples of such shared meanings in Karimnia 2018). For example, individual satisfaction, based on comfort in public space is an important factor observed to be weakly related to a high level of togetherness, which occurs through public use of space or in response to collective needs (Karimnia 2018). Togetherness therefore cannot be assumed to arise from passive engagement and presence in space.

From a sociological perspective, space is perceived as more 'public' when it tolerates differences and heterogeneous individuals (Lofland 1989; Young 1990). Sufficient evidence proves that the right to differences in public space provides plenty of opportunities for identity construction, social encounters, and practices of civility and conflict, which become evident in public space as a realm of expression (Varna and Tiesdell 2010). Through appropriations, such social encounters and engagement can contribute to the transformation of publicness. This is not to say that public space is only a territory of thrown togetherness and 'civil inattention,' which are insufficient conditions for public culture (Young 1990; Lofland 1989, 462; Amin 2008). The appropriation of space emphasizes the conditions that allow differences in ideas, people, desires, things, and demands to become public as forms of spatial practices based on the right to space and the right to use as well as to produce.

As the concept 'trial by space' emphasizes, the right to access is crucial for public space to have the expected impact. As Marcuse (2005) stressed, access to public space is becoming tightly controlled, resulting a sharp line between 'those who are entitled to enter and those who are excluded' (111). Based on social control grounds, not everyone is considered *the public* and publicness is more of a concern rather than a potential. The signs of control in urban spaces delimiting certain groups and behaviors, such as loitering or skateboarding, reveal the regimes of control in the everyday experience of space.

Appropriations can be enabled or restricted in different scales and types, through both material conditions such as the form and scale of space, and immaterial conditions such as managerial rules, land use regulations, and rent policies. Both physical configurations and regimes of control can affect the levels of presence and micro-actions. As Mitchell (1995) pointed out, the system plays an important role in excluding certain people from certain places. For example, the presence of homeless people in a park as 'representative of a community' is less about the design of the park. As a result, publicness is challenged and transformed by some groups and their actions in unexpected locations, illustrating a complex interplay between different relations of access to public space. The value of such appropriations lies in their interruptive characteristics and transformative power of expression towards more equality and inclusion in public spaces (for more about the conditions that allow or restrict public expression, see Kohn 2004; Low 2009; Hou 2010).

The consideration of people not only as the users but also as the producers of space was introduced by the 'everyday urbanism' approach, which explored forms of bottom–up and incomplete urban design and redefined both 'public' and 'space' through the lens of lived experience (Crawford 1999). As Crawford illustrates, everyday public spaces serve as public arenas where 'debates and struggles over economic participation, democracy and the public assertion of identity take place' (28). The publicness of these spaces therefore has 'in-between' characteristics in terms of spatiality, temporality, activities, management, and/or sense of publicness (Simões Aelbrecht 2016). The micro-actions in these spaces reflect macro socio–economic and political issues in the form of alternative practices with transformative power (Karimnia 2018).

As everyday urbanism suggests, urban design should enable appropriations by treating micro-practices as assets, applying them as situated knowledge through a local–expert partnership process to leverage the intentions underlying them. These practices (see examples in Karimnia 2018) do not offer a fixed model of public–private collaboration or type of public space, but rather illustrate a process through which actors and their socio-cultural and political needs and desires become public. Appropriations of urban spaces shape conditions (physical space, managerial rules, and operational network) that can favor equity and care via sharing space as an urban common for degraded needs. These practices not only shed light on invisible actors affected by larger systematic issues in their everyday urban practices but also give them positions from which to act and produce publicness. Public spaces should accordingly be seen as places giving visibility to such issues, groups, and constantly changing patterns. For example, Brighenti (2010) described the urban wall 'subject to both strategic and tactical uses' that is beyond an urban territory and can become a media, enable practices such as graffiti writing, which turn walls into surfaces of projection for visible traces and assertion. He discussed the processes by which a wall becomes part of an ongoing conversation, 'shared attention and the field of the distribution of immediate and mediated visibilities' (326).

We want to clarify that the appropriation of public space does not help designers and planners if it is seen as a separate means of production. Everyday public spaces should be discussed in relation to established public territories that allow or restrict emerging publicness. Appropriations help us assess publicness in relation to and beyond the use of space or, recalling Amin (2008), beyond intense togetherness as signifying success. Appropriations reveal ways of destabilizing the prevailing publicness as spatial practices that can contest what is defined as ordered, formal, or proper in public space and then restabilizing these practices. Therefore, tactics and strategies, permanence and impermanence, people and designers should all be seen as in constant interaction to transform publicness in urban spaces. Designing publicness of spaces requires such a dialectical understanding of appropriations to think of publicness beyond the territory of public spaces. Therefore, the process of designing publicness is fundamentally about *enabling*, through both the materiality of space and its socio–economic and political relations to the larger system, which we continue discussing in the following section.

Taking a dialectical approach to practice

By discussing the socio–spatiality of appropriations, we tried to address public space not as a designed public territory but rather as a 'spatial actant' with publicness as its consequence:

spatiality that 'brings about a certain effect in a certain situation or place' (Kärrholm 2007, 440). A territory designated 'public' through public ownership or public function is perceived as public through individuals' appropriations, which entail domination and taking 'ownership' for their own uses or expressions. Instead of designing publicness into a legally and functionally public territory, appropriations encourage us to design conditions for such democratic experiences that can stimulate, generate, afford, and sustain individual and collective spatial practices that support the voice of the powerless, and the visibility of the invisible ones. Design, therefore, is about the possibility of affording certain actions, promoting the visibility and availability of conditions for the perception of different alternatives.

A dynamic public realm accommodates appropriations as everyone's right to practice their being in space. In fact, appropriations narrow the gap between ideal or intended publicness and practiced publicness and suggest urban designers to take account of the diversity and possible visibility of different groups, their practices, and, accordingly, the possible conflicts. Appropriations expand design responsibilities to encompass political actions through which individuals' appropriations of space can be enabled or restricted. These 'creative political actions' can be undertaken through a set of 'moral decisions,' to impose extra control on the levels and types of publicness that accordingly rule out these possibilities of place, to expand knowledge and experiment with unexpected or unintended actors (e.g. varying in background, race, gender, and class), physical spaces (e.g. corridors, thresholds, and edges), and time (e.g. the multiple temporality of activities), or to support collective actions (Inam 2014).

One example of appropriation of space, and of a dialectical approach to designing publicness, was when southeast Asian migrants, primarily from the Philippines, working as domestic workers for Hong Kong families gathered every Sunday in the ground floor of the HSBC building in the city center. This space is publicly accessible and privately managed, but the presence and social encounters of the workers were not considered simply *public life* at first. They gathered, sat on straw mats, did their nails, and got haircuts, yet such meetings and gatherings were unexpected uses of this place by unexpected actors. This minority appropriated an accessible space to address their needs and opinions and engaged the (other) public in their struggles, at different levels of salience to consciousness. By discussing appropriations, we would like to identify the powerful actions that can transform the publicness of spaces. The choice of this space can be discussed in relation to its physical accessibility, being semi-open, and its central location, but mostly in relation to the workers' presence and actions, rooted in their dissatisfaction, in this unexpected space. Appropriations highlight the pseudo-publicness of privately-owned public space, systematic socio–economic exclusions in cities, and the unintended consequences of designed 'public' spaces. After unsuccessful efforts to relocate these domestic workers, their gathering in the city and intended activities were facilitated through adding temporary services such as toilets and adjusting the car traffic around the area. In this way, urban practice enhanced the publicness of space by enabling a less powerful and visible group, together with their ignored or derided rights, to become public and materialized in the city.

As the example above illustrates, the appropriation of public space helps us understand urban design beyond the spatio–temporal dynamics and the multiple intersecting material processes involved. It illustrates alternative practices through which design enables the visibility of others and their potentials to make change and transform publicness. This concerns urban designers' central position in the process, which is usually led by the architectural will 'to create,' 'to shape spaces so as to give them social utility as well as human and aesthetic/symbolic meanings.' and 'to give material form to the longings and desires of individuals and collectivities' (Harvey 2000, 199–201). The urban design field has been directed towards new knowledges and methodologies by which the complexity of urban issues, such as publicness and the power dynamics of 'making' processes, should be acknowledged. Yet, urban design practitioners have cut themselves off from reflective processes through which they could observe the unintended consequences of their practice. Instead of following the instrumental roles of 'making,' as giving shape to a product, appropriations can be assessed as experiments and help urban designers with their thinking, by altering their position and giving them the will and power to lead an uncertain process. Leading such process means arranging the underlying economic investments and preparing for the consequence of possible compromises, which requires navigating and reflecting on ambiguities that might be unclear from the beginning (Inam 2014; Verma 2011). This in fact concerns not only urban design knowledge but also its moral and reasonable decisions in listening to and observing differences and different ways of making.

In line with the above understanding, 'social and spatial relationships are dialectically inter-reactive, inter-dependent … social relations of production are both space-forming and space-contingent' (Soja 1980, 211). According to Low (2009), physical interventions in the landscape (or the part of it that is considered in urban design) 'attract opposition because they produce socio–spatial forms that reference deep and still unresolved or unresolvable conflicts among political economic forces, social actors and collectivities' (25). Urban design, to contribute to the transformation of publicness, should consider appropriations not as conflicts but as opportunities to reflect on actions that unintentionally, unexpectedly, or unknowingly restrict transformation process.

What we advocated, i.e. designing publicness through emergent appropriations, is a nexus between urban space and the intentions and politics involved in making it public. Such design is not a linear process, and appropriations can interrupt such a linear understanding in which physical space can become a condition of and for changing intentions. In this way, urban design is a process of enabling or constraining such becoming, and public space is the spatial representative of these relations. When a space is appropriated repeatedly in everyday use, this highlights a controlled condition in which dynamic tension can become stabilized, giving rise to visible desires in public space. Design makes this dynamic visible and becomes a social and political act of producing publicness.

Conclusion

In recent years, there has been tremendous effort to put the important notion of public space back on the agenda of research and practice (Haas and Mehaffy 2019). Dealing with the multidimensional factors that affect the publicness of space calls for a shift in designers' conventional thinking and in their engagement in the design process, to being more receptive of an inclusive and equitable approach to hearing, observing, and enabling everyone as publics. Appropriations are highlighted as alternative practices of producing publicness, illustrating the socio–political impact of design. They call for micro-practices, as design consequences, to be brought into planning and design processes.

We are aware that designing public space is not sufficient for cultural and social production, and that public space per se is not a social infrastructure. We need a practice of designing relationships that produce and transform publicness in terms of infrastructure. These practices, as appropriations reveal, are networks linking physical space, access, controls, interests, as well as cultural and social structures. Appropriations expose less visible conditions by which publicness is truly transformed. Such transformation takes time and invites designers to assume new responsibilities and understand their practice in terms of various sequences and outcomes.

We argue that appropriations have the potential to denote a critical politics rather than just policies for producing public space. The question to be addressed then is what conditions should be embedded into the built environment to ensure that publicness is produced, reproduced, and transformed. This is not a question we have attempted to answer here, although the authors' previous empirical work articulated specificities and identified modes of planning and design that engender the growth of these conditions.

To conclude, we call for the development of a field of study in which the design of publicness is an explicit focus, both as a locus of criticism of the way public space is used or misused as a justification for urban development, and as an active undertaking to understand how the right to appropriate space and produce publicness can be stimulated by urban design.

References

Amin, A. (2008) Collective Culture and Urban Public Space. *City*, 12(1): 5–24.

Brighenti, A. M. (2010) At the Wall: Graffiti Writers, Urban Territoriality, and the Public Domain. *Space and Culture*, 13(3): 315–332.

Carmona, M. (2014) The Place-Shaping Continuum: A Theory of Urban Design Process. *Journal of Urban Design*, 19(1): 2–36.

Carmona, M. (2017) The Formal and Informal Tools of Design Governance. *Journal of Urban Design*, 22(1): 1–36.

Carmona, M., S. Tisdell, T. Heath, and T. Oc. (2010) *Public Places, Urban Spaces: The Dimensions of Urban Design*. New York: Routledge.

Castells, M. (1983) *The City and the Grassroots: A Cross-Cultural Theory of Urban Social Movements*. Berkeley: University of California Press.

Childs, M. C. (2010) A Spectrum of Urban Design Roles. *Journal of Urban Design*, 15(1): 1–19.

Crawford, M. (1999) Blurring the Boundaries: Public Space and Private Life. In: *Everyday Urbanism*, J. Chase, M. Crawford, and J. Kaliski. (eds). New York: The Monacelli Press, pp. 22–35.

Cuthbert, A. (2010) Whose Urban Design? *Journal of Urban Design*, 15(3): 443–448.

Foroughmand Araabi, H. (2017) Multiple Expectations: Assessing the Assumed Roles of Theory in Relation to Urban Design. *Journal of Urban Design*, 22(5): 658–669.

Friedmann, J. (1988) Life Space and Economic Space: Contradictions in Regional Development. In: *Life Space and Economic Space: Essays in Third World Planning*, J. Friedmann. (ed). New Brunswick: Transaction, pp. 93–107.

Gehl, J. (1987) *Life between Buildings*. New York: Van Nostrand Reinhold.

Gibson, J. J. (1986 1979) *The Ecological Approach to Visual Perception*. Hillsdale: Erlbaum.

Goodsell, C. T. (2003) The Concept of Public Space and Its Democratic Manifestations. *The American Review of Public Administration*, 33(4): 361–383.

Haas, T. and M. Mehaffy. (2019) Introduction: The Future of Public Space. *URBAN DESIGN International*, 24(1): 75.

Hajer, M. and A. Reijndorp. (2001) *In Search of New Public Domain*. Rotterdam: NAI.

Harvey, D. (1975) The Geography of Capitalist Accumulation: A Reconstruction of the Marxian Theory. *Antipode*, 7: 9–21.

Harvey, D. (2000) *Spaces of Hope*. Edinburgh: Edinburgh University Press.

Hou, J. (2010) (Not) Your Everyday Public Space. In: *Insurgent Public Space: Guerrilla Urbanism and the Remaking of Contemporary Cities*, Hou, J. (ed). New York: Routledge, pp. 1–17.

Inam, A. (2011) From Dichotomy to Dialectic: Practising Theory in Urban Design. *Journal of Urban Design*, 16(2): 257–277.

Inam, A. (2014) *Designing Urban Transformation*. New York: Routledge.

Iveson, K. (2007) *Publics and the City*. Malden: Blackwell.

Karimnia, E. (2018) "Producing Publicness: Investigating the Dialectics of Unintended Consequences in Urban Design – Practices in Stockholm and Malmö" PhD diss., KTH Royal Institute of Technology.

Kärrholm, M. (2007) The Materiality of Territorial Production. *Space and Culture*, 10(4): 437–453.

Kohn, M. (2004) *Brave New Neighborhoods: The Privatization of Public Space*. New York: Routledge.

Lefebvre, H. (1991) *The Production of Space*. Oxford: Blackwell.

Lofland, L. H. (1989) Social Life in the Public Realm: A Review. *Journal of Contemporary Ethnography*, 17(4): 453–482.

Low, S. M. (2009) Towards an Anthropological Theory of Space and Place. *Semiotica*, 175: 21–37.

Madanipour, A. (1995) Dimensions of Urban Public Space: The Case of the Metro Centre, Gateshead. *Urban Design Studies*, 1: 45–56.

Madanipour, A. (2006) Roles and Challenges of Urban Design. *Journal of Urban Design*, 11(2): 173–193.

Madanipour, A. (ed). (2010) *Whose Public Space? International Case Studies in Urban Design and Development.* New York: Routledge.

Marcuse, P. (2005) The "Threat of Terrorism" and the Right to the City. *Fordham Urban Law Journal*, 32(4): 767–785.

Mehta, V. (2014) Evaluating Public Space. *Journal of Urban Design*, 19(1): 53–88.

Mitchell, D. (1995) The End of Public Space? People's Park, Definitions of the Public, and Democracy. *Annals of the Association of American Geographers*, 85(1): 108–133.

Moudon, A. V. (1992) A Catholic Approach to Organizing What Urban Designers Should Know. *Journal of Planning Literature*, 6(4): 331–349.

Nasar, J. L. (1998) *The Evaluative Image of the City.* Thousand Oaks, CA: Sage.

Nasar, J. L. (2011) Environmental Psychology and Urban Design. In: *Companion to Urban Design*, T. Banerjee and A Loukaitou-Sideris. (eds.). New York: Routledge, pp. 162–174.

Németh, J. and S. Schmidt. (2011) The Privatization of Public Space: Modeling and Measuring Publicness. *Environment and Planning B: Planning and Design*, 38: 5–23.

Punter, J. (2003) *The Vancouver Achievement: Urban Planning and Design.* Vancouver: UBC Press.

Simões Aelbrecht, P. (2016) 'Fourth Places:' The Contemporary Public Settings for Informal Social Interaction among Strangers. *Journal of Urban Design*, 21(1): 124–152.

Soja, E. W. (1980) The Socio–Spatial Dialectic. *Annals of the Association of American Geographers*, 70(2): 207–225.

Soja, E. W. (2003) Writing the City Spatially. *City*, 7: 269–280.

Varna, G. and S. Tiesdell. (2010) Assessing the Publicness of Public Space: The Star Model of Publicness. *Journal of Urban Design*, 15: 575–598.

Verma, N. (2011) Urban Design: An Incompletely Theorized Project. In: *Companion to Urban Design*, T. Banerjee and A Loukaitou-Sideris. (eds). New York: Routledge, pp. 57–69.

Young, I. M. (1990) *Justice and the Politics of Difference.* Princeton: Princeton University Press.

Zukin, S. (1991) *Landscapes of Power: From Detroit to Disney World.* Berkeley: University of California Press.

5

LANDSHAPE URBANISM
The topography of public space

Karl Kullmann

Introduction: landscape and urbanism

Although historically grounded in garden making, landscape architecture's primary motivation throughout the past century has been the design of public space. In recent years, this focus extended from parks and plazas into the wider urban fabric, as the landscape urbanism movement advocated for a more ecologically expressive approach to urban design. In contrast to traditional urban design methods that prioritized the built form, landscape emerged as an active agent in the structuring of urban space. However, an unproductive disciplinary standoff ensued, whereby prioritizing extensive urban landscape systems appeared incompatible with more traditional templates for compact, human-scaled cities.

Set within the wider trans-disciplinary project of reconciling landscape and architecturally based approaches to urban design (see Kullmann 2018), this chapter explores the agency of landform in shaping urban public space. Underlying topography establishes a common ground in the sense that landform is integral to the performance of both natural systems and urban form. The intimate connection between landscape and landform is reflected in the suffix *scape*, which is etymologically linked to *shape* (Casey 2002). The land*shape*, both pre-existing and designed, manipulates spatial experience and performance through its contours. Across scales, the shape of the land influences where and how cities are built, where public spaces are situated, and how we interact with those places. As a malleable medium, landform, like sculpture, qualifies as a 'plastic art,' albeit an art that we walk on, not around.

Topography and the city

Throughout urban history, establishing level ground is a driving factor in the founding of settlements, with the earliest cities of Mesopotamia situated amidst irrigated flood plains. Roman surveyors also favored level sites, for which the surveying staff (*Groma*) was plumbed to sit perpendicular to a horizontal surface (Rykwert 1976). Earthworks were typically engineered for hydraulic or defensive purposes, with the latter reaching its zenith in the star-fortification towns of Renaissance Europe.

Naturally topographic sites were reserved for refuge settlements, as was the case with the medieval hill towns built following the decline of Roman stability. Centuries later, cities founded during the Colonial Era also often compromised on the levelness criterion in order to take of advantage of strategically located natural harbors set into rugged coastlines. In these cases, cutting and filling was typically undertaken in order to carve functionality from an otherwise dysfunctional landscape (Leatherbarrow 1999). Particularly rugged regions were attributed lower scenic value or avoided as badlands.

Notwithstanding the practical challenges that terrain poses to city making, it is notable that cities founded in topographic settings routinely rank among the world's most distinctive. San Francisco is renowned for the dramatically contrasting manner through which the urban grid and hill parks amplify the underlying topography (see Figure 5.1) (Lipsky 1999). Around Sydney Harbor, protruding peninsulas support geologically expressive public spaces. In the favelas of Rio de Janeiro, unplanned public spaces cling precariously to the slopes of granite megaliths. And in Hong Kong, ladder-streets lead uphill to botanical gardens and micro-parks that are engineered onto the hillside (Kullmann 2017).

In the twenty-first century, these historically distinct narratives of level and hilly urban settings are increasingly convergent. As global urbanization accelerates, cities originally founded on level ground are pressured to expand into hilly hinterlands, thus becoming more topographic overall. From San Diego to Sao Paulo, Madrid to Nairobi, and from Nagasaki to Perth, the un-leveling of sprawling cities is evident across continents.

Topography and urban theory

Despite the persistent intersection of cities and landform, topography has generally remained a peripheral topic throughout the past century of urban theory. Within modernism's 'towers in the park' vision, the city was elevated onto *pilotis*, beneath which the greened public realm was

Figure 5.1 Contour map contrasting the street grid and underlying topography of San Francisco's Russian Hill and Telegraph Hill districts.

Image Credit: Karl Kullmann

to flow unhindered. However, when realized on a large scale through post-war reconstruction in Europe and urban renewal in the US, decoupling the city from the ground unintentionally devalued public space. Too often, modernism's pastoral landscape ideal devolved into flat, sprawling, automobile dominated wastelands (Ingersoll 2006; Jacobs 1961; Sennett 1990).

In reaction to the significant shortcomings of the modern project, two urban models more directly considered landform. Firstly, ecological planning specifically incorporated the criterion of topography through 'suitability-analysis' mapping techniques (see McHarg 1969). Although these methods proved effective at saving steeper terrain for public space and lowering residential densities, the commercial returns from developing desirable topographic locations often outweighed this reasoning. Moreover, by encouraging dispersed development, ecological planning arguably accelerated suburban sprawl (Hill 1992).

The second significant post-modern urban design model, which remains current, takes a more compact approach to cities. Traditional urban design focuses on reinstating the key formal qualities of pre-industrial settlements. With precedent cities most often sited on navigable waterways, traditional urban design is predisposed to a topographically based town-and-country dichotomy. As depicted in the influential Rural-to-Urban Transect, landform is typically positioned on the periphery as a scenic backdrop to the level urban core (see Duany 2002). Although applicable in certain ideal situations, transects through many metropolitan areas reveal more intricate relationships between urban and land morphologies (see Bosselmann 2011).

Topography and public space

In the nineteenth-century tradition, the size of public spaces strongly dictated landform strategies. At the smaller end of the spectrum, urban squares and neighborhood parks were ideally level, with levelness permitting ease of construction, circulation, servicing, and use. At the larger scale, public parks were free to undulate. The topographic park served the desire to bring nature into the city, providing a largely immobile population with a physically and psychologically invigorating respite from the industrial metropolis. Paris's vertigo-inducing Parc des Buttes-Chaumont, which occupies a former quarry site, and New York's Central Park, which reveals the city's postglacial bedrock, epitomize mid-nineteenth century land-shaping of public space to create a wild counterpoint to the city (see Figure 5.2).

The layouts of Parc des Buttes-Chaumont and Central Park draw on the traditions of the sublime and picturesque, and on the ancient history of the garden as an imagined representation of a distant aspirational place. However, as the influence of the picturesque faded into the twentieth century, the modern edict of form following function filtered from buildings to the outdoors. At first, this shift transformed the private garden, which was reconfigured from an immersive escape to a functional extension of the building's floor plan (Lewis 1993). By mid-century, automobile affordability meant that city dwellers were more able to venture beyond the city limits to real wild landscapes. It followed that city folk no longer required a simulacrum of that experience to be compressed into an urban park (Cronon 1995; Riley 1988).

With real nature now in reach through an automobile excursion, public space became less of a self-contained explorative experience, and more of an open stage for organized activities. For those activities, and the parking lots that serviced them, the ideal surface was level. While the sports field is a clear example of a level surface accommodating a set program, over time event programming became a consistent theme across the various scales and typologies of public space.

Figure 5.2 Constructed postglacial wilderness creates an explorative retreat from the city in Central Park, New York.

Image Credit: Erin Ching

Programming the public realm with events and activities revitalized many public spaces and benefited both citizens and City Hall. Grass-roots organizations seeking to revive community spirit leveraged event programming as a means of reclaiming blighted public space and instilling civic interaction and pride. But perhaps most tellingly, with neoliberal policies impacting the provision and maintenance of public spaces, organized events often provided a lucrative commercial base for economic self-sufficiency (Kullmann 2015).

It followed that even as expanding cities became more topographic overall, a progressive leveling out transpired across all scales of public space. This transformation reflected and reinforced a general homogenization of public space typologies in the contemporary city. Whereas urban squares, urban parks, and other more specialized types such as botanic gardens traditionally exhibited distinct identities and rituals, their events and supporting furniture and props became increasingly uniform.

The topographical and typological homogenization of public space brings unintended deficiencies. Constituting public space as a level stage places a heavy burden on perpetual programmatic novelty and on the support of temporary props and furnishings. If and when the planned events fall idle, leveled public space risks offering no compelling alternative strategy for catalyzing more spontaneous forms of site-specific engagement. The leveled landscape is divested of agency in the sense that rather than actively influencing the body through its contours, it becomes a passive stage for superimposed activities.

The topography of place

Incorporating, enhancing, or repatriating landform in the public realm suggests a range of potential benefits for urban life. From an experiential perspective, landform is a fundamental component of spatial cognition, orientation, landscape character, and place making (Tuan 1974). Indeed, the intimate connection between topography and home is embedded in the etymology of the Ancient Greek *topos*, which means *place*.

The topography of place is experienced sensorially. The primary topographic sense is kinesthetic, whereby landform and gravity interact with the body to influence balance and orientation. When encountering landform, we deploy proprioception to calibrate the positions and angles of the body in relation to itself and the surrounding physical environment. The secondary topographic sense is visual, whereby landform interacts with sightlines and depth perception to obscure and reveal the environment and enclose and open landscape 'rooms.'

Supplementing the principal senses of kinesthesia and sight, landform also interfaces with the senses of hearing and smell through the deflection or amplification of sounds and scents. And lastly, topography interacts with the most immediate of senses, which is the sense of touch. This interface is most likely to occur in the context of a textured surface, which requires feeling the way with one's feet. In particularly steep situations, the sense of touch potentially engages use of the hands. Moreover, touch receptors on the skin may be indirectly stimulated through exposure to or shelter from the wind, which is topographically shaped.

As perceived through the senses, landform plays a key role in shaping cognition and legibility of the urban environment. Following Kevin Lynch's emphasis of topographic gradients, land slope and morphology provide body-based orientational cues for directional differentiation when navigating the city (Lynch 1960). In addition to the general lay of the land, topographic gradients may include natural hydrological systems that flow through engineered urban landscapes. Alternatively, topographic gradients may leverage artificial landforms that serve as way-finding devices in otherwise flat urban environments with repetitive self-similar layouts.

Topographic motifs for public space

The topographies of public spaces fall into two distinct categories: *discovered* and *designed*. Discovered topography exists prior to the establishment of a public space and is likely to be the reason for an area remaining un-built and consequently designated as public space by default. Discovered topography results either from natural process such as soil erosion and accumulation, or as a by-product of human processes, such as quarrying, landfill, or infrastructural works. The stairway parks of San Francisco, which occupy street easements that are too steep for traffic, are an example of public space adapted to a discovered topography.

Designed topography is formed with intentionality to shape cultural experience and/or the ecological performance of a public space. Designed topography may be fabricated onto or into otherwise level sites to create new topographic gradients; it may be configured to invoke a historical topography that was lost to the city, such as a gulch or dune; or, it may be shaped to enhance or amplify a discovered topographic condition. Gasworks Park in Seattle, in which an expressive landform cloaks a landfill site, is an example of a designed landform amplifying an existing topography. Drawing on this and other examples of discovered and designed landforms, the following section explores three topographic motifs for land-shaping public space.

Motif 1: performance topographies

The first topographic motif challenges the convention of levelness as the foundation for flexible and vibrant public spaces. To be certain, a flat open space can accommodate the most diverse, manageable, and universally accessible range of events on its surface. It would, for example, be counterproductive to situate a weekend market on the side of a steep hillside instead of a level field, square, or promenade. Nevertheless, utilizing a large flat space often requires an organizational critical mass that may be less responsive to small-scale user-generated initiatives. For individuals or small groups, the apparently limitless programmatic possibilities of a flat expansive space may appear too intimidating to initiate engagement. As a consequence, people are more likely to remain on the edges of public spaces as bystanders rather than participants.

Topography offers cues, or starting points for catalyzing user-generated engagement with public space. Although a variable topographic surface is likely to accommodate fewer superimposed programs than a flat open space, it is more likely to encourage the invention of novel site-specific activities that are not dependent on organized events and supporting apparatus. As Lynch observes, shaping artificial topography into 'plastic uncommitted forms' potentially offers 'suggestive material for spontaneous action' (Lynch 1972, 112).

A topographical basketball court located in Volkspark Potsdam playfully illustrates this concept (see Figure 5.3). With the variable surface disrupting conventions that equate a level playing field with a fair game, players are obliged to spontaneously readapt to a site-specific version of basketball. Indeed, the un-level playing surface actually *levels* the field in the sense that informal mixed teams use the topographic variations to offset height differences amongst players. Through the interrelationship of the form of the ground and the activities it nurtures, the playing court shapes an interactive public space.

Figure 5.3 Variable playing surface encourages a site-specific version of basketball at Volkspark, Potsdam. Image Credit: Karl Kullmann

A spectrum of amorphous 'plastic' landforms potentially offers catalytic surfaces for user-engagement. Latent or active natural processes may directly shape or inspire the design of ground plane. Wind and water may carve out and mound up the ground, while seismic forces may combine with gravity to forge expressive geologies. Alternatively, topography may be shaped in the studio where it is folded from paper, sculpted from clay, or virtually modeled from topological algorithms. Whether simple or complex, or discovered or designed, landform is, in essence, comprised of convex and concave surfaces. These principal land-shapes influence human behavior, with concave forms tending to gather people together, and convex forms tending to be more dispersive.

When fully activated, the hollowed landform of the outdoor amphitheater exemplifies the gathering impulses of concave topography. However, when a large amphitheater configured for traditional performance/audience dynamics lies empty and un-programmed, it may be as unapproachable as a smooth level space. To circumvent this condition, concave forms need not be so deep or typologically specific. As is epitomized in the gently inflected form of Siena's historic Piazza del Campo, shallow concaved forms subtly encourage gathering, influence desire lines and shape other site-specific spatial behaviors. Melbourne's Federation Square offers a contemporary interpretation of this shape. Built over a capped railroad, the shallow concaved form enhances the square's magnetism as a political and cultural event gathering space (see Figure 5.4).

In contrast to the immersive gathering spaces associated with concave topography, *convex* landforms are inherently more dispersive. By obscuring sightlines between users, convex slopes foster a more individualized landscape experience that draws on the wider

Figure 5.4 Shallow concaved form enhances the square's potency as a central political gathering space on Federation Square, Melbourne.

Image Credit: Rebecca Finn

Figure 5.5 Convex and concaved shaped artificial topography at Gasworks Park, Seattle
Image Credit: Karl Kullmann

landscape and sky-scape. The amorphously shaped artificial topography of Seattle's Gasworks Park exemplifies convex public space (see Figure 5.5). The otherwise featureless convex slopes of a prominent hill act as a beacon within the park. Sitting, lying, rolling, and circulating are common activities on the slopes, as park users become both actors and audience in an expansive uncoordinated performance. And yet, even when the park is crowded with people, the convex slope contributes to the sensation of solitude and an expanded sense of personal space.

Motif 2: threshold topographies

The second topographic motif for public space applies landform as a framing device. Within the figure/ground hierarchy of traditional urban form, buildings typically clearly framed town squares and other hard-surfaced public spaces. Around the built edge, streets and doorways formed distinct thresholds through which to enter the public space. In the case of urban parks and other soft-surfaced public spaces, the perimeter fence typically substituted buildings as the framing device. A limited number of gates, which were locked overnight, tightly controlled access and clearly demarcated the park from the city.

In the modern city, public space was systematically de-framed. With modern towers set well back from property lines, the capacity for built form to tightly frame public urban squares was diluted. Concurrently, urban parks were defenced in a process that extended the centuries-long deconstruction of the archetypal garden frame (Kullmann 2016). In European cities, defencing enabled the publicization of royal hunting parks, while in the US it signified the democratization of civic space. Yet it was also symptomatic of a reduction in maintenance and custodianship of public space, with the unfenced urban park forced to fend for itself within a wider landscape of divestment.

With the aim of re-establishing legibility and typologies of public space, traditional urban design approaches have sought to reinstate figure/ground framing. Although productive in the case of urban squares, seeking to re-fence urban parks would be counterproductive for urban communities that seek fewer, rather than additional, physical barriers. Reconstituting the boundary fence also risks escalating the disjunction between the hyper-paced fabric of contemporary cities and the traditionally anchoring role of parks (Ingersoll 1997).

Topography has the potential to intervene in this framing problem, whereby weakly delineated public spaces dissolve into the city, while overly enclosed spaces are overlooked and underused. Reconceived in its purest sense, the frame operates as a semi-permeable threshold that filters permutations of physical movement, visual connectivity, aural information, and olfactory experience. Thresholds represent a third space that is distinct from inside and outside a public space. Within the frame's thickened threshold zone, the efficient rhythms of the city are slowed down and possibilities for interaction are amplified (Stevens 2007).

Reconceiving the frame as a topographically formed threshold potentially balances the goals of connectivity and distinctiveness in the public realm. When mounded up around the edge of a public space, the crest of an encircling landform creates an artificial horizon line. As occurs with the real horizon (as formed by the curvature of the earth), a topographic horizon encompasses a field of perception. However, whereas the real horizon tracks each person as they move around, an artificial horizon remains tied to the landform and fixed in place. As a result, the topographic horizon of an encircling landform is readily transcended as people move through the space.

The Esplanade in Fremantle, Western Australia illustrates the performance of a topographically formed threshold in a public space. A 120-foot diameter horseshoe shaped mound is the only topographic feature on an otherwise level harbor-side park (see Figure 5.6). The concaved internal space creates useful social facility as a meeting and performing place that is sheltered from the prevailing wind. At six-feet high, the mound's artificial horizon crests just above average eye-height, which is sufficient to visually obscure the interior hollow space from the outside. Physical permeability offsets this subtle visual enclosure, with immediate access and egress enabled in all directions. The openness that results is illustrated in the tendency for people crossing the Esplanade parklands to deviate to the mound, crest the threshold, loiter a while, and then resume their onward journey.

Motif 3: micro topographies

The third topographic motif for public space shifts scales from immersive forms that engage the body visually and kinesthetically, to a finer textural scale that is engaged through the sense of touch. In Classical thought, the immediacy of touch was considered anathema to distant reason and thus relegated to the lowest of the five traditional senses. Direct contact was further tarnished in picturesque ideology, where touch was connected to contamination and intervention. Looking was regarded as a more appropriate expression of the aspiration to un-intrusively comprehend untouched nature (Pollak 1998).

The privileging of vision at the expense of touch extended into the modern city, where the topography of the public realm was smoothed down to the micro, or material, scale. Dematerialization of the physical environment had profound implications for place making. Because touch requires something to be touched, it exists in a symbiotic relationship with materiality, which exerts friction back to the sensing body. Friction between the body and materiality

Figure 5.6 Horseshoe shaped mound creates a space framed by a topographic threshold on the Esplanade,
         Fremantle
Image Credit: Lucy Farley

amplifies attentiveness and enhances cues for engaging with our surroundings (Sennett 1998). Our predisposition towards tripping on small aberrations in otherwise flat pavements vividly illustrates our loss of attentiveness in frictionless, visually reliant environments.

To counteract the smoothing of modern cities, Richard Sennett calls for public spaces that initiate 'visceral resistance, commitment and expression' (Sennett 1998, 20). This is not to imply that public space become an adventure playground, but is a call for reincorporating micro-topographies in order to retrain our flat-complacency. Whereas we are attuned to quickly pass over smoothed spaces, the abundant friction of roughly textured surfaces slows us down and leads us to more mindfully engage the immediate environment.

The relegation and revival of the cobblestone encapsulates this narrative. Since Roman times, river-sourced cobblestones, and later quarried granite setts, served as a key surfacing material for public spaces. In the twentieth century, in all but the most historic districts, cobbled surfaces were systematically covered with a layer of asphalt or replaced with concrete or modular pavers.

So comprehensive was this material transformation, that by the twenty-first century, in-ground tactile markers for vision-impaired way finding were often the most expressive micro-topographies evident in the smoothed urban environment.

The more recent revival of the cobblestone repatriates friction in the public realm. Although the materiality of the cobblestone is initially transmitted visually, it is primarily imparted texturally via the sense of touch through the feet. In this regard, cobblestones function as inbuilt tactile markers, providing legibility and way finding cues in public space. The cobbled surface marks thresholds, frames spaces, and through its friction, materially communicates an area as a slower space.

Micro topography extends beyond textured surfaces to include small, but nevertheless influential, level changes. For example, stairways with lower than standard risers foster a gentler gait and engender a sense of lightness in the experiencer. Although it is true that stairs associated with major pedestrian thoroughfares must adhere to relevant codes, secondary circulation networks through public spaces can deviate from these norms. In the tradition of the garden path, small-level changes that follow non-standardized topography can recapture our attentiveness and recalibrate our proprioceptive norms.

Marlene-Dietrich-Platz, in Berlin's Potsdamer Platz precinct, illustrates the deployment of small level changes into an otherwise level urban fabric (see Figure 5.7). In this public theater forecourt, terraces with two-inch risers are arranged into a shallow amphitheater that descends slightly below street level. Despite the extremely compressed overall level change, the stepped micro-topography creates a niche landscape that performs as a meeting, gathering, observing, and resting space.

In addition to texture and small level changes, a third manifestation of micro-topography occurs through the process of wear. Wear refers to the abrasion or erosion of a surface or material through the friction of repeated use. Although generally interpreted as a negative process in architecture, wear is a more intrinsic and potentially constructive process in the landscape. In comparison with buildings, which are generally fully formed when new, landscapes remain works in progress that deform and adapt to shifting uses over time. Embracing wear challenges inherited notions that frame touch as exerting a negative impact on natural landscapes. Given that urban landscapes are often already highly modified, the act of touching may be reframed as a constructive process, whereby wear and deformation shape culturally legible and useful spaces over time (see Rothery 1912).

Figure 5.7 Deployment of diminutive level changes on Marlene-Dietrich-Platz, Berlin.
Image Credit: Tanya Eggers

Figure 5.8 Cobblestones amplifying a previously worn desire path at Park Rabet, East Leipzig.
Image Credit: Karl Kullmann

Constructive wear may take the form of desire paths, which are etched into the ground through the passage of many feet, thus reconfirming an informal route over time. In instances where these imprinted paths influence more formal circulation patterns, user behavior directly shapes the design of public space. This process is evident in Park Rabet in East Leipzig, where a worn desire path leading towards an adjacent access street was integrated into the landscape design. The desire path is minimally surfaced in loosely spaced granite sets to minimize mud and dust and maintain permeability, but otherwise left intact by the designers (see Figure 5.8).

Conclusion

From town squares to sports fields, the history of public space is in many regards a story of finding or making level ground. However, when taken to extremes in the quest for efficiency and convenience, leveling and smoothing out the city eliminates the niches and nuances that often grant public spaces their distinctive character. Moreover, notwithstanding the influence of neoliberal conceptions of economically justifiable use, public space is not akin to the floor plate of a commercial building: it is not compelled to be flat so as to maximize potential functionality and economic yield. On the contrary, as the common ground on which a sense of community and place evolve, public space is necessarily more amorphous as it simultaneously endures and adapts over time.

Shaping the ground amplifies the amorphous nature of public space and challenges assumptions that place the choreography of everyday life on a smooth, level stage (Cache 1995). Drawing on the parallel history of 'wild' urban park design, topographic urban design operates across scales. At the site scale, mounded landforms may create semi-permeable thresholds that enframe spaces without enclosing them. At the intermediate scale, concave landforms may subtly calibrate spaces for gathering, while convex forms may foster more contemplative experiences. And at the finer scale, micro topographies may take the form of small-level changes and rough materials that increase surface friction and slow us down.

By reacquainting us with our kinesthetic sense, the experience of topographic space catalyzes creative behavioral responses. When encountering topography, we quickly learn how to make best use of its contours: which route to take in order to moderate or accentuate gravity; where to situate ourselves in a niche; where to congregate; where to seek out a view or solitude, and which activities to adapt and invent. By acting on the body and shaping use through its contours, topography enhances the agency and possibility of public space.

References

Bosselmann, P. (2011) Metropolitan Landscape Morphology. *Built Environment*, 37(4): 462–478.

Cache, B. (1995) *Earth Moves: The Furnishing of Territories*. Cambridge: MIT Press.

Casey, E. S. (2002) *Representing Place: Landscape Painting and Maps*. Minneapolis: University of Minnesota Press.

Cronon, W. (1995) The Trouble with Wilderness. In: *Uncommon Ground: Rethinking the Human Place in Nature*, Cronin, W. (ed). New York: W. W. Norton & Co, 69–90.

Duany, A. (2002) Introduction to the Special Issue: The Transect. *Journal of Urban Design*, 7(3): 251–260.

Hill, D. R. (1992) America's Disorganized Organicist. *Journal of Planning Literature*, 7(1): 3–21.

Ingersoll, R. (1997) Landscapegoat. In: *Architecture of Fear*, Ellin, N. (ed). New York: Princeton Architectural Press, 253–259.

Ingersoll, R. (2006) *Sprawltown: Looking for the City on Its Edges*. New York: Princeton Architectural Press.

Jacobs, J. (1961) *The Death and Life of Great American Cities*. New York: The Modern Library.

Kullmann, K. (2015) The Usefulness of Uselessness: Towards a Landscape Framework for Un-activated Urban Public Space. *Architectural Theory Review*, 19(2): 154–173.

Kullmann, K. (2016) Concave Worlds, Artificial Horizons: Reframing the Urban Public Garden. *Studies in the History of Gardens and Designed Landscapes*, 37(1): 15–32.

Kullmann, K. (2017) Hong Kong, Grounded: Photographs of the Contact Zones between the Mountain and the Multilevel Metropolis. *Places Journal*, May. https://doi.org/10.22269/170502.

Kullmann, K. (2018) Design with (Human) Nature: Recovering the Creative Instrumentality of Social Data in Urban Design. *Journal of Urban Design*, 24(2): 165–182.

Leatherbarrow, D. (1999) Leveling the Land. In: *Recovering Landscape: Essays in Contemporary Landscape Architecture*, Corner, J. (ed). New York: Princeton Architectural Press, 171–184.

Lewis, P. (1993) American Landscape Tastes. In: *Modern Landscape Architecture: A Critical Review*, Treib, M. (ed). Cambridge: MIT Press, 2–17.

Lipsky, F. (1999) *San Francisco: The Grid Meets the Hills*. Paris: Editions Parentheses.

Lynch, K. (1960) *The Image of the City*. Cambridge: MIT Press.

Lynch, K. (1972) The Openness of Open Space. In: *The Arts of Environment*, Kepes, G. (ed). New York: Braziller, 108–124.

McHarg, I. (1969) *Design with Nature*. Garden City: Natural History Press.

Pollak, L. (1998) The Absent Wall and Other Boundary Stories. *Daidalos*, 67: 94–105.

Riley, R. B. (1988) From Sacred Grove to Disney World. *Landscape Journal*, 7(2): 136–147.

Rothery, G. C. (1912) *Staircases and Garden Steps*. Ithaca: Cornell University Library.

Rykwert, J. (1976) *The Idea of a Town: The Anthropology of Urban Form in Rome, Italy and the Ancient World*. London: Faber and Faber.

Sennett, R. (1990) *The Conscience of the Eye*. London: Faber and Faber.

Sennett, R. (1998) The Sense of Touch. *Architectural Design Profile*, 68(132): 18–22.

Stevens, Q. (2007) *The Ludic City: Exploring the Potential of Public Spaces*. London: Routledge.

Tuan, Y-F. (1974) *Topophilia: A Study of Environmental Perception, Attitudes, and Values*. New York: Columbia University Press.

6

SOCIAL JUSTICE AS A FRAMEWORK FOR EVALUATING PUBLIC SPACE

Setha Low

Introduction

This chapter proposes that distributive, procedural, and interactional dimensions of social justice as well as recognition of difference and an ethic of care provide a foundation for evaluating public spaces and a basis for social justice assessments of public infrastructure. The proposed social justice framework for assessing public space was developed based on twenty years of ethnographic research on parks, plazas and streets in New York City, Philadelphia, Jersey City, New York State, and San José, Costa Rica and a careful reading of the literature on the just city, right to the city and social justice planning.

An overview of the impact of neoliberal urbanism on public space and the resulting social exclusion of users sets the stage for the rationale of this chapter, that a social justice perspective would enhance the diversity, cultural recognition and social interaction among people who would otherwise not come into contact with one another. Two literature reviews follow: one on the "just city" and a second of urban design models that evaluate social inclusion in public space. Based on this literature, five dimensions of social justice are proposed to enhance diversity and equity in public space. These dimensions are then illustrated by data from Tompkins Square Park in New York City.

Neoliberal urbanism and social justice consequences in public space

Each year members of the Public Space Research Group (PSRG) undertake ethnographies of public spaces throughout Manhattan and Brooklyn using behavioral, movement and physical traces mapping, participant observation, interviewing, and archival documents to track the changes in the use and diversity of everyday streets, sidewalks, squares, and small parks. Over this period of documentation, the PSRG has found added restrictions, urban design modifications and increasing policing surveillance. These public space changes are conceptualized through an expanded notion of securitization that includes the affective, physical, spatial, legal, regulatory, and financial strategies used to reinforce governmental and corporate control (Soja 2010; Low 2019). Similar changes have also been documented through ethnographic research on plazas in San José, Costa Rica as well as in other cities throughout Latin America (Herbert and Brown 2006; Low 2017).

The findings include the emergence of a new "affective landscape" (Ramos-Zayas 2012) represented by hostile architecture and an emerging structure of a feeling based on the fear of others that is replacing an earlier climate of tolerance and appreciation of urban diversity. Benches with arms in the middle to prevent a person from lying down are becoming more common. Physical and spatial changes include the gating of previously open public squares and plazas as well as new forms of enclosure, such as the use of Jersey barricades and bollards to line streets and sidewalks to make them difficult to use.

These physical changes were reinforced in New York City by post 9/11 legislation that allows for stop-and-search by military personnel and police, and have encouraged zero-tolerance arrests and other more aggressive forms of securing space (Lippert and Walby 2013). In San José, there also has been an increase in aggressive policing with the addition of new mobile municipal police as a response to the rise of drug-related and organized crime.

New regulatory instruments such as Business Improvement Districts (BIDs) and public-private partnership such Privately Owned Public Spaces (POPS) have proliferated particularly in New York City, playing a central role in restructuring urban space and reconfiguring urban governance (Kayden 2000; Ward 2007; Németh 2010; 2011). In Costa Rica, BIDs and POPS are slowly developing, however, large police towers have been erected above the main pedestrian shopping street for greater control of crime and presence of undesirables. Both New York City and San José public spaces have added long lists of rules and regulations posted on squares, sidewalks, and plazas. In addition to these affective, spatial, physical and regulatory strategies, a more subtle "financialization of everyday life" is also increasing in both the United States and Costa Rica with moral overtones that make it harder for people with lesser means to participate in public life (Duneier 1999; Shepard and Smithsimon 2011; Fennell 2014; Low 2019).

Access and use of public spaces have suffered accordingly and is increasingly exclusionary for some users while benefitting others. Young men and women of color avoid policed and secured squares (Fiske 1998; Sorkin 2008; Cresswell 2015). Low-income residents read the symbolic cues of physical deterrents or ornate landscape features and avoid these restrictive landscapes (Kayden 2000; Miller 2007; Cresswell 2015). Middle-class individuals and tourists are happier with these highly controlled and managed public spaces, while teenagers avoid them because of their long lists of rules (Sorkin 2008; Staeheli and Mitchell 2008; Shepard and Smithsimon 2011). Homeless individuals are locked out of gated public spaces at night and by the increasing use of hostile architecture (Kayden 2000; Németh 2011). Individuals who are perceived as "Middle-Eastern" or "Arab," in New York City as well as illegal immigrants worry about the increased police profiling and surveillance cameras, and stop-and-search by ICE (Immigration and Customs Enforcement) makes many immigrants hesitant to leave home or spend time in public space (Ruben and Maskovsky 2008; Roy 2009; Newman 2011; Fassin 2013). In addition, street vendors in both New York and San José are having a more difficult time both because of the new regulations and the degree of policing and surveillance (Duneier 1999; Miller 2007; Chesluk 2008; Brash 2011).

The consequent lack of diversity in public space and the restrictions of these securitization practices call for a clearer agenda to press for better public space. Based on ethnographic research in parks, beaches, and plazas there is empirical evidence about what can be done to encourage diversity in public space and create a more inclusive public sphere (Low et al. 2005; Houssay-Holzchuch and Teppo 2009; Castells 2012). But to have a greater impact on design and governance practices and urban policymaking, a clearer moral and philosophically based argument and evaluative framework is necessary. A strategy is needed to illuminate the harmful long-term consequences and highlight the dangers of neoliberal urbanism in producing a more restrictive public realm.

The just city and public space

With the increasing securitization changes in public space, clearer arguments are needed to justify the criteria by which urban space can be considered unjust. Many well-known theorists have argued for equitable public space as well as for greater voice in the public sphere generally using two different formulations. The first group of theorists focus on "justice," defined as relative fairness and evaluated comparatively. For example, Iris Young (2001) insists on evaluating inequality in terms of social groups, since group-based comparisons are more likely to reveal patterns of structural inequalities that privilege some more than others (Mitchell 2003). Alternatively, theorists who focus on rights that are universal are invariably assessed at the individual level. For example, Henri Lefebvre's (1991) "right to the city" offers a utopian imaginary of a better society and the right to a better life, where individual rights to public space can be defended by judicial actions.

Axel Honneth (2004), on the other hand, poses his argument not on the elimination of inequality, but instead on the avoidance of humiliation or disrespect. His theory of justice is based on the degree of recognition that produces human dignity and the right to recognition for self-identity. Nancy Fraser (1990) brings these two lines of argument together in her single goal of "participatory equality" that includes both recognition and redistribution.

These theories of justice and rights have been drawn upon to address the concerns of planners, government agents, elected officials, and other policy makers. Addressing their practical concerns is essential because without clear articulation of what we are striving for, and some measure of accountability, it is difficult to struggle effectively. It was merely a legal oversight that allowed the privately owned public space of Zuccotti Park to be used for Occupy Wall Street, an exception within a security landscape that might not happen again.

The first of these planning models is Susan Fainstein's (2004) well-known "just city" formulation based on Lefebrve's (1991) "right to the city." A second model based on recognition and difference is offered by Ruth Fincher and Kurt Iveson (2007) who assert "the need for planning frameworks to take adequate account of the diversity of cities, by identifying and working with the many publics which inhabit them" (Fincher and Iveson 2008, 7). Their theory addresses the concerns of Young (2001) about evaluating inequalities in term of social groups, Honneth's (2004) use of recognition to promote human dignity, and incorporates Fraser's (1990) work. They provide a more relational approach to diversity (Fincher and Iveson 2008). A third approach is my work on social justice in public space derived from ethnographic research on urban parks, plazas, and gated communities (Low 2003; Low et al. 2005; Low and Smith 2006) and my critique of Fainstein's (2005) contention that diversity should not be at the center of a social justice analysis of public space.

The criticism of Fainstein's just city proposal is that she utilizes too narrow a definition of justice such that her utopian aims are only partially fulfilled. Similar to Marcuse (Brenner et al. 2011) who wants to go beyond a distributive theory of justice and concentrate instead on the dimensions of solidarity and difference, distribution solutions alone do not result in fairness in everyday life. Much closer to the formulation of Fincher and Iveson (2008), but developed from the social and organizational psychology literature, three dimensions of justice—distributive, procedural and interactional—are essential to address the multiple kinds of unfairness, injustice, and indignities that people suffer in public space. This formulation assumes that like Fincher, Iveson, Young, and Fraser's notions of diversity and difference are fundamental with regard to public space. Thus, the five dimensions of social justice attempt to avoid many of the limitations of the other formulations and to offer a broader perspective better fitted to the evaluation of public space.

Urban design models for the evaluation of social justice in public space

Within the field of urban design there is already evidence of a recognition of the importance of social justice in the design of public space and there are a number of evaluation models that have a social justice component. One set of models assess the "publicness" of the space in comparative terms (Varna and Tiesdell 2010; Nemeth and Schmidt 2011) while others focus how the sense of fear or fantasy influences urban design and management (Van Melik, Van Aalst, and Van Weesep 2007) or measure public space in terms of overall quality of life (Mehta 2014). Each of these models are addressed and considered as precursors of the social justice propositions and criteria developed in this article.

The desire to measure the publicness of space emerged from the increasing privatization and neoliberal restructuring of urban centers, particularly in rapidly gentrifying and global cities. In the US, New York, San Francisco, and Los Angeles have led the way in utilizing BIDs and POPS as strategies for providing "safe and secure" public spaces to attract customers for local businesses and to create aesthetically pleasing settings for corporate headquarters. Unfortunately, these newly developed public spaces do not necessarily attract a lively mix of residents and tend to exclude lower-income users, minorities, and immigrants. Researchers interested in understanding these differences and determining what design and management elements restrict use, have developed a series of urban design evaluative models.

For example, Varna and Tiesdell (2010) consider publicness a multi-dimensional concept and use five "meta dimensions—ownership; control; civility; physical configuration; and animation" (575) measured along a scale of publicness to come up with a "star model" of these dimensions. Applying their model to two public spaces in Rotterdam, they found that they could in fact discriminate the degree of publicness and whether it was due to management or design control.

Nemeth and Schmidt (2011), on the other hand, selected three core components of publicness: ownership, management, and uses/users. These components were understood as interacting with other factors along a public/government to private/corporate axis. Their instrument measured ownership and management employing an index that included: laws and rules governing the space, surveillance and policing present, design and image making, and access and territoriality. Applying their model to a range of New York City sites, they determined that more "private/corporate" public spaces were significantly higher in features that discouraged or controlled use than publicly owned spaces.

Both of these studies of publicness demonstrate how privatization decreases social diversity and kinds of use in public space. They set the stage for thinking more deeply about the principles and values—social justice or profit and business driven—that underlie design, planning and urban development proposals and implementation.

Producing "secured" urban spaces that are highly surveilled and policed, and "themed" spaces filled with events, fun-shopping, and cafes, are two other strategies cities use to produce spaces perceived as safe. Van Melik, Van Aalst, and Van Weesep (2007) argue that as public space becomes safer it also becomes more exclusive, as they put it, "safer and more comfortable, but for whom and at whose expense?" (40). Their study leads directly to the proposal that social justice principles need to be included in evaluating the design and management of public space.

The most comprehensive model for assessing public space is the Public Space Index (PSI) that categorizes the design, environmental qualities, user experience and activities that make up a people-friendly space (Mehta 2014). The PSI measures "the quality of public space by empirically evaluating its inclusiveness, meaningfulness, safety, comfort and pleasurability"

(57) through 42–45 variables ranked 0–3 and weighed based on observed local conditions and user interviews. Vikas Mehta, with a team of graduate students, applied the PSI in four downtown spaces in Tampa employing observations of site characteristics and interviews and surveys with people through a mixed methods research design. Visiting the selected sites six times spread out throughout the day and week they tested the PSI and found that it provided detailed information about the four spaces and insights into the overall characteristics and quality of the downtown areas. It was found to be useful for evaluating both the design and management of space and informing "citizens of the condition of equity, securitization and individual and group rights in their local public realm" (85).

Propositions for socially just public space

The following five propositions for evaluating more socially just public space have evolved over the past ten years through conversations with Kurt Iveson, presentations at professional meetings and previous publications that trace their development (Low 2013; Low and Iveson 2016). They draw upon the literature and foundational philosophical texts cited above, but at the same time have been informed by ethnographic fieldwork in plazas (Low 2000), gated communities (Low 2003), urban parks (Low et al. 2005), privatized public spaces (Low and Smith 2006), and most recently research with the Public Space Research Group and graduate students on the changing character of public space in New York City such as Moore Street Market (Low 2017) and Tompkins Square (Low 2019). Each is defined briefly and then illustrated through a rapid ethnographic study of Tompkins Square in New York City.

Public space and distributive justice

Distributive justice refers to questions of how the wealth and burdens of society should be distributed to achieve a just city. The discussion revolves around whether benefits and burdens should accrue to individuals equally, according to need, according to merit, or to those who are the least well off (Rawls 1971). Distributive justice draws upon equity theory that "argues that fairness means that people's rewards should be proportional to their contributions" (Tyler 2000, 118). To be more precise, it is that one's perceived ratio of reward/contribution should be equivalent to the ratio of the comparison to other (Whiteman et al. 2012).

Public space and procedural justice

Procedural justice refers to the way that the processes of negotiation and decision-making influence perceived fairness by individuals. While early research on social justice supported the findings that people felt most satisfied when outcomes were distributed fairly, subsequent research found that the favorability of an outcome is less crucial when the underlying allocation process is perceived as fair (Cropanzano and Randall 1993; Tyler and Blader 2003).

Public space and interactional justice

Distributional and procedural outcomes are not the only relevant issue when determining fairness or justice (Tyler and Blader 2003), and instead the way that a person is treated is equally important. Interactional justice is embedded in the relationships and interactions that take place in a specific situation or place. How a person is treated by others, and particularly by people in power contributes to a sense of fairness and is characterized by respect of the individual, propriety of behavior, and truthfulness in relationships (Bies and Moag 1986).

In addition to the three types of justice—distributive, procedural and interactional—two other social justice dimensions contribute to a more comprehensive analysis and evaluation of public space. The concept of recognition is drawn from the work of Fincher and Iveson (2008) and included in this framework because of its focus on cultural identity and implicit norms of behavior and practice that need to be accommodated to allow diverse people to use and feel comfortable in public space. The second addition is an ethic of care that was derived from multiple sources including Tronto (2013), Fennell (2014), Watson (2006), and Morgan (2010) and reflects the impact of Occupy Wall Street politics (Maharawal 2013).

Public space and recognition

The idea of recognition in public space addresses the systematic devaluing and stigmatization of some urban identities and ways of life in cities. Disputes over all sorts of urban issues are instigated by groups who argue that their particular values and needs ought to be taken into account. When such groups feel that their identities and ways of being in the city are unfairly denigrated or stigmatized, social justice becomes a matter of status and has an inter-subjective dimension: the pursuit of equality involves working against "cultural patterns that systematically deprecate some categories of people and the qualities associated with them" (Fraser 1990; Cresswell 2015; Iveson 2007).

Public space and the ethic of care

While an ethic of care is not well-developed as a concept in public space research, women's environmental activism is often described in maternalistic terms—as if motherhood and caring for the environment go hand in hand (Tronto 2013). Further, the core proposition of ecological integrity and social justice is based on the politics of care (Morgan 2010; Fisher 2012). Tom Hall and Robin Smith (2013) discuss the role that "kindness" plays in the "good city" and depict a caring city as more resilient and contributing to conviviality. An ethic of care in public space, in this sense, would focus on attending to others' needs, not just passively through recognition or interaction, but in pro-social and life-enhancing ways.

This framework for evaluating social justice in public space is illustrated through an evaluation of Tompkins Square in New York City.

Social justice and public space: Tompkins Square, New York City

This research was undertaken in response to the Department of Parks and Recreation Commissioner's desire to improve visibility, safety, and social inclusion by reducing the height of the fencing that surrounds New York City's urban parks. While a number of parks still have 8- to 9-foot fencing, particularly around children's playgrounds and the perimeter, best practices suggest that 4-foot fencing increases visibility, provides better access, and appears more welcoming to users. Park planners, however, found that while most community groups support fence height reductions, the residents who lived around Tompkins Square protested against this intervention, even after supporting it initially. In an effort to understand what was creating this change in opinion, the author and a group of graduate students decided to use a new methodology, the Toolkit for the Ethnographic Study of Space (TESS), to understand why the change of heart among residents and park users.

Tompkins Square, New York City

Tompkins Square Park is located on the Lower East Side of Manhattan bordered by East 10th Street to the north, Avenue A to the west, East 7th Street to the south, and Avenue B to the east. It was originally proposed in 1811, but not completed until 1834 and has always been a site of tension, beginning with its construction intended to stimulate urban development to conflicts about homelessness and the building of a tent city in the 1980s and the 1990s. The most recent conflict has been over the 2015 appearance of police towers to surveil users, part of a neighborhood gentrification strategy to reduce the lingering number of homeless people and drug users who remain.

The park contains clearly differentiated sections of grassy meadow, three playgrounds, a dog run, sports fields, skateboard area, and gardens separated by either a 4- or 8-foot iron fence creating a patchwork of subdivided spaces that only meet in a few wide circular areas that connect the sidewalks. These iron fences were first installed to protect gardens from animals and children in the 1860s. The most recent fences, however, were part of the renovations undertaken during 1930s under the direction of the Department of Parks for the City of New York. Robert Moses who was the commissioner at that time was committed to standardization and increased efficiency in public space. Tompkins Park was only one of the many small parks and plazas that were redesigned and reconfigured during this period with these goals in mind (Caro 1975; Corrigan 2018).

Methodology

The research team employed the TESS methodology (Low, Simpson, and Scheld 2018) that included behavioral, movement and physical traces mapping, participant observation, user interviews, and archival documentation. Data analysis was focused on the research question of why residents were opposed to lowering the children's playground fence and included a matrix of attitudes of different constituency groups as well as the content analysis of the maps and field notes. Emphasis was placed on the triangulation of data from the different methods that produced the following preliminary findings.

Findings

The research team found that there were multiple reasons that local users and residents were opposed to lowering the fence height around the children's playground related to various perceived dangers and the proximity of these problems. The most often mentioned reason was the perception by users that the location of the playground at the edge of the park next to a busy street and noisy traffic required that the park maintain the imposing barrier. The second reason was that the majority of interviewees were concerned about the number of homeless people who waited each day for a volunteer group to bring food near the entrance to the playground. They worried that they did not know the homeless individuals and sometimes they acted in ways that frightened them or their children. The homeless individuals who were interviewed near the playground expressed concerns about danger in the park late at night. Further, everyone interviewed including playground users, chess players, and the homeless individuals said that the nearby bathrooms were dangerous because of the drug users who used them to shoot up. Although only mentioned in a few interviews, the research team observed physical traces confirming that there is a rat population located near the playground that had been there for some time and remained active.

Overall, the children's playground was perceived as dangerous and fearful and the research team concluded that lowering the fence height was symbolic of this multiply determined negative perception of the place. For the users lowering height of the fence would have little impact on re-establishing a sense of security.

Social justice implications

While this study was not specifically undertaken to explore social justice, the data and findings are useful for illustrating how the social justice framework would work to enhance an understanding of the social, physical, and political situation in this public space and offer solutions to the overwhelming perception of danger and safety in this park.

First, in terms of distributive justice, there is not adequate space for all the users in this space. Indeed, it could be argued that it is the social injustice arising from a lack of affordable housing, food stamps, mental health facilities, and social-services support that creates a situation in which homeless individuals must wait each day for food near the entrance to the children's playground. In terms of a fair allocation of resources and public goods, the homeless individuals do not have their own adequate and safe spaces, and therefore are perceived as impinging on the playground. Another distributive problem might be whether the rat control resources of the city are adequate to cover this park or whether these resources focus on other wealthier or politically organized neighborhoods. Thus the distributive justice issue might not be occurring only within the spaces in the park but would require assessment of the just distribution of urban services and resources over all in the city and determining whether this park was receiving its fair share of rat control, police patrolling of the bathrooms, and social services for the homeless individuals rather than focusing on problems within the park.

In terms of procedural justice, the residents might be better able to accommodate their relationship with the homeless individuals and volunteer groups if the city began a process of negotiation about the use of the space. If all the users were able to meet and discuss the problem, and there was a fair procedure for voicing and remediating the conflicts, would this improve the social relations? Based on the procedural justice literature, greater cooperation and tolerance is an outcome when the process of decision-making is collective rather than only in the hands of one group (Whiteman et al. 2012).

There is clearly a problem with interactional justice as homeless individuals and others are not treated with respect by the children's caretakers and other users. However, the homeless men also feel that the city is not keeping them safe from whatever (or whomever) in inhabiting the park at night. In one interview with three homeless men, they said that they were fearful of spending the entire night in the park and left before midnight. A program of bringing park users together to get to know one another, or more targeted park supervision could work to impart rules and procedures for better kinds of social interaction. Such a program might also lessen the fear experienced by all the users.

Recognition of cultural patterns and practices is also missing in this park. Nearby Hester Street Playground has a variety of Chinese, Latinos, and White users and we observed that everyone talks, and their children play together as well as celebrate cultural holidays. In contrast at Tompkins Park there is more of a withdrawal of interaction, except within cultural groups. There are some young people at Tompkins who sunbathe in the grassy areas and share the space with homeless people who are also sleeping there, but in general there is not much recognition of cultural and behavioral difference except for complaints and little celebration of difference.

On the other hand, an ethic of care is being expressed by all the volunteer and social service groups who feed and help homeless individuals in this park. Also, the children's caretakers (parents and nannies) do watch out for one another, and a chess player also seemed to know who was in the park and keeps an eye out for danger. The role of the police in terms of care in the park, was mixed, in that they tracked some users but not others, but they were always present and seemed to recognize most of the park regulars including an artist, the chess players, a few men and women who drink in the park and some homeless individuals. Greater empowerment of park users to offer care and more caring based social services could contribute to the overall health and sense of social justice in the park.

The Tompkins Park example suggests that the proposed social justice framework adds to the assessment of the public space and offers solutions that are not only within the confines of the urban space, but that are also innovative in facing current park management challenges.

Acknowledgements

The author would like to thank Mitchell Silver, Commissioner, and Sarah Neilson, Chief of Long Range Planning of the New York City Parks Department for their support and for suggesting that we undertake this project. Sarah Neilson, a planner by training, gave us access to maps and insights into the kinds of questions we should focus on. The author would also like to thank Merrit Corrigan for access to her fieldnotes and unpublished paper that form the basis of my thinking on the topic. Other graduate students who worked in Tompkins Square included Benji Sullu, Anthony Ramos, and Elisandro Garza who gathered data that are reported as finding with Elisandro having completed the interviews with the homeless individuals in Spanish, and Anthony who interviewed the chess players and questioned the role of the police. The research team was critical to the success of the project and I want to acknowledge their important role in data collection. Without the help of all of these people this research could not have been completed. The conclusions, however, are my own.

References

Bies, R. J. and Moag, J. F. (1986) Interactional Justice. In: *Research on Negotiations in Organizations*, Vol 1, Lewicki, R. J., Sheppard, B.H. and Bazerman, M. (eds). Greenwich: JAI Press, pp. 43–55.

Brash, J. (2011) *Bloomberg's New York: Class and Governance in the Luxury City*. London: The University of Georgia Press.

Brenner, N., Marcuse, P. and Mayer, M. (eds). (2011) *Cities for People, Not for Profit: Critical Urban Theory and the Right to the City*. New York: Routledge.

Caro, R. (1975) *The Power Broker*. New York: Knopf.

Castells, M. (2012) *Networks of Outrage and Hope: Social Movements in the Internet Age*. Cambridge: Polity Press.

Chesluk, B. (2008) *Money Jungle: Imaging the New Times Square*. New Brunswick: Rutgers UP.

Corrigan, M. (2018) "We Want to Go in Our Cage and Feel Safe:" Tompkins Square Park and the Parks without Borders Initiative." Unpublished PhD diss., The Graduate Center, City University of New York.

Cresswell, T. (2015) *Place: An Introduction.* Oxford: Wiley Blackwell.

Cropanzano, R. and Randall, M. L. (1993) Injustice and Work Behavior: A Historical Review. In: *Justice in the Workplace: Approaching Fairness in Human Resource Management,* Cropanzano, R. (ed). Hillsdale: Lawrence Erlbaum Associates, pp. 3–20.

Duneier, M. (1999) *Sidewalk.* New York: Farrar, Straus and Giroux.

Fainstein, S. S. (2004) Cities and Diversity: Should We Want It? Can We Plan for It? *Urban Affairs Review,* 41(1): 3–19.

Fainstein, S. S. (2005) Planning Theory and the City. *Journal of Planning Education and Research,* 25(2): 121–130.

Fassin, D. (2013) *Enforcing Order: An Ethnography of Urban Policing.* Cambridge: Polity Press.

Fennell, C. (2014) Experiments in Vulnerability: Sociability and Care in Chicago's Redeveloping Public Housing. *City and Society,* 26(2): 262–284.

Fincher, R. and Iveson, K. (2008) *Planning and Diversity in the City: Redistribution, Recognition and Encounter.* Basingstoke: Palgrave.

Fisher, D. (2012) Running Amok or Just Sleeping Rough? Long-Grass Camping and the Politics of Care in Northern Australia. *American Ethnologist,* 39(1): 171–186.

Fiske, J. (1998) Surveilling the City: Whiteness, the Black Man, and Democratic Totalitarianism. *Theory, Culture and Society,* 15(2): 67–88.

Fraser, N. (1990) Rethinking the Public Sphere: A Contribution to Actually Existing Democracy. *Social Text,* 25/26: 56–80.

Hall, T. and Smith, R. (2014) Care and Repair and the Politics of Urban Kindness. *Sociology,* 49(1): 1–16.

Herbert, S. and Brown, E. (2006) Conceptions of Space and Crime in the Punitive Neoliberal City. *Antipode,* 38(4): 755–777.

Honneth, A. (2004) Recognition and Justice: Outline of a Plural Theory of Justice. *Acta Sociologica,* 47(4): 351–364.

Houssay-Holzchuch, M. and Teppo, A. (2009) A Mall for All? Race and Public Space in Post-Apartheid Cape Town. *Cultural Geographies,* 16(3): 351–379.

Iveson, K. (2007) *Publics and the City.* Oxford: Blackwell Publishing.

Kayden, G. (2000) *Privately Owned Public Spaces.* New York: John Wiley.

Kayden, J. S. (2000) *Privately Owned Public Space: The New York City Experience.* New York: John Wiley and Sons.

Lefebvre, H. (1991) *The Production of Space.* New York: Blackwell.

Lippert, R.K. and Walby, K. (2013) *Policing Cities: Urban Securitizaton and Regulation in a Twenty-first Century World.* New York: Routledge.

Low, S. (2000) *On the Plaza: The Politics of Public Space and Culture.* Austin: University of Texas Press.

Low, S. (2003) *Behind the Gates: Life, Security, and the Pursuit of Happiness in Fortress America.* New York: Routledge.

Low, S. (2013) Public Space and Diversity: Distribution, Procedural and Interactive Justice for Parks. In: *The Ashgate Research Companion to Planning and Culture,* Young, G. and Stevenson, D. (eds). Surrey: Ashgate, pp. 295–310.

Low, S. (2017) *Spatializing Culture: The Ethnography of Space and Place.* New York: Routledge.

Low, S. (2019) Domesticating Security. In: *Spaces of Security,* Low, S. and Maguire, M. (eds). New York: New York University Press, pp. 141–162.

Low, S. and Iveson, K. (2016) Propositions for More Just Urban Public Spaces. *City,* 20(1): 10–31.

Low, S. and Smith, N. (2006) *Politics of Public Space.* New York: Routledge.

Low, S., Taplin, D. and Scheld, S. (2005) *Rethinking Urban Parks: Public Space and Cultural Diversity.* Austin: University of Texas Press.

Low, S., Simpson, T. and Scheld, S. (2018) Toolkit for the Ethnographic Study of Space.

Maharawal, M. (2013) Occupy Wall Street and a Radical Politics of Inclusion. *Sociology Quarterly,* 54(2): 177–181.

Mehta, V. (2014) Evaluating Public Space. *Journal of Urban Design,* 19(1): 53–88.

Miller, K. F. (2007) *Designs on the Public: The Private Lives of New York's Public Spaces.* Minneapolis: University of Minnesota.

Mitchell, D. (2003) *The Right to the City: Social Justice and the Fight for Public Space.* New York: Guilford Press.

Morgan, K. (2010) Local and Green, Global and Fair: The Ethical Foodscape and the Politics of Care. *Environment and Planning A*, 42: 1852–1867.

Németh, J. (2010) Security in Public Space: An Empirical Assessment of Three Cities. *Environment and Planning A*, 42: 2487–2507.

Nemeth, J. and Schmidt, S. (2011) The Privatization of Public Space: Modeling and Measuring Publicness. *Environment and Planning B Planning and Design*, 38: 5–23.

Newman, A. (2011) Contested Ecologies: Environmental Activism and Urban Space in Immigrant Paris. *City and Society*, 23(2): 192–209.

Ramos-Zayas, A. (2012) *Street Therapists: Race, Affect, and Neoliberal Personhood in Latino Newark*. Chicago: University of Chicago Press.

Rawls, J. (1971) *A Theory of Justice*. Cambridge: Harvard University Press.

Roy, A. (2009) Civic Governmentality: The Politics of Inclusion in Beirut and Mumbai. *Antipode*, 41(1): 159–179.

Ruben, M. and Maskovsky, J. (2008) The Homeland Archipelago: Neoliberal Urban Governance after September 11. *Critique of Anthropology*, 28(2): 199–217.

Shepard, B. and Smithsimon, G. (2011) *The Beach Behind the Streets: Contesting New York City's Public Spaces*. Albany: SUNY Press.

Soja, E. W. (2010) *Seeking Spatial Justice*. Minneapolis: University of Minnesota Press.

Sorkin, M. (2008) *Indefensible Space: The Architecture of the National Insecurity State*. New York: Routledge.

Staeheli, L. A. and Mitchell, D. (2008) *The People's Property: Power, Politics and the Public*. New York: Routledge.

Tronto, J. C. (2013) *Caring Democracy: Markets, Equality and Justice*. New York: New York University Press.

Tyler, T. R. (2000) Social Justice: Outcome and Procedure. *International Journal of Psychology*, 35(2): 117–125.

Tyler, T. R. and Blader, S.L. (2003) The Group Engagement Model: Procedural Justice, Social Identity, and Cooperative Behavior. *Personality and Social Psychology Review*, 7(4): 349–361.

Van Melik, R., Van Aalst, I. and Van Weesep, J. (2007) Fear and Fantasy in the Public Domain: The Development of Secured and Themed Urban Space. *Journal of Urban Design*, 12(1): 25–41.

Varna, G. and Tiesdell, S. (2010) Assessing the Publicness of Public Space: The Star Model of Publicness. *Journal of Urban Design*, 15(4): 575–598.

Ward, K. (2007) Creating a Personality for Downtown: Business Improvement Districts in Milwaukee. *Urban Geography*, 28(8): 781–808.

Watson, S. (2006) *City Publics: The (Dis)Enchantments of Urban Encounters*. New York: Routledge.

Whiteman, D., Carpenter, N.C., Horner, M.T., Caleo, S. and Bernerth, J.B. (2012) Fairness at the Collective Level: A Meta-Analytic Examination of the Consequence and Boundary Conditions of Organizational Justice Climate. *Journal of Applied Psychology*, 97(4): 776–791.

Young, I. M. (2001) Equality of Whom? Social Groups and Judgments of Injustice. *The Journal of Political Philosophy*, 9(1): 1–18.

PART 2

Influences

The civic role of public space for the practice of democracy via discourse and debate has existed in the West since the Greek agora. Emphasizing the role of public space in democratic societies, Hannah Arendt, in *The Human Condition* (Arendt 1958), argues that public space provides the ability for people to come together, to discuss, and to recognize each other's presence, which is a crucial constituent of democracy. Politics has ubiquitously been recognized as the major influence on public space, and thinkers have long suggested that public space is a mirror, an expression of civil society, hence the civility of a society is enacted in public space. Yet, the making and use of public space is influenced by more than politics. The understanding, making, and use of public space in contemporary societies around the globe is shaped by a plethora of components and influences. The everchanging frameworks that impact and produce the meanings of public space are established by cultural factors, legal frameworks, environmental and geographic conditions, alterations of economic conditions, disciplinarization of the behaviors of specific ethnic groups, together with geo-political factors of the current new global order: migrations and flows of capital and labor. These ubiquitous and pressing issues can influence the interpretation, and therefore the type of uses in public spaces in different cultures for different groups and communities, in varying environmental, climate, geographical conditions, or because of religious, political, and established meanings.

The opening chapter by Jason Luger and Loretta Lees critically engages with questions of how to approach, and reframe, twenty-first-century urban public space. The authors explore how public space should be theorized when its occupation is governed and restricted by spatiotemporal realities that differ between one place and another, and promote a reflection on patterns and divergences to start forming a workable ontology, appropriate for a digital, neoliberal, authoritarian, and non-Western era.

Luisa Bravo reflects on the role public space plays in supporting a sustainable urban development. Her entry point is the New Urban Agenda adopted at the UN Habitat III conference in Quito (2016) that promotes public space as the human dimension of the urban context, able to provide opportunities of interaction and sharing, fostering civic identity, and social cohesion.

A place-oriented public health perspective on safety in public spaces is explored by Vania Ceccato in her chapter. A thorough bibliometric analyses for a 50-year period is explored linking public place, safety, and health issues. The chapter closes with a discussion of theoretical and practical potentialities as well as future challenges when adopting a place-oriented public health perspective to safety in public spaces.

Through case studies of public open spaces in Beijing, Hong Kong, and Taipei, Kin Wai Michael Siu, Yi Lin Wong, and Jiao Xin Xiao examine the inclusiveness of public open spaces and the current issue of exclusion of Visually Impaired Persons (VIPs). The authors discuss the need for synchronization between policy makers, designers, facility management teams, and users at different stages of work to offer truly inclusive public open spaces to visually impaired persons.

The chapter by Matthew Carmona draws from experiences in London to explore the design, development, use, and management of public spaces between 2008 and 2018—the austerity years—on the processes of shaping public spaces. The chapter suggests that austerity has been a period of significant innovation in the collective approach to public space.

Ceren Sezer's chapter introduces the concept of visibility as a useful tool to diagnose and assess inclusiveness in public spaces, which are understood as spaces that are open and accessible to all people, regardless of social, cultural, and economic differences. The author argues that visibility of distinctive urban groups on the street is a highly political issue and a key feature of socially inclusive cities.

Central urban parks and gardens were important public spaces in attempts to shape Trinidadian and Zanzibari society during British colonial regimes in the nineteenth and early twentieth centuries. Garth Myers uses examples to reiterate the ways in which imperialism and colonialism structured nature (literally and figuratively), and then shows the agency of ordinary people in reinterpreting the public spaces of postcolonial cityscapes.

Through a literature review and research in Southeast Asian cities, Tigor W. S. Panjaitan, Dorina Pojani, and Sébastien Darchen argue that theories on public space that apply to the West do not necessarily fit the context of Eastern cities. Despite these differences, the authors also recognize and explore converging trends in the nature of public space in both the West and the East due to neoliberalism, privatization, and globalization.

Su-Jan Yeo examines the case study of urban nightlife in Little India, a tourist nightspot, but also the locus of everyday (night)life for many unskilled and transient south Asian migrant workers in Singapore. The chapter illuminates tensions between state-driven and collectively-enacted actions that shape the use of public space after dark and define the degree of in/exclusions within it.

Tokyo, as other Japanese cities, has vibrant public spaces where social, cultural, and political lives unfold, even though they do not conform with foundational principles of the mainstream urban theories. Darko Radović explores the fact that Japanese, along many other non-Western languages, has no equivalent for the term "public," arguing for the need for local definitions, new, non-dogmatic readings and application of Lefebvre's right to the city, and for the right to difference.

Reference

Arendt, H. (1958) *The Human Condition*. Chicago: University of Chicago Press.

7

PLANETARY PUBLIC SPACE

Scale, context, and politics

Jason Luger and Loretta Lees

Introduction: place, platform, planet

Conceiving space as a static slice through time, as a representation, as a closed
system and so forth are all ways of taming it.

(Massey 2005, 59)

Public space is stretched, assembled, and practiced across different contexts in complicated
ways. It has never been easy in urban theory to categorize exactly what public space is:
public space has long been portrayed as both a material site for human encounter and the
production of material publics, as well as a broader 'sphere' (as framed by Habermas 1989)
in and across which publics form and co-relate. We argue it is time to re-think the nature
of public space, given 21st-century realities such as neoliberal privatization, emergent forms
and uses of space around the world, and the way digital networks and social media present
a new frontier for social and political encounters (if not physical encounters) and also new
methods of regulation, control and surveillance.

In some contexts, traditional notions of the public stretch into more private landscapes;
in others, the line between public and private are hard and demarcated. Lefebvre's (1968)
concept of the *right to the city*, in which accessible public space has always played a central
role, needs revisiting given both the rapid rise of digital networks and the surging currents
of authoritarianism and illiberalism in the global urban environment. Rights in one place are
not necessarily interpreted the same way in another, and there are unique cultural and religious characteristics attached to space across diverse contexts. Centrality, too, needs revisiting, given the rise of planetary suburbanization and changing relations between center and
periphery (an outdated dichotomy – see De Vidovich 2019).

Another important question is the degree to which politics produce public space and
how public space produces politics, given recent geopolitical shifts and the (re)turning global
proliferation of radical and reactionary political movements and moments, from pro-
democracy activism in Hong Kong to right-wing marches in the United States.

In many global contexts, the assembly of human bodies in public space comes with
a variety of caveats, restrictions, and performances. This is true in authoritarian settings, but
just as true in liberal-democratic urban settings, which, indeed, contain authoritarian elements

73

such as racist surveillance and state violence. Public space geographies everywhere are laden with moralistic assumptions and colonial/postcolonial legacies, the skeletons and residues of historical injustices and uneven power relations. These characteristics unite central London with Tiananmen Square, Red Square, or a street corner in Lagos or Luanda.

Nonetheless, public space remains vital for political formation and representation. Spontaneous eruptions and openings force their way through walls, borders, and enclosures; public spaces are reclaimed and re-appropriated. Protestors gather in public spaces from Hong Kong to Paris to Honduras, albeit using it in different ways and rallying around and against site-specific issues. In the United States and beyond, the #blacklivesmatter movement uses public space (connected to digital space) to call attention to the unjust surveillance and state violence aimed at black bodies in public space (e.g. Davary 2017). Women, LGBT groups and religious, racial, and ethnic minorities produce emancipatory public spaces at the same time that they face barriers to equal access and participation.

Decades of urban 'placemaking' agendas around the world, in which corporations and private companies have largely taken over the production of urban places, has given rise to semi-public squares, parks, and plazas. The notion of 'public' itself is open for discussion in some contexts: contrast Singapore, where the state is primary land-owner and producer of the built environment, against, say, a North American city like Houston in which the market has been free to terraform the urban landscape at will. Or a context like Mecca, where capitalism and religion intersect and access to public space is governed by religious doctrine. Is the Vatican a public space? Jerusalem?

The domestic sphere and the way that public and private intersect (and are produced by) domestic space, requires revisiting in a paradigm where public housing is increasingly replaced by private. In different cultures and contexts, domestic spheres can take on public roles and uses (e.g. the 'void decks' of east-Asian social housing). Likewise, public spaces like streets, squares, plazas, and parks come with context-specific rules, rights, and restrictions. Beaches and other natural areas in and bordering cities are enclosed and liberated. Finally, the advent of global digital infrastructure and the rise of the 'network society' and *smart city* ideas has provided, at least in theory, a planetary public space accessible to anyone with internet access or a smart-phone (Castells 1996). This space, however, is likewise rife with dynamic public-private divides and debates, and legal and political restrictions. 'Platform urbanism' (see Han and Hawken 2018) may be highly-networked and automated, modeled and algorithmic, but critical questions emerge about the relationship between *smart city* imaginaries and public space (*where in the smart city is public space? Can there be a 'public platform' urbanism?*).

The possibilities/impossibilities for emancipatory transformations to occur as a result of encounters in public space differ across scale, forms, and contexts (see for example Hou and Knierbein's 2018 global survey of public space, scale, politics, and territory). The popular understanding of public space as a sort of accessible agora implies the existence of a democratic structure; this democratic structure has not assisted the global poor, people of color, the homeless, or women, for example, equally. Surveillance and data collection as part of 'smart city' initiatives have blurred the line between democratic public space and dystopian police state. For the reality is that urban public space is not necessarily a space of and for democracy. For example, much of the state brutality, assault. and repression against black populations that have given rise to #blacklivesmatter happened in, and through, public space. In short, a *planetary* lens offers a fuller understanding of the complex nature of public space.

Framing public space/place

The urban theorist Henri Lefebvre (1974) argued that public space does not exist naturally but is an 'imposed normality' by power structures (i.e. the state, or state-market nexus) (23). Furthermore, space is only brought into existence and understood as being a full set of social relations, experiences, and encounters at a specific time: 'space is nothing but the inscription of time in the world. Spaces are the realizations, inscriptions in the simultaneity of the external world of a series of times, the rhythms of the city, the rhythms of the urban population … the city will only be … reconstructed on its current ruins when we have properly understood that the city is the deployment of time … of those who are its inhabitants' (Lefebvre 1967, 10). Space, therefore, is envisioned by Lefebvre as constantly being remade; each moment in time constitutes a different space. Furthermore, space is simultaneously local and global, comprised by and comprising social relations at intersecting scales, at different speeds. The way that public space is networked and relational giving rise to uneven power geometries, some of which are progressive, others repressive, is best explored by venturing through taxonomies of comparative public spaces in order to activate and bring conceptual understandings (like Lefebvre's) into view (see Massey 1993).

The concept of planetary urbanization (Merrifield 2013) has greatly widened what we might consider the urban to be, and thus, complicated the discussion on the location and texture of public space. Harvey (1989; 1996) has argued that modern capitalism transforms our understanding of distance, whilst Urry (2000; 2007) asserted that in the 'new mobilities paradigm,' there is a general blurred experience of space and geography. In the new 'informational city,' territory does not matter: borders and boundaries are only symbolic, and 'place' becomes less important than the flows occurring within and across it (Castells 1989; Merrifield 2002). However, the framing of public space, within digital, global, and multiscalar flows, as a very real, material *place* remains essential. A cosmopolitan and decolonized understanding of public space must take account of local context and site-specificity; subaltern lenses and locally constituted knowledge. Only then can global theory on public space truly be generated.

Public space is the site of the social, the political, and the self. This need not necessarily be a material 'place,' but the human body cannot – at least not yet – actually exist in cyberspace (beyond social media profiles and avatars). Thus, urban public space provides a gathering point for human bodies in a way that no other space can. Cresswell and Verstraete (2002) suggest that place is not only performative, but provides the very conditions for creative (and transgressive) social practices:

> place is made and remade on a daily basis. Place provides a template for practice – an unstable stage for performance. Thinking of place as performed and practiced can help us think of place in radically open and non-essentialized ways where place is constantly struggled over and reimagined in practical ways. Place is the raw material for the creative production of identity rather than an a-priori label of identity. Place provides the conditions of possibility for creative social practice. (25)

Public spaces can be as small as a single room, a street-corner, an alleyway – or, if one is willing to consider digital space as an extension of urban public space – then as large as the entire planet (even interstellar). The inter-scalar nature of public space and the relations occurring and intersecting at different speeds and via different circuits is one of the complexities plaguing any attempt to universalize a theory of public space, and this has not been lost on urban theorists. For example, Marston et al. (2005) built upon authors such as Neil Smith, Doreen Massey, and Neil Brenner, all of whom have addressed the complexity of scale and inter-scalar relations, by the radical proposal that 'scale' be abolished completely in favor of a 'flat' ontology, where 'the dynamic properties of matter produce a multiplicity of complex relations and singularities that sometimes lead to the creation of new, unique events and entities, but more often to relatively redundant orders and practice' (422).

However, we propose that attention to how public space is produced and experienced at different scales is crucial to building larger conclusions about the potential (and limitations) of each scale as an urban political, cultural, and social space. For larger processes and structures – inequality, racism, sexism, and bigotry; political encounters and resistances; capital flows – also occur at a variety of scales, speeds, and spatial form(s).

Urban sites have always been the foci of political formation and constructions of publics. For instance, the Athenian agora was always a gathering place for ideological exchange and dissent, and central urban sites remain fundamental to political movements (even as urban growth patterns sprawl and become diffuse). When mass protests appear on the news, more often than not, they are in centrally located urban spaces.

'Squares' are therefore associated with revolution, as is the central city, a gathering point that becomes synecdoche for social struggle. Marx associated the city with both industry and revolution. Lefebvre (1968) associated central Paris – that place of monuments, museums, avenues, and parks – as the 'site' of the widespread socialist activism of the late 1960s. Forty years later, Mayer (2009), Soja (2010), and Harvey (2012), among others, proposed again that the centrality of urban place remains crucial to activism. Enter Zuccotti Park, Central Hong Kong, Tahrir Square, Gezi, and the steps of St. Paul's Cathedral, all of which have been occupied for various causes.

The question then emerges: what does 21st-century public space *look* like, and is the central square still an appropriate archetype? Must contemporary public space be a material site, or can digital spaces also function as public space for representation? Finally, how have conceptions of the *centrality* of public space changed, and what does centrality now look like, in the digital age? What to make of the large (and growing) part of global urbanization that is suburban in nature? (Sub)urbanists such as De Vidovich (2019) are right to suggest that the 'center'/'periphery' bifurcation is outdated, since the majority of urban politics, processes and encounters happen away from the city-center. Suburban public space (and politics) are undertheorized, and vital to explore. If the central square is over-theorized, then the spaces of suburban politics are crucially under-theorized.

There are other factors not always considered in framing and comparing public space. The contextual differences (local histories, patterns of land ownership), cultural differences (racial, ethnic, religious practices or traditions), spatial and legal limits and restrictions, varying degrees of censorship (including the ways users of space self-censor, which is often culturally and locally-influenced), physical scale (public space in a village versus a mega-city), and even things such as weather and climate between these urban places need to be better explored (see Lees 1997, on the 'public' spaces in and around Vancouver's public library). In some contexts, protest is illegal; in others, it is partially legal; in other places, it might simply be too hot, too rainy, or too cold. In addition, protest need not look the same, or even take place in public places at all, which segues into another debate in the literature on activism. Authors do not agree over how important public space really is in terms of its relation to the urban encounter and activism in cities. Yet, regardless of how it now looks and feels, and where it is located in the digitally connected world, the symbolism of the 'center' remains a key recurring theme in the literature on public urban space.

Emerging discussions are increasingly re-framing the idea of the square (and the commons): building on the idea of the digital agora and the cyber-commons. Electronic or digital space – ranging from email to social media (Twitter, blogs, Facebook, and many others) – has opened a 'new frontier of public space' (Mitchell 1995, 122). This digital space may be even more crucial than material public spaces in some settings where the use and occupation of the built environment is particularly restricted or controlled (such as in an authoritarian city like Singapore), and where there is comparative freedom on the use of digital platforms. A new dialogue on digital urban space has given rise to robust explorations of various digital publics and digital political encounters and these explorations of digital public encounters are not limited to the west (Merrifield 2013). In fact, many studies have focused on the importance of digital public space in non-Western contexts, from Tunisia to postcolonial India (Bellin 2017; Datta 2018). The digital turn in urban studies has seen digital space approached through the lens of feminism (Fotopoulou 2017) and queerness, as in Miles' (2017) journey into LGBT digital (app-based) public social spaces. However, in other places, the digital sphere is more policed: in Iran or China, for example, internet activity is restricted and mis-use can result in imprisonment or even death. The internet is not completely free, anywhere, but offers different pathways for encounter and communication (Lees 1997).

In this framing, public space exists simultaneously in 'Tweets' and on the street (Gerbaudo 2012), with the digital and material being mutually-constitutive. Amin (2008) asked if it is still reasonable to 'expect public spaces to fulfil their traditional role as spaces of civic inculcation and political participation?' (5). Amin suggested that civic and spatial practice is no longer as connected to urban public space as it once was, or, indeed, to 'place' at all; that such civic practices are now 'shaped in circuits of flow and association that are not reducible to the urban' (6). Still, Amin conceded that public space/public urban place still has a role to play in the political encounter, 'allowing us to lift our gaze from the daily grind and as a result, increase our disposition toward the other' (6). Mitchell (1995; 2003) likewise argued for the importance of public spaces – particularly squares and central parks – as foci of publics and as spaces of representation for identity and politics.

But in many contexts, democracy and political representation operate very differently and public space may come attached with restrictions, or radically different interpretations and performances. For example, in the illiberal world, those spaces that are 'open' or 'closed' to publics necessitate a reconceptualization of where, and how, the formation of publics occur. Authoritarian cities, in particular, often feature a higher degree of state-influence around the use and occupation of the built environment when compared to oft-studied liberal-democratic cities. The cyber-agora plays a highly important role in more 'closed' contexts, as demonstrated by recent socio-political movements that had online dimensions (Hong Kong protests in 2019 are one compelling example).

Scott (2017) suggests that 'a vast extension of the common has taken place in these last couple of decades as a direct effect of the development of cyberspace' (9). In fact, Scott argues that, as a result of this cyber-extension, 'the scalar dimensions of many (but by no means all) segments of the common now extend far beyond any single urban area and in numerous instances are nothing less than global' (26–27). This cyber-extension has real implications for the encountering of networks and the networks of encounter in the urban realm.

Castells (2013) observed how the interrelationships of urbanized individuals are becoming more de-personalized as web-based exchanges increasingly supplant face-to-face interaction. Scott (2017), on the other hand, does not envision the cyber-common as supplanting interaction, but rather as complementing it and forming new methods and modes of communication, a theme echoed by other authors such as Merrifield (2013) and Gerbaudo (2012). In Scott's (2017) conceptualization, the 'digital common' is both global in scope but 'punctuated by strong localized articulations' and a 'complementary form of social reality subject to its own specific structural logics and ... accompanied by its own specific kind of effects' (26–27). Whether or not the digital/immaterial realm can ever become more important, or more influential than material sites – actually supplanting the material public square – is a provocative question and one worth exploring.

Global public space, local spatial politics

Singapore's Hong Lim Park, also known as 'Speakers' Corner' (see Figure 7.1), is a centrally located 0.94-hectare park where public gatherings occur. However, as seen below, the park is often devoid of human bodies, outside of the activities that are scheduled in the park. These activities and events and protests must be pre-approved by Singaporean authorities and are not permitted if they violate certain topics or themes. For example, an annual LGTBQ celebration known as 'Pink Dot' is permitted to take place, but only as long as it does not identify as a 'Gay Pride' event. Those who are not Singaporean permanent residents are forbidden from attending. Further complicating the park's status is the fact that the Singaporean government technically owns the park and the surrounding land. So, whether or not Hong Lim Park

Figure 7.1 Speakers' Corner/Hong Lim Park: Singapore hybrid public space?
Image Credit: Jason Luger

qualifies as truly public, or a hybrid space within Singapore's authoritarian state-society context, remains an open question.

As discussed previously, it is difficult enough simply to *locate* public space in the material urban environment or in digital forms, at a variety of scales. The next question becomes, what counts as public space, and what does not? Neoliberal urbanism, globalization, and hybrid state-society-space formations have blurred and blended the lines between public and private space and created an ambiguous urban grey area that is both public and private. Also varying greatly from one terrain to the next are locally constituted interpretations of public vs. private space and the different roles/hierarchies of these spaces according to local politics, culture, customs, and traditions. In the example above, Singapore's 'Hong Lim Park' is a hybrid space that, from the outside, looks like a traditional urban public square. However, the space comes with restrictions on assembly, expression and use that differentiate it from similar squares in liberal-democratic cities.

Huge swathes of territories come with varying degrees of authoritarian restrictions and limitations on the use of space, as well as differing relationships between the state, land, and the built environment. China is perhaps the biggest example of an authoritarian state-land nexus: the degree of state ownership of space in a context such as China complicates the line between public and private. In China, with its particular (state-capitalist) restrictions on private property rights – where is the actual line between public and private space? The same question could be asked of other territories where the state plays a stronger role, whether that be Cuba or, as an extreme, North Korea's totalitarian-built environment.

But even in less authoritarian contexts the publicness of public space is not easy. Anna Minton (2014) discusses the reason that Occupy London was outside the steps of St Paul's Cathedral and not in Paternoster Square:

> they couldn't put that protest in Paternoster Square, which is actually where the Stock Exchange is. They were called Occupy London Stock Exchange. But Paternoster Square is a privately owned estate and they were immediately ejected from there. In fact, the City of London today is a series of privately-owned estates. It wasn't at all before Big Bang [the day the London stock market was deregulated]. The small space outside St Paul's Cathedral is actually pretty much the only sizable publicly owned land that's left under the Corporation of London's jurisdiction. The Church owns some of it and some of it is owned by the Corporation. But that's the only effectively public space. So that's why they were there. (34)

If places like London's Paternoster Square are public (despite private ownership), then can public space be extended indefinitely to include any space where human bodies congregate and encounter? This would have huge implications for how places like churches or mosques, stores or restaurants, private universities and schools, offices or homes are theorized.

The home, traditionally differentiated as a private space, takes on public dimensions. This is true in some cultures where communal dining or socializing, even with strangers, is more common. In some contexts, housing is public (state owned) to a higher degree – Singapore, for example, where 85% of residents live in state housing. Singapore's public housing estates (known as HBD flats) have open 'void decks' on the ground floor, which form a communal space for public use. Meanwhile, in other contexts, traditionally public domestic spaces are being reformatted as private, or public/private hybrids (for example, the large-scale destruction and redevelopment of public housing into 'mixed-income' communities, mostly with private financing, in market-led urban contexts).

Where coming up with a single definition or fixed taxonomy of public space may be impossible, there are some unifying elements that run across all of these spatial forms, place formations, hybrids, and scales. These are, the shared concept of 'rights' – which come attached with notions of access, and centrality. Secondly, the fact that power operates across and from all public spaces as a governing and regulating force. Thirdly, all public spaces are political spaces, both formed by politics and giving rise to new politics.

De-centering the center: power, space, and identity

Henri Lefebvre (1968) proposed that access to public space – notably, the centrality of urban public space – was crucial to realizing the 'right to the city.' However, *where* centrality is, and *what* type of access to this centrality may or may not be possible, is a more complex discussion. As previously noted, the center/periphery dichotomy no longer reflects the reality of urban development patterns, which are diffuse and suburbanized. Many urban inner-cities, for example, are now gentrified (in the Global South and North), limiting access to the poor and working class; meanwhile new centers of encounter and politics form on the urban margins.

The reality is that many urban environments do not have traditional 'central' public spaces, and that not everyone has equal access to such central spaces, wherever they exist. People of color, for example, have long been excluded from, and denied access to, public space and public centrality. The field of black geographies points to examples where public space can be 'read' according to the layers of embedded presence, and absence, of black bodies: see for example Katherine Mckittrick (2014) on the absence/presence of blackness in urban spaces in Canada and the USA. Or A.M. Simone's (2016) explorations of a more global, unifying understanding of 'blackness,' in terms of those that are peripheralized in the global city – marginalized and denied access to the center. #Blacklivesmatter is a movement largely focused on the right of black bodies to access, without fear of racial assault or police brutality, public spaces (whether they be streets/sidewalks, parks, plazas, or indoors).

Women, too, have (over history) had a complicated relationship with public space. The Athenian Agora was not a place where women were part of the operation of politics, and women are still excluded from full representation in public spaces in many global contexts. Abuses and assaults against women are also frequently located in public space – #Metoo is one reaction against this. Homeless persons, refugees, LGBTQ persons, sex workers, the disabled, and others, too, are not fully represented in public space, or given access to centrality. Public space is exclusionary as much as it is a space for democracy; and as a produced space, is governed by certain uneven power flows and imbalances. Nor is digital public space (if such a space is public at all) a panacea: digital divides and exclusions exist, whether behind paywalls or degrees of censorship (e.g. China's 'great firewall'). Furthermore is the fact that despite the advent of the network society, large swathes of territory and publics remain unplugged and disconnected from cyberspace (Castells 2013).

Garbin and Millington (2018) highlight how central London has become a crucial site for representation for the Congolese diaspora, protesting about Congolese political developments thousands of miles away (see also Brown and Yaffe 2017 on the non-stop picket against apartheid in the 1980s). It is the symbolic global centrality of inner-London that gives tremendous meaning and representative power to the Congolese gathered in place – even if, as previously noted, much of the urban space in central London is no longer owned by the public. As a central, internationally visible urban space, Westminster becomes synecdoche not only for the Congolese struggle (about events across the globe), but also, linking into the wider #Blacklivesmatter global struggle and representative imaginary. Herein, geographically central urban space (whether truly public or semi-public) remains vital for the meeting of publics, the formation of the political, and the constructions of a shared, global black consciousness. As such, it is inextricably linked to the cyber-sphere; to global struggles, and to a variety of scales and multi-scalar relations. But it is also inherently local and site-specific, a very real and material assembly of bodies in the urban place that happens to be a few square kilometers in Westminster, London.

Indeed, the future of public space may be increasingly characterized by informal and highly networked relations between the global periphery and the center (wherever that center may be). Pieterse and Simone (2018) explore the rise of 'new urban worlds' in Africa and Asia, regions that will lead global urbanization in the 21st century. In Africa's pop-up and fast-growing cities, sometimes growing in the absence of formal planning and formally planned civic spaces, publics create new spaces through everyday social practice. The rise of 'DIY' urbanism and corresponding DIY public spaces can be seen as both a byproduct of neoliberalism's destruction of the traditional public realm, but also a series of creative and even emancipatory responses to the voids resulting from the privatization (and abandonment) of the public built environment. Douglas (2018) and Kinder (2016) show how cities from Los Angeles to Detroit feature publics that are reclaiming and producing urban space within the interstitial shadows left behind as traditional planning has broken down or abandoned large urban swathes. This pop-up urbanism is emancipatory at the same time that it is inadequate and unequal. Still, the networked and multi-scalar nature of contemporary public space as well as its increasingly DIY and informal nature are just two things to consider as we conclude with some beginnings of a possible outline for a new language of public space going forward that better encapsulates the contemporary paradigm.

Conclusion: towards a new ontology of planetary public space?

The publicness of place enables creative social practices to generate encounters, relations and politics that reverberate across scale, time, and space. Reconfiguring an ontology of urban public space to suit the 21st century is crucial in order to better converse with the diversity, scale, and implications of global geo-politics (Rokem and Boano 2017). These politics begin and end, as ever, in public space. Indeed, we may need to expand even further the conception of public space, to consider not only the space itself (whether in material or digital form): but also, all of the practices, structures, performances, and regimes that are attached to (and form) space (as Ruppert 2006, suggests). Through this *further-expanded, planetary* field, the possibilities and paradoxes of public space can be seen in full.

Thus, we conclude with some questions, and some suggestions, in order to begin forming an outline for what a new, cosmopolitan language of urban public space might consist of. In doing so we summarize several of the points addressed in the chapter. First, we suggest that any understanding of planetary public space should be both site-specific and spatial, taking into account the local characteristics as well as the global and multi-scalar flows that exist in and beyond territory. Only then can the actually existing power relations, uneven geographies, social networks, and political potential be explored in a full sense. This might mean, practically, for 'old' methods of observing and dissecting space (the ground-breaking public space ethnography of Elijah Anderson 2011 for example) to be joined with newer methods of digital research and digital ethnographies (e.g. Lane 2018; Aurigi 2016 on digital streets, digital cities).

Second, we suggest it may be time to revisit what constitutes publicness in the built and natural environment. The fixed city center or formally planned public square can no longer be a proxy or synecdoche for where publics form and encounter; these official central places have too often been privatized, enclosed, surveilled, or abandoned altogether. Public space is being reclaimed and activated in new ways and sometimes temporally fleeting moments, assembled and disassembled as needed, and not always occurring in the urban center. Given the possibilities to access publics through cyberspace, we ask whether it is time to extend public space into the home, the office, or really anywhere a smart phone can connect to the internet. Whether or not this represents a blending together of 'public space' with Habermas's 'public sphere' is a question we do not seek to answer, but pose nonetheless.

Finally, we propose that it is time to decolonize both public space and the construction of knowledge around public space. The vast majority of urban theory making stems from Europe and North America and thus is based on specific (and previously dominant) urban forms, power relations, and embedded socio-spatial networks. The rise of urbanization in the Global South, which both presents convergences and stark divergences from previous urban models, demands a re-think about how, where, and in what ways humans produce space and space produces humans.

Decolonizing the language of public space does not only need to happen in a geographic sense: for far too long, public space has been unevenly tilted towards use and control by the powerful and by global elites. The global poor, women, racial, ethnic, religious, and sexual minorities are demanding their voices be heard and presences felt in public space and public discourse. This requires that planetary public space be racialized, gendered, queered, and stripped of pre-existing assumptions and attributes taken as givens, or taken as natural. Herein lie several exciting opportunities for research and theory in approaching the complexity of the planetary urban.

References

Amin, A. (2008) Collective Culture and Urban Public Space. *City*, 12(1): 5–24.

Anderson, E. (2011) *The Cosmopolitan Canopy: Race and Civility in Everyday Life*. New York: W.W. Norton.

Aurigi, A. (2016) *Making the Digital City: The Early Shaping of Urban Internet Space*. London: Routledge.

Bellin, E. (2017) Networked Publics and Digital Contention: The Politics of Everyday Life in Tunisia. *Political Science Quarterly*, 132(2): 358–360.

Brown, G. and Yaffe, H. (2017) *Youth Activism and Solidarity: The Non-Stop Picket against Apartheid*. London: Routledge.

Castells, M. (1996) *The Information Age* (98). Blackwell: Oxford.

Castells, M. (2013) The Impact of the Internet on Society: A Global Perspective. In: *Ch@Nge: 19 Key Essays on How the Internet Is Changing Our Lives*, Gonzales, F. (ed). Bilbao: BBVA, pp. 127–148.

Castells, M. (1989) *The Informational City: Information Technology, Economic Restructuring, and the Urban-Regional Process*. Oxford: Blackwell.

Cresswell, T. and Verstraete, G. (eds). (2002) *Mobilizing Place, Placing Mobility: The Politics of Representation in a Globalized World*. Minneapolis: University of Minnesota Press.

Datta, A. (2018) The Digital Turn in Postcolonial Urbanism: Smart Citizenship in the Making of India's 100 Smart Cities. *Transactions of The Institute of British Geographers*, 48: 405–419.

Davary, B. (2017) # Black Lives Matter. *Ethnic Studies Review*, 37(1): 11–14.

De Vidovich, L. (2019) Suburban Studies: State of the Field and Unsolved Knots. *Geography Compass*, 13(12440): 1–14.

Douglas, G. (2018) *The Help-Yourself City: Legitimacy and Inequality in DIY Urbanism*. Oxford: Oxford University Press.

Fotopoulou, A. (2017) *Feminist Activism and Digital Networks: Between Empowerment and Vulnerability*. London: Springer.

Garbin, D. and Millington, G. (2018) 'Central London under Siege:' Diaspora, 'Race', and the Right to the (Global) City. *The Sociological Review*, 66(1): 138–154.

Gerbaudo, P. (2012) *Tweets and the Streets: Social Media and Contemporary Activism*. London: Pluto Press.

Habermas, J. (1989) *The Structural Transformation of the Public Sphere*, Trans. Burger, T. Cambridge: MIT Press.

Han, H. and Hawken, S. (2018) Introduction: Innovation and Identity in Next-Generation Smart Cities. *City, Culture and Society*, 12: 1–4.

Harvey, D. (2012) *Rebel Cities: From the Right to the City to the Urban Revolution*. Verso: London.

Harvey, D. (1996) *Justice, Nature and the Geography of Difference*. Cambridge: Blackwell.

Harvey, D. (1989) *The Condition of Postmodernity: An Enquiry into the Origins of Cultural Change*. Cambridge: Blackwell.

Hou, J. and Knierbein, S. (eds). (2018) *City Unsilenced: Urban Resistance and Public Space in the Age of Shrinking Democracy*. London: Taylor & Francis.

Kinder, K. (2016) *DIY Detroit: Making Do in a City without Services*. Minneapolis: University of Minnesota Press.

Lane, J. (2018) *The Digital Street*. Oxford: Oxford University Press.

Lees, L. (1997) Ageographia, Heterotopia, and Vancouver's New Public Library. *Environment and Planning D: Society and Space*, 15(3): 321–347.

Lefebvre, H. (1996 [1967]) The Right to the City and Theses on the City, the Urban and Planning. In: *Writing on Cities-Henri Lefebvre*, Kofman, E. and Lebas, E. (eds). Blackwell: Oxford, pp. 147–159 and 177–184.

Lefebvre, H. (1968) *Le Droit De La Ville* (2nd Edition). Paris: Anthropos.

Lefebvre, H. (1991 [1974]) *The Production of Space*, Trans. D. Nicholson-Smith. Oxford: Basil Blackwell.

Marston, S., Jones, J. and Woodward, K. (2005) Human Geography without Scale. *Transactions of The Institute of British Geographers*, 30(4): 416–432.

Massey, D. (2005) *For Space*. London: Sage.

Massey, D. (1995) *Space, Place and Gender*. Minneapolis: University of Minnesota Press.

Massey, D. (1993) Power-Geometry and a Progressive Sense of Place. In: *Mapping the Futures: Local Cultures, Global Change*, Bird, J., Curtis, B., Putnam, T. and Tickner, L. (eds). London: Routledge, pp. 60–70.

Mayer, M. (2009) The 'Right to the City' in the Context of Shifting Mottos of Urban Social Movements. *CITY* Special Issue, 13(2–3): 362–374.

Mckittrick, K. (2014) The Last Place They Thought Of. In: *The People, Place, and Space Reader*, Gieseking, J. J., Mangold, W., Katz, C., Low, S. and Saegert, S. (eds). New York: Routledge, pp. 309–313.

Merrifield, A. (2013) The Urban Question under Planetary Urbanization. *International Journal of Urban and Regional Research*, 37(3): 909–922.

Merrifield, A. (2002) *Metromarxism*. New York: Routledge.

Miles, S. (2017) Sex in the Digital City: Location-Based Dating Apps and Queer Urban Life. *Gender, Place & Culture*, 24(11): 1595–1610.

Minton, A. (2014) Privatising London: A Conversation with Anna Minton. In: *Sustainable London? The Future of a Global City*, Imrie, R. and Lees, L. (eds). Bristol: Bristol, pp. 29–42.

Mitchell, D. (1995) The End of Public Space? People's Park, Definitions of the Public, and Democracy. *Annals of the Association of American Geographers*, 85(1): 108–133.

Mitchell, D. (2003) *The Right to the City: Social Justice and the Fight for Urban Space*. London: Guilford.

Pieterse, E. and Simone, A. M. (2018) *New Urban Worlds: Inhabiting Dissonant Times*. London: John Wiley & Sons.

Rokem, J. and Boano, C. (2017) Introduction: Towards Contested Urban Geopolitics on a Global Scale. In: *Urban Geopolitics*, Jonathan R. and Camillo B. (eds). London: Routledge, pp. 17–30.

Ruppert, E. (2006) Rights to Public Space: Regulatory Reconfigurations of Liberty. *Urban Geography*, 27(3): 271–292.

Scott, A. J. (2017) City and Society. In: *The Constitution of the City*, Scott, A. J. (ed). Cham: Palgrave Macmillan, pp. 1–9.

Simone, A. (2014) *Jakarta, Drawing the City Near*. Minneapolis: University of Minnesota Press.
Simone, A. (2016) Urbanity and Generic Blackness. *Theory, Culture & Society*, 33(7–8): 183–203.
Soja, E. (2010) *Seeking Spatial Justice*. Minneapolis: University of Minnesota Press.
Urry, J. (2007) *Mobilities*. Malden, MA: Polity Press.
Urry, J. (2000) *Sociology beyond Societies: Mobilities for the Twenty-First Century*. London: Routledge.

8

PUBLIC SPACE AND THE NEW URBAN AGENDA

Fostering a human-centered approach for the future of our cities

Luisa Bravo

Introduction

In Europe and most of the Western world, there is consensus that public space contributes to the common good and that its quality makes economic, social, environmental, and cultural impacts (Ibeling 2015). In the Global South, that awareness is not yet fully established; often public space is poorly designed, maintained, or managed as a leftover space, while publicness is jeopardized by private interests, a political regime, or massive urbanization. The United Nations Human Settlements Program (UN-Habitat) has invested in public space through the Global Public Space Programme. Currently active in more than 30 countries the Program is strategically linked to the implementation of the New Urban Agenda (NUA), an action-oriented document pursuing a paradigm shift in sustainable development of cities and human settlements. While the concept of public space remains ambiguous and not actively champion—despite a growing citizen interest in active engagement in the decision-making process—investment in public space is an undervalued task for many local governments.

This chapter will, from a global perspective, present the contemporary complexities associated with urbanization and will discuss how the New Urban Agenda (NUA) and the Agenda 2030 for Sustainable Development are defining public space as a priority for the future of cities, establishing new models of governance, and renewing human-oriented thinking.

Public space in the age of massive urbanization

Our world is ever-increasingly urban; cities are rapidly expanding, and urbanization is one of the most urgent challenges for policy makers and city leaders. According to the United Nations, 55% of the world's population lives in urban areas, a proportion that is expected to reach 68% by 2050. For reference, only 10% lived in cities at the beginning of the 20th century. This growth, however, is a recent phenomenon. In 1990, ten urban concentrations were classified as megacities, and in 2014 that number rose to 28. It is expected to reach 43 by 2030 and projections show that another 2.5 billion people could be added to urban populations by 2050, overwhelmingly concentrated in Asia and Africa (UN 2018b).

In developing countries, the process of urbanization has become nearly uncontrollable, with the number of slum dwellers increasing nearly 200 million between 1990 and 2014 (UN 2016a). Though located primarily in sub-Saharan Africa, Pakistan, Mexico, India, and Brazil, informal settlements are also occurring in Paris, Berlin, and Rome. Large-scale migration has enabled the growth of those informal settlements across Europe (UNECE 2016). For example, Cañada Real Galiana, a slum located 15-minutes away from Madrid's city center, is home to around 30,000 residents, many of whom are long-standing, legal immigrants from the Global South (Anders and Sedlmaier 2017). As a result of conflicts, temporary refugee camps in the Middle East have turned into legitimate cities, despite the lack of real infrastructure. Around a quarter of the world's urban population lives in informal settlements presently, and in Davis' (2006) estimation that may increase to one-third. As urban environments are primarily human environments, urbanization is creating a novel set of human-centric issues.

Since it is a key enabler of human rights, empowering women, providing opportunities for youth, creating a sense of security and well-being, and promoting equity, inclusion, and democracy, public space is the crucial instrument to promote a human-oriented approach in cities. Public space is a common good, it is the social glue that can contribute to advancing mutual trust, cooperation, and solidarity. Well-designed and well-managed public spaces can provide opportunities for formal and informal economies by attracting investments, entrepreneurs, and services, enhancing the value of land and property, redefining urban environments with human vibrancy and livelihood, encouraging walking, cycling and play, and improving physical and mental health.

Surprisingly, the commitment to pursue a human-oriented urban planning strategy is not addressed in the urban agendas of politicians and local governments. Major concerns of cities and built environments are defined through different categories and expertise related to housing, services, infrastructure, environmental issues, and technological solutions that also generate investment and provide funding opportunities. Too often, public space is not listed as a primary question, but rather as a collateral component, mostly intended as design activities related to landscape urbanism or infrastructure facilities. While academic research on public space is well established through cross-disciplinary approaches and investigation, there is a lack of interest or an unprepared expertise in defining a comprehensive urban strategy for local implementation, that is built around humans and their life in the public domain. Many cities' public spaces, as the arena for human interaction, promote individualism and segregation as a form of existence, rather than inclusiveness, through neglect, under-design, underuse, and abuse. In many countries, particularly rapidly urbanizing and low-income, the proliferation of informal settlements and their exclusion from basic services make communities vulnerable and exposed to the risk of crime and violence, seriously impacting social cohesion and civic identity. In addition, the increasingly privatized public domain, and the lack of facilities to fully access, live, and enjoy public space often isolate communities, limiting their freedom to engage the public sphere (Sassen 2015).

We mainly refer to public space's cultural richness, identity, and diversity, but public space now addresses inequalities, poverty, and conflicts. Public spaces are meant to be open, inclusive, and democratic, but physical, social, and economic barriers pose challenges for many local communities and social groups. We also experience theoretical barriers to an open discussion on public space. Discourse is often limited to specific national or linguistic areas, and the dominance of examples from the Global West and North limit knowledge about global public space by imposing oversimplified views of design, management, and use (Bravo 2018). In the same way that the ideal public space promotes openness and inclusivity, *The Journal of Public Space*, a collaborative partnership between UN-Habitat and City Space Architecture, is entirely dedicated to being an open-knowledge platform, focused on learning from unheard countries and to discussing neglected and emerging topics that can be side-lined by mainstream publications.

UN-Habitat launched its Global Public Space Programme in 2012 to promote public space as an urban asset in more compact, connected, and socially inclusive cities. Moreover, it seeks to share knowledge, best practices, and methodologies for local governments through policy guides, capacity building, and advocacy work. The main objective of the Programme is to promote public space as a priority in the political agenda. This is accomplished by raising awareness about sustainable planning principles, addressing rapid expansion, retrofitting existing settlements towards more sustainable patterns, and reducing poverty and inequality. The Programme is active in more than 30 countries and has partnered with local organizations and communities to build pilot projects around the development, upgrade, and management of public space (Andersson 2016). The Programme applies a city-wide approach for city-wide impact and began in Nairobi, where the UN-Habitat is headquartered; these initial projects focused on spaces in Dandora, a low-income, high-crime residential neighborhood. The project implemented a series of small interventions aimed at reaching a large impact in the community and was developed through youth engagement and in cooperation with local organizations.

Public space as a key concept included in the New Urban Agenda

The UN-Habitat has established a database of public spaces in 289 cities in 94 countries, and in 2018, public space as an indicator was reclassified by UN Statistical Commission from Tier III to Tier II, meaning that there is consensus around the concept and established research methodology, but data are not regularly produced by countries. The methodology to identify the area of the city considered public space is based on three steps: 1) spatial analysis to delimit the built-up area of the city, 2) estimation of the total open public space, and 3) estimation of the total area allocated to streets. These assessments are directly tied to the motives of the NUA and the Agenda 2030.

The NUA is the new global mandate, that envisions an urban paradigm shift for the sustainable development of cities and human settlements over the next 20 years. This document was the result of two years of negotiations, engaging more than 100,000 people in drafting, shaping, and discussing its contents. Most notably, the NUA clearly defines its commitment toward a new form of development by promoting safe, inclusive, accessible, green, and quality public spaces, including streets, sidewalks and cycling lanes, squares, waterfront areas, gardens, and parks, that are multifunctional areas for social interaction and inclusion, human health and well-being, economic exchange, cultural expression and dialogue among a wide diversity of people and cultures, and that are designed and managed to ensure human development and to build peaceful, inclusive and participatory societies as well as to promote living together, connectivity, and social inclusion. (UN 2016c, 9)

The importance of public space is highlighted again by Agenda 2030 for Sustainable Development (2015). This document included 17 Sustainable Development Goals (SDGs), which operationalize inclusive and democratic principles. For example, SDG 11 is set out to "make cities and human settlements inclusive, safe, resilient and sustainable," and it includes a specific target related to public space, where "by 2030, provide universal access to safe, inclusive and accessible, green and public spaces, in particular for women and children, older persons and persons with disabilities".

As reported by United Nations Development Programme (2017), the United Nations Conference on Trade and Development estimated that achieving the SDGs will take between $5 and 7 trillion (US), with a $2.5 trillion investment gap in developing countries. The *Better Finance, Better World* report explained that achieving the SDGs could create $12 trillion of opportunities in the agriculture, energy, and health markets alone, as well as creating 380 million new jobs by 2030 (BFT 2018). Philanthropic capital and private financing are critical to this implementation; in 2019 the UN-Habitat Assembly convened a Business Leaders Dialogue to define the role of the global business community in advancing the actions required to implement the NUA and achieve the SDGs. The Dialogue was an opportunity to present UN-Habitat's business engagement strategy and to facilitate multi-sectorial collaboration opportunities. The investment in public space should be linked to profit activities for its implementation; partnerships with the private sector can be defined by shared principles, common goals, and should be grounded in an ethical conduct aimed at the public good.

Because of set and agreed upon indicators, the SDGs provide, for the first time, a platform where public spaces can be globally monitored. Now leaders and citizens, who have previously overlooked or undervalued the importance of public space in a multi-functional urban system, can refer to a specific target and indicator as an opportunity to be engaged in implementing and upgrading public space design and policies. As detailed in their 2018 report, the Global Public Space Programme's recent work has been devoted to creating public spaces in contested and high density neighborhoods, promoting urban culture, human rights, social cohesion, and inclusivity, elaborating guidelines through civic engagement, with a specific focus on children with disabilities, improving safety and mobility, integrating vulnerable groups through participatory planning, and empowering local communities and building peace in post-conflict scenarios, therefore anchoring public space in national policies. In several countries, UN-Habitat is working at the local level in planning debates and processes to inform national policy change and mainstreaming sustainable planning principles into national guidelines. These efforts are in support of countries' most vulnerable populations and seek to decrease the tensions between host communities and refugees.

In 2019, UN-Habitat published *The Silent Revolution of Public Spaces in Afghanistan* as a result of a collaboration with the Government of the Islamic Republic of Afghanistan. The Clean and Green Cities Programme, implemented in twelve Afghan cities, included the development and upgrade of 49 parks, with infrastructure, services, and public spaces. As stated in the introduction:

> The provision of public spaces in cities across Afghanistan is shaping the lives of the citizens, especially women and the poor, and the relationship with their cities. It is a social and cultural revolution that has empowered women, created jobs for the poor and improved the living standards of the citizens. It is a silent and peaceful revolution that has transformed the people and the city, building trust and confidence in public institutions.
>
> *(UN-Habitat 2019a, vi)*

The UN-Habitat's Global Public Space Programme regularly reviews its mission by engaging qualified professionals. In the past three years about 1,200 decision makers and stakeholders have been involved in planning charrettes and trainings. It has already produced a number of reports and publications, like the Global Public Space Toolkit (2015a), a practical reference for local governments aimed at enabling specific legislation also involving civil society, and is currently working on a compendium of inspiring practices and a guide for city leaders related to a city-wide public space strategy. The Programme also includes projects related to the use of digital technologies to increase levels of participation, civic engagement, and education of youth. One was developed through a partnership with Ericsson, where mixed-reality technology was used in urban and public space design, and another with Microsoft, which used the video game Minecraft (UN-Habitat 2015b; UN-Habitat. 2019b).

According to the United Nations, there are 1.8 billion people between the ages of 10–24 and nearly 90% live in less developed countries (UNFPA 2014). As a counterpoint, the number of older persons is expected to double by 2050 (United Nations 2017; 2019). The design of cities and public spaces should become more age friendly (WHO 2007). Additionally, more than one billion people in the world live with some form of disability, highlighting the urgent need to create more inclusive urban environments. Data indicate that in some countries more than 30% of persons with disabilities find transportation and public spaces inaccessible (United Nations 2018a). The challenge for public space is not just the quantity or location, but rather its full accessibility and safety (Pineda et al. 2017). The ambitious NUA and Agenda 2030's SDGs are set to create a future where urban citizens will define public space as urban commons, a fundamental human right, a stage for freedom of expression, civic empowerment, and human coexistence.

Investing in the human capital

In the Global South, post-conflict regions, or divided territories, where people may live in inadequate conditions and in extreme poverty, public space serves as an extension of small living environments, places where social interaction and mutual support can be established; it promotes community bonds and provides children with play spaces. Public space can also host informal economic activities along streets and on sidewalks. In the UN-Habitat (2013) report "Streets as Public Spaces and Drivers of Urban Prosperity," it is clear that "those cities that have failed to integrate the multi-functionality of streets tend to have lesser infrastructure development, lower productivity and a poorer quality of life" (iv). Cities are prosperous if they have prosperous streets, meaning that at least 30% of the city should be dedicated to the street pattern, not just for cars, but also for pedestrian, transit, and bicycle use. Cars, both in the Global North and South, are dominating urban environments, while walkability in many cities is in decline. The resulting pollution and climate change are harmful for human health; according to the World Health Organization (2012) air pollution is the cause of approximately 7 million premature deaths annually, while extreme weather events and natural disasters are resulting in millions of deaths globally.

People are at the core of the urban agenda. Wealthy cities should be the result of an investment in human capital and in a human-oriented approach to dealing with the complexity of urban planning and design. Although formally defined as "the knowledge, skills, competencies and attributes embodied in individuals that facilitate the creation of personal, social and economic well-being," (OECD 2011, 18), human capital is largely understood as a factor of economic production. Following the global financial crisis of 2008, for example, human capital, combined with Gross Domestic Product, was used as an indicator to assess a country's "welfare stateness" with alternative indicators such as the UN's Human Development Index and the OECD's Better Life Index, further highlighting the linkage between human capital and economic health (OECD 2011). Following the recent publication of the Human Capital Project (2018) the World Bank announced it would increase its human capital investments in Africa by 50% in the next funding cycle; including other financing, some $15 billion between 2021 and 2023 will be sent to the continent.

UN-Habitat's City Prosperity Index (CPI) is a tool to measure the sustainability of cities and it "considers that urbanization, as a process, should adhere to human rights principles, while the city, as an outcome, should meet specific human rights standards that need to be measured" (UN-Habitat. 2016, 3). CPI is defined by six critical dimensions and public space, along with health, education, safety, and security, is included in the quality of life indicator. The quality of public space depends not only on the overall quantity of public areas, but also from street networks, connectivity, walkability, and public transportation, as the ability to access a space is just as important as the space itself. The prosperity of a city depends on the human-oriented nature of its urban form and a good design of public spaces can have a positive impact on other CPI dimensions, such as infrastructure development, environmental sustainability, and high levels of equity and social inclusion. As Kim (2015, 10) notes, "intelligent design solutions require a deeper understanding of the design problem," and Tonkiss (2013, 177) argues, "cities are among the clearest of cases that design is never simply a technical process." While urbanization might have produced toxic urbanism, policy makers, community members, and developers produce outcomes together and therefore have the ability to re-evaluate and promote positive change.

The United Nations (2012) adopted the resolution "The Future We Want," which reaffirmed the importance of the Universal Declaration of Human Rights and "the role of civil society and the importance of enabling all members of civil society to be actively engaged in sustainable development" (8). In the same year, UN-Habitat launched "I'm a City Changer" to advocate for a global movement centered on individual, private, and public initiatives that empower the disadvantaged and work with communities to produce solutions that will improve their surroundings. Both the resolution and the campaign were based on a clear strategy – engage every urban citizen in sustainable development discourse, be open to grassroots solutions, and embrace the civil society as an agent of change.

As Brenner et al. (2012) argue, it is increasingly important to "underscore the urgent political priority of constructing cities that correspond to human social needs rather than to the capitalist imperative of profit-making and spatial enclosure" (2). If our guiding social and human values are not sustainable, then our cities become unsustainable too (Radović 2009). Now, more than ever, we need a proper understanding and consciousness based on human development, we need a different system of education based on public space culture for the activists, innovators, and leaders of the future, we need to invest in human capital, and most importantly, we need to become humanists and unfold a new human-oriented vision.

Design, policies, vision

The 9th World Urban Forum (WUF9) in 2018 was the first UN event dedicated to engaging global stakeholders on critical issues related to the implementation of the NUA and sought to define a common strategy for further steps. A certain criticism has been raised against the NUA for a limited implementation process, as well as thin financing and monitoring mechanisms. While it is a comprehensive document, implementing the NUA through adequate policies at the local level is difficult. WUF9 sought to address some of these concerns by seeking operational solutions to the following questions: "how can we effectively implement an urban agenda addressing urban prosperity, social welfare, and environmental protection, where public space and a humanistic approach could play a central role?" and "how can we translate key recommendations into tangible actions on the ground, so that they are able to improve the quality of life of urban residents and make public space inclusive and accessible for all?"

While a human-oriented vision is imperative for public space culture, we acknowledge that only through local government is the implementation of such policy and strategy possible, as they are directly responsible for regulating, designing, financing, building, managing, maintaining, preserving, and defending public space. United Cities and Local Governments has developed several initiatives to push local governments toward developing a public space policy framework, namely comprehensive strategies intended to localize key SDGs targets, rather than prescribe policy, governance, and environmental interventions (UCLG 2016).

Despite a long tradition of authoritarian politics, Brazil defined a new model grounded on a public space strategy: participatory budgeting. This program, promoted by Porto Alegre and adopted in 1989, was intended to be "a process of decision making based upon general rules and criteria of distributive justice, discussed and approved by regular, institutional organs of participation," aimed at redefining civic participation to the transformation and future of the city (de Sousa Santos 1998, 463). It was selected by the United Nations as one of the 40 urban innovations worldwide in 1996 and presented at the Habitat II conference. It has since been adopted by more than 2,700 governments worldwide. Following Porto Alegre's model, more governments are working to define an innovative and responsible approach that fosters participation in the decision-making process. Urban thinkers and doers have been working on redefining the urban paradigm by reclaiming public space for people; although there is not a universal recipe for this worthy endeavor, the commitment of civil servants in reimaging priorities for the future of cities and putting people at the core of the urban agenda is paving the way for a New Humanism. In fighting and protesting against the established system of power, they are implementing the New Humanist perspective that "is not only focused on the search for values – which it must also be – but oriented towards the implementation of concrete programmes that have tangible results" (Bokova 2010, 2).

Conclusion: a non-negotiable demand for a new humanism

The UN's commitment to inclusive development is oriented around people, and therefore, to public space; it is a call for country leaders to prioritize the needs of the most marginalized and disadvantaged, those facing dire poverty and discrimination. But this goal is not just for developing countries. Economic recession, rising unemployment, and homelessness are becoming realities in the West, while the lingering global financial crisis and unprecedented migration are simultaneously redefining the public realm as a politicized and contested space (Hou and Knierbein 2017).

In opposition to neoliberal Public–Private Partnerships (PPPs), a number of civic organizations are advocating for a Public–Private–People Partnerships (4P) approach (Marana, Labaka, and Sarriegi 2018; Perjo et al. 2016). This ensures that "people" are part of conversations and, in fact, must be included in the process to make cities and public spaces more sustainable. The architecture of the SDGs presumes that ordinary people, as well as institutions and governments, will work to achieve a better urban future. In order to accelerate the process of engagement, the Museum for the United Nations – UN Live, is working on a daring and visionary strategy that will "connect people everywhere to the work and values of the United Nations and catalyze global efforts towards achieving its goals" and "connect the hearts, the minds and the hands in one effort, building a bridge between awareness and action" (Edson 2018).

The active contribution of people, either through co-design or co-production (or both) is necessary for inclusive and equitable spaces, as they foster tolerance, conviviality, dialogue, and democratic exchange. Public spaces are where all citizens can claim their right to the city, which, as Harvey (2008) points out, "is far more than the individual liberty to access urban resources: it is a right to change ourselves by changing the city" (23).

As stated by the Barcelona declaration on public space, the right to the city is "a new paradigm that provides an alternative framework to re-think cities and urbanization" (Habitat III. 2016, 1). Public space, intended as the right to the city, is not an individual right, since it is shaped by a collective dimension that can influence the processes of urbanization. As such, it is no longer "me" or "them," it is "we;" the world, with all the overwhelming burdens, tensions, and paradoxes derived from urbanization, that the Global North can view from afar, is closer than we think. Even in our advanced Western societies, as Harvey (2008) argues, "the freedom to make and remake our cities and ourselves is one of the most precious yet most neglected of our human rights" (23). It is time to work together to strengthen individual and local efforts towards a greater outcome for the cause of public space, for us, the people, and for all those who are struggling daily for a better life and a better future (UN 2016b).

References

Anders, F. and A. Sedlmaier. (eds). (2017) *Public Goods versus Economic Interests: Global Perspectives on the History of Squatting.* New York: Routledge.

Andersson, C. (2016) Public Space and the New Urban Agenda. *The Journal of Public Space*, 1(1): 5–10.

Blended Finance Taskforce (BFT). (2018) *Better Finance, Better World. Consultation Paper of the Blended Finance Taskforce.* London: Business and Sustainable Development Commission.

Bokova, I. (2010) *A New Humanism for the 21st Century.* Paris: UNESCO.

Bravo, L. (2018) We the Public Space. Strategies to Deal with Inequalities in Order to Achieve Inclusive and Sustainable Urban Environments. *The Journal of Public Space*, 3(1): 163–164.

Brenner, N., P. Marcuse, and M. Mayer. (eds). (2012) *Cities for People, Not for Profit. Critical Urban Theory and the Right to the City.* London: Routledge.

Davis, M. (2006) *Planet of Slums.* London: Verso.

de Sousa Santos, B. (1998) Participatory Budgeting in Porto Alegre: Toward a Redistributive Democracy. *Politics & Society*, 26(4): 461–510.

Edson, M.P. (2018) Recorded Presentation. "Head, Heart, and Hands: Building a New Global Museum from the Bottom Up." UN Live. Accessed 15 June 2019. www.youtube.com/watch?v=3X95pFXQeyc

Habitat III. (2016) *Barcelona Declaration for Habitat III.* New York: United Nations.

Harvey, D. (2008) The Right to the City. *New Left Review*, 53(Sept–Oct): 23–40.

Hou, J. and S. Knierbein. (eds). (2017) *City Unsilenced. Urban Resistance and Public Space in the Age of Shrinking Democracy.* New York: Routledge.

Ibeling, H. (2015) *Middle Ground. In: Europe City. Lessons from the European Prize for Urban Public Space.* Zurich: Lars Müller Publishers.

Kim, A. (2015) *Sidewalk City: Remapping Public Space in Ho Chi Minh City.* Chicago: University of Chicago Press.

Marana, P., L. Labaka, and J.M. Sarriegi. (2018) A Framework for Public–Private–People Partnerships in the City Resilience-Building Process. *Safety Science*, 110: 39–50.

Organisation for Economic Co-operation and Development (OECD). (2001) *The Well-Being of Nations: The Role of Human and Social Capital*. Washington D.C.: World Bank Group.

Perjo, L., C. Fredricsson, and S.O. E Costa (2016) "Public–Private–People Partnerships in Urban Planning" [Working paper]. www.balticurbanlab.eu/materials/working-paperPublic-private-people-partnerships-urban-planning

Pineda, V., S. Meyer, and J. Cruz. (2017) The Inclusion Imperative. Forging an Inclusive New Urban Agenda. *The Journal of Public Space*, 2(4): 1–20.

Radović, D. (2009) *Eco-urbanity: Towards Well-Mannered Built Environments*. New York: Routledge.

Sassen, S. (2015) "Who Owns Our Cities – and Why This Urban Takeover Should Concern Us All." The Guardian, 24 November. Accessed 15 June 2019. www.theguardian.com/cities/2015/nov/24/who-owns-our-cities-and-why-thisurban-takeover-should-concern-us-all

Tonkiss, F. (2013) *Cities by Design: The Social Life of Urban Form*. Cambridge: Polity Press.

UN-Habitat. (2013) *Streets as Public Spaces and Drivers of Urban Prosperity*. New York: United Nations.

UN-Habitat. (2015a) *Global Public Space Toolkit from Global Principles to Local Policies and Practice*. New York: United Nations.

UN-Habitat. (2015b) *Using Minecraft for Youth Participation in Urban Design and Governance*. New York: United Nations.

UN-Habitat. (2016) *Measurement of City Prosperity. Methodology and Metadata*. New York: United Nations.

UN-Habitat. (2019a) *The Silent Revolution of Public Spaces in Afghanistan*. New York: United Nations.

UN-Habitat. (2019b) *Mixed Reality for Public Participation in Urban and Public Space Design: Towards a New Way of Crowdsourcing More Inclusive Smart Cities*. New York: United Nations.

United Cities and Local Governments (UCLG). (2016) *Public Space Policy Framework*. Barcelona: UCLG.

United Nation Population Fund (UNFPA). (2014) *The Power of 1.8 Billion Adolescents, Youth, and the Transformation of the Future: State of World Population*. New York: United Nation Population Fund.

United Nations (2015), General Assembly Resolution 70/01, Transforming Our World: The 2030 Agenda for Sustainable Development, A/RES/70/01. New York, United Nations. Available at: www.un.org/ga/search/view_doc.asp?symbol=A/RES/70/1&Lang=E

United Nations (UN). (2012) *The Future We Want*, Resolution 66/288. New York: United Nations.

United Nations (UN). (2016a) *World Cities Report 2016: Urbanization and Development. Emerging Futures*. New York: United Nations.

United Nations (UN). (2016b) *Leaving No One Behind: The Imperative of Inclusive Development, Report on the World Social Situation 2016, Executive Summary*. New York: United Nations.

United Nations (UN). (2016c) *General Assembly Resolution 71/256*, New Urban Agenda, A/RES/71/256*. New York: United Nations.

United Nations (UN). (2017) *World Population Ageing 2017*. New York: United Nations.

United Nations (UN). (2018a) *Disability and Development Report. Realizing the Sustainable Development Goals by, for and with Persons with Disabilities*. New York: United Nations.

United Nations (UN). (2018b) *World Urbanization Prospects 2018*. New York: United Nations.

United Nations (UN). (2019) *World Population Prospects 2019 Highlights*. New York: United Nations.

United Nations Development Programme (UNDP). (2017) *Impact Investment to Close the SDG Funding Gap*. New York: United Nations.

United Nations Economic Commission for Europe (UNECE). (2016) *Report on Informal Settlements in Countries with Economies in Transition in the UNECE Region*. New York: United Nations.

World Bank. (2018) *The Human Capital Project Brief*. Washington D.C.: World Bank Group.

World Health Organization (WHO). (2007) *Global Age-Friendly Cities: A Guide*. Geneva: World Health Organization.

World Health Organization (WHO). (2012) *Burden of Disease from the Joint Effects of Household and Ambient Air Pollution for 2012*. Geneva: World Health Organization.

9

SAFETY IN PUBLIC SPACES FROM A PLACE-ORIENTED PUBLIC HEALTH PERSPECTIVE

Vania Ceccato

Introduction

This chapter contributes to the existing body of literature by arguing in favor of a place-oriented perspective on public health when dealing with issues of safety in public spaces. By adopting this perspective, crime and other preventable harm in public spaces, such as fall incidents, are assumed to be issues that can be treated as endemic or epidemic infectious diseases. This aim is achieved by first scanning five decades of the international literature. Using VOSviewer, a software tool for constructing and visualizing bibliometric networks, clusters of research are generated that link safety in public spaces and health outcomes (Van Eck and Waltman 2011). Then, three current research strands are discussed in detail linking public place, safety, and health issues.

The mechanisms linking the quality of the public space with safety and health issues are not yet completely understood (Moore et al. 2018). However, the apparently simple choice that one makes when selecting a residential area to reside, has a major long-term impact on ones' way of living (Lorenc et al. 2013; Moore et al. 2018). This is because the type of residential area (the physical and social environments, the infrastructure, the layout of public spaces) affects ones safety and daily levels of physical activity, which have potential for adverse health consequences (Sandy et al. 2013). The type of residential area also affects how much time one spends in cars, whose use contributes to car crashes, pedestrian injuries, and not least, pollution (Ewing et al. 2014). Similarly, recent research shows links between the maintenance of public spaces, overall crime levels and health outcomes (Branas et al. 2018). Well-kept vacant land are inviting public places that make residents willing to keep an "eye on the streets" (Jacobs 1961), and engage in physical activities with health benefits.

Despite the growing interest on issues of public space, crime, and safety linked to public health, hitherto, there has not been any assessment on mapping the peer-reviewed literature that links these fields. Therefore, in the first part of this chapter, bibliometric indicators of published literature pertaining to public space, crime, and safety linked to public health are presented. Then, three of the most recent research strands are discussed in detail. The chapter closes with a discussion of potentialities and challenges when adopting a place-oriented public health perspective to safety in public spaces based on reflections of the current state of the field.

The current study

This chapter reports the results of a bibliometric search of the international literature over a period of 50 years that includes the potential links between the public space/place and fear, public space/place and crime, or public health and public space/place. The exclusion criteria limited the search to quantitative English literature published between 1968 and 2018. In VOSviewer a further selection was made excluding articles that were entirely concentrated in topics outside the scope of this article, such as addiction, public governance, and technological solutions. Retrieved data were exported from Scopus/PubMed to Excel for analysis and tabulation. The bibliometric search was performed using VOSviewer version 1.6.10 (Van Eck and Waltman 2011). These networks visualize the size and links between individual publications in SCOPUS and PUBMED. Two strategies were employed: (1) scanning the area with the keywords and (2) searching more specific and detailed studies (see Table 9.1).

Drawing on this literature review, three research strands of research will be discussed in more detail. The selection of these three areas was based on the nature of the articles and their eligibility over time. *Nature of articles*: Studies that combined issues of public space/place, crime, safety linked to public health with a clear spatial, multi-disciplinary place-based perspective were selected. *Eligibility over time*: The amount and increase of articles published from 1968–2018 in a specific topic indicated the stability of a theme (for instance, crime concentration as a topic increased since 1980s). Using visual thematic maps from VOSviewer, these articles were later aggregated into three groups (as discussed below) and others were excluded. The selection of articles is not meant to be extensive, but rather illustrative of the field.

Table 9.1 Results of the searches in PubMed and Scopus

Query Terms	PubMed results	Scopus results
1 ("public place" OR "public space") AND fear	12	386
2 ("public place" OR "public space") AND crime	10	260
3 (("public space") OR "public place") AND "public health"	90	823
4 (("public space") OR "public place") AND "public health" AND fear	2	10
5 (("public space") OR "public place") AND "public health" AND crime	11	14
6 (("public space") OR "public place") AND "public health" AND safety	13	60
7 (("public space") OR "public place") AND "public health" AND built environment	17	268

Results

Safety in public place as a public health problem: studies from 1968 to 2018

With the intention of illustrating both the nature of studies and their evolution through time in this field of research, a search of publications was performed using the Scopus database for the past 50 years. Figure 9.1 shows the output of five sets of keywords used as the basis for the bibliometric search (Table 9.1). The research found 1984 articles between 1968 and 2018 with duplicates that were later excluded resulting in a total of 1718 articles (1592 from SCOPUS and 126 from PUBMED). This search is illustrated through a combined set of figures showing mapping and clustering of the most frequently cited publications (see Figures 9.1a and 9.1b) that appeared in the field in the period 1968–2018.

The temporal evolution of the research field

Most of the retrieved documents were published in the last 20 years (Figure 9.1). As illustrated in Figure 9.1c, publications on the issues of *public space/place*, *crime* and *public health*, and *crime/fear of crime* and *public space/place* started in the late 1980s. Note that it is not until the late 2000s that topics linking *crime* and *fear* and *public space/place* started to be published, with about 25–30 publications per year. It took 20 more years for publications dealing with *crime* and *public space/place* to reach about 140 per year each, with a steadily increase since 2000s, with few episodes of increase in a variety of themes reflecting the importance of sustainability and governance issues (see Figure 9.1c). This was followed by publications dealing with *public place/crime and public health*, which constituted around 60 publications per year that were concerned with the interlinkages between physical environment, lifestyle factors (smoking, drug consumption) and health. The contribution of authors and institutions is predominantly dominated by North America and Western European countries.

The thematic evolution of the field

Term maps are used to illustrate the thematic evaluation of the research field (Figure 9.1a with center in "public place" while Figure 9.1b with center in "crime"). Colors indicate clusters of related terms; it helps in knowing the facets of a particular theme over a period of time. Note that five main themes are clustered in the 50 years. Note that overall, keywords related to *public space/place* were not dominant when the search for *public health* and *crime* was performed from 1968 to 2018. This topic appears only more recently among the retrieved publications.

Public place (linked to other keywords such as urban design, urban area) and public health (epidemic, healthcare) dominates the Figures 9.1a and 9.1b. The nature of public places varies from parks to streets to squares to entire parts of neighborhoods. The largest cluster (in green, Figure 9.1a) of publications links issues of public health and crime with institutional and policy terms such as *government*, *governance*, *law enforcement*, and *policy* but also tobacco, drugs, smoking. A strand of research from the 1970s and 1980s that linked the experience of victimization to health effects was, for instance, highlighted by Coakley and Woodford-Williams (1979). Figure 9.1d shows the density visualization map of keywords in literature from Scopus only with at least five repetitions of key words.

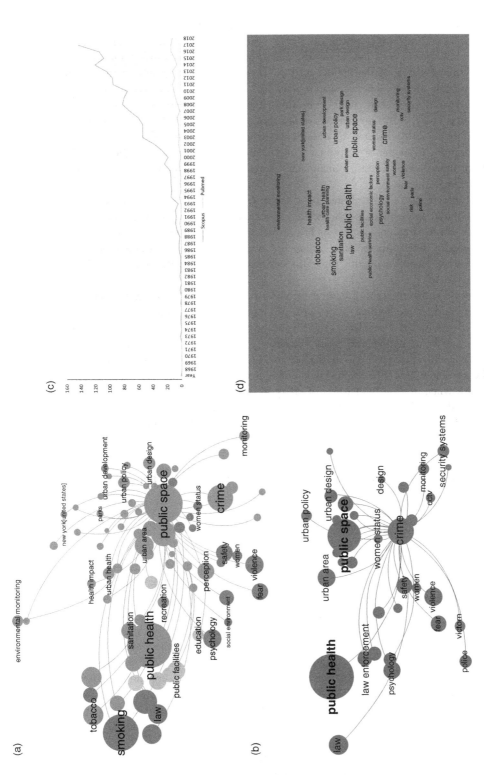

Figure 9.1 Term maps of the research field center around, (a) "public place" while, (b) with center in "crime," (c) publications linking crime and fear to public health and built environment, 1968–2018,(d) density visualization map of keywords in literature from Scopus only with at least five repetitions of key words based on the search 1968–2018

Image credit: Vania Ceccato

Publications that *simultaneously* deal with public health, safety, and public space (or at least, public health and safety related to public space/place) are still rare. The search in specific journals (for related public health, safety, and public space/place) indicates how scarce the subject was until the 2000s and how it became a popular subject matter more recently with the advent of new research areas like sustainability. An example of a strand of research that is somewhat stable over time links health to places via accidents and injuries. They often show conflicting results about the effect of the environment (e.g. Rees-Punia et al. 2018) because of differences in methodologies and the wide range of conceptualizations of elements of public spaces. Public spaces can vary highly (e.g. from vacant land, urban parks, forests, unmaintained urban places) that either have an effect on safety (negative or positive).

A better integration of these clusters of research is illustrated when the three sets of main keywords – public space/place, public health, crime and fear – were used (see Figure 9.1b). Public space is associated with words such as urban area, urban design, safety, women (here linked to gender issues and the intersectionality of safety). Recent research reviews with a public health perspective on this topic are summarized by Rees-Punia et al. (2018) and Moore et al. (2018).

This vast literature indicates a central focus on public spaces. These studies often offer a situational perspective with a micro spatio-temporal focus to the problem. Some of these public spaces are "enablers," which means they create opportunities for crime, injuries, or self-harm, while others are "stressors" or "mediators" that link to physical activity and health outcomes (e.g. Branas et al. 2018; Ceccato and Uittenbogaard 2016). Thus, the current state-of-the art research points towards a place-oriented perspective on public health when dealing with issues of safety in public spaces. From the set of publications described over 50 years, three examples of research areas (relating the public spaces, crime/fear of crime/safety/accidents and public health) are in the section below discussed more in detail.

Safety and public spaces: examples of a place-oriented perspective on public health

Crime concentrations in public spaces

The fact that the incidence of crime is associated with specific types of neighborhood environments has long been known in criminological literature (Sampson and Raudenbush 2004; Shaw and Mckay 1942). But only in recent decades have researchers recognized the need to link crime and disorder to micro-scale processes (Sherman et al. 1989; Weisburd 2015).

The biggest turning point was when researchers found out that crime, like many other types of observable phenomenon, tends to be highly concentrated in small geographic places, following a Pareto principle (Pareto 1964). In the late 1980s, these crime concentrations ("crime hot spots") were highlighted in particular in seminal studies by Pierce et al. (1988). Pierce et al. (1988) found that in Boston, Massachusetts, for instance, 3.6% of addresses produced 50.0% of emergency calls to police. In the 1990s, this evidence was corroborated by studies focusing on problem-oriented policing, in particular violent areas (Braga et al. 1999). Here these locations are public places, street corners, squares, interstitial spaces between buildings.

However, Weisburd and colleagues have been the most important advocates for the need to consider crime at a very detailed level in urban environments, in public places, such as street segments (Weisburd 2015, 2018). Weisburd (2015) referred to the "law of crime con-centration at places" to explain why place-based approaches are, as the author suggests, an efficient way to allocate crime prevention resources. The author produced the first cross-city comparison of crime concentration by street segments and also indicated the importance of police community collaboration to reinforce informal social controls at the micro-place level to improve public safety.

Some of these public places that concentrate crime are places where people converge, such as transportation nodes like bus stops or train stations (Ceccato et al. 2015; Loukaitou-Sideris 2012). Such places become hotspots of crimes at particular times because they may generate crime opportunities (crime generators, crime attractors, see Brantingham and Brantingham 1995) but they may also radiate crime to surrounding areas. This crime concentration has a direct effect on the quality of life of communities, affecting housing markets (by for instance, pulling property prices down, see Ceccato and Wilhelmsson 2012) and making public spaces less welcoming and walkable. In summary, if safety problems at particular public spaces are identified and addressed *before* crime occurs, harm can be reduced at that particular spot and there are overall benefits for society. As crime decreases and the quality of life improves, public resources can be utilized more efficiently and diverted to over communities.

The impact of public spaces on safety and health

An increasing body of research is showing positive associations between the quality of the urban environment, crime and safety, and overall health (e.g. Branas et al. 2016; 2018; Kondo et al. 2018a; 2018b; Mccormick 2017; Twohig-Bennett and Jones 2018). A particular strand of this literature has been investigating the importance of green areas on health. Twohig-Bennett and Jones (2018) found evidence of the beneficial influence of greenspace on a wide range of health outcomes, while Mccormick (2017) found similar results for children. Kondo et al. (2018b) found consistent negative association between urban green space exposure and mortality, heart rate, and violence, and positive association with attention, mood, and physical activity.

Another recent strand of research includes other factors of the urban environment and health. Rugel et al. (2019) found that public neighborhood nature was linked to a higher sense of community belonging. In turn, a high sense of community belonging was linked to reduced depression, negative mental health, and distress. Interestingly, Bornioli et al. (2018) found that walking in an urban environment promotes affective benefits to health despite the absence of nature. Research at this time also reports positive findings linking commu-nity-based initiatives and violence reduction. For instance, Shepherd (2007) argued that alcohol-related violence can be prevented through public health and community initiatives, in conjunction with other community agencies, including the police and local authorities that work with the quality of outdoor public spaces.

A third strand of research links urban maintenance and urban planning to less crime, in particular violent crime (Cowen et al. 2018; Freedman et al. 2018; Kondo et al. 2018a; Stacy 2018). Stacy (2018) suggested that demolitions decreased crime by about 8% (particu-larly homicides) on the block level, with the largest impact concentrated one to two months after the demolition occurs, while Freedman et al. (2018) found a decrease in robbery and assaults at the county level with new construction and rehabilitation centers. However, Cowen et al. (2018) found that neighborhoods with higher levels of walkability have greater levels of aggravated assault; they also found that increasing land-use diversity increases both aggravated assault and larceny. Also, Sakip and Salleh (2018) were able to identify the street-pattern characteristics that most influence snatch theft activities.

Finally, Branas et al. (2016) assessed how much abandoned buildings and vacant lots create physical opportunities for violence, by sheltering illegal activity and illegal firearms, and putting a price on such harmful effects. In another study, these authors show evidence that urban blight remediation programs can be cost-beneficial and reduce firearm violence. More recently, Branas et al. (2018) focused on particular blighted and vacant urban land as a source of threats to residents' health and safety, particularly in poor neighborhoods. Authors were able to create a process of standardized restoration of vacant urban land on a citywide scale and found that it significantly reduced gun violence, crime, and fear. They analyzed more than 500 sampled vacant lots that were randomly assigned into treatment and control study arms over 38 months. They found that those living near treated vacant lots reported significantly reduced perceptions of crime, vandalism, and safety concerns when going outside their homes, as well as significantly increased use of outside spaces for relaxing and socializing. Significant reductions in crime overall, gun violence, burglary and nuisances were also found after the treatment of vacant lots in neighborhoods below the poverty line.

In summary, improving the quality of outdoor public spaces may attract people, which is a positive quality but lots of people also means more crime opportunities (see e.g. Cowen et al. 2018). Similarly, desolated, ill-maintained public spaces are unattractive places; places that create the necessary conditions (anonymity) for crimes such as robbery and rape. More interestingly are recent findings from experimental studies. They show that fixing facades and improving the maintenance of once abandoned lots, reduce crime, increase residents' safety perceptions, boost the area's walkability, and positively affect residents' health. In conclusion, studies have so far studied how the quality of public space relates to safety and health. The relationship between these factors is sometimes unexpected; not always straightforward and should not be taken for granted.

Mobility and pedestrian falls in public places

Safe cities are walkable cities. People may avoid public spaces where they know they are at risk of falling or suffering other injuries. Although many more pedestrians are injured in falls than in pedestrian-vehicle collisions, most studies on accidents neglect pedestrian falls (which can be self-inflicted or in collision with a vehicle, bicycle, or another person), which leads to biased conclusions for transport and safety policies (Methorst et al. 2017). This is especially important for vulnerable groups in public transportation systems, in particular older adults, since they are overrepresented as victims among cases of pedestrian falls. Older adults avoid outdoor environments because of safety concerns, falling in wet conditions, or tripping or being hit by a vehicle at a crossing (Ceccato and Bamzar 2016). International studies have shown that older people may be at higher risk of under-recognized causes of traffic injuries, and to be able to better prevent them, it is necessary to know more about the nature of these traffic injuries, where and when they happen (Björnstig et al. 2005; O'Neill 2016).

Studies have more recently started to emphasize the detection of foci of accidents among older adults in public spaces, investigating the types of microenvironments (counts and by street segments) and visualizing these concentrations in space (Dai and Jaworski 2016; Dimaggio and Guohua 2011). A common finding of these studies is the concentrated character of these events. This is also corroborated by findings in Sweden, where 80% of falls among older pedestrians happen in 5% of the total mapped street segments, regardless of

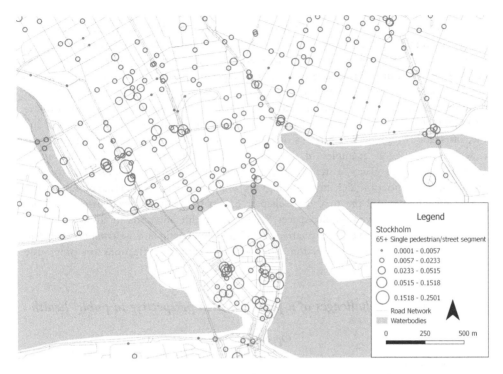

Figure 9.2 Hotspots of falls among older adult pedestrians in Stockholm mapped by street segment. Size of circles indicates the street segments with greatest concentrations of falls

Image credit: Vania Ceccato and Oscar Willems

type of municipality (Ceccato and Willems 2019). Figure 9.2 illustrates the hotspots for falls among older adult pedestrians in Stockholm mapped by street segment.

Falls take place in more densely populated areas and are associated with specific characteristics of the public places, e.g. mixed roads and bus stops but less frequently inclined streets. Among those events that are lethal, 63% happen in street or road segments and 20% in street or road intersections (Ceccato and Willems 2019). The most common places for elderly falls in relation to traffic accidents are a street or road segment followed by a pedestrian or bicycle path, sidewalk and then street or road intersection, often affected by environment conditions such as icy and slippery streets or uneven surfaces (Ceccato 2018).

Any new policy should also consider temporal patterns of these falls. In Sweden, for example, such incidents involving older adults often happen during daylight hours, on weekdays, and in the coldest months of the year, but there are regional variations. Awareness of these findings is essential for public health policies. Given the concentrated pattern of falls, a systematic maintenance of urban service across municipalities should be tailored to the local geography of falls to ensure elderly safety. As Ceccato and Willems (2019) shows, the great majority of falls are concentrated in a handful of locations in public spaces. If 5% of mapped street segments that concentrate 80% of falls among older pedestrians can be identified, then a reduction of falls and injuries can be ensured by interventions at these spots that deal with the problem. For instance, implementation of systematic maintenance services at these hotspots for falls in public spaces (e.g. snow plowing in pedestrian paths, fixing street holes and uneven surfaces, improving illumination), at particular times of the day and days of the week, also taking into account the incidence of falls by season. According to authors, a remaining challenge is that these concentrations of falls in public spaces are not always known by local authorities. Underreporting of cases is a problem; and when falls are reported, interventions to deal with the problem in street corners, parks, squares, may fail because of the poor cooperation between local actors to keep up with the maintenance in these public spaces.

Summarizing, the risks associated with the types of environment one is exposed to are not randomly distributed. These risks tend to follow particular patterns over space and time, often following people's patterns of routine activity. The geography of falls and traffic accidents are good examples of this, because they tend to show concentrations in a small percentage of places, a street corner, a park, a bus stop. Crime too tends to show a pattern of concentration and these findings are important because they have implications for research, policy, and practice. This chapter has so far attempted to illustrate linkages between issues of quality of public spaces, safety and health outcomes in studies adopting a place-oriented perspective. A normative set of guidelines split into *opportunities* and *challenges* is proposed in the next section based on this author's reflections on previously reported studies in this chapter.

Opportunities and challenges of a place-oriented perspective to public health

Opportunities

Some (public) places are riskier than others. Incidents of falls and other preventable acts, such as crime, are linked to the particular environmental conditions of a particular place and time that can be predicted and addressed, as they do not happen at random in urban space. In relation to crime in particular, Weisburd (2018) reminds that although city environments have been important to explain the geography of crime, place, as a micro unit of analysis (such as street corners or street segments), was something new and not well explored until recently. Yet, this realization has important implications in not only understanding why crime happens, but also how it can be prevented. This reasoning can also be applied to other events that happen in public spaces, such as traffic accidents and injuries.

Advantages of adopting a holistic approaches to risk. Research in the 1990s found strong associations between traffic accidents and crime (by crime type and different types of transportation means), suggesting that these events may share common causal processes (e.g. Junger et al. 1995). Tackling places with crime may also help avoid other incidents such as car crashes or pedestrian injuries because they share similar contexts (e.g. relate to particular nightlife environments, such as bars and restaurants serving alcohol) and specific patterns of routine activities.

The effectiveness of place-oriented interventions. Research has shown that structural interventions, which promote healthy behaviors by changing places may affect a broader population and may be more sustainable as compared to interventions that focus entirely on individuals. Evidence shows how the maintenance of urban public spaces can positively impact heart rate, crime, and general quality of life for residents.

Challenges

"Similar geography" does not mean "same cause." Research has so far struggled with methods, sample sizes, potential confounders, and the degree of generalizability of results. Longitudinal-oriented (randomized and controlled) intervention studies are needed to tease out mechanisms of a particular phenomenon or assess what specific changes to specific public places are necessary to alter injury and violence related outcomes (Kondo et al. 2018b).

Governance and the sectorial split. Public authorities dealing with urban risks, such as municipal health authorities and police, face a number of barriers to reach effective cooperation and ensure safe public spaces. Previous studies report that police officers, doctors, and urban planners rarely work together to improve areas with a concentration of problems.

The importance of context. Most of the international literature discussed in this chapter is dominated by case studies of large cities, often in North America or Europe; few are from smaller urban centers or countries from the Global South. The degree of generalizability of current findings has implications for research, but also for policy. A discourse around a place-oriented perspective on public health must be informed by place and context specific evidence.

Conclusions

In this chapter, a case for a place-oriented perspective on public health, when dealing with issues of safety in public places, is proposed. Relevant literature is inspected through the visualization of bibliometric networks spanning five decades of international literature associated with public space/place, crime/safety, and public health. Findings indicate that this area of research has, until recently, been scattered, reflecting different fields of research. Some work is beginning to overlap, however, one being the focus on place-based analysis. Based on these overlaps, an in-depth discussion of three recent research strands were presented. Regardless of whether the focus is on fall injuries or crime, evidence shows the importance of knowing more about the types of environments in which these harms occur, their potential determinants and longitudinal mechanisms, as well as their impact on neighborhoods and residents' health and overall quality of life. Despite the current evidence, this research area is in its infancy.

Finally, urban planners and public health professionals need clear-cut evidence on the impact of specific place-based interventions to help them address public health issues in practice, whether crime or fall injuries. A final argument is that a place-oriented perspective on public health has the potential to identify new solutions to safety problems in public spaces, but this knowledge must be made available to those who can make changes and, ultimately, save lives. Initiatives that link academia and practice are not only desirable but necessary, since research findings are still limited to a small group of professionals in universities. Translational knowledge in this area is therefore fundamental to reaching out to those who work with these issues in practice.

References

Björnstig, U., P.O. Bylund, P. Albertsson, T. Falkmer, J. Björnstig, and J. Petzäll. (2005) Injury Events among Bus and Coach Occupants: Non-Crash Injuries as Important as Crash Injuries. *IATSS Research*, 29(1): 79–87.

Bornioli, A., G. Parkhurst, and P.L. Morgan. (2018) The Psychological Wellbeing Benefits of Place Engagement during Walking in Urban Environments: A Qualitative Photo-Elicitation Study. *Health & Place*, 53: 228–236.

Braga, A.A., D.L. Weisburd, E.J. Waring, L.G. Mazerolle, W. Spelman, and F. Gajewski. (1999) Problem-Oriented Policing in Violent Places: A Randomized Controlled Experiment. *Criminology*, 37(3): 541–580.

Branas, C.C., M.C. Kondo, S.M. Murphy, E.C. South, D. Polsky, and J.M. MacDonald. (2016) Urban Blight Remediation as a Cost-Beneficial Solution to Firearm Violence. *American Journal of Public Health*, 106(12): 2158–2164.

Branas, C.C., E. South, M.C. Kondo, B.C. Hohl, P. Bourgois, D.J. Wiebe, and J.M. MacDonald. (2018) Citywide Cluster Randomized Trial to Restore Blighted Vacant Land and Its Effects on Violence, Crime, and Fear. *Proceedings of the National Academy of Sciences*, 115(12): 2946–2951.

Brantingham, P., and P. Brantingham. (1995) Criminality of Place - Crime Generators and Crime Attractors. *European Journal on Criminal Policy and Research*, 3(3): 5–26.

Ceccato, V. (2018) Patterns of Traffic Accidents among Elderly Pedestrians in Sweden. *Review of European Studies*, 10(3): 117–133.

Ceccato, V., and R. Bamzar. (2016) Elderly Victimization and Fear of Crime in Public Spaces. *International Criminal Justice Review*, 26(2): 115–133.

Ceccato, V., O. Cats, and Q. Wang. (2015) The Geography of Pickpocketing at Bus Stops: An Analysis of Grid Cells. In: *Safety and Security in Transit Environments: An Interdisciplinary Approach*, V. Ceccato and A. Newton. (eds). London: Palgrave Macmillan, pp. 76–98.

Ceccato, V., and A. Uittenbogaard. (2016) Suicides in Commuting Railway Systems: The Case of Stockholm County, Sweden. *Journal of Affective Disorders*, 198: 206–221.

Ceccato, V., and M. Wilhelmsson. (2012) Acts of Vandalism and Fear in Neighbourhoods: Do They Affect Housing Prices? In: *The Urban Fabric of Crime and Fear*, V Ceccato. (ed). Dordrecht: Springer Netherlands, pp. 201–213.

Ceccato, V., and O. Willems. (2019) Temporal and Spatial Dynamics of Falls among Older Pedestrians in Sweden. *Applied Geography*, 103: 122–133.

Coakley, D., and E. Woodford-Williams. (1979) Effects of Burglary and Vandalism on the Health of Old People. *Lancet*, 2(8151): 1066–1067.

Cowen, C., E.R. Louderback, and S.S. Roy (2018) The Role of Land Use and Walkability in Predicting Crime Patterns: A Spatiotemporal Analysis of Miami-Dade County Neighborhoods, 2007–2015. *Security Journal [published online]*.

Dai, D., and D. Jaworski. (2016) Influence of Built Environment on Pedestrian Crashes: A Network-Based GIS Analysis. *Applied Geography*, 73: 53–61.

Dimaggio, C., and L. Guohua. (2011) Roadway Characteristics and Pediatric Pedestrian Injury. *Epidemiologic Reviews*, 34(1): 46–52.

Ewing, R., S. Hamidi, and J.B. Grace. (2014) Urban Sprawl as a Risk Factor in Motor Vehicle Crashes. *Urban Studies*, 53(2): 247–266.

Freedman, M., E. Owens, and S. Bohn. (2018) Immigration, Employment Opportunities, and Criminal Behavior. *American Economic Journal: Economic Policy*, 10(2): 117–151.

Jacobs, J. (1961) *The Death and Life of Great American Cities*. New York: Vintage Books.

Junger, M., G-J. Terlouw, and P.G.M. Van Der Heijden. (1995) Crime, Accidents and Social Control Criminal Behaviour and Mental Health. *Victims & Offenders*, 5: 386–410.

Kondo, M.C., E. Andreyeva, E.C. South, J.M. Macdonald, and C.C. Branas. (2018a) Neighborhood Interventions to Reduce Violence. *Annual Review of Public Health*, 39(1): 253–271.

Kondo, M.C., J.M. Fluehr, T. Mckeon, and C.C. Branas. (2018b) Urban Green Space and Its Impact on Human Health. *International Journal of Environmental Research and Public Health*, 15(3): 445.

Lorenc, T., M. Petticrew, M. Whitehead, D. Neary, S. Clayton, K. Wright, H. Thomson, et al. (2013) Environmental Interventions to Reduce Fear of Crime: Systematic Review of Effectiveness. *Systematic Reviews*, 2(1): 1–10.

Loukaitou-Sideris, A. (2012) Safe on the Move: The Importance of the Built Environment. In: *The Urban Fabric of Crime and Fear*, V. Ceccato. (ed). Dordrecht: Springer Netherlands, pp. 85–110.

Mccormick, R. (2017) Does Access to Green Space Impact the Mental Well-Being of Children: A Systematic Review. *Journal of Pediatric Nursing: Nursing Care of Children and Families*, 37: 3–7.

Methorst, R., P. Schepers, N. Christie, M. Dijst, R. Risser, D. Sauter, and B. Van Wee. (2017) 'Pedestrian Falls' as Necessary Addition to the Current Definition of Traffic Crashes for Improved Public Health Policies. *Journal of Transport & Health*, 6: 10–12.

Moore, T.H.M., J.M. Kesten, J.A. Lopez-Lopez, S. Ijaz, A. Mcaleenan, A. Richards, and S. Audrey. (2018) The Effects of Changes to the Built Environment on the Mental Health and Well-Being of Adults: Systematic Review. *Health and Place*, 53: 237–257.

O'Neill, D. (2016) Towards an Understanding of the Full Spectrum of Travel-Related Injuries among Older People. *Journal of Transport & Health*, 3(1): 21–25.

Pareto, V. (1964) *Cours d'Économie Politique*. Geneva: Librairie Droz.

Pierce, G., S. Spaar, and B. Lebaron. (1988) *The Character of Police Work: Strategic and Tactical Implications*. Boston: Northeastern University, Center for applied Social Research.

Rees-Punia, E., E.D. Hathaway, and J.L. Gay. (2018) Crime, Perceived Safety, and Physical Activity: A Meta-Analysis. *Preventive Medicine*, 111: 307–313.

Rugel, E.J., R.M. Carpiano, S.B. Henderson, and M. Brauer. (2019) Exposure to Natural Space, Sense of Community Belonging, and Adverse Mental Health Outcomes across an Urban Region. *Environmental Research*, 171: 365–377.

Sakip, S.R., and M.N.M. Salleh (2018) "Linear Street Pattern in Urban Cities in Malaysia: Influence Snatch Theft Crime Activities." Paper Presented at the 8h Asia-Pacific International Conference on Environment-Behaviour Studies, The University of Sheffield, July 14 –15.

Sampson, R.J., and S.W. Raudenbush. (2004) Seeing Disorder: Neighborhood Stigma and the Social Construction of "Broken Windows". *Social Psychology Quarterly*, 67(4): 319–342.

Sandy, R., R. Tchernis, J. Wilson, G. Liu, and X. Zhou. (2013) Effects of the Built Environment on Childhood Obesity: The Case of Urban Recreational Trails and Crime. *Economics and Human Biology*, 11(1): 18–29.

Shaw, C.R., and H.D. Mckay. (1942) *Juvenile Delinquency and Urban Areas*. Chicago: University of Chicago Press.

Shepherd, J. (2007) Preventing Alcohol-Related Violence: A Public Health Approach. *Criminal Behaviour and Mental Health*, 17(4): 250–264.

Sherman, L.W., P.R. Gartin, and M.E. Buerger (1989) Hot Spots of Predatory Crime: Routine Activities and the Criminology of Place. *Criminology*, 27.

Stacy, C.P. (2018) The Effect of Vacant Building Demolitions on Crime under Depopulation. *Journal of Regional Science*, 58(1): 100–115.

Twohig-Bennett, C., and A. Jones. (2018) The Health Benefits of the Great Outdoors: A Systematic Review and Meta-Analysis of Greenspace Exposure and Health Outcomes. *Environmental Research*, 166: 628–637.

Van Eck, N.J., and L. Waltman (2011) Text Mining and Visualization Using VOSviewer. *Computing Research Repository (coRR) [published online]*.

Weisburd, D. (2015) The Law of Crime Concentration and the Criminology of Place. *Criminology*, 53(2): 133–157.

Weisburd, D. (2018) Hot Spots of Crime and Place-Based Prevention. *Criminology & Public Policy*, 17(1): 5–25.

10

INCLUSIVENESS AND EXCLUSION IN PUBLIC OPEN SPACES FOR VISUALLY IMPAIRED PERSONS

Kin Wai Michael Siu, Yi Lin Wong, and Jiao Xin Xiao

Introduction

The Greek origin of the word "public," *synoikismos*, means household units gathering together and worshipping different household gods (Sennett 1988). In the modern context, a public space refers to an area for all people to gather for different purposes. The space should have a capacity to allow different kinds of people who need each other to gather to form a public, such as neighbors using the same set of facilities or services. The combination of people and range of purposes result in the complexity of public spaces. This complexity includes change, dynamics, interactions, challenges, discoveries, exposures, uncertainties, compromises, and confrontations. However, if the notion that "public" is equivalent to "for all" or "belongs to all" (Hsia 1993a, 1994), and people with different cultural backgrounds, educational levels, and physical and sensory abilities should be able to access public spaces in their living environments easily, this poses a paradox: the "for all" nature has caused a "not for all" consequence in practice. The discussion of inclusiveness has attracted the close attention of designers and researchers in urban studies and public policy.

Public spaces allow all users to gather and meet others in the community; different groups of users occupy spaces and perform various activities. The people, activities, interactions, and atmospheres change from time to time, and public space is dynamic in nature. While the complexity of public spaces is often observable, and public space users can adjust themselves and alter their behavior in different situations and circumstances, users who have physical, sensory, or cognitive disabilities may not be able to do so by means of observation. Some may even be unable to understand or adjust themselves to their surroundings and they are thus excluded from the dynamic environment. One group unable to enjoy the inclusiveness and vivid interactions among different kinds of people is visually impaired persons (VIPs).

VIPs suffer from an obvious deficiency: they are unable to understand the immediate surroundings using first-hand visual information. They cannot perceive the unspoken attitudes, emotions, and gestures of others. They also cannot understand facilities out of their reach, and simple or small obstructions in public spaces are potential dangers to them. Tactile information regarding the usable environment and facilities in public spaces takes time to be assimilated and then being transformed into mentally visual representations.

Although governments and welfare associations have recognized the importance of inclusiveness in public spaces, and considerable progress has been made in current public space design to include more users with different needs, the so-called new inclusive designs do not comprehensively address the spatial difficulties of VIPs.

The quality and developmental and civic status of a city determine the inclusiveness of its public spaces. It is crucial to consider how inclusiveness is factored into the design of public spaces so that different kinds of users are able to enjoy the convenience and welfare provided by them. This chapter starts with a review of inclusiveness (and exclusiveness) of VIPs in China, followed by a comparative case study of public open spaces and related facilities in Beijing, Hong Kong, and Taipei. The word "open" in this chapter means outdoor environments such as outside seating areas and urban parks. The chapter also addresses the needs and difficulties of VIPs, discussing the need for synchronization between the open space provider (i.e. the government or private sector) and users in the design and offering of truly inclusive public open spaces to the visually impaired.

Inclusiveness (and exclusion)

Political theorist Robert Dahl (1971) considers inclusiveness to be one of the dimensions of democracy; inclusiveness refers to being "the proportion of the population entitled to participate" (Coppedge et al. 2008, 633). This suggests that inclusiveness represents how a group of people accepts diversity and includes others regardless of differences. Although this concept originated from political science, it is applicable in the context of public and inclusive design. Addressing inclusiveness through design means to design products, environments, or services "to meet the needs of a wide range of users, including both mainstream users and those with specific needs" (Goodman-Deane et al. 2014, 886). According to the British Standards Institute (2005), inclusive design is "the design of mainstream products and/or services that are accessible to, and usable by, as many people as reasonably possible … without the need for special adaptation or specialized design." Coleman et al. (2007) suggests that inclusive design is user-centered, population aware, and business focused. The design itself has to be functional, usable, desirable, and viable.

Inclusive design emerged when designers became aware that it was necessary to design for disabled and elderly people (Clarkson and Coleman 2015). Initially, it addressed the more varied physical needs of people. However, after years of development, inclusive design has shifted from a medical model to a social one in which people have the right to access different kinds of public services and spaces, or information and related services, via the internet or other digital media (Clarkson and Coleman 2015). Inclusive design has become a social concept suggesting how people with different needs might connect with each other.

While inclusiveness is explained through the provision of inclusive design, Keates and Clarkson (2003) believe that it is essential to consider design exclusion together with inclusive design. They suggested that examining how inclusive a design is does not show who and how many people can use the design. However, the results cannot help identifying how the design should be improved to be more inclusive; design exclusion is thus introduced to compensate for the inadequacy. Keates and Clarkson suggest that "by identifying the capability demands placed upon the user by the features of the product, it is possible to establish the users who cannot use the product irrespective of the cause of their functional impairment" (Keates and Clarkson 2003, 69). Lessening the demand will include more users with different needs. In other words, how a design does not include people with disabilities is a piece of essential information for re-design and improvement.

Visually impaired persons

There are approximately 1.3 billion people globally with different forms of visual impairment, including those with mild to severe impairment and blind persons. Thirty-six million of them are blind, with the majority of those 50 years old or above (World Health Organization 2018). The term visual impairment includes different levels of blindness (partial to total), different levels of distance and near vision impairment, and different kinds of color blindness. VIPs encounter great challenges in their daily lives compared with other differently abled people because they lack the most direct and immediate sense with which to perceive and understand their surroundings and the environment they live in. Although most existing inclusive public facilities and products are designed for people with physical disabilities, many do not cater for VIPs and therefore VIPs are included less in the public realm. A higher proportion of VIPs suffer from depression than those with good vision, creating additional mental-emotion barriers in their life (Evans et al. 2007).

Article 19 of the *Convention on the Rights of Persons with Disabilities* suggests that persons with disabilities should have equal rights to enjoy community services and facilities that are responsive to their needs in order for them to be included in the community (United Nations 2008). In this sense, the response of and actions taken by governments and other stakeholders has become crucial in establishing policies and building facilities for VIPs. Since their actions affect millions of people and their decisions govern how well VIPs can survive and live in their neighborhoods, inclusive policies and facilities are imperative for VIPs' daily quality of life.

VIPs in China

According to a World Health Organization (2010) report on global vision impairments, China accounts for 26.5% (about 75.5 million people) of the world's visually impaired population suffering from blindness and low vision. The percentage of VIPs in China is higher than the global proportion. Although the visually impaired population only makes up about 5% of China's population, the sheer number of people is concerningly large. Research has focused on the prevalence and causes of visual impairment nationally (e.g. Guo et al. 2017) and regionally (e.g. Tang et al. 2015), but most related research has discussed the issue from a medical perspective and concerns the physical health of VIPs.

In view of the needs and demands of VIPs and global concerns about disabled persons' rights, the Chinese government has enacted several policies and regulations to help VIPs to live a better life in society. Despite these initiatives and increasing concern, however, the life of VIPs in China is still constrained. For example, when an interviewee from the China Association of the Blind in Beijing was asked where the VIPs in China would go if they wanted to take a walk, he, a VIP, replied that VIPs often took their walks at home, back and forth from one end to the other. Tending not to go out and unable to go to open areas near their homes, VIPs' sphere of activity is very limited. This raises the question of how public facilities can assist VIPs in enjoying their life in the community.

Methods

Based on the concept of inclusive design and design exclusion, and the needs of VIPs, public open spaces in three representative cities in China—Beijing, Taipei, and Hong Kong—were chosen for a comparative case study in 2017. Through examining the inclusiveness of the current public facilities, the study investigated VIPs' ability to use facilities, and revealed how design should be improved. In addition to field studies, interviews were conducted to gain more information from the users directly.

Field study: the three cities

Beijing, Taipei, and Hong Kong are densely populated cities with similar social and cultural contexts. Although the values and living styles of people in these three cities may be different, they are all Chinese in culture. Their inclusive policies and regulations for persons with disabilities have been established in the last few decades, but their development and implementation approaches differ. A comparative analysis of their similarities and differences provide a rich data source for similar contexts (Siu et al. 2018).

Table 10.1 shows the districts, their population densities, and the number of public open spaces visited in each city by the research team. The districts visited have relative higher population densities than other districts, providing more potential cases related to the design exclusion of public open space facilities. In total, researchers in the team visited 45 public open spaces, where they took photos, and briefly observed the users.

Interview: participants

Thirty-six VIPs, aged between 18 to 75 years, were recruited by local VIP organizations across the three cities. Unstructured interviews were conducted, during which the VIPs were asked to share their experiences of using facilities in public open spaces in their neighborhoods; the unstructured nature of interviews helped the VIPs to relax and be more willing to talk about their daily lives. Each interview lasted approximately an hour and major topics of discussion included the difficulties VIPs encountered in public open spaces, their experiences and feelings while using the facilities, and suggestions to improve existing facilities. The interviews were recorded and transcribed for data analysis.

Table 10.1 Districts, population densities, and number of public open spaces visited in each city

Cities	Districts visited	Population densities (persons/km²)	Number of public open spaces visited
Beijing	Xicheng, Dongcheng and Chaoyang	24,372 21,881 and 7,530	15
Taipei	Da'an and Zhongzheng	27,283 and 20,975	15
Hong Kong	Kwun Tong and Kowloon City	55,204 and 37,660	15

Inclusive facilities for VIPs

The findings from field visits to the selected public open spaces are summarized below. Representative photos for each city were chosen from the extensive photographic documentation and are included in this chapter to aid the discussion. Access to the identified public open spaces and the entrances of and facilities in the spaces are discussed here.

Beijing

One of the significant spatial features in Beijing is the *hutong* (胡同 in Chinese). A *hutong* is the name for a narrow alley between traditional residential buildings. As the old areas were rebuilt with new buildings and houses replacing the traditional structures, some *hutong* became accessible to not just residents and pedestrians, but also vehicles. This modernized *hutong* connects different public open spaces within residential areas, however, there was no provision for inclusion, such as tactile paving thereby limiting the range of activities performed by VIPs.

In addition, rental bikes were piled up on streets and other obstacles blocked the streets (see Figure 10.1), and since these obstacles were not permanent, it was observed that VIPs were unable to adapt to the frequent change of the environment. Similarly, dangerous experiences discouraged them from going outdoors to open public spaces. The lack of access facilities and poor management thus resulted in an unwelcoming public environment for VIPs.

Several seating areas were found in the districts visited in Beijing. The areas were often small and surrounded by residential buildings, but had no facilities that catered for the needs of VIPs. In the seating area shown in Figure 10.2, no tactile paving was installed to guide VIPs to the area, and a curb raised the area up, meaning there was no level access. In addition, recycling facilities were not designed for VIPs or other users with disabilities. Overall, the environment was not favorable for VIP use, and the needs of VIPs were obviously not considered.

Taipei

Facilities designed to accommodate VIPs were rare in Taipei, with most of the inclusive facilities for people with physical disabilities or the elderly. Public open spaces were similar to those identified in Beijing, for example, Figure 10.3 shows a seating area in a small public open space surrounded by buildings. In this instance, no facilities for VIPs were observed, and the stone road in the small garden in Figure 10.4 was not designed with VIPs in mind either. The public open space did not have any facilities for VIPs. It seemed that VIPs did not have any public open space for leisure in the neighborhood; whether VIPs knew of this place and were able to access it for relaxation was unknown.

Further observations in Taipei indicated that ramps were often built at the entrances of public open spaces, yet those were for wheel-chair access, not built for VIPs. As highlighted in Figure 10.4, the entrance of a seating area we visited in Taipei, the ramp edge was not in good condition, creating a dangerous threshold and a trip hazard for VIPs. In addition, perforations in the drain cover would easily trap a VIP's probing cane.

Hong Kong

One of the unique features of Hong Kong's public open spaces is an array of locations and sizes. Although public parks are large enough for people to spend extended leisure time, most

Figure 10.1 Piled rental bikes blocking the street

Image Credit: Kin Wai Michael Siu, Yi Lin Wong, and Jiao Xin Xiao

Figure 10.2 Environment in a sitting-out area in Beijing
Image Credit: Kin Wai Michael Siu, Yi Lin Wong, and Jiao Xin Xiao

Figure 10.3 Seating area in a small public open space surrounded by buildings in Taipei
Image Credit: Kin Wai Michael Siu, Yi Lin Wong, and Jiao Xin Xiao

Figure 10.4 Entrance to a seating area in Taipei
Image Credit: Kin Wai Michael Siu, Yi Lin Wong, and Jiao Xin Xiao

of the seating areas are very small and sometimes situated between vehicular roads. Despite the locational disadvantage, Hong Kong's public open space had more inclusive facilities built for VIPs as compared to Beijing and Taipei. For example, most of the seating areas included tactile paving. Despite this, the implementation was lacking, as this only appeared at the entrances of seating areas, in front of or on the last step of staircases, and ramps (see Figure 10.5). There was no tactile paving to guide VIPs to or from other places.

Although tactile paving was installed at the entrance at the base of the staircases and ramp, it was not easy for VIPs to find without a connecting path because most entrances to the public open spaces were narrow. VIPs who were not familiar with the environment would not be able to go to the public open spaces without specific guidance or assistance

Figure 10.6 shows a public open space in Hong Kong. The paving was old and flat, without any obstacles or objects that could cause falls, a number of seats were provided for people to relax, and there were provisions for users to play Chinese chess. However, there was little color contrast in the design of the public open space, creating difficulty for VIPs to distinguish the facilities from the background. In addition, neither a tactile map nor tactile paving were found in the area.

Table 10.2 summarizes the findings related to inclusive facilities for VIPs in public open spaces in Beijing, Taipei, and Hong Kong. Most of the public open spaces in the three cities did not have any facilities for VIPs. While some had provided facilities, these were inadequate and did not cater to the specific needs of VIPs. Those facilities considered the needs of people with physical disability primary, with sensory disability a secondary concern.

Figure 10.5 Tactile guided path in front of staircases in a public open space in Hong Kong
Image Credit: Kin Wai Michael Siu, Yi Lin Wong, and Jiao Xin Xiao

Figure 10.6 Environment in a public open space in Hong Kong
Image Credit: Kin Wai Michael Siu, Yi Lin Wong, and Jiao Xin Xiao

Table 10.2 Inclusive facilities for VIPs in public open space in Beijing, Taipei, and Hong Kong

City	Path leading to open space	Entrance of open space	Facilities in open space
Beijing	*	*	*
Taipei	*	*	*
Hong Kong	*	**	**

Note: * no such facilities/very inadequate; ** facilities provided but inadequate; *** facilities provided and adequate

Difficulties encountered by VIPs and the need for the improvement of open space facilities

The comments from VIPs participating in the interviews are highlighted in Table 10.3.

Based on the data collected from the field studies and the interview findings, possible improvements are suggested below.

- Governments should be proactive when addressing the needs of VIPs when developing policies. VIPs are rather passive in making requests and finding solutions to tackle their difficulties, or more seriously, they do not understand that their difficulties can be addressed and solved through better design.
- Apart from designing facilities for the specific needs of VIPs, governments should install devices in street furniture and public facilities so that a signal could be sent to VIPs' smartphones when they enter public open spaces, as most VIPs use smartphones with headphones when they travel outside.

Table 10.3 VIP interview comments

City	Difficulties and needs
Beijing	• There were too many obstacles on the street, and the street environments changed frequently. VIPs did not dare to go out. • VIPs did not know the seating areas and other public open spaces in their neighborhood. • No helpers or volunteers would help them to go outside.
Taipei	• Although the government did not provide facilities for access, VIPs did not think it was a concern. • The community that the VIPs lived in was very helpful and inclusive. • Helpers and volunteers from non-governmental organization (NGOs) often offered help to VIPs. VIPs became dependent and would not try to go outside alone.
Hong Kong	• There were many other users in the public open spaces, and it was inconvenient for VIPs to go there. • There was no tactile information about public open spaces for VIPs. • They would only go to places with which they were familiar, such as the shopping mall near their home. • Although VIPs were very independent, and some would go outside alone, most preferred staying at home.

- Additional axillary products need to be designed. NGOs and VIP organizations can be included in the education process by notifying their clients about the new way of gaining information. Tactile and aural information design is an important aspect in improving VIPs' daily life quality, so there is a real need for innovative work.

- NGOs or VIP organizations have to take the initiative to link VIPs and government and establish a feasible and convenient channel for communication in policy making, implementation, and management stages. Their role in connecting different parties and obtaining welfare benefits at the three levels of policy work has to be clear. Direct and face-to-face encounters are the best way for VIPs, as VIPs can experience and feel others' presence.

- It is important to remember that not all blind persons are blind at birth and those individuals can serve as community facilitators. More inter-community connectivity, such as learning different ways to enjoy community life, can be done through organizing different kinds of social meeting sessions with other "sight-experienced" VIPs. Policy makers and designers can facilitate discussions around how spaces can better cater to and be explicitly designed for VIPs.

- VIPs have different types of impairments, and each person has specific difficulties related to light and color. A spectrum of VIP experiences should be included in any inclusive events, as the larger the number of VIPs a more holistic picture about their difficulties, needs, and preferences can be gained.

Other general suggestions include:

- Governments should provide inclusive facilities and cater to the needs of VIPs, but research is needed to further understand their needs before any implementation can occur. VIPs should be invited to participate in the research and design process to ensure their needs are directly identified and addressed.

- Channels to convey information to the public about the needs of VIPs in open space should be developed. Education is imperative to prepare the non-VIP public to use inclusive facilities, while also allowing more flexibility for VIPs.

- As it is highly recommended that VIPs be independent and enjoy their lives with minimal assistance, helpers and volunteers should also teach them how to be independent. This is the only way that the VIPs can live with dignity long-term.

- NGOs or community centers should organize different social activities or form interest groups for VIPs to build their communities.

- Helpers and volunteers from NGOs or community centers could introduce VIPs to new places, allowing VIPs to have more options to experience their neighborhood.

- From the interview findings, it is obvious that older VIPs have had more difficulties than younger VIPs. Some of the services provided may have to merge with those for the elderly; other VIP activities should be suitable for persons with different disabilities cultivating a more inclusive community.

Towards a truly inclusive public open space: synchronization

To make the previously suggested improvements feasible, effective communication between users, policy makers, designers, and public open-space management teams is essential. These parties have to work and coordinate closely, so that an inclusive environment can be provided for VIPs. The information held by each party should be synchronized so that all parties are at the

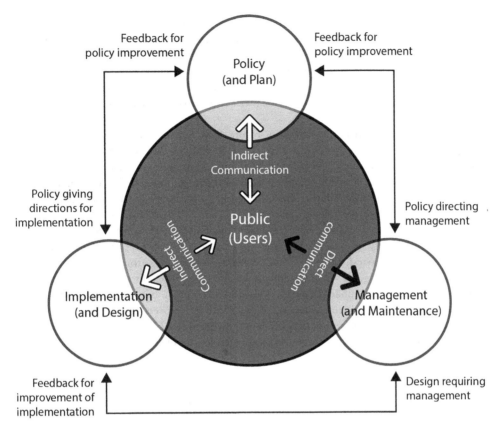

Figure 10.7 Policy-implementation-management model
Image Credit: Siu, 2009; Siu, Lu, & Xu, 2009

same stage of the design and developmental processes. The policy-implementation-management (PIM) framework (see Figure 10.7) describes the importance of synchronization and how it works (Siu 2009; Siu et al. 2009).

The PIM model presents three levels of work to establish a public facility and also the relationships among them and the public (users). Directions, aims, and plans are developed at the policy level. The model provides implementation with directions and leads management. At the implementation stage, the plans are put into practice and executed pragmatically. Designers are involved in this stage, and they have to consider different factors and design optimal methods and approaches to address the needs of the public. Through the implementation process, designers are able to provide feedback to the government for the improvement of the related policy. The establishment of public facilities does not stop at the implementation stage, as the design formulated at the implementation stage requires appropriate management and maintenance to sustain the facilities or the service. Subsequently, at the management level the management team is responsible for managing and maintaining the design, and responds to changes in society and deals with the public directly. The management team is therefore able to provide feedback to the government and designers to improve the policy and the design. This also means that management reflects the satisfactory level of users' experience directly and implies whether the satisfaction or the dissatisfaction would be prolonged. The public (users) is the center of the model: the ultimate goal of the model is to benefit users and facilitate the provision of better lives. The three levels are interconnected, and they provide information and mutual feedback to each other.

When VIPs are identified as a group of users, the model would have to be adjusted at the pragmatic level so that VIPs' needs can be addressed. Figure 10.8 shows one of the possible adjustments that the policy makers, designers, and management teams have to make to reach the group of VIPs.

Different from the PIM model shown in Figure 10.8, the role of VIPs as users is clearly identified. This group of people has a relatively low mobility because of their inability to understand the world through vision. Most of them seldom commute alone, explore their communities, and obtain new information related to the neighborhood. They are rather passive in seeking out new possibilities. Therefore, policy makers, designers, and management teams have to take proactive roles to reach VIPs. When VIPs are concerned, they can no longer stay in standard working zones (circles in Figure 10.7); they have to expand their services out proactively to reach VIPs and understand their needs (the pear shapes in Figure 10.8). In the process, they also have to include other users from the public, so that the community writ large are able to understand the needs of VIPs and the changes that would be made to the public facilities. The communication between VIPs and policy makers, designers, and management teams are crucial for the process being truly inclusive, as a clear agreement about the needs of VIPs needs to be made. NGOs or VIP organization mediating or intervening in this discussion is essential to stimulate discussion and smoothen the process.

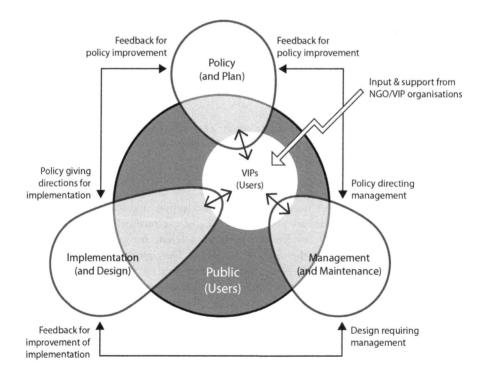

Figure 10.8 One of the possible adjustments in the policy-implementation-management model for VIPs

Image Credit: Kin Wai Michael Siu, Yi Lin Wong, and Jiao Xin Xiao

How much each party should extend their work to reach VIPs depends on different circumstances and level of work. Figure 10.8 shows an example of a possible adjustment that the three parties could make. In the figure it is clear that more work has to be done at the level of implementation to understand VIPs. The adjustment that each party makes is not static but volatile and the parties have to be flexible and responsive to changing situations. In this regard, the mutual feedback from policy makers, designers, and management teams is vital, as, first, the three parties have to arrive at a clear consensus on the direction of facility provision, and second, without effective communication among the parties, unsuccessful work may be carried out across different levels of the design and implementation processes. The success of a public facility depends on the public's satisfaction: the three levels have to work collaboratively; synchronization is essential to achieve this goal.

Conclusion

Public spaces, by definition, are open and designed to provide for all public gatherings and communication. People with different cultural backgrounds, educational levels, and physical and sensory abilities should be able to easily access the public spaces in their living environments. Inclusiveness is an important concern in this regard, however, people with disabilities are often excluded from public spaces through design. VIPs comprise one of the main groups that encounter great difficulties because they are unable to perceive the world visually; even small damage to facilities can cause great danger to VIPs. To address that issue, this chapter discussed the concept of inclusiveness and exclusion, and presented the situation of VIPs in China. Beijing, Taipei, and Hong Kong were selected as places to investigate the how facilities in public open spaces were inclusive and the difficulties encountered by local VIPs.

This chapter argued that synchronization should be a feature of the communication process between different parties in the design and development of facilities in public open spaces. These parties include users, policy makers, designers, and facility management teams. The emphasis should be put on users, and all comments and feedback from different parties should be valued equally. In many cities in China, it is still difficult to achieve this, as most public open space and facilities are provided by the government and the needs of other parties are less valued. This chapter should therefore serve as a reference for other cities that are beginning to develop truly inclusive communities, and the findings and framework presented here will be useful for those developments.

Acknowledgements

The authors would like to acknowledge the Humanities and Social Sciences Prestigious Fellowship Scheme (RGC 35000316) of the Research Grants Council, and Internal Competitive Research Grant and Endowed Professorship in Inclusive Design of The Hong Kong Polytechnic University for the data collection and the preparation of the paper. The authors thank the Architecture and Building Research Institute, China Association for the Blind, China Disabled Persons' Federation, Hong Kong Blind Union; Taiwan Foundation for the Blind and The Hong Kong Society for the Blind for providing a lot of useful information.

References

British Standards Institute. (2005) British Standard 7000-6:2005: Design Management Systems – Managing Inclusive Design – Guide.

Clarkson, P. J. and Coleman, R. (2015) History of Inclusive Design in the UK. *Applied Ergonomics*, 46: 235–247.

Coleman, R., Clarkson, J., Goodman, J., Hosking, I., Sinclair, K., and Waller, S. (2007) Part 1 Introduction. In: *Inclusive Design Toolkit*, Clarkson, J., Coleman, R., Hosking, I. and Waller, S. (eds). Cambridge: University of Cambridge, pp. 1–4 to 1–55.

Coppedge, M., Alvarez, A., and Maldonado, C. (2008) Two Persistent Dimensions of Democracy: Contestation and Inclusiveness. *The Journal of Politics*, 70(3): 632–647.

Dahl, R.A. (1971) *Polyarchy: Participation and Opposition*. New Haven: Yale University Press.

Evans, J.R., Fletcher, A.E., and Wormald, R.P.L. (2007) Depression and Anxiety in Visually Impaired Older People. *Ophthalmology*, 114(2): 283–288.

Goodman-Deane, J., Ward, J., Hosking, I., and Clarkson, P.J. (2014) A Comparison of Methods Current Used in Inclusive Design. *Applied Ergonomics*, 45: 886–894.

Guo, C., Wang, Z., He, P., Chen, G., and Zheng, X. (2017) Prevalence, Causes and Social Factors of Visual Impairment among Chinese Adults: Based on a National Survey. *International Journal of Environment Research and Public Health*, 14: 9.

Hsia, C.J. (1993) *Space, History and Society (Paper collection 1987–1992)*. Taipei: Taiwan Social Research Studies-03.

Hsia, C.J. (1994) *Public Space [in Chinese]. Public Art Series 16*. Taipei: Artists Publisher.

Keates, S. and Clarkson, P.J. (2003) Countering Design Exclusion: Bridging the Gap between Usability and Accessibility. *Universal Access in the Information Society*, 2(3): 215–225.

Sennett, R. (1988) The Civitas of Seeing. *Place*, 5(4): 82–84.

Siu, K.W.M. (2009) Quality in Design: Policy, Implementation and Management. In: *New Era of Product Design: Theory and Practice*, Siu, K.W.M. (ed). Beijing: Beijing Institute of Technology Press, pp. 3–15.

Siu, K.W.M., Lu, J.Y., and Xu, P. (2009) Design Standard of Tactile Guide Paths in China, Policy, Implementation and Management. In: *Proceedings of the 6th International Conference on Integrity-Reliability-Failure*, Gomes, J.F.S. and Meguid, S.A. (eds). Porto, Portugal: Edicões Instituto de Engenharia Mecanica e Gestao Industrial.

Siu, K.W.M., Wong, Y.L., and Xiao, J.X. (2018) Reliability and Failure of Policy Implementation of Inclusive Design: Case Studies of Open Space in Beijing, Taipei and Hong Kong. In: *Proceedings of the 6th International Conference on Integrity-Reliability-Failure*, Gomes, J.F.S. and Meguid, S.A. (eds). Porto, Portugal: Edicões Instituto de Engenharia Mecanica e Gestao Industrial.

Tang, Y., Wang, X., Wang, J., Huang, W., Gao, Y., Luo, Y., and Lu, Y. (2015) Prevalence and Causes of Visual Impairment in a Chinese Adult Population: The Taizhou Eye Study. *Ophthalmology*, 122(7): 1480–1488.

United Nations. (2008) *Convention on the Rights of Persons with Disabilities (CRPD) – Article 19*. Accessed 3 July 2019. www.un.org/development/desa/disabilities/convention-on-the-rights-of-persons-with-disabilities/article-19-living-independently-and-being-included-in-the-community.html

World Health Organization. (2010) *Global Data on Visual Impairment*. Accessed 3 July 2019. www.who.int/blindness/GLOBALDATAFINALforweb.pdf

World Health Organization. (2018) Blindness and Vision Impairment. Accessed 3 July 2019. www.who.int/news-room/fact-sheets/detail/blindness-and-visual-impairment

11

PUBLIC SPACE, AUSTERITY, AND INNOVATION (THE LONDON CASE)

Matthew Carmona

Introduction

Through an overview of the period 2008 to 2018, this chapter reflects on the impact of the austerity years on the processes of shaping public spaces. The chapter draws from the experiences of London to explore processes of the design, development, use, and management of public spaces during this period, many of which are likely to be applicable to other major cities that have shared an austerity mandate. The evidence suggests that we have witnessed a period of significant innovation side by side with major challenges in the collective approach to public spaces. Episodes of changing practice are used to illustrate the argument and cumulatively reveal a distinct and significant impact from austerity, although not necessarily in the manner that might have been expected.

Since the financial crisis of 2008 governments, municipalities, developers, and ultimately communities have had to rapidly adapt to a new more uncertain reality. Some argue that the period has marked a new and distinctive era with its own political-economy, governance, and societal norms (Bramall 2013). For others the period is simply a continuation, perhaps even a deepening, of the neo-liberal project (Bone 2012); a project that has frequently been underestimated as regards its capacity for transformative and adaptive change (Peck and Tickell 2002); and which continues to thrive (albeit evolving), particularly within its heartlands, such as England.

Winston Churchill argued that we should 'never let a good crisis go to waste.' Reflecting Churchill's maxim, in times of crisis some evidence exists of a flowering of ideas and practices relating to public spaces as the most quintessentially shared part of our built environment. Smog filled Victorian cities, for example, saw a flowering of public parks, Europe's bomb-ravaged post-war cities saw the new forms of expansive Modernist public space take root, and the economic shocks of the 1980s led to private corporations rediscovering the commercial value of traditional public spaces for example at London's Canary Wharf.

Whilst others have written about what they see as an 'austerity urbanism' this discourse often focuses more on the governance and political impacts of austerity in urban areas, rather than on its spatial implications (e.g. Peck 2012; Mayer 2013). Therefore, looking back over the past ten years, this paper asks, has a 'good crisis' gone to waste? Whilst every city is unique, and London is certainly different to other parts of Europe – where recession, alongside austerity, has been far deeper and more sustained in its impacts – London's size, diversity of governance practices, and early embrace of the austerity model makes it an interesting exemplar to explore (Christodoulou and Lada 2017).

The framework provided by the place-shaping continuum framework articulated in Carmona (2014a) is used to structure the discussion. The paper represents a journey across that continuum, from design and development, to the use and management of public spaces, eventually drawing out conclusions about the shifting power relationships that have taken place during the austerity years.

The London case: agnostic to advocate

Analysis begins with a brief exploration of the changing and varying political and policy context for public spaces and how this has responded to the drive for austerity. As a political and policy concern, issues relating to the provision and quality of public space have been on the rise, globally. De Magalhães and Carmona (2009 111) explain: 'From civic, leisure or simply functional spaces with an important, but to some extent discrete, part to play in cities and urban life, public spaces have become urban policy tools of a much wider and pervasive significance,' often at the forefront of policy debates around livability, sustainability, social inclusion, economic competitiveness, place image, and culture.

Whilst some question the contemporary relevance of public spaces 'in an age of urban sprawl, multiple usage of public space and proliferation of sites of political and cultural expression,' this has not generally been taken up by politicians (either nationally or locally) for whom public spaces are typically a binary concern (Amin 2009). Either they are off the agenda completely – a public space agnostic view – and therefore of little relevance beyond their management cost and maintenance liability, or they are a major opportunity with far reaching economic, social, cultural, health, and environmental potentials – a public space advocate view.

London demonstrates this binary approach well. The 1980s and 90s were a time of disinvestment and decline as regards the publicly owned streets and spaces of the city, which were largely viewed, in a purely managerial sense, as traffic arteries. This changed after 2000 and the election of the first London Mayor – Ken Livingstone – when public spaces moved decisively up the policy agenda. Livingstone argued that the quality of public space had a direct impact on the city's beauty, sustainability, connectivity, and safety – and therefore on its attractiveness to investors – and promised to create or upgrade 100 public spaces over five years. For Londoners this was undeniably a period when a noticeable new embrace of public space was apparent as café culture came to the city, but the complexities of delivering public spaces schemes meant that Livingstone's early ambition proved rash and only five schemes were realized by the end of his tenure (Carmona 2012).

Boris Johnson, the second mayor, came to power in May 2008 just as the grip of austerity began to tighten. His early emphasis was on leafy outer London, and he actively embraced austerity, even before it began to bite nationally (after 2010). Consequently, Johnson quickly set about dismantling the high-profile public space programs of his predecessor, including a plan to re-design Parliament Square, he disbanded Design for London (the Mayoral design arm), and announced the closure of the London Development Agency; previously a major source of project funding. Yet Johnson also inherited responsibility for delivering the London Olympics in 2012 and stimulating the housing market in London that had crashed spectacularly in 2009. These priorities, alongside his political concern for outer London (notably their struggling mixed-use shopping streets – high streets), meant that, despite contrary signs at the start,

Figure 11.1 Windrush Square, Brixton: Boris Johnson also benefitted hugely from the legacy of work on
public spaces conducted during the Livingstone era, as, like the gift that keeps on giving,
public space projects continued to mature throughout the austerity years

Image credit: Matthew Carmona

he also quickly embraced a strategic public spaces role for the city, continuing to invest heavily
throughout the austerity years (see Figure 11.1) (Carmona 2012, 38).

The third mayor – Sadiq Khan – has promised to continue the focus on public spaces
although with a stronger environmental emphasis directed at reducing pollution, clutter, and
congestion, and improving design (Khan 2016). Begun under Johnson, but significantly bol-
stered from 2016 under Khan, much of the discussion on London's streets and spaces is
increasingly seen in health terms, with the launch of Healthy Streets for London in early
2017 formalizing a range of initiatives into a long-term plan and directing the £2.1 billion
(€2.4 billion/$2.8 billion) of streets spending over the Mayoral term 'towards delivering
against the Healthy Streets Indicators' (TfL 2017).

London has therefore witnessed a journey from an agnostic (managerial) perspective to
a role as advocate with regard to the merits of public space investment, sometimes for classic
entrepreneurial reasons, but increasingly for social and environmental ones (Biddulph 2011).
Whilst this belief seems to have persisted during the austerity years, the discussion of the
changing political and policy context indicates that it has also been evolving as part of the
larger neo-liberal project that Peck and Tickell (2002) characterize (through gritted teeth) as
the 'commonsense of the times' (381). Thus, just as the wider political-economy of the city
has continued to evolve, so to have approaches to public space.

The discussion that follows pulls this apart and, by identifying trends in the design, development, use and management practices of London, determines what are the trends that can be detected during the austerity years.

Design

Space as spectacle

In London, playing on Boris Johnson's love of spectacle (he funded, for example, the Emirates Air Line cable car), the crisis years saw developers and others see the potential of public spaces as places of spectacle. In 2011, for example, the Mayor was quick to champion Gensler's ultimately unrealized plans for a floating boardwalk along the Thames, plans that were criticized for 'privatizing' parts of the river. He also approved revised plans for the now realized Walkie Talkie tower in the City of London that, in exchange for planning permission, delivered London's first ticketed 'public' space at the top of the tower (see Figure 11.2).

Elsewhere, Johnson strongly backed the Garden Bridge Trust with £60 million (€67 million/$79 million) of public money to deliver its garden bridge across the Thames; a scheme (now defunct) that some argue would have been ostensibly a private tourist

Figure 11.2 For example, the new 'public space' – Sky Garden – on top of the Walkie Talkie tower negotiated as part of the planning permission. Free to enter, but you need to book in advance, obtain a ticket, and pass through security

Image credit: Matthew Carmona

attraction rather than a public space (Minton 2014). The Trust strongly contested such assertions, although Khan (following a brief enquiry, and rather cowardly in the view of some) was quick to withdraw public support on coming to power. Each, for good or ill, extend traditional notions of public space whilst creating or envisaging new ways of experiencing the city with associated knock-on public interest and/or commercial benefits (Cho et al. 2016).

Investing in the ordinary

Whilst spaces of spectacle continue to be produced, the other side of the coin has been a new focus on 'everyday' spaces. London represents a case-in-point where many of the city's streets have suffered from decades of priority being given to traffic over people (Gehl Architects 2004). Whilst there have been individual examples of re-prioritizing space in favor of pedestrians (e.g. Kensington High Street), only recently has the approach been mainstreamed by Transport for London. Thus since 2014 London's streets have been re-classified against a nine-part matrix that, to varying degrees, recognizes each as a 'place' as well as a corridor for 'movement,' and this will now define design standards across the city. Gradually (very gradually), a re-balancing of ordinary street space is occurring (TfL 2013; 2016).

The challenges of densification

Public spaces are also playing a key role in the everyday processes of growth and densification across the city. With its tightly drawn growth boundary, London exemplifies this and successive iterations of the London Plan have supported densification in areas well served by public transport and in the city's strategic opportunity areas; a quid pro quo for which has been the provision of high-quality public spaces.

The results have sometimes proved controversial, including the 2009 re-design of Chelsea Barracks where a scheme by Richard Rogers became mired in controversy following a damming intervention by Prince Charles and was subsequently replaced by traditional terraces and mansion blocks around a series of garden squares (Adams 2010). In 2017 a battle raged over rival plans to redevelop the huge Mount Pleasant sorting office with, on the one hand, contemporary medium and higher-rise blocks and a linear park against, on the other, traditional mansion blocks and a central classically designed square. Elsewhere much larger regeneration projects at Stratford and Nine Elms are being implemented around the city's first significant new urban parks since Victorian times. And the regeneration (aka replacement) of post-war housing estates are substituting unloved 'indeterminate land oozes,' as Jane Jacobs (1961) once christened them, with diverse 'contained' green infrastructure, that often blurs the boundaries between public and private (GLA 2015, 33). Densification means that public spaces are having to 'work a lot harder' than they have before in the sorts of low-density cities that predominate in the UK.

Development

Big business, 'private' spaces

Traditionally public space is thought of as a public good, paid for, delivered, and managed by the public purse, but London demonstrates that this is frequently not the case. In London, as home-grown funding for development became scarcer in the austerity years, footloose international money flooded in to fill the gap, notably from the Middle East and China (Pitcher 2013). This is shaping many of the largest development projects in the city, including three new Westfield shopping centers (built since 2008) and many new high density and often high-rise housing and office developments.

In such places arguments rage around whether associated 'private' public spaces are too commercial, corporate, securitized, sanitized, and exclusionary in feel, and therefore not really public at all, with some arguing (notably *The Guardian*) that the austerity years have seen a gradual escalation of such 'privatization' processes (Garrett 2015; 2017). In reality the resulting privately owned and managed public spaces continue to be as varied (in experiential terms) as their purely public and pseudo-public counterparts, and, like the other cities, public spaces are shaped by a negotiation between commercial interests and regulatory policies and practices (Carmona and Wunderlich 2012).

Valuing the temporary and exploratory

Arguably, debates relating to the privatization of public space are so hotly pursued precisely because such interventions, for good or ill, are so permanent. Yet increasingly both public authorities and private developers are interested in temporary interventions that bring sites into socially beneficial and/or profitable use whilst they are waiting for development to start or (in the austerity years) for the economy to recover (Ferreri 2015). In London such 'meanwhile' uses are also seen as means to shape perceptions of the emerging place during the development process, for example the swimming pond and skip garden that featured amongst the program of temporary spaces animating yet to be developed parts of the huge Kings Cross redevelopment.

Reflecting on this move to the temporary, Tonkiss (2013, 315) warns:

> As useful as meanwhile uses can be, it is important to note how quickly the pop-up can become the tear-down, and the fine margin that at times separates the pioneer use from the urban land-grab, or the creative incubator from the developer demonstration project.

Others argue that in challenging times, these sorts of interventions 'enable new types of creative conversations to happen between parties traditionally considered in opposition' and beyond the short-term impact of the meanwhile use itself, they can lead better development outcomes over the long-term (Kamvasinou 2017, 205). In London temporary interventions have been used to encourage: new ways of using the city, such as traffic free days on Regent Street; new ways of seeing the city, including the sea of ceramic poppies that slowly grew at the Tower of London throughout 2014; new revenue opportunities, through the more intensive use of underutilized spaces such as car parks for farmers markets at the weekend; or for the testing out of new ideas (see Figure 11.3).

Case by case negotiation

Case by case negotiation is the key in London, with major developments subject to bespoke planning agreements. Typically, such negotiations encompass a wide range of 'public goods' from schools to roads, and streets to social infrastructure. In such complex negotiations there is a danger that public space issues are given inadequate attention and

Figure 11.3 Exploratory cycleway layouts (to the right of the image) in London's Bloomsbury, implemented at minimal cost prior to more permanent investments (to the left of the image) being made

Image credit: Matthew Carmona

that spaces are then either sub-standard when completed or long-term rights and responsibilities are inadequately resolved. Carmona (2014b) argues that there is need for the mayor to adopt a clear and simple charter of public space rights and responsibilities to cover the whole city and this has now been promised in the latest (2017) revised London Plan. However, as each of the 33 boroughs remain autonomous in these negotiations practice (and interest in such matters) continues to vary and developers remain in a strong bargaining position.

Infrastructure as place-making

Within this context, although public expenditure was dramatically cut back during the austerity years, expenditure on new public space projects faired relatively well, in large part because of its association with expenditure on the 2012 Olympic Games, and latterly on public transport. Across England, whilst funding to local government reduced on average by 25% between 2010 and 2014 new transport infrastructure spending in London has been shielded by the need to address London's growth and an historic backlog in infrastructure investment (NAO 2014).

During this period also, national government has been on a journey. In the past it viewed such infrastructure as costly bits of technical kit to be delivered at minimum cost to

Figure 11.4 The Queen Elizabeth Olympic Park, Stratford – post-games
Image credit: Matthew Carmona

the public purse and with little concern for the local impact. Only now is a realization dawning that such investments are pieces of city building with huge place-making potential well beyond the infrastructure itself (Savills 2015, 6–7). In London, the commitment to carefully design a new public realm around the 40 new Crossrail stations (at the time of writing) soon to be opening across the city represents the clearest demonstration of this thinking (see Figure 11.4). At the same time the range of new and enhanced spaces associated with the Olympics have been transformational in parts of East London, demonstrating the potential of infrastructure-led public space.

Use

From utilitarian to leisure and specialist uses

Whilst the gradual re-balancing of space (already refereed to) in favor of pedestrians and cyclists is likely to be a long-term trend, other pressures are also acting to change how public spaces are used. Particularly pervasive is the impact of the internet on almost every aspect of life, and most particularly on shopping habits and on the consequential viability of traditional mixed (shopping) streets (Carmona 2015a). In London, one consequence is that many local shopping streets are slowly moving from a utilitarian retail or service (e.g. banking) function for everyday needs to a leisure and more specialist range of functions, including catering to the tastes of new immigrants to the city (Wrigley and Lambiri 2014, 19). Coffee shops are spreading like wildfire and pop-up coffee venders are appearing on every busy corner leading many to wonder when 'peak coffee' will be reached (Haughton 2015).

The changes have increased in pace since the crash of 2008 when spending power fell dramatically and increasingly moved online (ONS 2016). All this is changing both why people visit key public spaces in the city and how they function.

A tale of two cities (and different populations)

If, for some users, public spaces have become places of leisure, for others this is not the case. London, as a global city, has long been considered a magnet for the sorts of creative classes that cities are increasingly intent on attracting. Arguments have raged about the gentrification impacts of these populations and about provision for those at the opposite end of the social spectrum. The differential impact can be seen in the state of the city's traditional retail streets. Whilst some (in affluent areas) thrive, others (in less affluent areas or on busy trafficked roads) have been struggling to adapt to the new realities of the online marketplace, large multi-national discount retailers, and gentrified populations seeking the different (leisure) experience already discussed, and who are prepared to travel.

The fate of many of London's traditional street markets is a strong bellwether of this and have either declined and all but disappeared during the austerity years or have had to adapt and find a new income as part of the leisure economy (Jarvis 2015). The contrasting fates of Borough and Petticoat Lane markets, both on the edges of the City of London, but serving very different populations has been tracked for over ten years and reveals the stark story of adapt and survive or fail to adapt and go under (Kim 2017). The rise in visible homelessness has also been dramatic, with changes in migration patterns across Europe, austerity-led reductions in benefits entitlements in the UK, and cuts in services for the homeless and those with mental health difficulties, all leading to significant rises in rough sleepers (see Figure 11.5) (Crisis 2016). Associated rises in begging and arrests (up 90% according to some estimates) are also contributing to stark and very public contrasts between haves and have-nots (Watts 2014).

Reasserting democratic space

Since 2008 demonstrations against austerity politics in Europe and beyond have reconfirmed the critical democratic role of public space. In London, protest and demonstration remains a regular occurrence, with the 'Occupy' and 'Stop the War' camps of 2011 and 2012 forcefully re-asserting the historic role of public space for demonstration and political purposes (Tonkiss 2013). Eventually the legal limits of such protests were tested in the courts and the camps began to disappear as legislation effecting Parliament Square was redefined (via the Policy Reform and Social Responsibility Act 2011) and elsewhere rights to protest and of association under the 1998 Human Rights Act were shown not to extend to the right to occupation. Consequently, the early light touch policing and tolerance of such activities, which were largely peaceful, has, since 2012, been replaced with more active and rapid intervention as and when deemed necessary.

By contrast, the rights and wrongs of the 2011 riots, which affected large parts of London in a far more dramatic and disturbing manner, remain contested. Under Mayor Boris Johnson they nevertheless helped to drive significant public funds in the direction of the most affected areas, much of which was focused on improvements to the physical built environment of the spaces that had been targeted.

Figure 11.5 The rise in rough sleepers represents a direct and highly visible impact of austerity
Image credit: Matthew Carmona

Management

Caring for the everyday physical fabric (public and private)

In London, the first cuts to be made when public finances got tighter were to the budgets for managing the city's public spaces; cuts that have fallen disproportionately on those living in the most deprived areas (Hastings et al. 2015). Whilst at first this largely went unnoticed, the impact is cumulative as revealed in official statistics relating to the dramatic increase in potholes and associated accidents on the city's streets (Williams 2016). As decline in the public physical fabric sets in, so potentially does the closely associated well-being of citizens who now experience an uglier and seemingly less loved environment, leading to a stronger desire amongst commercial interests to manage new spaces themselves (London Assembly 2011).

Underpinning some of these trends, across Europe, are the New public management approaches, applied to public spaces, that are so strongly associated with the politics and practices of neo-liberalism (De Magalhães and Carmona 2007). These play into the debates (already highlighted) regarding access to and rights over privately owned public spaces and the tendency for management regimes to be over officious in their enforcement of privately defined management codes. Concern that 'In an age of austerity … Budget pressures on local authorities are increasingly leaving the management of public space in the hands of developers' led in 2011 to an enquiry by the London Assembly into such practices (London

Assembly 2011, 9). Whilst the inquiry concluded that greater attention should be given in policy to ensuring a consistent consideration of management issues at the time planning permission is given, it also argued that 'private ownership or management of public space is not, in itself, a cause for concern'.

Space as empowerment

Across England, a second narrative of community empowerment and localism has accompanied that of austerity. Whilst some see this as a cynical attempt to paper over the cracks left by cuts in public services, community and local action has certainly flourished in some of the gaps (Hambleton 2011). Notably this has included significant numbers of community groups now involved in green space management across London, such as the Rocky Park community garden in Bethnal Green where volunteers have transformed an unloved 'hangout' into a productive space for flowers and vegetables (Bawden 2016). Elsewhere, crowd funding is being used to innovate new projects as diverse as The Line art walk along the River Lee (see Figure 11.6) and the proposed Coal Line urban park utilizing disused coal sidings in Peckham. At a larger scale the voluntary '20's plenty for us' campaign is encouraging communities to demand slower speed limits on residential roads, and, so far, has convinced nine of London's 33 boroughs to adopt 20mph speed limits across their entire local roads network (20's Plenty for Us 2016).

Figure 11.6 The Line art walk, from the O2 to the Queen Elizabeth Olympic Park, Damien Hirst's *Sensation* (2003) as unintended plaything

Image credit: Matthew Carmona

Management by cappuccino (sponsorship and advertising)

Beyond enlisting the direct help of citizens, tighter budgets for the management of public spaces have left local authorities increasingly looking to external sources of income to help fill some of the gaps, notably by exploiting commercialization opportunities. Whilst some hysterical reports have raised the potential for local authorities to close and sell off facilities such as parks, or otherwise give them over to private interests, such as companies in the personal fitness industry, there is no incidence of this actually happening beyond a single case of the tree tops in Battersea Park being rented out for adventure climbing (BBC News 2015; Plimmer 2016). Instead, commercialization activities are typically small scale and include renting pitches to the sorts of mobile coffee vendors already referred to. They also encompass sizable and longer-term outdoor advertising contracts as councils seek to cross-fund the management of public space through selling advertising rights on public buildings, bus shelters, and on free-standing hoardings (Outsmart 2016).

From 2010 the 'Boris Bike' cycle rental scheme (named after the former Mayor Boris Johnson) brought private sponsorship more visibly and ubiquitously into the public realm with sponsorship initially by Barclays and latterly Santander Banks. This mirrors schemes that have now become commonplace in Europe, although quickly new subversive technologies look likely to make these fixed infrastructure investments (with bike stands across central London) redundant as multiple private bike hire companies have (in 2017) begun operating in the city, with GPS tracked bikes that can be picked up and left anywhere without any of the heavy upfront infrastructure or management costs.

Figure 11.7 Barriers installed following the London terror attacks of 2017
Image credit: Matthew Carmona

The security question

A final management concern relates to the question of security that has long featured as a consideration in London with its history of IRA attacks from 1971 through 2001, and radical Islamist attacks from not long afterwards to 2017. The installation of security-measures (physical barriers) in a few high-profile public spaces has been the subject of much debate with a degree of target hardening around prominent buildings such as the Houses of Parliament and (recently) on bridges over the River Thames (see Figure 11.7); although also a determination that such issues will not impact on access to the city's key iconic venues (Syal and Asthana 2017). Unfortunately, on this most confounding of management issues, the recent vehicle attacks in London have demonstrated how profoundly difficult it is to address such concerns without impacting the democratic nature of Western public space.

Conclusion

Evolving power relationships

Terry Farrell (2017, 137) advances the thesis that 'there is only one thing worse for urban design than a recession, and that is a boom' with the inevitable emergence of cash-rich developers wanting to make a fast buck whilst the going is good. Clearly there is a contradiction here. Whilst recessions slow things down and offer greater scope for the careful consideration of projects, it is also far harder to get projects built, particularly if, as has been the case since 2008, larger economic challenges are matched by cuts in public sector funding. In fact, it was only the start of the period covered by this chapter that actually saw technical recession in London, after which the private sector continued to steam ahead with a vengeance whilst resources were being systematically removed from the public sector.

In London, the austerity years do not seem to have substantially altered the balance between players responsible for or affected by public space. Roberts (2017) goes so far as to suggest that, for London, austerity simply represented 'business as usual.' Public space has always been diverse in its ownership, management, and use, and continues to be so, although the period has seen a renewed emphasis on very large projects of different types. Typically, these have been driven forward by single developers who increasingly retain ownership, either themselves or via the long-term management organizations, of key spaces. Whilst this might be viewed as a move to more privately owned and managed spaces, in reality these sites were never in public ownership in the first place, or if they were, were often badly degraded and used only by those who had little choice but to do so (Carmona 2017).

The opportunities and threats caused by the rapidly changing economic and social environment that is London seems to have led to a period of rapid innovation across the place-shaping continuum used to structure this paper, perhaps reflecting London's ongoing role as city with a tradition of public space experimentation and innovation. Whether, ultimately this will lead to better or worse, more or less democratic public space is yet to be seen. The evidence collected here suggests that even single cities can exhibit many, sometimes quite contrasting practices and outcomes.

Place-making plus

Looked at another way, this might be framed as a move from a 'place-making' to a 'place-making plus' position, where public space investment is more explicitly being used to help deliver on wider public policy goals – health, economic, social and environmental – whilst, conversely, investments in urban infrastructure are being used to create better public spaces. Yet even in a city which is now in a public space advocate mode, public spaces need to be nurtured and protected against pressures (if and where they exist) to undermine the key qualities that give their essential sense of 'publicness.' These relate variously to such issues as transparency in the provision, rights, and responsibilities associated with space; addressing the needs of haves and have-nots simultaneously; finding new innovative means to breathe life into some of the most challenging and relentlessly sub-standard city spaces, such as declining mixed retail streets; and the need to balance democratic rights and inclusiveness with threats from extremism, unfettered economic exploitation (where it exists) and the impact of cuts (where they persist) in the prosaic but vital stewardship of public spaces.

Over this the neo-liberal project persists, in London still obvious in its expression. It is interesting then that within this context, the evidence gathered suggests that (in toto) the direction of travel, if not every outcome, is more positive than negative (the crisis, it seems, has not gone to waste); and, despite the financial pressures, commitment to public space and its critical social role in society, has been maintained.

References

20's Plenty for Us. (2016) *20's Plenty for London Update.* www.20splenty.org/20_s_plenty_for_london_update_sep_16.

Adams, S. (2010) "Prince of Wales's Emotional Chelsea Barracks Letter Revealed." *The Telegraph*, 24 June. www.telegraph.co.uk/culture/culturenews/7850091/prince-of-waless-emotional-chelsea-barracks-letter-revealed.html.

Amin, A. (2009) "Collective Culture and Urban Public Space." *Public Space*, 2 June. www.publicspace.org/en/text-library/eng/b003-collective-culture-and-urban-public-space.

Bawden, T. (2016) "Britain's Public Parks Threatened with Privatisation as Cuts Stretch Council Budgets." *The Independent*, 27 June, www.independent.co.uk/news/uk/home-news/britains-public-parks-threatened-with-privatisation-as-cuts-stretch-council-budgets-10349408.html.

BBC News. (2015) "London's Parks 'Could Become Inaccessible to the Public.'" 21 June. www.bbc.co.uk/news/uk-england-london-33205239.

Biddulph, M. (2011) Urban Design, Regeneration and the Entrepreneurial City. *Progress in Planning*, 76 (2): 63–103.

Bone, J. (2012) The Neoliberal Phoenix: The Big Society or Business as Usual. *Sociological Research Online*, 17(2). www.socresonline.org.uk/17/2/16.html.

Bramall, R. (2013) *The Cultural Politics of Austerity*. Basingstoke: Palgrave Macmillan.

Carmona, M. (2012) The London Way: The Politics of London's Strategic Design. *Architectural Design*, 82(1): 36–43.

Carmona, M. (2014a) The Place-Shaping Continuum: A Theory of Urban Design Process. *Journal of Urban Design*, 19(1): 2–36.

Carmona, M. (2014b) "Neo-Liberal Public Space, in London (and Beyond)." *Urban Design Matters*, 12 January. www.matthew-carmona.com/2014/01/12/neo-liberal-public-space-in-london-and-beyond/.

Carmona, M. (2015a) London's Local High Streets: The Problems, Potential and Complexities of Mixed Street Corridors. *Progress in Planning*, 100: 1–84.

Carmona, M. (2015b) Re-Theorising Contemporary Public Space: A New Narrative and a New Normative. *Journal of Urbanism*, 8(4): 374–405.

Carmona, M. (2017) The Publicisation of Private Space. *Town & Country Planning*, 86(11): 494–496.

Carmona, M. and Wunderlich, F. (2012) *Capital Spaces, the Multiple Complex Spaces of a Global City*. Oxford: Routledge.

Cecil, N. (2016) "Sadiq Khan Plans to Create Ultra Low Emissions Zone in Central London by 2019." *Evening Standard*, 10 October. www.standard.co.uk/news/london/sadiq-khan-plans-to-create-ultra-low-emission-zone-in-central-london-by-2019-a3364961.html.

Cho, I.S., Heng, C.K., and Trivic, Z. (2016) *Re-Framing Urban Space: Urban Design for Emerging Hybrid and High-Density Conditions*. New York: Routledge.

Christodoulou, C. and Lada, S. (2017) Urban Design in the Neo-Liberal Era: Reflecting on the Greek Case. *Journal of Urban Design*, 22(2): 144–146.

Crisis. (2016) *The Homelessness Monitor: England*. www.crisis.org.uk/data/files/publications/homeless ness_monitor_england_2016_execsummary_v1.pdf.

De Magalhães, C. and Carmona, M. (2007) Innovations in the Management of Public Space, Reshaping and Refocusing Governance. *Planning Theory and Practice*, 7(3): 289–303.

De Magalhães, C. and Carmona, M. (2009) Dimensions and Models of Contemporary Public Space Management in England. *Journal of Environmental Planning and Management*, 52(2): 111–129.

Farrell, T. (2017) There Is Only One Thing Worse for Urban Design Than a Recession, and That Is a Boom. *Journal of Urban Design*, 22(2): 137–139.

Ferreri, M. (2015) The Seductions of Temporary Urbanism. *Ephemera*, 15(1): 181–191.

Garrett, B. (2015) "The Privatisation of Cities' Public Spaces Is Escalating. It Is Time to Take a Stand." *The Guardian*, 4 August. www.theguardian.com/cities/2014/dec/30/what-i-want-from-our-cities-in-2015-public-spaces-that-are-truly-public-boris-johnson-london.

Garrett, B. (2017) "These Squares Are Our Squares: Be Angry about the Privatisation of Public Space." *The Guardian*, 25 July. www.theguardian.com/cities/2017/jul/25/squares-angry-privatisation-public-space?mc_cid=8237a7a204&mc_eid=aa61c2d78d.

Gehl Architects. (2004) *Towards a Fine City for People, Public Spaces and Public Life*. London: Copenhagen, Gehl Architects.

Greater London Authority. (2015) *Natural Capital, Investing in a Green Infrastructure for a Future London*. Green Infrastructure Task Force Report.

Hambleton, R. (2011) A Jekyll and Hyde Localism Bill. *Town & Country Planning*, 80(1): 24–26.

Hastings, A., Bailey, N., Bramley, G., Gannon, M., and Watkins, D. (2015) The Cost of Cuts: The Impact on Local Government and Poorer Communities, Joseph Rowntree Foundation. www.jrf.org.uk/sites/default/files/jrf/migrated/files/Summary-Final.pdf.

Haughton, J. (2015) "Have We Reached 'Peak Coffee'? Eight Stirring Facts and Stats." *Chartered Management Institute*, 30 April. www.managers.org.uk/insights/news/2015/april/have-we-reached-peak-coffee-eight-stirring-facts-and-stats.

Jacobs, J. (1984 [1961]) *The Death and Life of Great American Cities: The Failure of Modern Town Planning*. London: Peregrine.

Jarvis, G. (2015) "End of the Road for London's Traditional Street Markets? Meet the Last Stallholder in Hackney's 'Waste.'" *The Guardian*, 17 April. www.theguardian.com/cities/2015/apr/17/decline-london-street-market-hackney-waste.

Joseph Rowntree Foundation. (2015) The Cost of the Cuts: The Impact on Local Government and Poorer Communities. www.jrf.org.uk/sites/default/files/jrf/migrated/files/summary-final.pdf.

Kamvasinou, K. (2017) Temporary Intervention and Long-Term Legacy: Lessons from London Case Studies. *Journal of Urban Design*, 22(2): 187–207.

Khan, S. (2016) "A Greener Cleaner London." www.sadiq.london/a_greener_cleaner_london.

Kim, S. (2017) "London's Traditional Markets; Managing Change and Conflict in Complex Urban Spaces." PhD diss., UCL.

Lehrer, U. and Laidley, J. (2008) Old Mega-Projects Newly Packaged? Waterfront Redevelopment in Toronto. *International Journal of Urban and Regional Research*, 32(4): 786–803.

Lipton, S. (2004) "Introduction." In: *The Value of Public Space: How High Quality Parks and Public Spaces Create Economic, Social and Environmental Value*, Woolley, H., Rose, S., and Carmona, M. (eds.). London: Cabe Space, p. 2.

London Assembly. (2011) *Public Life in Private Hands Managing London's Public Space*. London: Greater London Authority.

Mayer, M. (2013) First World Urban Activism: Beyond Austerity Urbanism and Creative City Politics. *City*, 17(1): 5–19.

Minton, A. (2014) "What I Want from Our Cities in 2015: Public Spaces That Are Truly Public." *The Guardian*, 30 December. www.theguardian.com/cities/2014/dec/30/what-i-want-from-our-cities-in-2015-public-spaces-that-are-truly-public-boris-johnson-london.

National Audit Office. (2014) *The Impact of Funding Reductions on Local Authorities*. www.nao.org.uk/wp-content/uploads/2014/11/impact-of-funding-reductions-on-local-authorities.pdf.

Office for National Statistics. (2016) *Retail Sales in Great Britain: Oct 2016.* www.ons.gov.uk/businessin dustryandtrade/retailindustry/bulletins/retailsales/oct2016.

Oldenburg, R. (2000) *Celebrating the Third Place.* New York: Marlowe & Company.

Outsmart. (2016) "Clear Channel Wins New Bus Shelter Ad Contract." www.outsmart.org.uk/news/ clear-channel-wins-tower-hamlets-bus-shelter-advertising-contract.

Peck, J. (2012) Austerity Urbanism, American Cities under Extreme Economy. *City,* 16(6): 626–655.

Peck, J. and Tickell, A. (2002) Neoliberalizing Space. *Antipode,* 34(3): 380–404.

Pitcher, G. (2013) "International Money Boosts London Property Market." *Architects' Journal,* 18 January. www.architectsjournal.co.uk/news/international-money-boosts-london-property-market/8641408. article.

Plimmer, J. (2016) "UK Parks Endangered by Council Spending Cuts, MPs Warned." *Financial Times,* 13 October. www.ft.com/content/39164258-8634-11e6-a29c-6e7d9515ad15.

Roberts, M. (2017) Urban Design, Central London, and the 'Crisis' 2007–2013: Business as Usual? *Journal of Urban Design,* 22(2): 150–166.

Savills. (2015) *Spotlight London Infrastructure: Connecting Opportunities.* London: Savills World Research.

Strebel, I. and Silberberger, J. (2017) *Architecture Competition: Project Design and the Building Process.* Abing-don: Routledge.

Syal, R. and Asthana, A. (2017) "Parliament Security to Be Reviewed by Police and Authorities." *The Guardian,* 23 March. www.theguardian.com/uk-news/2017/mar/23/parliament-security-reviewed-after-attack-michael-fallon.

Tonkiss, F. (2013) Austerity Urbanism and the Makeshift City. *City,* 17(3): 312–324.

Transport for London (TfL). (2013a) *Better Streets Delivered, Learning from Completed Schemes.* www.urban designlondon.com/wordpress/wp-content/uploads/better-streets-delivered-web-version.pdf.

Transport for London (TfL). (2013b) *Street Types for London.* www.tfl.gov.uk/info-for/boroughs/street-types?intcmp=24919.

Transport for London (TfL). (2016) *Better Streets Delivered 2: Learning from Completed Schemes.* Forthcoming.

Transport for London (TfL). (2017) *Healthy Streets for London.* www.content.tfl.gov.uk/healthy-streets-for-london.pdf.

Watts, M. (2014) "90 Per Cent Rise in Beggar Arrests Is Helping London Drug Addicts, Charity Says." *Evening Standard,* 25 July. www.standard.co.uk/news/uk/90-per-cent-rise-in-beggar-arrests-is-help ing-london-drug-addicts-charity-says-9629182.html.

Williams, D. (2016) "Londoners Win £4 Million in Pothole Payouts in One Year." *Evening Standard,* 13 April. www.standard.co.uk/news/london/londoners-win-4m-in-pothole-payouts-in-one-year-a3224306.html.

Wrigley, N. and Lambiri, D. (2014) *High Street Performance and Evolution: A Brief Guide to the Evidence.* www.thegreatbritishhighstreet.co.uk/pdf/gbhs-highstreetreport.pdf.

12

VISIBILITY IN PUBLIC SPACE AND SOCIALLY INCLUSIVE CITIES

A new conceptual tool for urban design and planning

Ceren Sezer

Introduction

Cities can be characterized by the qualities of their public spaces. Potentially, urban public spaces can be open and accessible – both physically and perceptually – for all, no matter people's differences of age, gender, cultural background, economic status, ethnicity, or belief. Public space offers opportunities to see and be seen, to observe and be observed, to be noticed and be recognized. By being a ground for everyday life in the city, public spaces reveal the ways that they are used and appropriated by urban dwellers. Consequently, public spaces are potentially able to promote socially inclusive cities, allowing for diversity of voices and users from all social groups.

The inclusive features of public spaces are under pressure as a consequence of trends toward the commodification of urban development and the privatization of urban spaces (Madanipour 2010; Nikšič and Sezer 2017). Market-oriented growth models have revalorized public space as centers of urban leisure and consumption, leading to trends toward commodification. City authorities, planners, and developers have used public space as a marketing tool to create attractive places and brand cities through appealing images (Lash and Urry 1994). "Urban culture has become in itself a commodity," an object of "touristic gaze," which can be experienced the most in cities' public spaces (Fainstein 2007, 4; Urry 2002; Hall and Rath 2007; Madanipour 2010). Commodified spaces include historical city centers, with their iconic museums, festive marketplaces, concert halls, waterfronts, and characteristic architecture (Zukin 1995; Janssens and Sezer 2013).

Similar transformations have also been observed in streets of immigrant neighborhoods like Chinatowns and African and Turkish neighborhoods in major western European cities. Immigrant streets manifest cultural differences through their shops, restaurants, religious places, and organizations with distinctive names, signs, and unusual products and spatial practices. They offer an "authentic" experience of the city, exhibiting cultural features of minorities. Seeing an opportunity in this, municipalities and city commentators have promoted them as economically

137

valuable places for tourism (Hall and Rath 2007). This has led to the commodification of immigrant cultures for the purpose of cultural consumption and tourism, triggering processes of commercial and residential displacement in central neighborhoods (Diekmann and Smith 2015; Sezer and Fernandez Maldonado 2017; Sezer 2018).

Public spaces have been a matter of attention in academic and policy circles and, in many cases, considered as a solution to overcome the increasing social fragmentation in cities, promoting tolerance and social cohesion (Sezer 2019). The UN Habitat's (2016) New Urban Agenda commits itself to promoting public spaces as drivers of sustainable development, encouraging open and accessible spaces to foster socially inclusive cities. However, the evaluation of inclusive public spaces and its implications for urban design and planning have been disregarded. This chapter addresses this gap by introducing the concept of visibility as a useful tool to diagnose and assess inclusive public spaces, especially in the context of urban transformation processes.

The concept of visibility refers to the visual perception of the observable features of distinctive urban groups in public space, which give evidence of how these groups engage with, shape, and construct public space. The visibility of distinctive urban groups on the street manifests the rights of these groups to participate in public life of the city, which is a key feature of socially inclusivity. Consequently, the presence and changes in visibility of urban groups in public space is highly political, raising issues in relation to just and unjust urban conditions. The main question of this chapter is: how can the visibility aspect of public space be a useful conceptual tool to diagnose and assess inclusiveness of public space?

The following section presents the theoretical underpinnings of the chapter. The third section outlines the empirical research focused on the presence and change in the visibility of distinctive Turkish amenities in Amsterdam in 2007 and 2016 in the context of urban transformation. The final section presents reflections and conclusions.

Inclusive public space, visibility in public space, and immigrant amenities

Inclusive public spaces

The ideal of inclusive public space is undeniably linked to the rise of the modern city and public life. Public life, in an inclusive public space, is the political ground for "recognizing and affirming diverse social groups by giving political representation to these groups and celebrating their distinctive characteristics and cultures" (Young 2000, 240). But in real life, public space has never been entirely free and inclusive, nor was it ever equally available to all. The politics of public space determines "who and what come to count as being truly 'public' and/or 'political' as well as how and where they can come to count" (Lees 1998, 232). Since ancient times, "various social groups – the elderly and the young, women and members of sexual and ethnic minorities – have, in different times and places, been excluded from public space or subject to political and moral censure" (Jackson 1998, 173). Likewise, movement and migration of people have generated conflicts and contestation between newcomers and old residents, and individuals and institutions (Hou 2013; Sezer, forthcoming).

Nevertheless, inclusive public spaces offer multiple opportunities for negotiation and exchange, providing mechanisms for the recognition and expression of the voices and perspectives of vulnerable groups. This perspective on inclusive public spaces is clearly associated with the right of citizens to the participation in and appropriation of their shared urban environment (Purcell 2002).

Reviewing the literature on the inclusivity of public space, Akkar (2005) has developed an inclusivity model based on four interrelated dimensions of access:

Social access, the perceived openness of public spaces, or the feeling of belonging and welcoming for different social groups. This is fundamentally related to the sociability in public spaces, binding people to people, and people to places, providing a sense of familiarity, belonging, and safety (Mehta 2019).

Access to activities, the presence of a balanced diversity of activities and social groups. A series of indicators to promote an active public life in public spaces have been advanced, including variety of land uses, patterns of opening hours, presence of street markets, spaces that enable people watching, open street facades, etc. (Montgomery 1998; Mehta 2013).

Physical access, the bodily access of public spaces by different individuals and groups including those physically, sensory, or mentally limited and impaired groups. Absence of physical barriers, availability of various modes of access, and connectivity of public spaces improve physical access (Carr et al. 1992; Hamraie 2017).

Access to information, discussions and intercommunications, availability and accessibility of information about events and activities and "the design, planning, development, management, control and use processes on public spaces" (Akkar Ercan and Memlük 2015, 197).

Visibility in public space and immigrant amenities

In a broad sense, visibility can be understood as the state of being able to see or be seen. Visibility in public space is understood here as the visual perception of the observable features of individuals and groups, which gives evidence of their lived experiences and how they engage with, shape, and construct the built environment within the course of everyday life (Brighenti 2007; Cancellieri and Ostanel 2015; Hatuka and Toch 2017; De Backer 2019). Visibility in public space can be expressed in different ways, through corporal manifestations such as clothing and hair styling; or distinctive cultural events, gatherings of urban groups, such as religious and cultural festivals; or the distinctive amenities of immigrant neighborhoods.

The concept of visibility has received attention in early writings about the modern city, which celebrate urban social life as a ground for surprises, excitement, and stimulation (Wirth 1938). Visibility is specifically associated with public spaces, as they are identified as spaces of "appearance" (Arendt 1958). Sennett (1990) has argued that visibility in public space is a fundamental aspect of urban social life, as it offers opportunities to city inhabitants to experience individuals and groups from different life choices and backgrounds. In this way, visibility and inter-visibility in public spaces may generate unease between different urban groups, yet they are essential for developing a sense of civility among city inhabitants (Sennett 1970; Loukaitou-Sideris and Ehrenfeucht 2009). From this perspective, visibility is considered a key feature of public spaces, which characterize their "public" character as open and accessible (Arendt 1998 [1958]; Sennett 1990; Lofland 1998; Brighenti 2007).

The role of visibility has been blurred since the advent of Information Communication Technology and the related mass circulations of images (Aitken and Lukinbeal 1998; Brighenti 2007; Hatuka and Toch 2017; Sezer, forthcoming). But visibility is still useful to study the ways that urban groups shape public spaces for their own needs. This is particularly relevant in the city quarters in which immigrants have settled and established their businesses and social connections. In particular, immigrant amenities offer various opportunities to observe and experience immigrant cultural expressions (Göle 2011; Nell and Rath 2012; Hall 2015). Their different ways of visibility can provide empirical evidence for assessing the level of inclusivity of public spaces. Visibility can be then related to the four dimensions of accessibility from Akkar's (2005) inclusivity model of public spaces:

For *social access*, the visibility of immigrant amenities provides multiple opportunities for promoting *encounters and civility* among immigrants, as well as immigrants and other groups. Immigrant amenities promote regular contacts among immigrants, helping them to develop their sense of belonging to a place and community (Carr et al. 1992; Kusenbach 2006; Akkar Ercan and Memlük 2015; Sezer 2019). In this way, they may generate civility, mutual respect, and recognition between immigrant groups and other groups, without neglecting their differences (Sennett 1970; Young 1990).

In terms of *access to activities*, the visibility of amenities provides empirical evidence for the *use and user diversity* of public spaces. Immigrant commercial amenities are often small-scaled independent businesses, which add diversity of uses at street level, as opposed to chain stores. Their specialization and urban location influence the diversity of their clients (e.g. residents, visitors, and tourists).

In terms of *physical access*, the visibility of immigrant amenities is inherently linked to the *physical setting* of public spaces. Physical access also relates to visual accessibility, in terms of legibility—by which people can understand their environment—and visual permeability, allowing the visual experience of public spaces (Bentley et al. 1985). Space syntax tools analyze street networks according to their physical and visual accessibility, relating them to walkability, safety, and sociability (Dalton and Bafna 2003).

In terms of *access to information, discussions and intercommunications*, the visibility of immigrant amenities may be helpful as a means to realize immigrants' will for *participation and appropriation* of public spaces. The use of public space for their needs of amenities manifests their participation in the symbolic, physical, social, and political construction of public space (Sezer, forthcoming).

Empirical research: fieldwork, context, and findings

Fieldwork and context

This section addresses the transformation of immigrant neighborhoods in the context of urban renewal processes. It focuses on the visibility of immigrant amenities in immigrant neighborhoods of Amsterdam, studying their presence and changes at a street level in the period between 2007 and 2016. The visibility of immigrant amenities is used as a proxy for the presence and appropriation of public spaces by immigrant groups through their distinctive signs, languages, specialized products, cuisines, and unconventional uses. Turkish amenities are selected as the object of study, as Turkish immigrants are one of the largest immigrant groups in Amsterdam and generally considered a vulnerable population group due to their high welfare dependency. The inclusion of Turkish immigrants into mainstream society has been widely addressed both in public and academic debates (Yücesoy 2006; Crul et al. 2012).

Since the 1980s, the Amsterdam Municipality has been marketing its image with the slogan Iamsterdam (Iamsterdam 2016). The city has become a major tourist destination and a large real-estate bubble has gained momentum since 2013 (UBS 2019). Real-estate market trends and local planning policies and practices have had a significant role in the city's transformation processes, especially in inner-city neighborhoods, leading to gentrification

(Uitermark et al. 2007). Immigrant neighborhoods, more specifically their public spaces, have been drastically affected by these trends.

To approach the concept of visibility from four perspectives presented above—*encounters and civility; use and user diversity; physical setting*; and *participation and appropriation*—this section analyses the social and spatial characteristics of immigrant amenities. Social characteristics refer to the opportunities that amenities offer for sociability among immigrants, and between immigrants and other groups, which is relevant to the first theme, *encounters, and civility*. Spatial characteristics at the city level refer to the occupation of public space by immigrant amenities in terms of their location (center, outskirts), and their spatial distribution in relation to immigrant residential concentration areas. At the neighborhood level, it focuses on the ways that immigrants are using and changing streets according to their needs. The last three themes relate to the spatial characteristics of immigrant amenities.

The empirical research has been conducted in five successive steps:

1. Data collection about the location and general observation of immigrant amenities at city level: the locations of Turkish-related immigrant amenities, identified through their language, products, and cuisines, were mapped. This was used to identify the clusters of amenities in specific streets, and later for the selection of two cases for in-depth analysis. The observations were carried out by walking and cycling from October to December 2007.
2. Data collection about the social characteristics of selected amenities at neighborhood level: the objective was to identify whether and how these amenities promote social encounters between the immigrant community and other groups (the public realm), or within the immigrant community (the parochial realm). Observations, photography, and audio recording were used to collect data about these social characteristics. Seven in-depth and 40 unstructured interviews were conducted with owners and visitors of the amenities. For all interviews, the questionnaires included open-ended questions related to the frequency of visits and time spent in the amenities; location of socialization with friends in Amsterdam; country of origin of clients; etc.
3. The selection of two streets, which could provide relevant information about the presence and changes in the visibility of immigrant amenities: these streets were chosen according to their location (inner-city/outskirts); changes in the residential concentration of immigrant population of the neighborhoods in which they are located (increasing/decreasing); types of users that the street is catering to (residents/visitors/tourists). For the latter, a map was produced showing overlapping amenities associated with residents, citywide visitors, and tourists. This was done to evaluate the attractiveness of streets for tourism-related urban development.
4. Identification of the changes in the two selected streets in terms of diversity and vitality: to measure diversity, the observations included the functions and time schedules of all existing street amenities, including Turkish amenities. To do so, two rounds of fieldwork were conducted in November 2007 and November 2016. To measure street vitality, the presence of people at different times of the day and night, and different types of street uses were counted, recording people's age and gender, precise location, and the activities they engaged in, such as gathering and lingering. Both observations were carried out during weekdays (Wednesdays) and weekends (Saturdays), during the busiest time: morning, afternoon, and evening peak hours. The data were mapped, annotated, and photographed.

5. Data on the presence of and changes in immigrant amenities in the selected streets were mapped and documented in field visits done in November 2007 and November 2016. The fieldwork activities included the observation, photographing, and annotating the changes. Additionally, a set of six interviews were conducted in 2016 with shop owners to learn about their experiences in relation to the changes in their businesses in the context of urban transformations. The notes from these interviews included the approximate age, and gender of the shop owners; the duration of their businesses; the strategies that they used to adapt their businesses to the on-going urban transformation process; and the future prospects of their businesses.

6. The final step interpreted the research findings about the changes in the presence and absence of the immigrant amenities in relation to the inclusiveness of public spaces.

Findings

Regarding spatial characteristics, this study found that most Turkish amenities were located in post-war housing estates built on the outskirts of Amsterdam. In inner-city neighborhoods, these amenities were located in affordable housing areas developed around the beginning of the 20th century and post-war social housing areas. The Turkish amenities were mostly commercial amenities catering to locals located on the main and secondary shopping streets. There were also some exceptional examples, which were amenities catering to citywide users. These were specialized shops, such as Turkish music shops and restaurants offering regional cuisines. Communal amenities, including mosques and teahouses were not very visible at the street level, as they were mostly located on the backstreets of the neighborhood. Figure 12.1 presents some examples from Turkish amenities and Figure 12.2 presents the streets in which these amenities clustered.

At the outskirts of the city, amenities were predominantly commercial and mostly located on main and secondary streets of residential neighborhoods. They mainly catered to locals, although some restaurants attracted citywide customers. Communal amenities occupied larger areas, clustering a few other amenities, such as immigrant organizations, teahouses, and shops.

At the neighborhood level, the study found that Turkish amenities presented significant differences in terms of their personalization and legibility, clearly distinguishing between communal and commercial amenities. Commercial amenities shared personalized façades and interiors, with large name boards, window displays, and a profusion of products. Their entrances were clearly legible, suggesting their functions, as they stimulate regular and informal uses, such as gathering and chatting, which promoted the visibility of these amenities.

Communal amenities may be less legible if they belong to a small community and lack financial means to rent/construct their own buildings with certain architectural styles. Among these amenities, mosques promoted a large variety of unplanned public space uses, such as men gathering in front of mosques before and after prayer times, which promoted the legibility of these amenities and increased their visibility. Figure 12.2 presents some of the general characteristics of Turkish amenities in terms of their personalization and legibility.

The study revealed significant differences between commercial and communal amenities in terms of their social characteristics as related to their capacity to promote social contacts within the immigrant and larger community. Analysis of parochial and public realms found

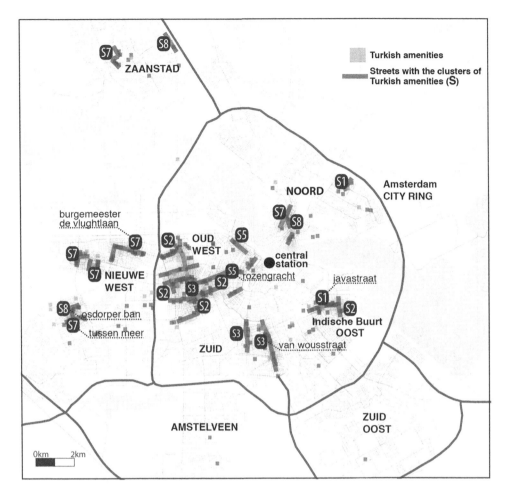

Figure 12.1 The streets with the clusters of Turkish amenities in Amsterdam and the location of the case
streets, Burgemeester de Vlugtlaan and Javastraat

Image credit: Ceren Sezer

that commercial amenities promoted both. This was most obvious in Turkish food shops
and restaurants, which stimulate daily chats between immigrant owners and their clients.
Supported by long working hours and their attractive fusion cuisines, these interactions promoted both public and parochial realms.

Turkish communal amenities, especially mosques, which are used almost exclusively by
Turkish people, were central in their parochial life. Within this parochial realm, communal
amenities were gendered spaces, as well as differentiated according to different political and
religious views. By contrast, Turkish secular cultural organizations, such as those teaching
the Turkish language, were vivid places of encounters between Turkish and other groups,
promoting the public realm.

The study focused on two commercial streets with clusters of Turkish amenities in
Amsterdam to study how they changed in terms of diversity and vitality, and how Turkish
amenities relate to these changes in the context of urban transformation between 2007 and
2016. These streets are Javastraat and Burgemeester de Vlugtlaan (see Figure 12.1).

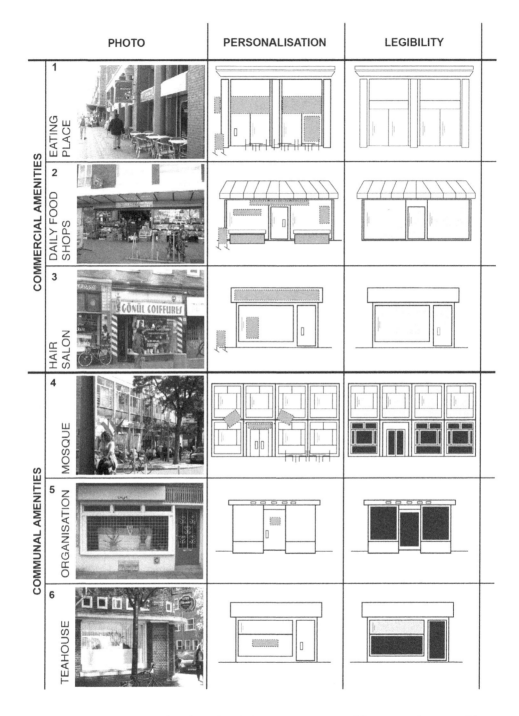

Figure 12.2 Neighborhood level analysis of Turkish amenities in terms of their personalization and legibility at street level

Image credit: Ceren Sezer

Javastraat

Javastraat is the main shopping street of the Indische Buurt neighborhood, which was built in the beginning of the 20th century as a social housing area for skilled workers. In the 1980s the neighborhood became gradually multicultural, and in 2007, the local municipality and housing associates developed a neighborhood renewal program for the improvement of housing and retail streets. This program visioned Javastraat as a trendy shopping street of Amsterdam.

In 2007, Javastraat was characterized by immigrant shops—specifically Turkish—which increased beginning in the 1980s after the arrival of Turkish migrants to the neighborhood. Besides few national chain shops and supermarkets, shops in Javastraat were mainly independent businesses, such as bakeries, groceries, clothing shops, and eateries. More than half of these independent businesses were Turkish, specializing in Turkish products, such as food and clothing for Islamic fashion. The advertisement boards of these shops, with different languages and their appealing windows with a wide variety of products, enriched the street diversity in Javastraat. Figure 12.3 illustrates the variety of Turkish amenities in black dots in Javastraat.

In terms of vitality, Javastraat was an attractive destination both for residents and visitors of the neighborhood. Different kinds of street use include sitting, standing watching, gathering,

Figure 12.3 Turkish amenities in Javastraat in 2007

Image credit: Ceren Sezer

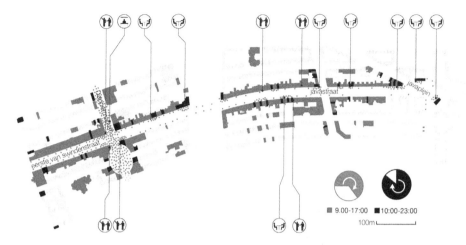

Figure 12.4 User intensity and user activities in Javastraat, 2007

Image credit: Ceren Sezer

and socializing. Dappermarket, a local street market offering cheap food and clothing products contributed to street vitality especially during its opening hours (9am–4 pm). After the market was closed, Javastraat was less active, except around few shops and eateries, which were Turkish and open late. The presence of Turkish women, who were clients of these shops, and Turkish men visiting teahouses located in Javastraat, generated a vivid street life. Figure 12.4 presents the street amenities and their general opening hours, and street user intensity and user activities in Javastraat.

By 2016, Javastraat had changed from an immigrant street into a street attractive to citywide visitors and tourists. The population profile of the neighborhood changed due to increasing housing prices. The transformations include a new street design with widened sidewalks, new sitting furniture, and a decrease in parking. There was a significant increase in the number of trendy eateries and pubs, while the number of immigrant shops decreased. Two-thirds of Turkish amenities were closed or converted into other businesses. For example, one of the Turkish bakeries changed its products and became specialized in organic food. As a few shopkeepers stated, owners were not included in the decision-making process about Javastraat's transformation; they were only invited to the meetings by the municipality to be informed about the urban renewal project.

In terms of vitality, some of the new businesses in Javastraat, especially cafes, restaurants, and pubs, have longer opening hours that changed the active street use. The presence of young customers in these new premises gave a new look to the street and increased the use of the street both during the day and at nighttime. Widened sidewalks, reduced car parking, and new street furniture promoted street use.

Burgemeester de Vlugtlaan

Burgemeester de Vlugtlaan is the main shopping street of Geuzenveld-Slotermeer neighborhood, which was built in the 1950s. In the 1970s, the neighborhood was attractive for young middle-class families; however, its demographic composition changed with the arrival of immigrant groups. In 2000, the local government and developers initiated a new urban renewal program, which recognized the historical value of the neighborhood and in 2007 nominated it as a "Municipal conservation site." A new museum was established to nominate a neighborhood planner and designer. These interventions have improved the image of the neighborhood and increased its cultural value.

In 2007, Burgemeester de Vlugtlaan was characterized by immigrant shops. Half of the 40 businesses situated on the street were Turkish, including food, furniture, and clothing shops. A Turkish mosque, located in a receded location, clustered with a Turkish grocery shop, a kick-box salon and organization for children and women. Figure 12.5 presents the variety of Turkish amenities in black dots along Burgemeester de Vlugtlaan.

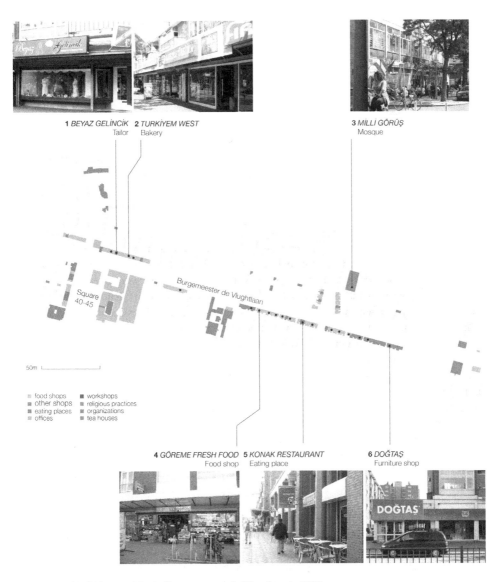

Figure 12.5 Turkish amenities in Burgemeester de Vlugtlaan in 2007

Image credit: Ceren Sezer

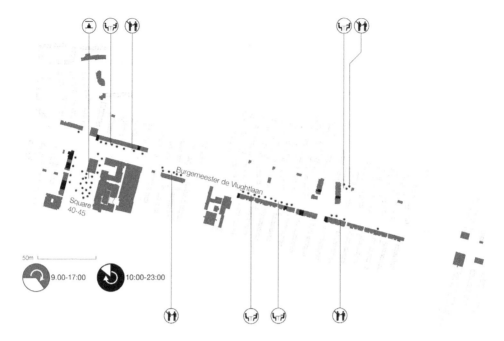

Figure 12.6 User intensity and user activities in Burgemeester de Vlugtlaan, 2007

Image credit: Ceren Sezer

In terms of vitality, the daily market, which was open from 9 am to 4 pm, had an influence on the use intensity and the user behavior of the street. At night, the street was almost empty except for the parts around the Turkish cafés and restaurants where youngsters gathered until late in the night. Similarly, the Turkish mosque brought vitality to the street, especially before and after prayer times. The mosque was a gendered place, almost only used by men. As Figure 12.4 presents the street amenities and their general opening hours, and street user intensity and activities on Javastraat, Figure 12.6 highlights the same along Burgemeester de Vlugtlaan.

In 2016, changes in the diversity and vitality of Burgemeester de Vlugtlaan were not significant in comparison to Javastraat; the Turkish shops were still the dominant feature of the streetscape. There were a few new Turkish eateries, which brought a new look to the street with their modern design and the museum increased street use and intensity during both day and nighttime. Additionally, the opening of a Turkish restaurant specializing in halal fast food enhanced street vitality, especially at night, and brought an extra buzz to the street life.

Conclusions

The findings of this research have been useful to explain fundamental issues to answer the main question: how can the visibility aspect of public space be a useful conceptual tool to diagnose and assess inclusive public spaces?

The empirical assessment of the immigrant amenities along two streets in Amsterdam from 2007 to 2016 shows diminishing Turkish amenities in the inner-city case study street, which has led to a significantly lower visibility of Turkish migrants. As such, Javastraat has lost its image as a "characteristic" immigrant commercial street and transformed into a street dominated by trendy cafes and restaurants, missing most of its distinctive cultural features.

As the visibility of distinctive urban groups in public space is strongly linked to their presence and involvement in everyday life of the city, the diminishing visibility of the Turkish amenities suggests a negative impact on some key features of inclusive public spaces, including diversity, participation and appropriation, social encounters and civility. Additionally, the physical features of the built elements shaped by contextual (e.g. geographical and political settings) and spatial factors (location, centrality as well as personalization and legibility) influence the inclusive features of public spaces.

Regarding the diversity of inclusive public spaces, this study provides empirical evidence on the ways that Turkish amenities enrich the diversity of the streets in terms of their types, distinctive functions, open hours and user groups, specifically residents and visitors. Turkish shops and restaurants, which are very often small-scale independent businesses, cater to immigrants, but may have a varying clientele depending on their location and specialization. Turkish mosques manifest their cultural differences through their own architectural styles and spatial practices, and cater to specific groups, mostly Turkish migrants.

The research shows that Turkish amenities promote social encounters amongst Turkish immigrants as well as with other groups. The function and location of amenities determine different types of social encounters. Turkish amenities located in the inner city tend to generate more social encounters between Turkish people and other groups, than those amenities located in the outskirts of the city. Turkish communal amenities tend to promote social encounters amongst Turkish immigrants, since they target a very specific user group.

What is still not certain is whether these social encounters generate forms of civility between Turkish immigrants and others. Further research to examine the role of immigrant amenities to promote civility in terms of mutual recognition and respect would be necessary to complement and support these findings.

The analysis presented in this chapter shows that visibility can be operationalized by studying the spatial and social characteristics of immigrant amenities in public spaces. Measuring and documenting the spatial (at city and neighborhood level) and social (social life of parochial and public realm) characteristics of immigrant amenities, the visibility of culturally distinctive groups in public space can be compared in a synchronic and diachronic way.

The results have important implications for urban planning, design theory, and practice. In terms of theory, the research claims that visibility is not an abstract concept; it has practical implications for assessing the qualities of inclusive public spaces. In this way, the research adds to urban design theory by introducing a new conceptual tool to study and assess the inclusive features of public spaces.

The implications for practice are also significant, as the findings indicate that urban renewal and planning schemes' decisions and interventions, play a fundamental role in the formation of inclusive features in public spaces. Comparing the findings of the two case studies showed that urban planners and designers should be aware and well informed about their own role in urban transformation processes in culturally sensitive neighborhoods. The study also shows that the visibility of distinctive groups can and should be considered to design public spaces that contribute to create and sustain inclusive practices in the built environment.

Acknowledgements

The author would like to express her sincere appreciation to the Editors for their generous comments and support during the review process.

References

Aitken, S. and Lukinbeal, C. (1998) Of Heroes, Fools and Fisher Kings: Cinematic Representations of Street Myths and Hysterical Males in the Films of Terry Gilliam. In: *Images of the Street: Planning, Identity, and Control in Public Space*, Fyfe, N. (ed.). New York: Routledge, pp. 141–159.

Akkar Ercan, M. and Memlük, N.O. (2015) More Inclusive than Before? The Tale of a Historic Urban Park in Ankara, Turkey. *Urban Design International*, 20(3): 195–221.

Akkar, M. (2005) Questioning Inclusivity of Public Spaces in Post-Industrial Cities: The Case of Haymarket Bust Station, Newcastle upon Tyne. *METU Journal of Faculty of Architecture*, 22(2): 1–24.

Arendt, H. (1998 [1958]). *The Human Condition*. Chicago, IL: University of Chicago Press.

Bentley, I., McGlynn, S., Smith, G., Alcock, A. and Murrain, P. (1985) *Responsive Environments: A Manual for Designers*. London: Architectural.

Brighenti, A.M. (2007) Visibility: A Category for the Social Sciences. *Current Sociology*, 55(3): 323–342.

Brighenti, A.M. (2010) The Publicness of Public Space: On the Public Domain. *Quaderno*, 49(March): 1–56. Trento: Università Degli Studi Di Trento.

Cancellieri, A. and Ostanel, E. (2015) The Struggle for Public Space: The Hyperdiversity of Migrants in the Italian Urban Landscape. *City*, 19(4): 499–509.

Carr, S., Francis, M., Rivlin, L.G. and Stone, A.M. (1992) *Public Space*. Cambridge: Cambridge University Press.

Crul, M., Schneider, J. and Lelie, F. (eds.). (2012) *The European Second Generation Compared: Does the Integration Context Matter?* Amsterdam: Amsterdam University Press.

Dalton, R.C. and Bafna, S. (2003). The Syntactical Image of the City: A Reciprocal Definition of Spatial Elements and Spatial Syntaxes. 4th International Space Syntax Symposium, London.

De Backer, M. (2019) Regimes of Visibility: Hanging Out in Brussels' Public Spaces. *Space and Culture*, 22(3): 308–320.

Diekmann, A. and Smith, M.K. (eds.). (2015) *Ethnic and Minority Cultures as Tourist Attractions Book*, Vol. 65. Bristol: Channel View Publications.

Fainstein, S.S. (2007) Tourism and the Commodification of Urban Culture. *The Urban Reinventors*, 2: 1–20.

Göle, N. (2011) The Public Visibility of Islam and European Politics of Resentment: The Minarets –Mosques Debate. *Philosophy and Social Criticism*, 37(4): 383–392.

Hall, C.M. and Rath, J. (2007) Tourism, Migration and Place Advantage in the Global Cultural Economy. In: *Tourism, Ethnic Diversity and the City*, Rath, J. (ed.). New York: Routledge, pp. 1–24.

Hall, S.M. (2015) Migrant Urbanism: Ordinary Cities and Everyday Resistance. *Sociology*, 49(5): 853–869.

Hamraie, A. (2017) *Building Access: Universal Design and the Politics of Disability*. Minneapolis, MN: University of Minnesota Press.

Hatuka, T. and Toch, E. (2017) Being Visible in Public Space: The Normalisation of Asymmetrical Visibility. *Urban Studies*, 54(4): 984–998.

Hou, J. (2013) *Transcultural Cities: Border-Crossing and Placemaking*. New York: Routledge.

Iamsterdam. (2016) "Diversity in Amsterdam." Accessed May 2019. www.iamsterdam.com/en/local/about-amsterdam/people culture/diversity-in-the-city.

Jackson, P. (1998) Domesticating the Street, the Contested Spaces of the High Street and the Mall. In: *Images of the Street: Planning Identity and Control in Public Space*, Fyfe, N. (ed.). New York: Routledge, pp. 176–191.

Janssens, F. and Sezer, C. (eds.). (2013) Special Issue: Marketplaces as Urban Development Strategies. *Built Environment*, 39(2): 165–316.

Kusenbach, M. (2006) Patterns of Neighbouring: Practicing Community in the Parochial Realm. *Symbolic Interaction*, 29(3): 279–306.

Lash, S. and Urry, J. (1994) *Economies of Signs and Space*. London: Sage.

Lees, L. (1998) Urban Renaissance and the Street: Spaces of Control and Contestation. In: *Images of the Street: Planning Identity and Control in Public Space*, Fyfe, N. (ed.). New York: Routledge, pp. 236–253.

Lofland, L.H. (1998) *The Public Realm: Exploring the City's Quintessential Social Theory*. Hawthorne: Aldine de Gruyter.

Loukaitou-Sideris, A. and Ehrenfeucht, R. (2009) *Sidewalks: Conflict and Negotiation over Public Space*. Cambridge, MA: MIT Press.

Madanipour, A. (ed.) (2010) *Whose Public Space? International Case Studies in Urban Design and Development.* London: Routledge.

Mehta, V. (2013) *The Street: A Quintessential Social Public Space.* New York: Routledge.

Mehta, V. (2019) Streets and Social Life in Cities: A Taxonomy of Sociability. *Urban Design International,* 24(1): 16–37.

Montgomery, J. (1998) Making a City: Urbanity, Vitality and Urban Design. *Journal of Urban Design,* 3(1): 93–116.

Nell, L. and Rath, J. (2012) *Ethnic Amsterdam: Immigrants and Urban Change in the Twentieth Century.* Amsterdam: Amsterdam University Press.

Nikšič, M. and Sezer, C. (2017) Special Issue: Public Space and Urban Justice. *Built Environment,* 43(2): 161–304.

Purcell, M. (2002) Excavating Lefebvre: The Right to the City and Its Urban Politics of the Inhabitant. *GeoJournal,* 58(2–3): 99–108.

Sennett, R. (1970) *The Uses of Disorder: Personal Identity and City Life.* New York: W. W. Norton.

Sennett, R. (1990) *The Conscience of the Eye: The Design and Social Life of Cities.* New York: Knopf.

Sezer, C. (2018) Public Life, Immigrant Amenities and Socio-Cultural Inclusion. *Journal of Urban Design,* 23(6): 823–842.

Sezer, C. (2019) Visibility of Turkish Amenities: Immigrant Integration and Social Cohesion in Amsterdam. In: *Public Space Design and Social Cohesion,* Aelbrecht, P. and Stevens, Q. (eds.). New York: Routledge, pp. 220–241.

Sezer, C. (forthcoming) Visibility as a Conceptual Tool for the Design and Planning of Democratic Streets. *Space and Culture.*

Sezer, C. and Fernandez Maldonado, A.M. (2017) Cultural Visibility and Urban Justice in Immigrant Neighbourhoods of Amsterdam. *Built Environment,* 43(2): 193–213.

UBS. (2019) "UBS Global Real Estate Bubble Index 2019." www.ubs.com/global/en/wealth-manage ment/chief-investment-office/life-goals/real-estate/2019/global-real-estate-bubble-index-2019.html.

Uitermark, J., Duyvendak, J.W. and Kleinhans, R. (2007) Gentrification as a Governmental Strategy: Social Control and Social Cohesion in Hoogvliet, Rotterdam. *Environment and Planning A,* 39(1): 125–141.

UN-Habitat. (2016) "Habitat III, New Urban Agenda." www2.habitat3.org/bitcache/ 99d99fbd0824de50214e99f864459d8081a9be00?vid=591155&disposition=inline⊕view.

Urry, J. (2002) *The Tourist Gaze.* London: Sage.

Wirth, L. (1938) Urbanism as a Way of Life. *American Journal of Sociology,* 44(1): 1–24.

Young, I. (1990) *Justice and the Politics of Difference.* Princeton, NJ: Princeton University Press.

Young, I.M. (2000) *Inclusion and Democracy.* Oxford: Oxford University Press.

Yücesoy, E.U. (2006) *Everyday Urban Public Space: Turkish Immigrant Women's Perspective.* Amsterdam: Het Spinhuis.

Zukin, S. (1995) *The Cultures of Cities.* Oxford: Blackwell.

13

(POST?)-COLONIAL PARKS

Urban public space in Trinidad and Zanzibar

Garth Myers

Introduction

This chapter focuses on public spaces in the capital cities of two former British colonies, Trinidad and Zanzibar, discussing colonialism's central urban parks and gardens in Port of Spain and Zanzibar. It examines the respective colonial botanical gardens and main public park of each city, and the roles these played in shaping colonial society. This is followed by a discussion of alternative landscapes and urban environments produced by the colonized peoples in both places, including into the post-colonial era. One certainly sees here the role of British colonialism in producing urban public space and nature in its interests. Yet ordinary residents often reframed landscapes into something quite different. While colonial mindsets endured into the post-colonial period, they were refracted into different urban cultures of nature in the two cities. The post-colonial stories of urban public space in Trinidad and Zanzibar offer us a manifestation of how divergent urban contexts grew into post-colonial urbanisms—heterogeneous yet comparable—in the 21st century. The chapter begins with colonial Trinidad, followed by colonial Zanzibar, and then the alternative rewriting of public space in both settings.

Spaces of colonial nature in Port of Spain (Trinidad)

The two most significant park spaces of Port of Spain, Royal Botanic Garden (see Figure 13.1) and Queen's Park Savannah (see Figure 13.2), occupy adjacent parcels north of the city center. From its creation in 1820 through the 1950s, the Royal Botanic Garden served colonial research interests, but its grounds also served the symbolic and aesthetic interests of colonial urban society. Government House, the residence of the colonial governor, looked out over the gardens from their eastern edge. The global scope of its tree collection sent the clear message of Britain's imperial reach; by 1895, plants from all around the empire as well as the Americas were there, and a guide for tourists glorified "the beauty of the scene" in the Garden as well as the view from it of the Queen's Park Savannah and the hills of Laventille and St. Ann, with their "luxurious tropical verdure" (Hart 1895, 26).

As Stuempfle (2018) writes, "Government House clearly asserted authority and grandeur within the Botanic Gardens," for its observational position overlooking the Savannah and the city (69). The Savannah District was the heart of colonial Trinidad in political and

Figure 13.1 Royal Botanical Garden, Kapok Valley, Port of Spain
Image credit: Garth Myers

Figure 13.2 Queen's Park Savannah, Port of Spain
Image credit: Garth Myers

socio-natural terms. The Savannah hosted horse-racing, cricket, football, rugby, hockey, and golf matches from the 1820s on, along with military reviews and parades; it was the landscape for the reproduction of the colonial power structure. By 1916, the Savannah was surrounded by the Botanic Garden, Government House, and the most elite residences of Port of Spain as well as the colony's central cultural institutions. This district was "a low-density residential and recreational area in which there was a premium on order, control, peacefulness and cultivated nature … an alternative to the commotion of downtown" and a "high status" area for "the city's most prominent inhabitants" (Stuempfle 2018, 70).

The botanist Robert Orchard Williams (commonly called R.O.) served as Curator of the Royal Botanic Garden from 1916 to 1933, and then as Director of Agriculture in Trinidad (1938–1945) and Zanzibar (1946–1949), authoring guides to the "useful and ornamental plants" of both colonies. In his biography of his father, R.O. Williams, Jr. (1990) described the Queen's Park Savannah as "insulating" colonial elites "from the squalid town and waterfront;" he considered the Savannah the "only relieving feature" of Port of Spain (24). The Savannah had been an estate, granted by the then Spanish colonial regime in 1782 to the Peschier family, French settlers via Grenada (Williams 1990; Costelloe 2017). Williams (1990) claimed that "fortunately the regulations laid down and promises made to the Peschier family had been strong enough to keep [the Savannah] free from urban development" and that the "fashionable residential areas developed around it" as a consequence (24). Yet the Savannah is alive in the local cultural and spiritual imagination as the venue for the annual Carnival, the central performative ritual for Trinidadian cultural identity. In R.O.

Williams' years as curator, his son wrote, "traditionally the famous pre-Lent carnival procession took place on the encircling road" (24). But this was an innocuous rewriting of a tumultuous narrative. Carnival celebrations had had a violent political history already in Trinidad by the time Williams arrived. The colonial era was marked by the steady takeover of Carnival by the creative culture of the African-origin working-class masses living east of the Savannah in Belmont, East Dry River, Laventille, and East Port of Spain. *Jab jab*, *jab molassie*, *cowband*, *bat*, and *devil mas* performances (all forms of Carnival masquerade involving seemingly violent or dark imagery tied to African cultures and spiritualism seen as threatening to the colonial order) originating in East Port of Spain, the "dirty, ugly, noisy, threatening and potentially violent" side of the town, manifested the Carnival derided by elite Trinidadians and colonial (and post-colonial) governments. These became central to the holiday as expressions of "protest and rebellion against the harsh and unfair conditions which marked [the] daily existence" of the masses (Matthews 2016, 187–88).

During the interwar years, the colonial regime attempted to control, or even eliminate, the working-class Carnival by moving it to the controlled space of the Savannah. Throughout these years, competing variations of the celebration of Carnival occurred downtown and in the Savannah, but slowly and steadily as the 20th century progressed—strikingly after Independence—the more popular, more working-class, and more African elements of Carnival became transcendent on the Savannah.

Spaces of colonial nature in Zanzibar

The comparable colonial parks of Zanzibar, Migombani Botanical Garden (see Figure 13.3) and Mnazimmoja Park (see Figure 13.4), tell a somewhat different colonial story. When Migombani was created in the 1870s, the surroundings were rural, lying beyond Ng'ambo, the "Other Side" of Zanzibar Town that was only beginning to expand. Like the rest of Ng'ambo, Migombani's land in the 1870s consisted of the *viunga* (literally, the attached things) of town elites, overwhelmingly Omani overlords. From the 1870s through the 1940s, as Ng'ambo lands became urbanized, land control shifted toward south Asian immigrant businesspeople, or was dedicated as *waqf* land (an Islamic form for the inheritance of property) upon the death of the Omani owners. Migombani had been the property of a prominent family of the al-Harthi clan, Omani rivals to the royal al-Busaidi family who ruled Zanzibar until the establishment of the British Protectorate in 1890. Even into the 1930s, when the part of Migombani not included in the Botanical Garden had been dedicated as the waqf of Aziza binti Salim al-Harthi, this waqf plantation had some 400 clove and coconut trees on 17 acres (UK Public Record Office Colonial Office file CO 618/59/1, 9 February 1934). The al-Harthis had ceded the rest of Migombani to the British for the Botanical Garden in 1873 as British influence expanded and overtook American, German, and French imperial agents. By the time Zanzibar became a British Protectorate, the Botanical Garden was firmly established, but it steadily ceded scientific significance to a colonial agricultural research station farther from town. By the time of the arrival of R.O. Williams as Director of Agriculture in the mid-1940s, Migombani was surrounded on its northern edge by the first planned suburbs of Ng'ambo, the Holmwood estates. To the west lay the Ziwani Police Barracks and the as-yet undeveloped Kilimani hill. Farms, dairies, and pastures took up most of the eastern edge, and the sea comprised the southern edge.

Thus, in contrast to Trinidad's Royal Botanic Gardens, Migombani was far from centrally located in the urbanism, geographically or politically. By the 1940s, it operated largely as a commercial supplier of ornamental plants for the expatriate elite and the burgeoning tourism industry. The Director of the Parks and Gardens Department of Zanzibar

Figure 13.3 Migombani Botanical Garden, Zanzibar
Image credit: Garth Myers

Figure 13.4 Mnazimmoja Park, Eid celebrations, Zanzibar
Image credit: Garth Myers

Municipality in 1992, Mohamed Mzee (1992) nevertheless spoke wistfully to me of the era of Williams' tenure overseeing Migombani as a part of the Agriculture Department. He lamented its 1951 transfer to the Zanzibar Town Council, which never developed the capacity to maintain the Botanical Garden (Mzee 1992; Zanzibar National Archives file AK 7/15, 26 April and 25 October 1948).

Mnazimmoja [One Coconut Tree] Park is a more central park for colonial Zanzibar, somewhat comparable to Queen's Park Savannah. Mnazimmoja consists of two large segments. What became by the early 20th century the Mnazimmoja Recreation Grounds was, at the beginning of the Protectorate, the backwater, literally, of the Pwani Ndogo, the tidal basin and creek which separated the Stone Town peninsula—what was called Zanzibar Town until the 1960s—from Ng'ambo, the historic African and Swahili side of the city that was the "Native Location" in the colonial era (Myers 1994). Further south, a thin barrier sand bar built up over time into a semi-forested strip of land separating Pwani Ndogo from the Indian Ocean and connecting Stone Town with the island of Unguja. This strip was the first Mnazimmoja Park, but an 1871 cyclone removed all but one coconut tree, giving the park the name it still has (Christie 1876). In the first decade of the 20th century, the British developed part of the strip into a golf course that wrapped around the southeastern corner of Pwani Ndogo. In 1899, the colonial regime began to fill in the southern lagoon of Pwani Ndogo that by the 1870s had been used, at low tide, as an informal picnic area for the Town elites and as the festival area for the end of Ramadhan (Bissell 2011). The reclaimed land opened to the public as the Mnazimmoja Recreation Ground in 1915 and was declared free for "the use and enjoyment of the public forever" (Zanzibar National Archives file AB 40/42; Meffert 1991). In combination, these two portions of Mnazimmoja held golf, tennis, outdoor squash, cricket, and football matches, along with holiday celebrations, as the largest public space in the city, throughout the colonial era.

Since slavery was not outlawed until 1897, many of Ng'ambo's first residents were enslaved Africans, freed slaves, servants, and peasants, from mainland cultures and formerly rural peoples from Zanzibar's three Swahili-speaking indigenous communities (Myers 1994; Fair 2001). Most residents of Stone Town in the early Protectorate were from Arab, south Asian, or European communities. The European quarter—as designated and developed by the British between the 1890s and 1940s—occupied the southern triangle of the Stone Town peninsula, directly adjacent to what would become Mnazimmoja. As the colonial regime made use of and attempted to reproduce and codify into laws the separation of Stone Town and Ng'ambo, they filled in the Pwani Ndogo to make the boundary less liquid (Bissell 2011). In 1913, a colonial sanitation report made the political aims of the park explicit: Mnazimmoja "should remain an open space and a neutral zone between the European and the native quarter of the town" (Zanzibar National Archives file AB 2/264, 9). Like the Queen's Park Savannah, Mnazimmoja and the landscape around it became associated with the colonial era's elite. By the 1940s, the park's western edge contained the British Residency, the High Court, the massive town hospital, the showcase Peace Memorial Museum built in 1925, and housing for colonial officials. Casuarina trees marked both sides of the road separating the golf course from the recreation ground. On the border of the park in Ng'ambo's Kikwajuni neighborhood, a string of middle-class Swahili, Comorian, and Indian residences were constructed beyond the elegant *allee* of saman trees planted by the colonial regime in the 1930s.

The African-Swahili Kikwajuni community's relationship to the golf club became apparent with the formation of its first football club, the Caddies, in the 1920s. A second Kikwajuni club, the New Kings, came into being in the 1920s (Fair 2001). Both clubs practiced on segments of the grounds not turned over to the golf course or cricket pitch, and football's popularity led the colonial regime to add "three additional football pitches … adjacent to the makeshift ground at Mnazi Mmoja" (Fair 2001, 234). The neighborhood's women utilized the grounds for practice sessions of popular dance groups. And both men and women used the Mnazimmoja Park for the celebration of the two main Muslim holidays (Bowles 1991). As with the Queen's Park Savannah, then, Mnazimmoja served as a zone for the reproduction of the colonial order, even as it steadily became a site for countering that order. It was the town's central public space, and both its buildings and planned uses reinforced and symbolized the colonial vision. While they lacked the explicitly counter-colonial punch of Afro-Trinidadian alternative Carnival celebrations, the Afro-Swahili informal uses for the park manifested alternative visions for the use of public space, whether in playing football on the cricket grounds or dancing on the picnic grounds. The lack of border fencing, along with the open beach along the south edge of the park, enhanced informal access to a more significant extent than the Trinidad parks (where the Botanical Garden was fenced in and the Queen's Park Savannah encircled by busy roads).

Alternative landscapes of post-colonial urbanism

Trees and parks were certainly essential agents in the colonial plans in both former colonies, along with the urban nature-landscape more broadly. But what became of those colonial intentions is quite varied. In part this had to do with the lackadaisical manner of implementation for colonial plans, or what Bissell (2011) proposed as a sort of calculated failure. But it also had to do with the highly varied and stratified responses to the colonial projects in different colonies. This is well illustrated in the post-colonial stories of parks, gardens, avenue trees and landscapes of Trinidad and Zanzibar, where neither society had had a prior custom for urban parkland public spaces.

Powerful people in colonial and post-colonial cities have wanted to use parks and plants to articulate their vision of their power. Plants were shipped across the British colonial tropical world to give botanical life to that vision. The demise of colonialism left cities with plants and parks that expressed the colonial vision, and post-colonial states often sought to continue speaking with shaped natural space and living things – and sometimes in a colonial manner: in Trinidad, after all, it is still the *Royal* Botanic Gardens and the *Queen's* Park Savannah. But the people of Trinidad and Zanzibar have often found different ways of re-deploying plants and parks for different purposes.

In the contemporary Royal Botanic Gardens, much of the space is grassy picnic park land. The city has built a strip of gazebos to encourage more families to use the park for picnics. The picnic area portion has clear fencing around it on three sides, the fourth edge being the forested segment that eventually ends with the fence for the Emperor Valley Zoo. The zoo took over another segment of the Botanic Garden, as did the national presidential headquarters above the Botanical Garden. Despite the park being dramatically changed from Williams' era, this remains a model park for an orderly state, or one that wishes to project such an image – as in the celebrations planned for its 200th anniversary. In the Queen's Park Savannah, too, the order of a colonial world seems to remain beyond just the name – other than in the Carnival season, when it is occupied by the spirits of alternative cultural visions.

Yet, in the neighborhoods surrounding the Savannah, alternative landscapes and socio-natures abound. Greater Port of Spain displays numerous examples of local people re-thinking the landscape. One prime example exists in the heavily African-influenced east end of Port of Spain. East Dry River, Laventille, Sea Lots, Belmont, and San Juan were rife with "African cultural retentions" from the early 19th century on (Gift and Kiteme 2013, 95). For example, many urban dwellers of African origin utilized "sou-sou" micro-credit systems in small groups, a retention from Yoruba *esusu* financial cooperation institutions (Cummings 2004, 63). Part of this narrative ties into the history of Africans in Trinidad. Almost 40 percent of the Trinidadian African community came from the Bight of Benin and held many cultural elements in common; unlike most Caribbean islands, "the majority of enslaved blacks in Trinidad" at the time of emancipation "were born in Africa" (Gift and Kiteme 2013, 96) and still spoke African languages. Igbo, Rada (Beninois), and Mandinke cultural practices endure into the present, most notably in music and dance, but also in spiritual and religious practice (Carr 1989; Besson 2011; 2012). So many Yoruba speakers resided in Laventille and East Dry River that the hilly area was collectively known as Yoruba village (see Figure 13.5). Belmont, which was known as Freetown in the 19th century, had Igbo, Kongo, Mandinke and Rada communities alongside Barbadian migrants (Cummings 2004; Stuempfle 2018). These are the poorest parts of Port of Spain, associated with high crime since the 1870s. Yet there is still a substantial natural environment in these hills, and the West African spiritual and healing practices as well as the musical instruments used in many songs originating in the northeast of the city depend upon specific woods and essential oils found in the hills – or even plants grown in the barrack-yards (Cummings 2004). Ties to Africa today extend from the Afrikan Oils store just down the street from the Queen's Park Savannah to the Success Laventille Networking Committee with its "*Ujamaa* [Swahili for Family-ness] Newsletter" that promotes "self-knowledge, traditional African practices, cultural self-assurance, and a view of themselves as heroic survivors rather than as victims" (cited in Gift and Kiteme 2013, 101).

Figure 13.5 Laventille Hills, Port of Spain
Image credit: Garth Myers

The colonial regime had banned African drums in the 1920s, perhaps reducing direct ties to appropriation of wood from local trees – but this only gave rise to the greatest musical gift Trinidad has given the world, the steel pan, invented in the yards of East Port of Spain using old oil barrels (Cozier 2012). Calypso and steel pan music and West African-inspired masquerade became the heart of Trinidad's Carnival celebrations on the Queen's Park Savannah, alongside stick-fighting rituals with their direct parallels in Africa – including in Zanzibar, where they are central to the *Mwaka Kogwa* celebration of the Swahili Shirazi "Persian" New Year (Brereton 2009 [1981]; Cummings 2004). The neighborhoods on the East Side have used these cultural elements – and, in a very tangible sense, the cultural landscape – "to convey a political vision, one of resistance to oppression and a refusal to negate the cultural continuities that they were able to create for themselves" (Gift and Kiteme 2013, 106) from the 19th century to the present. "Reciprocal relationships" were essential to Afro-Trinidadian communities in Port of Spain, their lively usage of plants and parks in stark contrast to the "subdued" business district of today as the downtown elite cleared out the "village" dynamics of the city (Cummings 2004, 177).

The hills north of the Botanic Gardens belong to the middle-and-upper class suburb of St. Ann's, but these hills are also home to the community afforestation and environmental education organization, Fondes Amandes (see Figure 13.6). Led and run largely by women activists, Fondes Amandes seeks to stabilize the hillsides, but more significantly to raise consciousness and the quality of life for poor women in and around the city. Theatre space and outdoor classrooms are the venues for the group's work, combining pride in Trinidadian connections to nature with a liberatory agenda.

Figure 13.6 Fondes Amandes Community Re-Forestation Project, St. Anns
Image credit: Garth Myers

In Zanzibar, the avenue trees, parks, and gardens of the colonial era have taken on new meanings, too, perhaps more dramatically – both in the revolutionary single-party era (1964–1992) and the post-revolutionary era since then. The state has changed the dynamic of its uses of trees, parks, and gardens; so, too, have ordinary residents. The post-revolutionary order has brought physical redevelopment and redeployment of the Forodhani Gardens in front of the former Omani Sultan's Palace and the administrative headquarters of the Protectorate at the Beit al Ajaib (House of Wonders) in the interest of tourism. Yet Forodhani retains, underneath, strong local uses and meanings beyond tourism; it is alive with ordinary people every night, gathering to socialize with friends and family. By contrast, Victoria Garden in the British enclave of Vuga, where the colonial Legislative Council once met, has never returned to the position of political centrality colonialism attempted to give to it. The old LegCo building is a secondary storage facility for the municipal parks and gardens department. It symbolized the political process that had excluded Zanzibar's clear Afro-Swahili electoral majority from political power and the conditions which led to the popular revolution, and it simply was not a big enough structure to house the revolutionary regime's House of Representatives. That body met first (1984–2008) at a former social hall bordering Mnazimmoja but on the Ng'ambo side and, since 2008, in a massive new structure on the southeastern edge of town.

Mnazimmoja (see Figure 13.7) lost its golf course with the revolution, but then more recently lost its significance as a revolutionary-era center for other sports activities. It has retained its utility as a central ground for the celebration of Ramadhan's end, and its far southwest corner, Suleiman Maisara, is still a key site for celebrations of the anniversaries of

Figure 13.7 Mnazimmoja Park looking out toward Peace Memorial, Vuga, Zanzibar
Image credit: Garth Myers

the revolution, even as its tennis courts have withered from lack of use or upkeep. And the biomass of the park is disappearing. The grass has been eaten into for informal parking areas. A 2017 drainage project meant digging up the park down its center, and it had not yet really recovered by January 2019. The *mitiulaya* (saman) trees planted in the 1930s remain on the northern side of the park, in all their grandeur, but many of the *mivinje* (casuarinas) bordering the south edge have been cut. The Peace Memorial, the Golf Club, and other colonial structures have all faded into states of disrepair.

The fading of Mnazimmoja looks like intensive maintenance by comparison with what has happened to the Migombani Botanical Garden. A key node of botanical research in the western Indian Ocean region for a time, many of Migombani's historic trees are gone (with the prominent exception of one extraordinary kapok tree – see Figure 13.8). By January 2017, an environmental officer for the Revolutionary Government of Zanzibar put it this way in an interview: "Migombani Botanical Garden has died." Even there, however, a grassroots community group emerged from the haphazard settlement north of the park, Jang'ombe, to proclaim that rumors of Migombani's death were premature, seeking external funding to secure its borders and gain income from admission, to feed their families and to

Figure 13.8 Kapok tree, Migombani Botanical Garden, Zanzibar
Image credit: Garth Myers

plant trees to revive the garden. While Migombani has no fences, the community group has at least one member stationed by the main road to converse with anyone who seeks to enter the park area. Yet its northern half is mostly—as it has been for thirty years—an informal football pitch for Jang'ombe residents.

Perhaps the most fascinating narrative among parks and gardens in the colonial city belongs to Jamhuri Gardens, just adjacent to Mnazimmoja. This space was originally a physical part of the Pwani Ndogo, just north of and nearly contiguous with Mnazimmoja. While the southern end of the Pwani Ndogo was filled to create Mnazimmoja, the middle stretch surrounding Kisiwandui (Smallpox Island) was filled in 1931 to create what is now called Creek Road, the physical dividing line between Stone Town and Ng'ambo. However, a drainage canal was created, along with a circular pool at the south end just past what had been the small island in the tidal inlet. This pool, meant as an element of beautification of the landscape, slowly faded, and then was gradually filled in through the early revolutionary regime. In the early 1990s, the Zanzibari revolutionary regime of Dr. Salmin Amour created an ornamental garden in the sunken remains of the circular pool. This park has been well maintained, and the revolutionary government added a children's playground to the western quarter of the circle in the 2000s. What is more remarkable is how readily ordinary residents have taken to the use of this park. It is home to informal Islamic and civic education groupings and day-to-day political discussions.

The Zanzibar government's *2015 Ng'ambo local area plan* (DoURP 2016), nicknamed Ng'ambo Tuitakayo [The Ng'ambo We Want], proposed a new "Central Park" that would link Mnazimmoja to the unprotected southern side of the remains of Migombani, and a "green corridor" from Jamhuri Gardens eastward through the center of Ng'ambo. These ideas might be implemented, but the reality is that ordinary people are already remaking and reframing these parks and corridors. The Zanzibari novelist, MS Abdulla, created the Sherlock Holmes of Swahili fiction with his namesake character, the detective Bwana MSA; Bwana MSA had a thick book he regularly consulted as his guide in solving cases, *Kinyume cha mambo*, "The opposite of matters." This title still evinces a core kernel within Zanzibari culture: its joyously defiant, ornery embrace of the opposite of matters, of contradiction. A city full of signs telling you what not to do there, where you've done exactly that. It is a revolutionary democracy that everyone knows is neither revolutionary nor democratic. Its pious Muslim identity is inseparable from its historic and still pervasive openness to otherness and difference, its incredible tolerance of diversity and stark impiety. One Kikwajuni resident dismissed the 2015 local area plan as "Ng'ambo WAitakayo" – "the Ng'ambo THEY want." Boundaries of parks, where they are even noted, are meant to be crossed. This opposite-of-things cultural energy defines a strong core of Zanzibar's socio-nature, in seeming contrast to Trinidadian culture – but even there, where the air is thick with a post-colonial propriety that is not really so "post," a vein of gleeful irreverence finds its way into the cultural bedrock. This is encapsulated in the work of the Lordstreet Theatre Company created by playwrights, Tony Hall and Errol Fabian, dedicated to fostering "play", so that the transgressive spirit of the alternative *J'ouvert* (daybreak) Carnival in the city's public spaces – an anti-colonial alternative, even in a royal botanic garden, next to the Savannah of the Queen's Park.

Conclusion

In Trinidad, the story has commonalities with Zanzibar, with some divergences. The Trinidad and Zanzibar cases highlight the variability of, but also the potential for, post-colonial cultures in decolonizing the urban landscape. Ultimately, the patterns in cities living in the aftermath of British colonialism specifically appear in the repeated urban forms and spaces for nature that the colonial regimes sought, both in terms of botanical gardens and central parks. Across Britain's other sub-Saharan African and Caribbean holdings, similar parallels played out, even to the point of similar realms for divergence. Prita Meier (2016) sees the cities of the Swahili coast as having the contradiction of their essence emerging from a constantly changing array of globalized "elsewheres" even while their architecture rests on a projection of unchanging stone permanence. This gives rise to a churning ongoing effort to reconcile "the need for mobility and mixing on one hand and fixity and rootedness of the other hand" (3). In some senses, this is a contradiction one can see in the Caribbean, and in other post-colonial settings. There is in both of the small cities of this chapter both a feeling of—an insistence on—endurance and permanence and at the same time a sense that what makes these cities exist is their set of flows and links with elsewhere. The cities can be found on their transversal lines.

References

Besson, G. (2011) "The Rada Community." *The Caribbean History Archives*. Accessed 14 April 2018. www.caribbeanhistoryarchives.blogspot.com/2011/12/rada-community.html.

Besson, G. (2012) Personal Communication with Author, 12 June 2012. Port of Spain.

Bissell, W.C. (2011). *Urban Design, Chaos and Colonial Power in Zanzibar*. Bloomington: Indiana UP.

Bowles, B.D. (1991) The Struggle for Independence, 1946-1963. In: *Zanzibar under Colonial Rule*, Sheriff, A. and Ferguson, E. (eds.). London: James Currey, pp. 79–106.

Brereton, B. (2009 [1981]). *A History of Modern Trinidad 1783–1962*. Trinidad: Terra Verde Resource Centre.

Carr, A. (1989). *A Rada Community in Trinidad*. Port of Spain: Paria Publishing.

Christie, J. (1876). *Cholera Epidemics in East Africa*. London: Macmillan.

Costelloe, C. (2017) "Within the Walls of Peschier Cemetery: 200 Years of Family History." *Guardian* Trinidad, 13 August, Page B3.

Cozier, C. (2012) Personal Communication with Artist Christopher Cozier, 10 June 2012.

Cummings, J. (2004). *Barrack-Yard Dwellers*. St Augustine: University of The West Indies.

Department of Urban and Rural Planning (DoURP), Zanzibar. (2016) *Ng'Ambo Local Area Plan, Zanzibar*. Revolutionary Government of Zanzibar And African Architecture Matters.

Fair, L. (2001). *Pastimes & Politics: Culture, Community and Identity in Post-Abolition Urban Zanzibar, 1890–1945*. Athens: Ohio University Press.

Gift, S. and Kiteme, O.K.O. (2013) Freedom of the Spirit and African Cultural Retentions: The Case of East Port of Spain, Trinidad and Tobago. In: *Trajectories of Freedom: Caribbean Societies, 1807–2007*, Cobley, A. and Simpson, V. (eds.). Kingston: University of the West Indies Press, pp. 95–108.

Hart, J.H. (1895). *Visitors' Guide to the Royal Botanic Gardens Trinidad*. Trinidad: Government Printing Office, Port of Spain.

Matthews, G. (2016) Elevating the Masses through the Masquerade: George Bailey's Afrocentric Mas in Trinidad's Carnival. In: *In the Fires of Hope, Volume 2: Essays on the Modern History of Trinidad and Tobago*, Mccollin, D. (ed.). Kingston: University of the West Indies and Ian Randle Publishers, pp. 186–197.

Meffert, E. (1991) *Architectural Notes about Zanzibar Town*. Unpublished.

Meier, P. (2016). *Swahili Port Cities: The Architecture of Elsewhere*. Bloomington: Indiana University Press.

Myers, G. (1994) Eurocentrism and African Urbanization: The Case of Zanzibar's Other Side. *Antipode*, 26(3): 195–215.

Mzee, M. (1992) Interview with Mohamed Mzee, Director of the Department of Parks and Gardens, Zanzibar Municipality, 20 January.

Stuempfle, S. (2018). *Port of Spain: The Construction of a Caribbean City, 1888–1962*. Kingston: University of The West Indies Press.

UK Public Record Office File CO 618/59/1: Housing and Town Improvement in the Native Town, Zanzibar.

Williams, R.O. Jr. (1990). *A Plantsman's World: The Biography of Robert Orchard Williams*. Mona, Jamaica: Typescript Available at The University of the West Indies Library, Mona.

Zanzibar National Archives File AB 2/264: Report on Sanitary Matters in the East Africa Protectorate, Uganda And Zanzibar, By Professor W. J. Simpson.

Zanzibar National Archives File AB 40/42: Secretariat: Mnazi Mmoja Recreation Grounds.

14

GLOBAL HOMOGENIZATION OF PUBLIC SPACE?

A comparison of "Western" and "Eastern" contexts

Tigor W. S. Panjaitan, Dorina Pojani, and Sébastien Darchen

Introduction

Public space reflects the city or nation in which it is embedded – its socio-cultural, economic, and political conditions. In the broad region that has traditionally been considered as the "West" (i.e. Western Europe, North America, and Australia), public space has been defined by its openness and accessibility to everyone – thus reflecting Western democratic values of freedom and equality. By contrast, in absolutist or totalitarian nations public spaces have been typically designed to symbolize the power of governments over people and overwhelm users with sheer size (Pojani and Maci 2015).

Views about the nature of public space and the challenges that it faces are, for the most part, formulated based on studies and observations from the perspective of developed Western countries. The present chapter adopts a different vantage point. Public space is analyzed from the lens of Southeast Asian cities, in contemporary definition consisting of Vietnam, Laos, Cambodia, Thailand, Myanmar, Malaysia, Indonesia, Singapore, Philippines, as well as several small countries or island-nations in Oceania. Through a review of the available literature and our own experience conducting research in the region, this chapter challenge the view that theories on public space which apply to the West also fit the context of "Eastern" cities (i.e. cities in Southeast Asia). The chapter advocates for urban analyses and planning approaches sensitive to local cultures.

The chapter views public space as a physical manifestation of the public sphere, and therefore inseparable from it. But physical public space is vital to citizenship and cannot be replaced with an abstract and fluid public sphere such as that envisioned by the German philosopher Jürgen Habermas in the 1960s (Habermas 1991). More recently, concerns have been raised that the advent of virtual space might efface physical space and eventually lead to new forms of urbanism. But it is also evident that new technologies have reinforced the role of traditional cities, which concentrate business headquarters and the creative professional class (Carmona 2010a). For subaltern societies with limited access to digital media, public space may be the only public sphere available.

To place public space issues in context, this chapter opens with a theoretical discussion of the public sphere, as conceptualized in the "West" and the "East," prior to moving to a discussion of physical space. This chapter argues that, traditionally, differences in the public sphere between Western and Eastern cultures have been physically translated into specific features and characters in the respective public spaces. In the East, a different economic level, culture, political regime, gender balance, and role of religion in society, as well as the colonial legacy, have combined to produce distinctive results, which significantly differ from Western models. Therefore, a different approach is needed to study public space in this context. However, the chapter also recognizes and explores converging trends in the nature of public space in both the West and the East, which are due to neoliberalization and privatization, coupled with globalization and a decline of the traditional public sphere in the West.

The public sphere

Starting with the Enlightenment in the 17th century, a notion took hold in Europe and the United States that society should enjoy the freedom to deliberate about public affairs and formulate public solutions (Reinelt 2011). A civic culture and critical political discourse arose outside of state or religious control. Habermas (1991) was among the first to theorize the concept of "public sphere" (*Öffentlichkeit*), as it pertains to the West. He ideally defined it as an open and democratic forum (physical or abstract) where everyone, regardless of status, has equal opportunities and rights to discuss issues affecting society. The essential characteristics of the public sphere in Habermas' conceptualization are its critical nature, its universal access, and its focus on public issues.

Some critics have contended that in the West, this idealized and universalist version of the public sphere has never existed. The public sphere was monopolized by the elites to legitimize their own power, while other key groups, such as the poor, the uneducated, women, people of color, migrants, ethnic and sexual minorities, and criminals, were excluded or had their voices muted. Even if a unifying public sphere were attainable, postmodernist critics have questioned its desirability. They have argued for the existence of multiple public spheres and counter-publics, in which "alternative" cultures find expression and fight against discrimination thus threatening social stability and bourgeois values (Gholamhosseini et al. 2018).

But while in theory postmodernism emphasizes multi-culturalism and acceptance of diversity, in practice it has not led to more participation in public life in the West. On the contrary, individuals appear to place more emphasis on their private life, family, and intimate friends. Critics note a sense of obliviousness, numbness, and disinterest in public affairs. Part of the blame for this state of affairs is laid at the foot of the internet, which allows for entertainment, errands, and even work to be undertaken online from home. Second, the content conveyed through contemporary electronic media is distorted by the power elites and the owners of capital. More recently, public opinions are increasingly manipulated through sound bites, fake news, echo chambers, public defamations, and the like. This discourages public participation in the Habermasian sense. It is becoming clear that the virtual public sphere requires as much regulation as the physical public space in order to guarantee freedom and equity of access, and maintain a level playing field.

In sum, in Western contexts, public sphere issues include fragmentation and dematerialization. In the East, the public sphere differs from the West in several ways, delineated below.

Diversity and fragmentation

Southeast Asia is a multi-cultural, multi-ethnic, multi-linguistic, and multi-religion region. Historically, it has been much more diverse than Europe, North America, and Australia.[1] Such diversity is the product of migration and international trade, but also colonization, both by Western maritime powers and by neighbors.[2] Given the extreme diversity and fragmentation of the region, there has never been a pretension of a unifying public sphere, such as that envisioned by Habermas for the West.[3]

Collectivist values

Notwithstanding a turbulent past, coexistence of different groups within the respective Southeast Asian countries is relatively congenial. All countries tend to score high on power distance and traditional values of collectivism (Hofstede et al. 2010). Individuals submit to the collective in exchange for loyalty, support, and protection. Central issues such as gender, ethnicity, religion, and citizenship are managed within the framework of dialogue and strategic negotiation in order to maintain social stability and inter-group harmony, all while preserving a multitude of cultural identities (Clammer 2002; Salmenkari 2014). Southeast Asian nations tend to be face-and-shame cultures – *mianzi* in Mandarin (Ho et al. 2004; Hofstede et al. 2010).[4] By contrast, in the Habermasian public sphere there is an expectation of debate and even confrontation, based on individual ideas and ideological positions. There are significant exceptions, however – both recent and historic.[5] In the West too, especially in Europe, inter-group conflicts are significant and evident in public space and the public sphere.

Rigid social hierarchy

Harmonious community relations in Southeast Asia are likely due to still rigidly hierarchical and patriarchal social structures, which prescribe fixed rules of behavior and speech based on one's position in the social matrix. Hierarchy is based on status but also on age and gender. Society is envisioned as a pyramid, at the apex of which sits the ruler. Similarly, the family is envisioned as a pyramid topped by the most elderly members, and women are subordinate to men. Traditionally, hierarchical relationships in the region have been encouraged by Hindu-Buddhist philosophy[6] – although in some countries (Indonesia and Malaysia), the advent of Islam introduced a more egalitarian spirit.[7] However, even here social distances based on inheritance or acquired wealth remain much more prominent than the West; hence more adherence to behavioral rules in public spaces. Eccentricity is much less tolerated or encouraged.

Subordination of women

In terms of gender, patriarchy is also much more pronounced than in the West. Feminism has made little headway in Southeast Asia.[8] Countries influenced by communism have provided an opportunity for women to engage in the public sphere, with women in influential positions gaining authority over lower-ranked men. However, they are still subject to the paternalist structures of their class, and their status has practically imposed a double burden on them: be active participants in public life while also complying with traditional duties within the family (Sangwha 1999). More currently, in Islamic countries Muslim women have been using revisionist interpretations of the religion (e.g. the *piety movement*) to press for rights and promote their presence in the public sphere (Rinaldo 2008). While this has "feminized" the public sphere, gender equality (in public space and elsewhere) is a still-distant prospect (Brenner 2011).

Strict public/private life separation

Another point of difference between the West and Southeast Asia is the latter's strict separation between the public and private sphere. Duos such as "inside-outside," "us-them," and "family-society" are viewed as dichotomies rather than continuums. In Confucian tradition, this is known as *nei-wai*. It is also discussed in terms of the opposition between *si* (selfishness) and *gong* (public spirit). *Nei-wai* applies to gender roles but also more broadly to the rules for interaction between those who are insiders and those who are outsiders. The "inside" is the feminine, intimate, and nurturing domain and the outside is the masculine domain, characterized by strength, dominance, assertiveness, and egotism (Sangwha 1999). The public sphere is seen as embodying masculine traits and therefore characterized by dissensus rather than consensus. The openness, mutual trust, and confidence needed for deliberation and agreement on public matters are only possible in smaller groups, the leaders of which then go on to deliberate with their peers following the pyramid of social hierarchy. Value conflicts are solved not through rational persuasion but through compromises that take all parties into account. This "representative" model departs from the Western model of an overall public sphere governed by rules of behavior to which all are subject, and where people must speak directly to those whose interests and views differ from their own (Salmenkari 2014). Public/private life separation also contrasts to the Western propensity to publicly perform private thoughts and emotions – a culture which is increasingly rewarded in social media and "reality TV" programming.

Crucial role of religion

Religion retains a key role in Southeast Asian public life. This is another important distinction with the West, in which the public sphere is envisioned as separate from religious thought. While reason and logic supplanted religious dogma in Europe starting with the Enlightenment, in Southeast Asia religious values continue to pervade all aspects of culture and public discourse. In more traditional Southeast Asian societies, people's behavior, dress, and speech in public are strongly influenced by religious norms and traditions (Brenner 2011). In some countries, including ones that are in theory secular, a particular religion is intrinsically linked to the national identity – for example Buddhism in Thailand and Cambodia and Islam in Malaysia and Indonesia (Fauzia 2013). In predominantly Islamic countries, there are attempts (discursive, at least) to apply *sharia* principles of social justice and wealth redistribution. In the Philippines, politicians have been known to adopt certain policies[9] mainly to find favor with the Catholic Church (Azada-Palacios 2013). The Buddhist doctrine of egolessness (*anattā*, no-self) is adhered to in Thailand, Cambodia, and Myanmar (Clammer 2002).[10] This contrasts sharply with the Freudian (or human-centric) concept of "ego" or "self" which is promoted in the West and is equated with the rational mind or individuality. In countries with Hindu minorities, *ahimsa* (the Hinduist ethos of non-violence) has political connotations as well – which, however, are not supported by nationalistic parties such as Indian People's Party (BJP). Even in communist and atheist countries such as Vietnam (and China), the influence of religion is evident. For example, to celebrate major life events, such as weddings and funerals, temples are preferred to the public spaces created by the government (Yongjia 2013).

Delayed penetration of mass communication

Another reason why an idealized public sphere concept has not formulated in Southeast Asia in parallel with the West is a delayed penetration of mass communication technologies. The telegraph and the printing press did not reach this region until the mid-19th century via European colonizers, the British in particular.[11] Through the press, locals could finally learn about events occurring beyond their immediate community, city, or even country, and contribute their own news and opinions. This served to raise indigenous people's awareness to fight against colonization and increase their political bargaining power (Ballantyne 2016). It also led locals to adopt more Western mores and appearances. Consequently, increasing public critical attitudes toward the government made the media an object of government censorship (Limapichart 2009). In the present, the press and other mass media continue to face restrictions at various levels.

State authoritarianism

All Southeast Asian nations achieved independence after the end of WWII. For several decades afterwards, the region contended with the Cold War, the Vietnam War, and anti-colonial struggles, initially led by communist parties. By the mid-1960s, outside mainland Southeast Asia the communists had been largely defeated. However, liberation from communist totalitarianism did not lead to democracy. On the contrary, democratic experiments in Indonesia and the Philippines failed, paving the way for subsequent authoritarian regimes, and some areas, such as the southern Philippines, southern Thailand, and Indonesia experienced (and continue to experience) episodes of regional and sub-nationalist armed conflicts. Singapore retained an authoritarian government. The Indochina peninsula, encompassing Vietnam, Cambodia, Laos, and Myanmar, remained under the control of communist dictatorships or military juntas. Conditions such as state control of the media and censorship of free speech precluded the consolidation of an open, Western-style public sphere (Huat 2008; Morgenbesser 2016). In Indonesia too, there have been attempts to restrict freedom of speech after the 1998 reform movement.

Advent of the internet

In the contemporary era, the advent of the internet and electronic media has affected the public sphere in Southeast Asia, as in the West. The anonymity afforded by the internet has provided the confidence needed for participation, discussion, criticism, and debate thus enhancing democratic practices (Salmenkari 2014). Recognizing the power of the internet in promoting democracy in the region, several governments have sought to control and censor it—for example, by banning sites reporting on citizen protests or criticizing public institutions—thus in effect attempting to restrict the public sphere.

As in the West, the internet has also served as a tool to spread fake news and engage in public defamations. For example, one could argue that the (mis)information campaign (online) and the organization of protests (in public space) that brought down the Governor of Jakarta, Ahok, in 2017 (who is ethnic Chinese and Christian) were products of the internet, while traditional newspapers may not have been complicit (Ismail 2016).

Public space

The public sphere finds its physical manifestation in public space. In Western democracies since Classical Antiquity, public space, at its core, has been characterized by its (a) publicness and (b) political connotations. The first trait, publicness, refers to equity, physical and psychological accessibility and openness, and freedom from group or government dominance. In an ideal public space, all citizens are at liberty to join in debates on public affairs. While activities can be contemplative, Western public space sustains democracy in that it holds the potential for citizens and civil groups to tout their existence and express their aspirations to one another. The politically charged nature of public space means that friction and conflict can and do occur.

Southeast Asia presents a different picture. First, purposely designed, representational public spaces in the manner of the Greek *agora* or the Roman *forum* did not exist in pre-colonial times. Traditionally, "public" space has been the street as well as any communal space left over between buildings, villages, or farming fields. Colonial spatial planning introduced the concept of land sub-division, which mandated a clear demarcation of property lines (Gunawan et al. 2013). In the more "formal" parts of cities, these regulations detracted from the spontaneity of public space. In more informal districts, common spaces remain highly active, vibrant, and multi-functional in nature and accommodate a range of practical activities and socio-economic functions: buying and selling, praying, arts-and-crafts, cooking and eating, spiritual-religious ceremonies (which in the more secular West have long moved indoors), and even services such as haircuts and ear cleaning (see Figure 14.1). On the other hand, the European concept of *boulevardier* or *flâneur* was never present in Southeast Asia.

Figure 14.1 Street barber in Bangkok
Image credit: Tigor Panjaitan, Dorina Pojani, and Sébastien Darchen

User-produced Southeast Asian public spaces are also, for the most part, managed by users. Sometimes intense use of street space is contested – for example, when different functions such as pedestrian circulation and food vending overlap in scarce space – but such conflicts are resolved though informal negotiation and deliberation, without government involvement (see Mehta 2018 and Figure 14.2). Typically, government efforts to ban or curtail activities that are intrinsic to local cultures, such as street vending, result in major public outcry (Dunlop 2017). In contrast, public space in the West is produced by governments (or, increasingly, private entities). Supra-authorities (the police or private security) are expected to mediate any conflicts that may occur, thus relieving users of responsibility but also, in a sense, disenfranchising them (Wang 2003; Sien 2003).

Some "planned" public spaces appeared in Southeast Asia under the influence of outsiders – for example, Fatahillah Square in old Jakarta, modelled after the Dutch *grote markten*; Bangkok's Ratchadamnoen Klang Avenue, inspired by 1930s Art Deco monumentalism; or Merdeka Square in Kuala Lumpur, a former cricket field left behind by British Residents. In some cases, entire city centers (e.g. Rangoon, Hanoi, Penang, Luang Prabang, and Bandung) were planned and laid out based on a classic European grid pattern punctuated by squares, parks, and fountains. However, these were re-appropriated by local cultures and narratives (Chifos and Yabes 2000; Sien 2003; Daniere and Douglass 2009; Oranratmanee and Sachakul 2014).

The patterns of use of public spaces are different in Southeast Asia compared to the West. In keeping with the hot and humid tropical climate and the nightlife culture, many

Figure 14.2 Street food in Kuala Lumpur

Image credit: Tigor Panjaitan, Dorina Pojani, and Sébastien Darchen

public spaces are more heavily used after dark than during the day. In terms of design, markets are often enclosed, covered, or tucked in narrow alley networks in order to provide shelter from the bright sun. Also, the functions of space are much more temporary and adaptive than in the West. For example, many streets, which are vehicular during the day, are closed to traffic starting at sunset and turn into markets and eateries, with movable barriers and food stalls placed along the roadbed (see Figure 14.3a and Figure 14.3b). Given strong kinship and community ties, these spaces are extremely convivial and gregarious (Yuen and Chor 1998; Oranratmanee and Sachakul 2014).

In more religiously-observant parts of Southeast Asia, religion is a strong regulator of public space interactions and conduct. The values of courtesy and public order take precedence over personal freedom (Evers and Korff 2000). In Hinduist Bali, the layout of settlements or even individual houses is prescribed by religious texts, as human designs are considered to be responsible for balancing the cosmic energies.

Finally, a crucial difference between Western and Southeast Asian public space is the latter's lack of civic and political functions. Although Southeast Asian countries are beginning to incorporate democratic principles into their political systems, political activities in public space are still either choreographed by the governments or take place under their watchful eye. In line with subaltern hierarchies, governments are positioned as the guardians of social harmony. Citizens are not free to congregate for the purpose of protesting or expressing dissent, lest social accord is compromised. In theory, only government-approved political activities, which comply with certain regulations, are allowed – although violent protests can and do occur from time to time (Clammer 2002; Douglass 2008).

Figure 14.3a Surabaya street is crowded by vehicles during the daytime
Image credit: Tigor Panjaitan, Dorina Pojani, and Sébastien Darchen

Figure 14.3b The street becomes a culinary center at night

Image credit: Tigor Panjaitan, Dorina Pojani, and Sébastien Darchen

Specific public space design approaches are followed with the express purpose of controlling urban populations. With the exception of sports fields and government complexes, the creation of large and open public spaces, which allow for crowds to gather, is avoided. A few larger spaces created by post-colonial governments, such as the Heroes Monument field in Surabaya, Patuxai Park in Vientiane, or Vimean Ekareach in Phnom Penh, serve as plateaus for featuring prominent national monuments and other state symbols (see Figure 14.4). Similar to European socialist countries during the Cold War (Pojani and Maci 2015), staged demonstrations take place in those to celebrate Independence Day and other national holidays (Lim 2007; Douglass 2008; Gibert 2013; Gunawan et al. 2013).

Contemporary homogenization of public space?

Notwithstanding the noted differences in the concept of public space between West and East, some commentators have advanced a theory that public space is being homogenized at a global level. Cities are under pressure to create standardized consumption environments that bring in revenues and are perceived as safe and attractive for white-collar residents and tourists. Meanwhile, designers, developers, and clients are not tied to particular places but operate across countries, and as a result, design templates are replicated from place to place with little thought to context, history, and culture. This homogenization theory is examined in the context of Southeast Asia through a theoretical framework adapted from Carmona (2010a 2010b). The two main components of the framework are: "under-managed space" and "over-managed space" (see Table 14.1).

Figure 14.4 Patuxai Park in Vientiane
Image credit: Tigor Panjaitan, Dorina Pojani, and Sébastien Darchen

Table 14.1 Public space issues in Western settings

Under-managed space	*Over-managed space*
Litter	Physical exclusion (street curbs, moving traffic, darkness)
Uncollected garbage	Psychological exclusion (fear, suspicion, subtle visual clues, tension, social conflict)
Graffiti	Financial exclusion (entry fees)
Ugly buildings	"Undesirables" (the homeless, smokers, drinkers, skateboarders)
Automobile traffic	Private security, cameras, regulation, symbolic restrictions
Car-related paraphernalia (railings, overpasses, gas stations, etc.)	Activities fragmented by age, ethnicity, race, occupation, sexual orientation
Spaces abandoned to market forces	Segregation (red light districts, ghettoes, gated communities)
Public retreat to private cars	Privatization (rental chains, malls, privatized road infrastructure, public space rented out for commercial events, corporate plazas, BIDs, theme parks, pseudo-historic quarters, "thirds-spaces")

Under-managed space

Southeast Asia fares worse than the West in terms of public space neglect and car invasion. In inner cities, litter, stench, and signs of vandalism owe to lower public resources for public space maintenance but also to a different attitude toward public space and law enforcement. For example, street vendors, cars, and motorbikes take over sidewalks, hampering pedestrian movement (see Figure 14.5). However, sometimes this is acceptable to local residents because of the

Figure 14.5 Traffic congestion and restricted pedestrian space in Manila
Image credit: Tigor Panjaitan, Dorina Pojani, and Sébastien Darchen

hierarchical and negotiating culture described earlier. These types of cultural practices can be a significant barrier to a well-functioning public space (see Stead and Pojani 2017).

Changing lifestyles and socio-economic milieus play a role in public space neglect. While the "old city" trading areas remain busy during working hours, some turn dead in the evenings. In the not-very-distant past, these spaces bustled with life until late at night. The diminishment or even abandonment of nightlife in the old city has to do with a growing separation of residential and commercial functions. In the industrious port cities of the past, merchants lived and worked in their *shophouses*,[12] which are now treated only as a workspace. For living, they prefer quiet and comfortable residential neighborhoods, away from the crowds and noise of the city center. The development of modern business and shopping centers, or even entirely new CBDs, has taken away some of the traditional customer base of shophouses, thus reducing the attraction of old cities, and activating a cycle of neglect (Ismail and Shamsuddin 2005). To some extent, this mirrors the experience of mono-functional Western CBDs (particularly American ones).

Another type of under-managed space is that encompassed by informal settlements – e.g. Indonesia's *kampung* or Thailand's *khlong* housing (see Figure 14.6). Typically, these are located on marginal land in urban outskirts or the inner city and may lack basic urban infrastructure and services. However, the public spaces therein (such as streets and alleys) may be intensely used for economic production and socialization. While some commenters praise the vernacular aesthetic and picturesque authenticity of informal settlements, to others this

Figure 14.6 Neglected informal settlement in Yangon
Image credit: Tigor Panjaitan, Dorina Pojani, and Sébastien Darchen

type of urbanity is shocking (see Pojani 2019). However, most informal settlements in Southeast Asia are immune from the extreme violence of Western ghettos because social hierarchies and religious controls apply here as in other residential spaces.

The effects of car invasion are also widespread and devastating. Due to rapid urbanization, Southeast Asian countries find themselves burdened with sprawling and motorizing megacities, without adequate mass transit systems. While streets and other public spaces accommodate a mass of pedestrians, car and motorcycle gridlock is commonplace too. The typical policy response to the growth in motorization and congestion has been the construction of additional road infrastructure, which has then displaced the social, economic, and cultural functions of road corridors. Both users' choices and governments' poor decisions (or neglect) are responsible for the invasion of public space by automobiles in Southeast Asia. Comprehensive and integrated long-term visions for urban sustainability and livability are rare (Stead and Pojani 2017).

Over-managed space

All the over-management issues raised with regard to the West have emerged in Southeast Asia in a major way. All countries have seen an increase in disposable incomes but also more economic polarization and physical segregation (Stead and Pojani 2017). Moreover, gated communities, both high- and low-rise, have become ubiquitous in cities ranging from

Manila to Vientiane (Sajor 2003; Shatkin 2008). In a hierarchical, heterogeneous, post-colonial, and status-conscious society, these are sought after by those with financial means and tolerated by the rest.[13] A growing sense of insecurity, intolerance, and suspicion of others play a role too (Douglass 2008; Hogan 2012; Tedong et al. 2014). For example, fear of crime is a main reason cited by wealthier Malaysians for retreating into private gated communities, although Malaysia has among the lowest crime rates in the world (Douglass 2008; Tedong et al. 2014). This preference for exclusivity has trickled down to the lower strata of society. It is not uncommon for poor urban *kampung* to be gated as well (Tedong et al. 2014).

With the advent of malls (see Figure 14.7), commercial activities are becoming increasingly segregated by class and income (Connell 1999; Leisch 2002; Firman 2004; Harun and Said 2009; Pomeroy 2011; Lim 2013; Robertson 2018). Spending most of one's time in air-conditioning is a crucial marker of status in a tropical climate (heat and humidity cause perspiration and foul smells, which locals frown upon). The apolitical environment of consumer-oriented spaces is acceptable to local populations that are accustomed to strong behavioral controls, both while indoors and outdoors (Tedong et al. 2014). In Islamic parts of Southeast Asia, malls (and indoor spaces more generally) offer a layer of symbolic protection for women.

Next to the standard Western-style malls, a new building typology has emerged in Southeast Asian cities: "total lifestyle megamalls," which, in addition to retail space, include sleek housing and services, and are often integrated with transit stations. Muang Thong

Figure 14.7 Open concept mall in Bandung

Image credit: Tigor Panjaitan, Dorina Pojani, and Sébastien Darchen

Figure 14.8 Singapore's Chinatown

Image credit: Tigor Panjaitan, Dorina Pojani, and Sébastien Darchen

Thani in Thailand, Grand Phnom Penh International City in Cambodia, Lippo Karawaci and Bumi Serpong Damai in Indonesia are only but a few examples (Shatkin 2008; Lim 2013; Tedong et al. 2014). In a few cases, concept malls are created which resemble pedestrianized public streets but are under private ownership and management (e.g. Greenbelt Mall in Manila).[14] However, the presence of modern malls has not led to an abandonment of traditional markets.

In addition to big-ticket items, privatization is also creeping up in more subtle ways. For example, some governments share the cost of public space projects, such as parks, with private developers. These co-funded spaces end up filled with (sometimes unaffordable) commercial activities in order to ensure a return on investment for the private partners. As in the West, some formerly productive ethnic enclaves, such as Chinatown in Singapore, have turned into tourist playgrounds (see Figure 14.8). Consumerist values seem to have taken hold everywhere.

Conclusion

Southeast Asia is so diverse and rich in history and culture that any generalizations and theorizations are difficult. However, based on the foregoing review and analysis, this chapter concludes that, in the contemporary era of globalization and internet-based communication, public space production and consumption in the region has similarities to the West. Neglect of "formal" and "informal" public space, and invasion by automobiles and motorcycles is ever-present. With governments in retreat from physical planning, the private sector plays an ever-growing role in creating and managing new forms of public spaces, the most conspicuous of which is, of course, the shopping mall. In the era of globalization, public space worldwide is becoming more homogeneous.

However, Western-based frameworks of public space management do not entirely fit the Southeast Asian context. These must be adjusted to include other aspects intrinsic to the local culture, society, and politics, which affect both public space and the public sphere. At a general level, to be successful, research studies or urban design projects based in Southeast Asia must take the following factors into consideration: ethnic fragmentation and diversity; persisting state authoritarianism; collectivist as opposed to individualistic values; rigid social hierarchies; strong patriarchy and subordination of women; critical role of religion in society; strict separation of public and private life, and a lack public space for political protest. Local wisdom must be called upon to guide policy and development.

Notes

1 In Indonesia alone, more than 700 languages are spoken, there are more than 600 ethnicities, and about six main religions are practiced, including Islam, Protestantism, Catholicism, Hinduism, Buddhism, and Confucianism (in addition to myriad local beliefs).

2 E.g. Chinese dominion of Vietnam (111 BC–AD 39) and the capturing of Taiwan by the Ming Dynasty in the 1700s.

3 For example, the cultural clashes between Buddhist Thais (the majority) and Thai-Malay Muslims in Southern Thailand lead to demands for special autonomy.

4 The concept of "face" is crucial for maintaining social prestige and achieving upward mobility, while "loss of face" can threaten those (Ho et al. 2004). In Indonesia, the spirit of community-based togetherness, collectivity, and mutual support is captured by the term *gotong-royong*. To affirm *gotong-royong* and practice gratitude (*kenduri*), male community members gather for afternoon meals, which usually are accompanied by religious rituals and praying for health, peace, rain, or other common needs.

5 Including the brutal military campaign against Rohingya Muslims in Myanmar, the discrimination or even violence against people of Chinese descent in Indonesia (*Chindos*), the ethnic-religious cleansing during Khmer Rouge's reign of terror in Cambodia, and persecution of Uyghurs in China.

6 In some ways, social hierarchy facilitated the colonization of the Southeast Asian region, as local kings were replaced by foreign rulers while the remainder of the social pyramid remained unaltered.

7 Based on Islamic doctrine, people in a position of authority or superiority are expected to take care of their subordinates or those less fortunate. This extends to the *sultan*, who cannot simply demand respect but is also presented with obligations towards his subjects.

8 An interesting fact that illustrates the degree of separation between men and women is *nüshu*, a script derived from Chinese characters that was used exclusively by women in the Hunan province of southern China since the Song and Yuan dynasty (13th century). Similarly, transgender women in Malaysia (*maknyah*) use a secret language, *Bahasa Seteng* (literally "half-language"), to protect their identity.

9 E.g. suspension of the death penalty under former President Gloria Macapagal-Arroyo.

10 The principle of egolessness assumes that humans are only but a speckle within nature; hence any self-aggrandizing stances are unwarranted.

11 Initially, these technologies were under the exclusive province of the white ruling elite to help regulate trade relations and keep the colonies under control. However, over time, as the end of the century drew nearer, indigenous people were increasingly able to access and use the technological devices of the era. In the 1880s, Singapore was already a printing press hub that supplied Islamic and Indo-Malay communities, as well as Christian missionaries, with books, newspapers, and pamphlets (Frost 2004).

12 Vernacular buildings, two-three stories high, with a shop on the ground floor and a residence above the shop.

13 Residential segregation based on class is hardly new to Southeast Asian cities. The colonizing elites set the example by separating from the rest of society into white-only zones. Segregation based on ethnicity (for example, Chinese, Arab, or Indian) was also common in the colonial era. This type of sorting was handy in controlling rebellious locals. In Indonesia, indigenous status (*pribumi*, literally "islander") was considered as a key distinction – and remained so in the post-colonial period (Tedong et al. 2014).

14 Formerly isolated countries, such as Cambodia, are poised for a further mall boom to catch up with the rest of the region (Kimsay 2013). In other countries, such as Malaysia, they are starting to decline in popularity due to retail glut – for example, there are 255 malls in the Klang Valley alone, which encompasses Kuala Lumpur and Putrajaya (Achariam 2017). This is the result of a shortsighted economic agenda centered on consumerism.

References

Achariam, N. (2017) "So Many Malls, So Few Shoppers." *The Malaysian Insight*, 31 March.

Azada-Palacios, R.A. (2013) "Hannah Arendt and the Possibility of Creating a Space for Religion in the Philippine Public Sphere." Paper Presented at Conference Religion, Secularity and the Public Sphere in East and Southeast Asia, 7–8 March, Asia Research Institute, Singapore.

Ballantyne, T. (2016) *Orientalism and Race: Aryanism in the British Empire*. Amsterdam: Springer.

Brenner, S. (2011) Private Moralities in the Public Sphere: Democratization, Islam, and Gender in Indonesia. *American Anthropologist*, 113(3): 478–490.

Carmona, M. (2010a) Contemporary Public Space: Critique and Classification. Part One: Critique. *Journal of Urban Design*, 15(1): 123–148.

Carmona, M. (2010b) Contemporary Public Space: Critique and Classification, Part Two: Classification. *Journal of Urban Design*, 15(2): 157–173.

Chifos, C. and Yabes, R. (2000) *Southeast Asian Urban Environments: Structured and Spontaneous*. Tempe: ASU Center for Asian Research.

Clammer, J.R. (2002) *Diaspora and Identity: The Sociology of Culture in Southeast Asia*. Selangor: Pelanduk Publications.

Connell, J. (1999) Beyond Manila: Walls, Malls, and Private Spaces. *Environment and Planning A*, 31(3): 417–439.

Daniere, A. and Douglass, M. (2009) *The Politics of Civic Space in Asia: Building Urban Communities*. London: Routledge.

Douglass, M. (2008) Civil Society for Itself and in the Public Sphere: Comparative Research on Globalization, Cities and Civic Space in Pacific Asia. In *Globalization, the City and Civil Society in Pacific Asia*, Routledge, pp. 45–67.

Dunlop, N. (2017) *"Will Bangkok's Street Food Ban Hold?"* *The Guardian*, 27 August.

Evers, H.D. and Korff, R. (2000) *Southeast Asian Urbanism: The Meaning and Power of Social Space*. Münster: LIT Verlag.

Fauzia, A. (2013) "Faith and the State: A History of Islamic Philanthropy in Indonesia." PhD diss., University of Melbourne.

Firman, T. (2004) Demographic and Spatial Patterns of Indonesia's Recent Urbanisation. *Population, Space, and Place*, 10(6): 421–434.

Gholamhosseini, R., Pojani, D., Mateo-Babiano, I., Johnson, L. and Minnery, J. (2018) The Place of Public Space in the Lives of Middle Eastern Women Migrants in Australia. *Journal of Urban Design*, 24 (2): 269–289.

Gibert, M. (2013) "Urban Transition and Public Space in Vietnam: A View from Ho Chi Minh City Street." Paper Presented at 8th Asian Graduate Forum on Southeast Asian Studies, Singapore, 24 July.

Gunawan, S.R., Nindyo, S., Ikaputra, I. and Bakti, S. (2013) Colonial and Traditional Urban Space in Java: A Morphological Study of Ten Cities. *Dimensi*, 40(2): 77–88.

Habermas, J. (1991) *The Structural Transformation of the Public Sphere: An Inquiry into a Category of Bourgeois Society*. Boston: MIT Press.

Harun, N.Z. and Said, I. (2009) "The Changing Roles of Public Spaces in Malaysia." Paper Presented at National Landscape Seminar, International Islamic University, Malaysia, 25–26 March.

Ho, D.Y., Fu, W. and Ng, S.M. (2004) Guilt, Shame and Embarrassment: Revelations of Face and Self. *Culture Psychology*, 10: 64–84.

Hofstede, G., Hofstede, G.J. and Minkov, M. (2010) *Cultures and Organizations: Software of the Mind*. London: McGraw Hill.

Hogan, T. (2012) Manila's Urbanism and Philippine Visual Cultures. *Thesis Eleven*, 112(1): 3–9.

Huat, C.B. (2008) Southeast Asia in Postcolonial Studies: An Introduction. *Postcolonial Studies*, 11(3): 231–240.

Ismail, N.H. (2016) "How Jakarta's First Chinese Indonesian Governor Became an Easy Target for Radical Islamic Groups." *The Conversation*, 7 November.

Ismail, W.H.W. and Shamsuddin, S. (2005) "The Old Shophouses as Part of Malaysian Urban Heritage: The Current Dilemma." Paper Presented at *8th International Conference of the Asian Planning Schools Association*, Penang, 11–14 September.

Kimsay, H. (2013) "City Poised for Shopping Mall Boom." *The Phnom Penh Post*, 28 August.

Leisch, H. (2002) Gated Communities in Indonesia. *Cities*, 19(5): 341–350.

Lim, M. (2007) Transient Civic Spaces in Jakarta Demopolis. In *Globalization, the City and Civil Society in Pacific Asia: The Social Production of Civic Spaces*, Routledge, pp. 211–230.

Lim, W.S. (2013) *Public Space in Urban Asia*. Singapore: World Scientific.

Limapichart, T. (2009) The Emergence of the Siamese Public Sphere: Colonial Modernity, Print Culture and the Practice of Criticism (1860s-1910s). *South East Asia Research*, 17(3): 361–399.

Mehta, V. (2018). Space, Time and Agency on the Indian Street. In: *The Palgrave Handbook of Bottom-Up Urbanism*, Springer International Publishing, pp. 239–253.

Morgenbesser, L. (2016) *Behind the Façade: Elections under Authoritarianism in Southeast Asia*. Albany: SUNY Press.

Oranratmanee, R. and Sachakul, V. (2014) Streets as Public Spaces in Southeast Asia: Case Studies of Thai Pedestrian Streets. *Journal of Urban Design*, 19(2): 211–229.

Pojani, D. (2019) The Self-Built City: Theorizing Urban Design of Informal Settlements. *Archnet-IJAR: International Journal of Architectural Research*, 13(2): 294–313.

Pojani, D. and Maci, G. (2015) The Detriments and Benefits of the Fall of Planning: The Evolution of Public Space in a Balkan Post-Socialist Capital. *Journal of Urban Design*, 20(2): 251–272.

Pomeroy, J. (2011) Defining Singapore Public Space: From Sanitization to Corporatization. *Journal of Urban Design*, 16(03): 381–396.

Reinelt, J. (2011) Rethinking the Public Sphere for a Global Age. *Performance Research*, 16(2): 16–27.

Rinaldo, R. (2008) Envisioning the Nation: Women Activists, Religion and the Public Sphere in Indonesia. *Social Forces*, 86(4): 1781–1804.

Robertson, H. (2018) "The Pristine Exclusivity of Cambodia's 'Imported Cities.'" *Nextcity*, 13 Feb.

Ruddick, S. (1996) Constructing Difference in Public Spaces: Race, Class, and Gender as Interlocking Systems. *Urban Geography*, 17(2): 132–151.

Sajor, E. (2003) Globalization and the Urban Property Boom in Metro Cebu, Philippines. *Journal of Urban Affairs*, 34(4): 713–742.

Salmenkari, T. (2014) Consensus and Dissensus in the Public Sphere: How East Asian Associations Use Publicity. *Studia Orientalia Electronica*, 2: 16–36.

Sandercock, L. (1997) *Towards Cosmopolis: Planning for Multicultural Cities*. London: Wiley.

Sangwha, L. (1999) The Patriarchy in China: An Investigation of Public and Private Spheres. *Asian Journal of Women's Studies*, 5(1): 9–49.

Shatkin, G. (2008) The City and the Bottom Line: Urban Megaprojects and the Privatization of Planning in Southeast Asia. *Environment and Planning A*, 40(2): 383–401.

Sien, C.L. (2003) *Southeast Asia Transformed: A Geography of Change*. Singapore: Institute of Southeast Asian Studies.

Stead, D. and Pojani, D. (2017). The Urban Transport Crisis in Emerging Economies: A Comparative Overview. In *The Urban Transport Crisis in Emerging Economies*, Springer, Cham, pp. 283–295.

Tedong, P.A., Grant, J.L., Wan Abd Aziz, W.N.A., Ahmad, F. and Hanif, N.R. (2014) Guarding the Neighborhood: The New Landscape of Control in Malaysia. *Housing Studies*, 29(8): 1005–1027.

Wang, D. (2003) *Street Culture in Chengdu*. Palo Alto: Stanford University Press.

Yongjia, L. (2013) Turning Gwer Sa La Festival into Intangible Cultural Heritage: State Superscription of Popular Religion in Southwest China. *China: An International Journal*, 11(2): 58–75.

Yuen, B. and Chor, C. (1998) Pedestrian Streets in Singapore. *Transportation*, 25(3): 225–242.

15

RIGHT TO THE CITY (AT NIGHT)

Spectacle and surveillance in public space

Su-Jan Yeo

Introduction

In the twenty-first-century globalized world, time is expanding through a technological epoch by which contemporary cities remain in a wakeful state of 24/7 hyper-connectivity; thus, emboldening planning strategies aimed at boosting urban productivity and consumption after dark. Concurrently, space is contracting through the rapid pace of urban development and dense confluence of an ever-diverse urban society; thereby, elucidating one's right to the city (or lack thereof).

This right to the city is further compounded at night as various actors—from public sector agencies and corporate entities to ordinary citizens—seek to shape the nocturnal culture of public space. And, in doing so, raising crucial questions: How does the night change our affective experience of public space? What actions, both state-driven and collectively-enacted, define the use of public space after dark? Who is in/excluded in the city at night through such actions, and what are the planning implications of these in/exclusions? This chapter will explore public space and the right to the city at night by way of two linked strands. The first, spectacle, highlights the opportunistic gains of the nighttime economy for urban regeneration, placemaking, and street vitality. The second strand, surveillance, points to the contentious practices of monitoring people in public space as a means to curb instances of anti-social behavior and transgressive activities, especially after dark.

The ethnic neighborhood of Little India in Singapore offers a unique cultural nightlife experience that attracts locals, tourists, and south Asian foreign workers (many of whom are employed in low-skilled, low-waged manual jobs). On any ordinary night, Little India's commercialized streets transform into sites of rest and revelry as people gather to shop, socialize, and soak in the urban spectacle. Conversely, these very same streets also became sites of riot at night when, in 2013, a fatal bus accident involving the death of an Indian migrant worker sparked a violent conflict between hundreds of foreign workers and the police. In the aftermath of the Little India 'riot,' surveillance was increased through additional installations of lighting, streetlamps, and police cameras in public space. By examining these two disparate vignettes of public space at night, this chapter aims to illuminate the complex entanglement between exerting surveillance, on the one hand, and promoting spectacle, on the other, while advancing the city at night as a common good and a right for all.

Public space and the nocturne

Urban life is shaped as much by time as by space. In a twenty-first century globalized world that is increasingly networked through digital and wireless technologies, time is no longer viewed as a constraint but rather a resource to be exploited in pursuit of economic modernization (Melbin 1987; Kreitzman 1999). The conventional unit of a 'nine-to-five' lifeworld is increasingly challenged by the postindustrial temporality of '24/7 capitalism' which, in turn, is transforming the nature of urban life in the contemporary city (Crary 2013). Moreover, in the competitive climate of this twenty-first century marketplace, the deliberate stimulation of new economic growth sectors in the lexicon of lifestyle and culture has produced myriad spaces of consumption which come alive at night. In effect, the contemporary city is an 'entertainment machine'—one in which '[e]ntertainment is the work of many urban actors' (Lloyd and Nichols Clark 2001, 358).

The production of nightlife as a form of entertainment involves multiple actors ranging from public sector agencies and corporate entities to ordinary citizens. At one level, planning authorities are developing policies and (de)regulations that boost urban productivity and consumption in the hours of darkness (Wolifson and Drozdzewski 2017). Urban rejuvenation schemes toward aestheticizing and programming the city after dark underscore a political mandate, which seeks to position nightlife as a symbolic expression of modernity, urbanity, and prosperity. The prevalent '24-hour' urban strategy, in particular, is illustrative of this attentive shift in policy- and plan-making which aims not only to manipulate the temporal dimension of the city but also the visual and experiential elements associated with urban life beyond the daytime realm. Here, the 'nightscape' is illuminated and 'imagineered,' often with great financial investment, to further attract capital, business, and tourism. And, in today's era of digital media, such efforts to produce and circulate symbols of 'global city' aspirations are reaching audiences on a worldwide scale.

In this global city context, the built environment is shaped more and more through state-led modalities of civic boosterism, urban entrepreneurialism, and engagement with the private sector (Harvey 1989); thus, effectively permeating neoliberal governance and planning of the 'nighttime economy' (Lovatt and O'Connor 1995; Hobbs et al. 2000; Hollands and Chatterton 2003; Shaw 2015). The co-creation (and, in some instances, co-branding) of nightlife entertainment districts by planning authorities and private sector contractors—through, for example, instruments such as design competitions, requests for proposals, and tenders—is a common practice by which to supply lifestyle activities and cultural experiences for the masses. However, there are also criticisms that point to the exclusivities and inequities engendered by such (carnivalesque and privatized) spaces of consumption as being less than inviting to all and, therefore, promulgating 'social divisions and lifestyle segmentations' (Hollands 2002, 168; see also May 2014).

At another level, ordinary citizens who traverse the nocturnal city through their nighttime activities in public space—mingling on the street, people-watching on the pavement, resting in the park, and so on—might readily identify with the duality of the nocturnal city as a distinct spatio-temporal milieu that is simultaneously characterized by laissez-faire spontaneity and routinized habituation (Yeo et al. 2016). The perceived sense of freedom, on the one hand, and the reverence for structure, on the other, informs individual and collective practices and behaviors in public space at night. As such, the plural meanings and orders attached to public space in the hours after dark have direct implications for planning and urban design. More interestingly, the prevalence of economic migrants working, living, and playing in the contemporary city—in other words, the nouveau stranger 'who comes today and stays tomorrow'—have been explored in relation to rising discords and dissensions associated with urban 'super-diversity' and 'difference' (Simmel 1950, 402; Vertovec 2007; Valentine 2013).

The potential for discords and dissensions emerges when the 'planned' nightscapes and the 'multiple publics' that gather in these formally-arranged spaces paint a disjointed picture of a normative night out, where 'normative' becomes the acceptable version of nightlife as explicitly shaped by the state through its regulations and controls. And, in this way, elucidating one's right to the city at night (or lack thereof). Here, the 'right to the city' refers to a new paradigm for urban inclusion that has taken the international stage through the Global Platform for the Right to the City initiative and defined as:

> the right of all inhabitants, present and future, temporary and permanent, to use, occupy and produce just, inclusive and sustainable cities, villages and settlements, understood as a common good essential to a full and decent life.
>
> *(GPR2C 2019)*

The city at night, on the one hand, comprises sites of spectacle through processes of urban regeneration and placemaking that promote spaces of mass leisure where street vitality is often times a crucial element. On the other hand, the increasingly overt surveillance of these streets after dark as a means of curbing 'anti-social' behavior and 'transgressive' activities might encourage liberation (and inclusion) of some groups yet engender alienation (and exclusion) of 'Others.' The following case study of Little India points to the paradox between spectacle and surveillance; and, in so doing, identifies key lessons toward advancing the city at night as a common good and a right for all.

Little India: a tale of two nights

Singapore's urban core is a legacy of the island-nation's past as a former British colony when planning policies of the time instituted racial districts for the Chinese, Malay, and Indian communities. In 1989, these historically and architecturally significant districts—named Chinatown, Kampong Glam, and Little India, respectively—were gazetted in a conservation master plan by the Urban Redevelopment Authority (Singapore's government agency on land use planning and conservation). Over the course of three decades, a combination of state-led and market-driven actions has contributed to the transformation of Chinatown, Kampong Glam, and Little India.

First, the conservation of vernacular shophouse buildings and road networks has helped to ensure that these heritage areas retain their distinct urban form amid the high-rise, high-density built environment of contemporary Singapore. Second, conservation status has boosted cultural tourism through the positioning and promotion of these heritage areas as unique visitor attractions. Third, conservation guidelines applied to these heritage areas allow for instances of adaptive re-use which, in recent years, has facilitated the introduction of new commerce activities alongside older businesses. The effect of retail gentrification has most notably transformed Chinatown and Kampong Glam into lifestyle enclaves for the middle class and creative industries (Ho and Hutton 2012). Little India by contrast is still visibly an ethnically-bound space articulated and exemplified by the material culture and social practices of an Indian sub population (Hee 2017). In this regard, Little India is often described as imbuing an 'authentic heritage' whose appeal attracts locals and tourists alike in search of an Indian-oriented experience in Singapore (Baker 2016).

Sites of spectacle: the production of carnivalesque play

Hear the sound of temple bells. Taste delicacies rich in flavour. Feel the softness of silk draped around you. See a burst of colours unfold before your eyes. Smell the fragrance of spices that tell of the aura of a culture.

(LISHA 2017)

The excerpt above, from the Little India Shopkeepers and Heritage Association website, is lyrical in tone and yet it underscores the very real spectrum of sensorial experiences that can be encountered routinely, day and night, in Little India and, above all, unequaled with experiences elsewhere in Singapore. Here, formal business establishments (restaurants, mini-marts, and shops providing specialty foods, garments, and services) co-exist with informal vendors (selling trinkets, long-distance calling cards, and fresh coconut among other small commodities) and in proximity to several places of worship (for Hindus, Muslims, and Buddhists)—altogether generating a habitual rhythm of interactions and exchanges from day to night. The famous Mustafa Centre in Little India is a large shopping complex known not only for its range of bargain goods but also for its 24-hour operations. There are also several eateries that remain open past midnight, drawing patrons in search of a late-night snack or supper. These semi-public spaces of consumption contribute to Little India's nighttime economy which then creates vitality on the streets after dark.

In this regard, Singapore is ranked one of the safest cities in the world where, according to a 2018 Gallup report, 94% of residents feel safe walking alone at night in their neighborhood (SPF 2018). As such, it is not uncommon to see women, families with young children, and the elderly participating in the nightlife of Little India. Moreover, Little India's compact built form of low-rise shophouses and narrow walkways has limited carrying capacity and, as such, nudges the flow of pedestrians and ground-level activities to spill over on the streets. On Sundays, in particular, the added presence of up to 100,000 south Asian migrant workers congregating in Little India (MHA 2014, 17) leads to significant crowding in public space, as shall be discussed further in the following section.

When the sun sets, Little India's ambient quality intensifies through a shambolic arrangement of lights that have functional and decorative utility. These light installations illuminate commonplace elements, casting attention to ordinary façades and fenestrations as well as makeshift stalls and mundane merchandise. The cumulative lighting after dark engulfs the spatially and ethnically bound district of Little India in a luminous glow, creating a temporal spectacle that can only be experienced at night. A more spectacular level of night lighting is staged annually in Little India for a period of two months during the Hindu festival of Deepavali (Festival of Lights). Here, the elaborate light-up ceremony involving the ornamental display of 2 million colored light bulbs transforms the arterial streets of Little India into 'a fantasyland of colourful arches and stunning lights' (Little India Shopkeepers & Heritage Association (LISHA) 2017; Heng 2018). The 'image and meaning' of this particular street setting are described by Mehta (2013) as a 'celebration street' for it provides the arena upon which to observe and participate in the open display of identity and community (18–19). During the Deepavali celebration, Little India also serves as a site for festival bazaars, cultural performances, and organized activities—many of which are scheduled at night and situated in outdoor public spaces such as open fields and pedestrianized streets.

The night presents visitors to Little India with myriad opportunities for carnivalesque play; yet, more often than not, the enactment of carnivalesque play at night tends to draw out distinctions of civilized/desirable and uncivilized/undesirable behaviors. In other words, '[c]ertain sections of the city may desire and/or stand to benefit from the city as carnival, but others may see it as a re-assertion of threat' (Lovatt and O'Connor 1995, 132). As Hubbard (2013) argues, this attention toward (in)appropriate forms of carnivalesque play at night can invoke public sentiments and social stereotypes that cast certain social groups as being more problematic and abject in their embodiment of nighttime leisure 'with their bodies becoming the focus of surveillance and media scrutiny' (277).

Sites of surveillance: the Little India 'riot' and its aftermath

Little India is a popular leisure space for south Asian migrant workers. In 2018, migrant workers occupying low-paid, low-skilled jobs in the construction sector represented approximately 20% (280,500 individuals) of the total foreign workforce (1,386,000 individuals) in Singapore (MOM 2019). The city-state is dependent on the inflow of migrant workers to help realize the economically driven aspirations that underpin much of urban development in contemporary Singapore. Little India, therefore, plays an important role as a hub of affordable and familiar goods and services which accommodate discerning needs and preferences of migrant workers from south Asia—and, more significantly, as the locus of collective experiences shared by a particular group of ethnic identities.

At night, and especially on Sundays (typically a 'rest day' for migrant workers), the streets and open fields of Little India fill with the corporeality of male migrant workers as they shop for weekly groceries and personal supplies, remit money to family members in their home countries, socialize with compatriots and, for the most part, pass time in these public spaces. For these migrant workers, Little India is not only a space for nighttime leisure where they partake in the act of being both spectacle and spectator but, in so doing, their forms of carnivalesque play (carom playing near an alley, picnicking on a vacant open field, newspaper reading on the curb, perusing streets vendors, and so on) reinforce kinship and routine. Yet, migrant worker identities (and welfare) present a fragmented landscape of contested and negotiated issues relating to the transience, vulnerability, and integration of this population within Singapore society (see also Harrigan and Koh 2015). There is distinction in the status and, by extension, treatment of '"foreign talents" in white-collar jobs' vis-à-vis 'unskilled foreign labor' that manifest through state structures and policies pertaining to migration and economic development in Singapore (Kaur et al. 2016, 28). From this viewpoint, migrant workers are seen as playing temporary yet necessary roles in 'dirty, difficult and dangerous' jobs that are unattractive to Singaporeans (30). Furthermore, the common reference to migrant workers as a transient segment of society reinforces differential treatment and underscores the lack of social relations between them and Singaporeans.

On Sunday 8th December 2013, an otherwise ordinary night in Little India turned for the unexpected through a two-hour altercation triggered by a fatal road accident involving a migrant worker of Indian nationality who died under the wheels of a private bus. A crowd of migrant workers in the vicinity at the time of the accident reacted angrily by attacking the bus. When the Singapore Police Force and Singapore Civil Defence Force arrived at the scene, the mob had grown in size and the unrest became more turbulent with dissidents overturning and setting ablaze emergency vehicles. Local mainstream media reporting of the event dubbed it a 'riot' and, according to its press coverage, the mob is estimated to have massed around 300 rioters (Lim and Sim 2014). In the context of Singapore, where strict laws promote a high degree of public order, the magnitude of this 'riot' is unparalleled and represents only the second but severest incident of public unrest in over than 40 years. More interestingly, the initial unfolding of this 'riot' on the street and its prolongation for two hours in the very same location points to a different image and meaning of the street from the earlier vignette of a 'celebratory street.' Here, the 'protest street' facilitates expressions of dissent by protestors seeking to expose their oppressions, grievances, and demands in a highly visible space (Mehta 2013, 19).

Mainstream media coverage on the Little India 'riot' shone a light on the circumstances of low-waged, low-skilled migrant workers in Singapore, thereby opening out the discourse from the familiar purview of policymakers to the overtly public conversations of ordinary Singaporeans. These conversations unearthed the rising discontents and displeasures felt by Singaporeans regarding the increasingly unavoidable presence of migrant workers in the public spaces of everyday life, including the 'strongly ethnicised "weekend enclaves"' that emerge on a temporal basis in places such as Little India (Yeoh et al. 2017, 645). Representation of the Little India 'riot' by local mainstream media also problematized such 'weekend enclaves' vis-à-vis the night as an especially precarious time of deviance, disorder, and danger. In this way, prompting authorities to not only respond to requests for greater security in public spaces and gathering areas where migrant workers often congregate but also to seize the policy windows which opened to 'encourage some workers to stay at their dormitory rather than travel out to a congregation area [such as Little India]' by working with operators and vendors to provide services and amenities within the dormitory (Ministry of Home Affairs (MHA) 2014, 65).

In the aftermath of the Little India 'riot,' several measures were implemented to deter anti-social behavior and transgressive activities. For example, the number of police cameras installed throughout Little India increased from 113 to 250 in the 12 months after the Little India 'riot' with plans for 88 more cameras by the following year, while police patrol was also stepped up. Furthermore, 100 streetlamps were erected to increase lighting at 42 locales which include alleys and back lanes (Sim 2014). Additionally, the designation of Little India as a liquor control zone introduced a ban on alcohol sales during specified days and times when large numbers of migrant workers are known to congregate in this spatially and ethnically bound district. It is this top-down 'taming' of the night through surveillance and control of public space that deserves closer examination from a planning and urban design perspective.

Right to the city (at night)

The dark of night affords an occasion to transform (directly or indirectly) the aesthetic, ambient, and affective qualities of public space through, most notably, the presence or absence of lighting and security. In this sense, the city at night engenders a myriad of human practices and activities with social, economic, and political dimensions. When examining the city at night from the standpoint of leisure and, more specifically, within the contextual case of Little India, two opposing orientations arise amid the notion of 'leisure.' First, the view that '[l]eisure must break with the everyday' as a means of liberation from habitual obligations is perpetuated in advanced capitalist societies (Lefebvre 2014, 55) and finds expression in the production of nightlife for the masses. The case of Little India is an illustration of government and business actions toward promoting an ethnic neighborhood as a tourist destination and nightspot for cultural activities and entertainment (in other words 'spectacles') worthy of attracting/distracting the gaze. This 'gaze,' as Williams (2008) argues, will vary from day to night and its 'deployment' is a barometer of power and hegemony (519).

Second, within the structural and 'general need for leisure' framework, there is also what Lefebvre (2014) describes as 'differentiated concrete needs,' which are increasingly determined and stratified by social group (54–55). The congregation of south Asian migrant workers in Little India, particularly at night and on Sundays, must be acknowledged as a perpetual activity that is more than mere mass leisure. For this marginalized group, Little India is no more a nightspot than it is a locus of 'everyday (night)life' which satisfies their needs for kinship and routine especially given their transient status in Singapore (Yeo, Hee, and Heng 2012). Their embodied and gendered practices and behaviors in public space— from the informal 'occupation' of streets and pavements to the social 'loitering' in open spaces—reinforces identity and shared experiences yet elucidates difference. It is this state of difference and '[t]he "othering" of strangers' that can lead to heightened levels of suspicion, thus making 'surveillance and enclosure of the commons even more compelling' (Banerjee and Loukaitou-Sideris 2013, 349).

The treatment of nightlife as a homogeneous form of mass leisure is in itself problematic because the nuanced outcomes of nightlife production and consumption in public space liberate as well as alienate. As Williams (2008) asserts, '[n]ight spaces organize and mediate the societal meanings and uses of the darkness: where to be with whom, and why, as well as what to do and how to do it during the "when" of darkness' (526). Herein lies a complex entanglement amid the organization and mediation of, on the one hand, nightlife spectacle and, on the other hand, nightlife surveillance. In both regards, there is an implicit relation to public order and control; and, by extension, the making of inclusionary and exclusionary nightlife.

The key lessons arising from the Little India case study in Singapore are threefold— these lessons are intended to serve as policy considerations for planning and designing urban nightscapes that are inclusive and convivial for all. First, it should be acknowledged that most urban agendas today not only espouse a 24/7 ethos but also promote standardized experiences of nightlife. As such, it is imperative to turn our attention toward better understanding the varied uses of public space at night beyond prevalent norms and assumptions by asking:

> When will the site be used: daytime, nighttime, seven days a week, twenty-four hours a day? Are there differences between day and night use or between workdays and weekends? Should the design address the issue of evening or night use? Are there potential conflicts among the foreseen uses? When? Why? How can these conflicts be solved?
>
> *(Palazzo and Steiner 2011, 40–41)*

Second, expanding public receptivity for the varied possibilities of public space use after dark is critical if nightlife is to be viewed holistically as a positive urban asset. Places existing at the fringe can be accepted as pleasantly lively, economically vibrant, and socially sustainable at night—even if somewhat liminal—by way of fair media representation. Third, more imaginative thinking is needed to generate alternative methods to the contentious practice of surveillance. People-centric solutions of managing crowd behavior and promoting public safety can be explored through, for example, innovative lighting design and interactive installations as well as community-derived programming ideas for public space use at night. In short, the enjoyment of nightlife as a convivial spectacle that shapes and is shaped by public space depends as much on cultivating attractive attitudes toward living with difference as it does on curbing unattractive attitudes that pose threats to safety and inclusion.

The contemporary city in an age of globalization constitutes 'multiple publics' and public space is an amenity by which people from all walks of life can partake—individually, with friends, or among strangers—in uninhibited modes of nighttime leisure that include promenading, talking, mingling, resting, and people-watching among many other forms of urban life. As demonstrated by this chapter, these nightly interactions and exchanges, however ordinary and mundane, enrich the lived experiences of urban dwellers especially those whose sense of community and attachment to place are temporally embedded in the city after dark. Public space is a common good both day *and* night; and, hence, fundamental to the social health and well-being of an urban society. The right to the city at night, therefore, is a right to be enjoyed by all.

References

Baker, J.A. (2016) Little India Isn't Messy – It's Authentic. *The Straits Times*, 19 December. Accessed 8 March 2019. www.straitstimes.com/singapore/little-india-isnt-messy-its-authentic

Banerjee, T. and Loukaitou-Sideris, A. (2013) Suspicion, Surveillance, and Safety: A New Imperative for Public Space? In: *Policy, Planning, and People: Promoting Justice in Urban Development*, Carmona, N. and Fainstein, S.S. (eds.). Philadelphia, PA: University of Pennsylvania Press, pp. 337–355.

Crary, J. (2013) *24/7: Late Capitalism and the Ends of Sleep*. New York: Verso Books.

Global Platform for the Right to the City (GPR2C). (2019) "The Right to the City Will Be Included for the Very First Time in an UN Document." Accessed 1 April 2019. www.right tothecityplatform.org.br/espanol-el-derecho-a-la-ciudad-sera-incluido-por-primera-vez-en-un-doc umento-de-la-onu/.

Harrigan, N. and Koh, C.Y. (2015) "Vital Yet Vulnerable: Mental and Emotional Health of South Asian Migrant Workers in Singapore." Accessed 1 April 2019. https://ink.library.smu.edu.sg/soss_research/1764/.

Harvey, D. (1989) From Managerialism to Entrepreneurialism: The Transformation in Urban Governance in Late Capitalism. *Geografiska Annaler: Series B, Human Geography*, 71(1): 3–17.

Hee, L. (2017) *Constructing Singapore Public Space*. Singapore: Springer.

Heng, M. (2018) "Little India to Celebrate Its 30th Year of Deepavali Light-Up." *The Straits Times*, 12 September. Accessed 1 April 2019. www.straitstimes.com/singapore/little-india-to-celebrate-its-30th-year-of-deepavali-light-up.

Ho, K.C. and Hutton, T. (2012) The Cultural Economy in the Developmental State: A Comparison of the Chinatown and Little India Districts in Singapore. In *New Economic Spaces in Asian Cities: From Industrial Restructuring to the Cultural Turn*, Daniels, P.W., Ho, K.C. and Hutton, T.A. (eds.). London: Routledge, pp. 220–237.

Hobbs, D., Lister, S., Hadfield, P., Winlow, S. and Hall, S. (2000) Receiving Shadows: Governance and Liminality in the Night-Time Economy. *The British Journal of Sociology*, 51(4): 701–717.

Hollands, R. (2002) Divisions in the Dark: Youth Cultures, Transitions and Segmented Consumption Spaces in the Night-Time Economy. *Journal of Youth Studies*, 5(2): 153–171.

Hollands, R. and Chatterton, P. (2003) Producing Nightlife in the New Urban Entertainment Economy: Corporatization, Branding and Market Segmentation. *International Journal of Urban and Regional Research*, 27(2): 361–385.

Hubbard, P. (2013) Carnage! Coming to a Town Near You? Nightlife, Uncivilised Behaviour and the Carnivalesque Body. *Leisure Studies*, 32(3): 265–282.

Kaur, S., Tan, N. and Dutta, M.J. (2016) Media, Migration and Politics: The Coverage of the Little India Riot in the Straits Times in Singapore. *Journal of Creative Communications*, 11(1): 27–43.

Kreitzman, L. (1999) *The 24-Hour Society*. London: Profile Books.

Lefebvre, H. (2014) *Critique of Everyday Life*. London and New York: Verso.

Lim, Y.L. and Sim, W. (2014) "Little India Riot: One Year Later – The Night that Changed Singapore." *The Straits Times*, 6 December. Accessed 5 January 2019. www.straitstimes.com/singa pore/little-india-riot-one-year-later-the-night-that-changed-singapore.

Little India Shopkeepers & Heritage Association (LISHA). (2017) Accessed 7 March 2019. www.littlein dia.com.sg.

Lloyd, R. and Nichols Clark, T. (2001) The City as an Entertainment Machine. In: *Critical Perspectives on Urban Redevelopment (Research in Urban Sociology, Volume 6)*, Gotham, K.F. (ed.). Bingley: Emerald Group Publishing Limited, pp. 357–378.

Lovatt, A. and O'Connor, J. (1995) Cities and the Night-Time Economy. *Planning Practice & Research*, 10 (2): 127–134.

May, R.A.B. (2014) *Urban Nightlife: Entertaining Race, Class, and Culture in Public Space*. New Brunswick, NJ: Rutgers University Press.

Mehta, V. (2013) *The Street: A Quintessential Social Public Space*. New York: Routledge.

Melbin, M. (1987) *Night as Frontier: Colonizing the World after Dark*. New York: Free Press.

Ministry of Home Affairs (MHA). (2014). *Report of the Committee of Inquiry into the Little India Riot on 8 December 2013*. Accessed 7 March 2019. www.mha.gov.sg/docs/default-source/press-releases/ little-india-riot-coi-report—2014-06-27.pdf.

Ministry of Manpower (MOM). (2019) *Foreign Workforce Numbers*. Accessed 8 March 2019. www.mom. gov.sg/documents-and-publications/foreign-workforce-numbers.

Palazzo, D. and Steiner, F.R. (2011) *Urban Ecological Design: A Process for Regenerative Places*. Washington, DC: Island Press.

Shaw, R. (2015) 'Alive after Five:' Constructing the Neoliberal Night in Newcastle Upon Tyne. *Urban Studies*, 52(3): 456–470.

Sim, W. (2014) "Little India Riot: One Year Later - The COI Checklist: What's Being Done." *The Straits Times*, 6 December. Accessed 5 January 2019. www.straitstimes.com/singapore/little-india-riot-one-year-later-the-coi-checklist-whats-being-done.

Simmel, G. (1950) The Stranger. In: *The Sociology of Georg Simmel*, Wolff, K.H. (ed.). Glencoe: The Free Press, pp. 402–408.

Singapore Police Force (SPF). (2018) *Annual Crime Brief 2018*. Accessed 1 April 2019 www.police.gov. sg/news-and-publications/statistics.

Valentine, G. (2013) Living with Difference: Proximity and Encounter in Urban Life. *Geography*, 98(1): 4–9.

Vertovec, S. (2007) Super-Diversity and Its Implications. *Ethnic and Racial Studies*, 30(6): 1024–1054.

Williams, R. (2008) Night Spaces: Darkness, Deterritorialization, and Social Control. *Space and Culture*, 11(4): 514–532.

Wolifson, P. and Drozdzewski, D. (2017) Co-Opting the Night: The Entrepreneurial Shift and Economic Imperative in NTE Planning. *Urban and Policy Research*, 35(4): 486–504.

Yeo, S.J., Hee, L. and Heng, C.K. (2012) Urban Informality and Everyday (Night) Life: A Field Study in Singapore. *International Development Planning Review*, 34(4): 369–390.

Yeo, S.J., Ho, K.C. and Heng, C.K. (2016) Rethinking Spatial Planning for Urban Conviviality and Social Diversity: A Study of Nightlife in a Singapore Public Housing Estate Neighbourhood. *Town Planning Review*, 87(4): 379–399.

Yeoh, B.S., Baey, G., Platt, M. and Wee, K. (2017) Bangladeshi Construction Workers and the Politics of (Im)mobility in Singapore. *City*, 21(5): 641–649.

16

THE STRANGE IDEA OF THE PUBLIC

No, *hiroba* (広一場) is not public space; so, what?!

Darko Radović

Introduction

There is a consensus that a successful city has to have vibrant public spaces, where social, cultural, and political lives unfold, where everyday life gets permeated by the highest expressions of urbanity. That consensus is based on a well-developed, coherent body of theories that enshrine the ideals of urbanity, socio-cultural, and the political character of cities, at the core of which is the idea of *public*. This essay poses a question: what if that is not true? The main aim is to advance discussions about the (im)possibility of translation and equivalence in cross-cultural research, while introducing peculiarities about the place of public life and space in Japanese culture and, in particular, the city of Tokyo.

The reasons for that question lie in the fact that Japanese language has no words to express the concept of "public," which is at the very core of dominant definitions of urban quality (Radović 2010). That lack indicates an absence of, or at least a very unusual situation with the very concept of public in Japanese culture. Neither transcribed *paburiku* [パブリック], nor indigenous *kōkyō* [公共] encapsulate the meaning of "public." *Paburiku* is just a phonetic approximation, which exposes the limitations of *katakana* syllabary script. Katakana is used for adopted foreign terms, and its use here explicitly labels the notion of "public" as foreign. In *kōkyō*, the key ideogram refers to an official, governmental power, even the princely estate, which is profoundly different from the *public* and, thus, transparently wrong. From that, another iteration of our dilemma arises: having no words necessary to express what is public, does Japanese culture have means to generate and live public (space)?

1

The above-mentioned consensus on what makes one city good is based on an ideal(ized) vision of *polis*, the mature city of ancient Greek civilization defined as "the locale where the density, the specific *gravity of discourse* [emphasis added] are greatest" (Steiner 1973, 322). The civilizational value of *polis* is profound. "In his early work, the first book of *Politics*, Aristotle declares how it is evident that the *polis* belongs to the things that exist by nature, and that man is by nature a creature intended to live in a *polis*" (Downey 1973, 316). Cities, thus, encapsulate what it means to be human.

In the steps of Henri Lefebvre, Sennett (1990) elaborates how the city is "a place that per-mits" and "encourages the concentration of differences," regardless of the physical size of the settlement (127). We all know numerous examples of both tiny but distinctly urban settlements, and cases where vast, pseudo-urban conglomerations lack even basic properties of urban life (Bogdanović 1973; Radović 2010). Aristotelian *koinonia* best encapsulates the complex relation-ship between humans and cities, commonly described as *urbanity*. *Koinonia* was understood as "the basic link [...] between the *polis* and the citizen", that "signifies active and purposeful rela-tionship and implies the existence and use of energy" (Downey 1973, 318).

Lefebvre's (1996) critical concept of *the right to the city* projected ancient ideals into the 20th century and beyond. For a metaphilosopher, and for us here,

> the right to the city is like a cry and a demand ... which cannot be conceived of as a simple visiting right or as a return to traditional cities. It can only be formu-lated as a transformed and renewed right to urban life...as long as the 'urban', place of encounter, priority of use value, inscription in space of a time promoted to the rank of a supreme resource among all resources, finds its morphological base and its practico-material realization.
>
> *(158)*

The fact that the city gave names to both politics (*polis*) and civility (*civitas*) places a particular kind of politics—the Greek ideal of democracy—at the core of what good urbanism, in terms of thinking, making, and living cities, should be about. That is where the unsettling questions asked above come to haunt us.

2

In order to address those questions we need to zoom out, from urban theory and the dilemmas that it creates to broader, philosophical levels of thinking. François Jullien opens *The Strange Idea of the Beautiful* (2016), which is devoted to untranslatability of the term *beautiful* into Chinese *mei*, 美 (an equivalent to the Japanese derivative *bi*), by pointing out how "East" never asked a question which Plato has put at the center of Western thought: "I am not asking what 'beautiful is' but *what is* 'the beautiful'" (8). That shift entitles Jullien to propose how whole European "philosophy was born from this added article and is promoted in this displacement" (9). Shadowing his exegesis of the consequences of "detaching the beautiful from what is beautiful," we can think of *the public* and a variety of phenomena that are *public*, of *the urban* and *urban* (9). For Socrates, "unable to be a thing, gold or marble, the beautiful will consequently become a notion" (10). Transposed into our topic, it emerges how, equally *unable to be a physical phenomenon* ("a thing, gold or marble"; materials which, incidentally, were common to classical public spaces) *the* public will consequently become a notion (as etymologically derived from the Greek *ennoia* and Latin *notionem*, "concept, conception, idea, notice") to, eventually, spread globally as simply "Western," through various exercises of power.

The meaning of *the public* is precisely defined there. The question is: what happens to the complex meaning of that culturally significant notion and its rich web of connotations when the word signifying it gets transcribed and appropriated as *paburiku*, or mistranslated as *kōkyō*?

For centuries, linguistic appropriations from Chinese cultures and language to Japanese were seamless. Those cultures are marked by an underlying similarity. But, transplantation and importation of profoundly alien concepts is difficult and poses a number of questions, especially when the original meanings are clearly defined. Once again, a simple *détournement* of Jullien's (2016) thoughts about beauty will help us explain that those who—in China and in Japan—immediately and without further elaboration resort to the (European) notion of the "public" in order to deal with their tradition are, inadvertently, in thrall to an illusion, on the cultural level, that is analogous to an anachronism in the historical order. It is *as though* "a cultural refinement manifested at the highest … *must have* [emphasis added] the concept of the public to crown it" (24).

Historians point out how "prior to the Meiji era (1868–1912), the Japanese language did not have a word for 'private'… The right of privacy was formally recognized in legal ruling for the first time by Tokyo District Court in 1964…" (Hidetaka and Tanaka 2001, 107). Therefore, in legal terms, "the *Western concept* [emphasis added] of 'public' and 'private' did not exist in Japanese culture until late in this (20th) century" (106).

3

While these fundamental, imported cultural concepts could not be translated, they were, and they still continue to be, imposed and self-imposed upon cultures of *the Other*. In opposition to that, Berman's *L'Epruve de l'etranger* demands translation to be grounded in "the desire to open up *the Stranger as Stranger* to his own space of and in language [and] to recognize *the Other as Other* [emphasis added]," which beautifully encapsulates what the ethics of cross-cultural encounter needs to be(come) (Worton 1998). The crucial point is the importance of *non-equivalence*. In her very opening sentence of *Text, Typology and Translation*, Trosborg (1997) clarifies that "equivalence can hardly be obtained in translation across cultures and languages, and it may not even be a desirable goal" (vii).

It should not be necessary here to further explain the relevance of translation theory to urbanism and the complex of themes associated with *the public*, especially in the times of rampant globalization. As embodiments of culture, cities can and they indeed should be understood as

Figure 16.1 Japanese roji in Nezu, Tokyo: spatial expression of Japanese residential culture
Image credit: Darko Radović

the "most *immoderate* of human texts," arguably "the supreme work, the work of works" (de Certeau 1984; Lefebvre 2016, 241). While having something substantial in common, the simplest lived experiences in what the locals may call *città, ciudad, ciutat, город, grad, град,* المدينة, เมือง อง, *Stadt, ville,* 市, 市内 differ profoundly. What gets communicated in other languages only exposes an ultimate untranslatability of such, complex, culture-specific terms laden with particular meaning. One could argue that cities are the complexity itself. At their finer social and physical scales, that complexity does not necessarily get reduced, but reframed. It makes perfect sense to expect that *lived experiences* of a narrow Japanese *roji* and the labyrinthine character of Thai *soi* are as profoundly different as their textual expressions in Japanese *kanji* 路地 and Thai *aksorn* ซอย are. While both 路地 and ซอย get translated into English as "lane," when explored as actual places, arenas of culture-specific practices, the irrelevance of any imposed, single "equivalent" becomes obvious. While the similarity exists at the basic level of physical urban morphology; *roji, soi* and lanes are all, indeed, linear, relatively narrow spaces bordered by buildings, lived experiences between those walls, when placed in Tokyo, Bangkok, or London, are incomparable. To claim the opposite would be equal to arguing how, when served in similar glasses, sake, whisky, wine, and beer all taste same.

4

In order to address Japanese (urban) culture and its peculiarities in relation to the concepts of *public* and public *space*, both of which are decidedly political constructs, we need to establish minimal grounding in geographic and historic conditions shaping this country.

In terms of geography, its defining condition is of its unique geological instability and extreme vulnerability to natural catastrophes. Japanese mentality and culture are profoundly influenced by the sense of imminence and unavoidability of disasters, the calamity which will come. Due to the limits of an essay, that fundamentally important theme will stay on the margins of our argument but it, nevertheless, has to be taken into account. In terms of history, we will emphasize several moments of equally fateful, even catastrophic magnitude. Those moments hold clues of direct relevance to the perceived and actual uniqueness of Japanese culture and language and, within that, the place and an ultimate absence of *public* in Japanese urbanity.

The first of such events happened during the pre-modern, feudal Tokugawa period (1603–1868) when, Shogun Tokugawa Iemitsu, threatened by increasingly aggressive colonial pressures and Christian missionary aspirations, declared *sakoku*, a "closed country." An interesting, and for our discussion significant paradox, marks that period. While we tend to associate urbanity with openness and connectivity, during these two centuries of self-imposed isolation from the rest of the world, the literacy rate in Japan "topped England's, and the country matched or bettered many other Western nations in education" (Feifer 2006, 47). During the *Pax Tokugawa* Edo (later renamed Tokyo) pulsed with an intense creative energy, "thanks to its relatively educated population eager to spend its leisure time and money on entertainment. Music, poetry, painting, theater (including for puppets), ceramics, and couture flourished so vigorously that their production and pursuit sometimes seemed feverish" (48). That illustrates how radically different urbanities are possible and that what may seem essential in some, even world-dominating cultures, does not necessarily apply, or *translate* at all into the cultures of the *Other*.

While the internal act of latent violence caused by hermetic closure of Japan, in 1867 an external, American threat caused the dramatic end of the Shogunate. The ensuing Meiji Era has defined all subsequent efforts (not) to open Japan to the World (Harootunian 2006; Taggart 2015). During that period Japan became "a country of excessive importation" and, importantly, the *"kingdom of translation"* (Tatsumi 2006, 171). As new social conditions started to acquire their physical, urban form, the measure of success, both in the eyes of the common Japanese and, in particular, in the eyes of their cultural and financial elites, became defined externally. Obsession with catching up and overcoming the "West" generated an urge not to reject, but to conceptualize own modernity (Harootunian 2006). The ensuing Taisho Era (1912–26) was an effort to dialecticize the resulting contradictions, hope, and gloom of the Japanese *Belle Époque*. Jinnai (1995) reminds how "nearly all the urban spaces we enjoy today—the avenues, street corners, plazas and parks—were built during this period" (18). The Taisho list of cultural imports was almost as long, and as impossible as the one crafted by Meiji elites. It also included a number of concepts untranslatable to Japanese language and culture – such as public, democracy, rights, citizen, avenue, piazza, and park. The perceived signs of modernity continued to be imported as gestures, invoking strong desire to break up with "backwardness." The culture in which streets had (and still have) no names was importing empty signifiers, Western terms and forms to be impregnated with retro- or neo-Japanese contents. In political terms, for internal use that was a promise of smooth transition into the new order, while towards the world the *Empire of Signs* projected the well-choreographed simulacrum of equivalence (Barthes 1982). A double reality created by renaming and repackaging demanded no fundamental change.

The violence and barbaric end of the World War, which in itself was a failed attempt at mis-conceptualized modernity, were followed by "the long postwar" (Harootunian 2006). With their decision to preserve the Emperor and the dynasty, "the United States literally undermined the very reforms it had implemented to eliminate prewar fascism and to put into place the foundations of a genuine social democratic structure" (2). That is evident in an acute lack of evolution of the Japanese political system up to the present day and, consequently, the lack of politicization of urban space. The key conservative rules and practices of traditional, top-down, patriarchal power system stayed intact. Embedded in strict social norms and customs, those rules ensured and enhanced repressive, internalized sense of shame and efficient self-control of the subjects. While the doors were opened wide to technical modernization, in social terms slow evolution was preferred over the possibility of any, let alone revolutionary change. An uneasy pretense of continuity lived despite the radical redefinition of power (the "blue-eyed shogun"). Over centuries of latent, internal, and external violence, the population has developed a "mental history of servility," which "helped naturalize the paradox of the emperor's ultimate sovereignty and limitless irresponsibility in Japanese culture" (Tatsumi 2006, 24).

This is where we can return to the investigation of (the absence of) the *concept* of public in Japanese society and the spatial repercussions (of that absence) and focus on examples that will help concretize this discussion and place it on the actual ground.

5

Jiyūgaoka is a precinct of Tokyo known for its distinct urban character. Only few years after the devastating Kanto Earthquake in 1923, suburban land subdivision and settlement started to emerge there. The first major development was Jiyūgaoka-gakuen High School, which brought name to the whole area. Term Jiyūgaoka, translated as *Liberal*, or *Freedom Hill*, was referring to new educational spirit of the short-lived Taishō democracy. As in many other parts of the rapidly expanding metropolis, urbogenetic sparkle came with new railway line. The opening of the Station in 1927 has triggered a pattern of growth characteristic to Tokyo. Three years later its significance grew further when Tokyu and Ōimachi Lines established an interchange there. Following rapid development after the Second World War, central Jiyūgaoka took its present form around 1970.

Today, this precinct ranks among top residential areas of Tokyo. Fine living areas are within walking and easy cycling distance from the Station, and an extremely commercialized but finely grained, unusually authentic urban center. Due to sensitive local community management (*machi zukkuri*, literally town-making), central Jiyūgaoka resists the onslaught of bigness, which became common to all important railway hubs in Tokyo. Its famed charm and an enviable quality of everyday life reach beyond the safe and comfortable, towards truly enjoyable lifestyles. This multifaceted quality has attracted young women and children, the subtle presence of tourists, an evident passion for groomed dogs, a booming café and tea culture, carefully organized and managed open-space programs—such as pedestrian-friendly weekends, festivals—and residential areas of distinct environmental quality. Our emphasis here will be on only two parts of central Jiyūgaoka, descriptions of which commonly employ term "public" – the Station square and one street which combines all of the above-listed qualities: Kuhonbutsugawa Ryokudō,

5.1

Japanese planning laws recognize only two kinds of open space – roads and parks. Those are managed by the authorities, primarily for post-disaster evacuation and other emergencies. The rest of the territory is almost entirely privately owned – including streets, lanes, and other kinds of open spaces, which would elsewhere be associated with public. Following these blunt laws, as the square in front of Jiyūgaoka Station was not a park, it had to be "road." Therefore, despite its piazza-like shape, this urban form without a cause has been given over to stationary taxis, a bus stop and slow vehicular traffic. Its official name is Jiyū-gaoka *Hiroba*. The characters forming term 広場 prosaically translate as "wide open area," referring to the physical condition rather than any social or formal property of the actual space (Kuma et al. 2015, 8).

Japanese efforts towards theorising their own urbanity often compare *hiroba* with piazza, coming up with conflicting findings about their relationship. Sand (2013), who uses these two terms interchangeably, offers a useful overview of the key texts dealing with that matter. He explains how in 1960 Takamasa and Kōichi contrasted the Greek agora as

> 'a healthy interpretation of humanity within the community of citizens, however limited,' with Japan's ancient capitals, whose urban form expressed 'Oriental despotic rule lacking communal solidarity.' This trait, they wrote, became yet more pronounced in the feudal cities that emerged in Japan during the subsequent medieval and Tokugawa periods. To compensate, postwar architects designed 'citizens' plazas' (*shimin hiroba*), most often adjacent to new municipal and prefectural office buildings. Later architects observed critically that these plazas were seldom used by ordinary citizens.
>
> *(43)*

Takamasa and Kōichi raised this issue in the year when political demonstrations, which marked much of the 1950s in Tokyo, reached their peak. Those events briefly took the list of imports away from the elites, with Tokyoites bypassing common misappropriation through mistranslation by taking political practices directly into the streets:

> Citizens protested in *places they understood as public* [emphasis added] property, either because it had been granted to them by the state or because they treated it as *their own by right as the citizens* [emphasis added] of a democratic polity.
>
> *(Sand 2013, 46)*

This protest against *Anpo*, the U.S.–Japan Security Treaty, led to resignation of the Prime Minister Kishi, but without achieving any actual change.

In the wake of *Anpo* protests, a number of architects fervently sought Japanese equivalents to Western *piazzas*. These often-contradictory efforts tried to balance irreconcilable desire for expressing uniqueness and universality. While generating interesting artistic and architectural experiments, they lacked urbo-political weight. Itō Teiji and his colleagues, thus, emphasized *kaiwai* ("activity space"), in which they saw the uniqueness of ways in which "ordinary people appropriated space spontaneously ... *Kaiwai* thus *supplemented citizen politics with an aesthetics of the everyday* [emphasis added]" (Sand 2013, 49). That statement is of critical importance for our argument that appropriation and requalification of imported concepts was open to any contents – as long as they were not political. In the years to come, aesthetization was to gradually fully replace politization, the very essence of paradigmatic *polis*.

After the *Anpo* moment, the protests in Tokyo continued gaining power towards new and, as it turned out, final peaks in 1968–69. In the lead-up to this renewed attempt to claim open spaces for public purposes, the ideas of Henri Lefebvre emerged in Japan, simultaneously with his action in France. His history of the Paris Commune was translated into Japanese in 1967, and,

> one aspect of LeFebvre's (sic!) work in particular struck a chord for student activists in Tokyo, as it had for students in Paris: the idea that the Commune had in its essence been a 'grandiose fête,' in which class barriers were broken down by spontaneous action in the streets. As in Paris, the student movement in 1968 Tokyo combined concrete demands for university self-government with a romantic rejection of bureaucracy and rationalism.
>
> *(Sand 2013, 51)*

It is important to remind that the list Lefebvre's keywords—including the crucial term *right*, as in his *right to the city*—remained fundamentally untranslatable to Japanese language, while enacted and tested in *praxis*.

The peaks of protests at the University of Tokyo and West Shinjuku Station were forceful, globally up to date and, when it comes to the demands of protesters – tragically inconsequential. But they have left an undesired, all-important mark on the Japanese politics of space. In their wake, the resilient Japanese ruling class decided that such unruliness was never to happen again. The notion of "public space" started to be increasingly shifted towards the "feel-good" quality. "Writer and participant Sekine Hiroshi put it pithily in an article published less than a year later ...: 'In short, all that had happened is that an imitation *hiroba* became a real *hiroba*, so the establishment used its power to suppress the *hiroba*, and confessed anew that this *hiroba* was an imitation.'" (Sand 2013, 57). The remote ideal of the citizen was waning, in favor of the new reality of the consumer.

These events have shaped Jiyūgaoka *Hiroba* of today. Formally "public," *hiroba* imitates *piazza*, while in reality it is private (only 10x10 meters there belongs to Meguro municipality). Activities resemble those in public realm, but they are carefully managed,

(a)

(b)

Figure 16.2 Jiyūgaoka *Hiroba*: on normal day, when occupied by vehicles (a), and pedestrianized, with preparations for strictly choreographed and controlled events (b)

Image credit: Darko Radović

choreographed and obsessively supervised by private stakeholders and their *machi zukkuri* organizations. When closed to traffic, the residents do not even traverse the piazza-like space. If not instructed how to use it, that space remains only the Station forecourt. With neither sense of belonging nor the right to freely take over, that space is *hiroba* but not a piazza.

5.2

The second space of Jiyūgaoka to briefly investigate here will be its sinuous, 2.2 kilometers long Kuhonbutsugawa Ryokudō. Central segment of the leafy Promenade is the spine of a vibrant shopping area, while the rest supports quiet, predominantly residential precincts. Of critical importance in the development of that quality was the fact that the Promenade was built on the top of Kuhonbutsu River. That river used to be an official boundary between the towns of Seta-gaya and Meguro, in both of which all land was, predictably, private. While in 1974 the reason for turning the river into a culvert was to avoid hazards associated with the badly kept waterflow (and, thus, association with "backwardness"), the accidental result was emergence of one of the most successful promenades in Tokyo. The newly established space belonged to neither Setagaya nor Meguro-ku; being neither a park nor a road, it could not be subjected to the existing rules. The flattened topographical gap became a gap in both property and legal terms, virtually non-existent and left to be managed by two adjoining communities. Their act of strategic foresight was to plant cherry trees along the full length of that improbable space. The new environmental conditions were unique, with qualities that instantly attracted visitors. As the majority of those were coming to Ryokudō on bicycles, the community managers chose not to fence the strip of cherry trees off, but to introduce two, almost uninterrupted lines of benches instead (funded by the adjoining small businesses, as everything "public" in Jiyūgaoka is). Despite the long-standing cultural prejudice against sitting out, the resulting quality of the Promenade, yet again, went beyond expectations. These two fateful moves—introduction of aesthetically and environmentally import-ant greenery and urban furniture, a rare commodity in residential Tokyo—made Kuhonbutsugawa Street of today one of the most pedestrian-friendly spaces in the metropolis, the backbone of an interesting lifestyle which blends local sensibilities with wanton cosmopolitanism – in a Japanese, consumption-driven way (Radović and Boontharm 2014).

Spaces are always socially constructed but, once conceived, they can also become the agents of societal change. The potential embedded in design, accidental qualities and an equally important meticulous care stimulate desire for creative and free expression. That causes hints at emergence of a highly subversive, latent *desire for public* (Radović and Boontharm 2014). If that is the case, the quality that seems to be emerging here would demand appropriations which, as in Lefebvre's (1996) powerful definition of the *right to the city*, are "the right to the *oeuvre* (participation) and appropriation (not to be confused with property but use value) …" (20). In other words, an appreciation of urban conviviality, of the universal, Nancean imperative for *being-with*, as the main constituent of any urbanity, might be emerging in Kuhonbutsugawa Street (Nancy 2000). Such residual culture bubbles here, amidst the mono-culture of extreme consumerism, surrounded by confronting super-ficiality and an insatiable appetite for fakeness.

In summary: while Jiyūgaoka *Hiroba* decidedly belongs to the Taisho import list, Ryo-kudō is formidably authentic. Despite its recent origin, it is indigenous, of this place and culture in an original, non-traditionalist way. The flashes of spontaneity make it appear proto-public. While the relationship between control and freedom remains firmly in favour of control (as this *is* Japan), here the tools and practices of control often get exposed in all

(a)

(b)

Figure 16.3 Central Kuhonbutsugawa Ryokudō: during the cherry blossom season, at its aesthetic best (a), and full of people, even during a rare snowy day in Tokyo (b)

Image credit: Darko Radović

Figure 16.4 Central Kuhonbutsugawa Ryokudō: the view from former café LaManda, one of early incubators of authentic, yet cosmopolitan urban flavour

Image credit: Darko Radović

of their banality, challenged by subtle bottom-up disobedience (as with ubiquitous bicycles that, despite obsessive prohibitions, still get parked everywhere). As opposed to the most of Tokyo, this unique street also offers spaces of non-consumption. To reiterate again, the only form of politics allowed is that of (smoothly imported) neoliberal consumerism. Kuhonbutsugawa Ryokudō is aesthetically pleasing, enjoyable and – lost in translation. It imitates, but it imitates well.

The term "imitation," used by Sekine above, precisely captures the "*katakana* character" of Japanese cultural importation. Katakana efficiently exposes alien concepts as foreign. Mispronounced, they get conceptually reduced to form devoid of essence which, in the case of *public, piazza,* and *rights* is the socio-political charge itself. In Japan,

> sociological and spatial studies of the city turned to the *sakariba,* or 'flourishing place,' as a key feature of Japanese urban culture. *Sakariba,* a native term in use since the Tokugawa period, referred to entertainment districts and marketplaces, and was thus defined by consumption as well as by the tendency of people to gather.
>
> *(Sand 2013, 61)*

The resurrection of terms and concepts from feudal Japan remains the most common feature of Japanese (pseudo)theorizing of their urban and architectural spaces – *kaiwai, sakariba, oku, en* ... , all of them always mysteriously untranslatable to the West. The denial of the lack of

Figure 16.5 Central Jiyūgaoka and its fake gondola, in an appropriately fake canal, next to the fake piazza, in a block-sized, scaled-down "Venice"

Image credit: Darko Radović

political evolution seeks satisfaction in (the myth of) uniqueness, the equation of cultural sustainability with conservation and the reinvention of old ways (Dale 2005). That practice helps keep the established power in place and, in return, gets full support from it. In 1968–69, "the battle for a more encompassing citizen control of central space in Tokyo had been lost to the police and, by extension, to the invisible hegemony of capitalist mass society" (Sand 2013, 61).

6

In conclusion, it is fundamentally important to understand that an absence of equivalents for untranslatables of relevance for our discussion of *the public*—such as *culture*, *philosophy*, *aesthetics*, *logic*, *rights* and—the idea(l) of *the citizen* in Japanese (and in many other non-Western languages) does not point at any lack or deficiency (Cassin 2014). The bottom line is that these important, Western concepts are among those "in which other cultures have shown hardly any interest, to the extent that often they do not even have name for it" (Jullien, 2014, 8). Their absence only highlights the possibility, inevitability, and the beauty of difference (Radović 2003). Franco Ferrarotti wisely warned that, when facing irreducible complexities of *the Other*, it is better "not to understand, rather than to color and imprison the object of analysis with conceptions that are, in the final analysis, preconceptions" (Dale 1986).

Therefore – no, *hiroba* (広場) is not public space; so what?!

References

Barthes, R. (1982) *Empire of Signs*. New York: Hill and Wang.

Bogdanović, B. (1973) Town and Town Mythology. *Ekistics*, 42(253): 240–242.

Cassin, B. (2014) *Dictionary of Untranslatables*. Princeton, NJ: Princeton University Press.

Dale, R. (1986) *The Myth of Japanese Uniqueness*. Oxford: University of Oxford Press.

de Certeau, M. (1984) *The Practice of Everyday Life*. Berkeley, CA: University of California Press.

Downey, G. (1973) Aristotle as an Expert on Urban Problems. *Ekistics*, 42(253): 316–322.

Feifer, G. (2006) *Breaking Open Japan*. New York: Smithsonian.

Harootunian, H. (2006) *Japan after Japan*. Durham, NC: Duke University Press.

Hidetaka, T. and Tanaka, M. (2001) Japanese Public Space as Defined by Event. In: *Public Places in Asia Pacific Cities*, Pu, M. (ed.). Dordrecht: Springer, pp. 107–118.

Jinnai, H. (1995) *Tokyo: A Spatial Anthropology*. Berkeley, CA: University of California Press.

Jullien, F. (2014) *On the Universal, the Uniform*. Cambridge: Polity Press.

Jullien, F. (2016) *The Strange Idea of the Beautiful*. London: Seagull Press.

Kuma, K., Jinnai, H. and Suzuki, T. (2015) *Hiroba: All about "Public Spaces" in Japan*. Tokyo: Tankosha Publishing.

Lefebvre, H. (1996) *Writings on Cities*. Cambridge: Blackwell.

Lefebvre, H. (2016) *Metaphilosophy*. London: Verso.

Nancy, J.-L. (2000) *Being Singular Plural*. Stanford, CA: Stanford University Press.

Radović, D. (2003) Celebrating the Difference. In: *Modernity, Tradition, Culture, Water*, King, R., Panin, O. and Parin, C. (eds.). Bangkok: Kasetsart University Press, pp. 65–71.

Radović, D. (2010) Right to the City – On Bridges and the Essence of Public Life. In: *Intentcity – The Political City*, Maturana, B. and McInneny, A. (eds.). Melbourne: Architects for Peace, pp. 7–29.

Radović, D. (2014) *Subjectivities in Investigations of the Urban*. Tokyo: flick studio.

Radović, D. and Boontharm, D. (2014) *In the Search of Urban Quality*. Tokyo: flick studio.

Sand, J. (2013) *Tokyo Vernacular Common Spaces, Local Histories, Found Objects*. Berkeley, CA: University of California Press.

Sennett, R. (1990) *The Conscience of the Eye*. London: Faber and Faber.

Steiner, G. (1973) The City under Attack. *Ekistics*, 42(253): 322–328.

Taggart Murphy, R. (2015) *Japan and the Shackles of the Past*. Oxford: Oxford University Press.

Tatsumi, T. (2006) *Full Metal Apache*. Durham, NC: Duke University Press.

Trosborg, A. (1997) *Text Typology and Translation*. Philadelphia, PA: John Benjamins Publishing.

Worton, M. (1997) Speaking (on) Theory: Teaching and Translation – Or Teaching as Translation. *Interstices*, 2(Spring/S) pp. 55–70.

PART 3

Types

We encounter public space every day—many types of public space. In the city, we find several typologies and many faces of public space—parks and playgrounds, streets, plazas and squares, waterfronts, greenways and trails, markets and memorials, the predominantly corporatized, commoditized, and homogenized public spaces of downtown, the vestigial spaces associated with public infrastructure, the residual spaces between buildings, and so on. These typologies vary, from the classical taxonomies to the newly recognized and often "loose" public spaces of the postindustrial city. Examining public space through the lens of a typological classification shows the immense range of the role of public space in the life of the city. There have been several past efforts in classifying public space. Yet, as the opening chapters in this section show, existing classifications are neither exhaustive nor meticulous. As cities evolve and transform, needs in and roles of public space change. Equally importantly, as the design, use, and management of public space evolves, new typologies emerge; existing ones become obsolete or adequately transform to create a need for reconsidering past classifications. The chapters in this section address this need and revisit the typologies and classifications of public space, while also introducing new ones and showcasing how existing typologies need to be reexamined in the context of political, cultural, and economic changes.

The first three chapters further our understanding of the classification of public space. Karen Franck and Te-Sheng Huang highlight that type is a common organizing feature of public space research but past categorizations of types as a descriptive or analytic tool are largely ad-hoc, depending primarily upon the researchers' academic lens. They point out the analytic advantages of devising and employing classifications of types that comprise mutually exclusive and exhaustive categories. The authors argue for a consistent use of employing design, use, *or* management of public spaces as criteria to create a better typological classification. Using case studies, the authors demonstrate how carefully defined and rigorously applied classifications of public space can reveal important similarities and differences with respect to their design, use, and management and help us with a more thorough understanding of public space.

Understanding what goes on in public spaces and how people use public space provides a critical view to the life of the city. The use of public space is also a good measure of publicness, as it indicates the absence of certain activities and users. Sverre Bjerkeset and Jonny Aspen propose a comprehensive set of categories to identify and record how people use public spaces. Through case studies of two neighborhoods in Oslo, they examine outdoor spaces that, in principle, are accessible to all. The authors demonstrate how classifying public space by use can benefit policy, planning, and design, as well as help foster dialogue across scholarly disciplines and fields of practice.

The definitions of public space often revolve around the multiaxial matrix of ownership, control, and access. Yet, access and control are better at making visible the agency and experience of users in space. In other words, the degree of public access and public control are the manifestation of how people experience the publicness of public space. Kim Dovey and Elek Pafka present a typology of public space focused on publicness, defined as the characteristics of space that create possibilities for public life. By thoroughly mapping three contrasting urban neighborhoods in Melbourne, Australia, the authors examine contemporary public space that delivers the varying conditions of publicness and privatization produced by a range of entities and actors in the globalized neoliberal context. The approach presents six broad overlapping categories of publicness that help distinguish and understand these spaces from the varying rights to access and public control through appropriation. The typology highlights intersections and overlaps and the resulting ambiguities of types regarding publicness. Equally important, their work shows that publicness can, in fact, be mapped and made visible.

As a physical manifestation of the public sphere, public space is the geography of politics, and sociability. In their chapter, Els Leclercq and Dorina Pojani ask if urban designers and planners can create a physical setting that fosters publicness, as well as sociability and aesthetic value. Their work adds to the growing body of literature on the evaluation of public space. The authors undertake a systematic visual assessment of three different projects in Liverpool, UK – an entirely private development, a public–private partnership, and a public project taken over by a community organization. Although all three are ostensibly public, their research finds stark differences and complex, dichotomous, overlapping, and polarizing interactions of publicness, sociability, and aesthetics. The authors urge designers and planners to pay keen attention to programming, access, and aesthetic rules so as to not override diversity and inclusivity, the core values of publicness.

Quentin Stevens' chapter draws on Latour's Actor-Network-Theory to examine six sets of diverse actors, tangible and intangible, that come together to facilitate new kinds of temporary urban open spaces. Using examples of artificial city beaches installed on formerly-industrial riverfronts in Germany, Stevens highlights the complex and dynamic relationships and processes through which these various actors come together. The author highlights how diverse forms of power help stabilize or transform these relationships and serve to re-imagine, re-develop, and re-purpose urban spaces to bring new public benefits.

Examining an existing type of public space in the current context of aging urban populations, Anastasia Loukaitou-Sideris explores the potential of neighborhood parks to offer health and recreation benefits for older adults. Six desirable characteristics for neighborhood parks are articulated, based on past literature and focus groups with older inner-city adults in Los Angeles. The chapter proposes design guidelines around ten themes that seek to satisfy these desirable characteristics.

Ken Greenberg argues that in spite of all the political, social, cultural, economic, and technological challenges, the desire for public space has returned as we realize that the commons are essential to our well-being as social creatures. Using the Bentway in Toronto, Canada, as an example, the author describes the public space as being amenable to diverse programming and varied uses. By describing the process in detail, the author shows that this was only possible as the result of a concerted, participatory effort oriented around the reimagining of the public, rather than a continued focus on the private. The Bentway shows how re-stitching the city's fabric, linking previously discrete places, symbolically represented the de-siloing, and partner-based initiative of current governance models can produce such a space. Suggesting breaking down barriers between top-down and bottom-up processes, the author argues that public space can no longer be viewed solely as "government" space, but rather civic (our) space.

Using strategic examples of contemporary skyscrapers in New York and London, Vuk Radović examines the privately owned public spaces (POPS) in contemporary commercial skyscrapers. The author presents a historical overview of the genesis of the "open-space" skyscraper as a manifestation of early-American colonialism, followed by examples of contemporary skyscrapers with such spaces. He studies a range of POPS including ones that are external to, within the physical boundary of, or publicly accessible spaces within a complex of buildings. The findings show that irrespective of the physical location of these public areas, the results are invariably homogenous and highly regulated, with limited public access. The author ends by providing the reader with an open-ended set of solutions that transcend the issue of public-space-as-externality and instead provide space and experience within the skyscraper which are truly public.

In the final chapter in this section, Renia Ehrenfeucht brings attention to the challenges in representing diverse histories and experiences in public space. The author examines the debate around public monuments to highlight how these are enacted and debated by diverse actors. Her chapter shows how communities and societies represent and attempt to reconcile deeply irreconcilable perspectives about their histories, people and events with little resolve. Ehrenfeucht argues that this inconvenient challenge to public space is also a necessary step towards building inclusive societies.

17

TYPES

Descriptive and analytic tools in public space research

Karen A. Franck and Te-Sheng Huang

Introduction

As socially and physically constructed, as described, occupied and regulated, the world is unavoidably composed of *types* of places (Franck and Schneekloth 1994). Indeed, the title of this book and the titles of many of the chapters in it refer to a type – "public space" – thereby distinguishing that kind of space from others that are, in some way, not that. In addition, the naming of more delimited public space types, such as street, park, plaza, or gated community, is largely unavoidable in research about public space.

The classifications of places according to type is used for describing as well as for explaining, prescribing, and regulating the environment (Franck and Schneekloth 1994). These may be existing classifications or newly developed ones. In this chapter we adopt the idea of type as a lens for interrogating the diverse and growing body of research about public space in different social-science disciplines, primarily in the US and the UK. Then we point to several ways researchers have employed existing or new classifications of types of public space as descriptive and analytic tools. We are particularly interested in the analytical uses of type. We demonstrate how, in a particular study, classifications of cases of public space, according to both more general and more fine-grained categories, can reveal important similarities and differences with respect to the design, use, and management of particular cases that would otherwise remain obscure.

Existing types as a lens for viewing public space research

A brief review of urban public space research through the lens of type suggests that researchers often focus on a particular type of space and that the types of interest to researchers and the perspective taken toward them have changed over time.

In *Death and Life of Great American Cities*, Jane Jacobs (1961) drew from her observations of the Hudson Street sidewalk outside her home. In *Social Life of Small Urban Spaces*, William Whyte (1980) reported his observations on the design and use of privately owned, outdoor plazas and adjacent sidewalks. In *Politics of Park Design*, Galen Cranz (1982) described the evolution of urban parks from 1850 through the 1960s. Francis et al. (1984) considered various kinds of "community open space" in New York City developed by local residents. Another early work, Jan Gehl's *Life Between Buildings* (Gehl 1987), focused on public spaces adjacent or in close proximity to residential buildings – largely sidewalks and squares. Although the particular type selected for study framed these works, types were rarely an explicit organizing feature of the work with some exceptions. Through her historical research Cranz (1982) identified four types of parks based on their design and management features. Subsequently, she and Boland (2004) identified a fifth type. Clare Cooper-Marcus and Carolyn Francis' (Cooper-Marcus and Francis 1990) book about what they call "urban open space" in San Francisco is organized around a series of types of such spaces. In their comprehensive book, *Public Space*, Carr et al. (1992) present a list of indoor and outdoor public spaces that include parks, squares, streets, and also greenways, market places, memorials, playgrounds, malls, and waterfronts and their case studies cover all these types.

This first generation of social research about contemporary public space addressed primarily outdoor spaces, most of which are publicly owned and publicly managed (with the exception of Whyte's plazas, Francis et al.'s community gardens and Carr et al.'s marketplaces). Following this first generation of work, many researchers continued to study types of publicly owned and publicly managed exterior spaces, as indicated by books about: a particular plaza (Low 2000), a particular sidewalk (Duneier et al. 2000), sidewalks more generally (Loukaitou-Sideris and Ehrenfeucht 2009; Blomley 2011), a particular waterfront (Campo 2013), several streets and sidewalks (Mehta 2014), parks (Low et al. 2005), memorials (Stevens and Franck 2015), and community gardens (Hou et al. 2009). The topic of these works is a particular case, a particular type, or several types of outdoor public space. College campuses have received only some attention (Cooper-Marcus and Francis 1990; Day 1999b).

Unlike outdoor ones, publicly owned, indoor spaces (e.g. public libraries and museums), with some exceptions, have not received much attention from public space researchers. The interior public spaces of greatest interest to researchers have been those that are privately owned and privately managed venues such as malls and corporate atria. These seem to be of interest because, by virtue of their ownership and management, they are not as accessible as traditional public spaces, do not allow political activities, and hence present a hybrid and often problematic circumstance of being public but not completely so. Notably, the types of consumerist space that are viewed favorably by researchers are farmers markets, other kinds of outdoor markets and market halls (Carr et al. 1992; Watson 2009) possibly because these businesses are smaller scale, are more likely to be local than the chain stores of malls and supermarkets, and in general have a more convivial and more informal atmosphere, functioning sometimes as community centers (Project for Public Spaces 2016). Indeed, it is precisely the small scale, convivial atmosphere, and eminently local features of what Ray Oldenburg (1989) named the "third place" that makes places of social interaction in communities, including cafes, coffee shops, bars, general stores, bookstores, and hair salons.

Starting in the 1990s, the perspective of researchers studying public space and commentators shifted from a neutral, if not optimistic, one to a more critical one that focused on the excessive control exerted by owners and managers, and accordingly so did the type of space studied. Starting with Mike Davis' *City of Quartz* (1990) and Michael Sorkin's *Variations on a Theme Park* (1992) writers pointed out how public spaces in the U.S., even those under public ownership, such as streets and sidewalks, were coming under stricter and more discriminatory controls that limited access to the space, sometimes as part of business improvement districts and sometimes by harsh police tactics aimed at the homeless.

Attention was also directed at the types of public space that had proliferated in the latter part of the 20th century: airports, malls, the interior spaces of large, privately owned, mixed-use complexes that incorporated space accessible to the public, such as the IDS

Center in Minneapolis, and skywalks linking such complexes. Many are places of consumption and often used as evidence of the privatization and commercialization of public space more generally. Subsequently gated residential communities came under similar scrutiny (Kohn 2004; Low 2004).

In the 2000s, indoor and outdoor privately owned public spaces (POPS) in New York City came under close scrutiny. These "bonus spaces," which private companies and other organizations provide in order to gain the city's permission to incorporate additional floor area, are privately owned and managed but intended for public use, many with design features and amenities promoted by William Whyte based on his research in the 1970s and now required by the Department of City Planning. Kayden et al. (2000) documented in systematic detail the location, key design features, conditions and use of the 503 bonus spaces in Manhattan, providing a critical analysis of the widely varying degrees of accessibility and attractiveness of these spaces. That is, many of the spaces are actually inaccessible or provide reasons, such as seating, for people to occupy them. This work has served as a useful guide for subsequent research (Smithsimon 2006; 2008; Miller 2007; Németh 2007; 2009; Schmidt et al. 2011; Németh and Schmidt 2011; Huang 2014a; 2014b; Huang and Franck 2018) and for efforts to make these spaces more accessible and more attractive to the public undertaken by a partnership between the Municipal Arts Society and Advocates for Privately Owned Public Space (apops.mas.org).

In suburbs without traditional retail streets or public squares, the mall became the preeminent "public space" in the community and yet does not permit the kinds of activities, particularly political ones, allowed, and required, by free-speech rights (Kohn 2004). The critique of privately created and owned public spaces often intended for consumption, such as malls, may be warranted given the control that is exerted over them by activities and occupants. This critique is only possible, however, because in modern times very different types of public space do exist.

The benefits retail spaces hold for women suggests the importance of distinguishing between the needs of different user groups, something that is infrequently done in public space research. In their broad-brush critiques of consumer-oriented spaces as being insufficiently public or overly restrictive, researchers have tended to overlook whether and how such types of spaces may meet particular needs of specific kinds of occupants or for pursuing specific kinds of activities, In fact, women continue to seek out the convenience, safety, and amenities of retail and eating venues in many countries, frequenting cafes, bookstores, and shopping centers in the US or malls in Cairo where they can meet friends without the necessity of male accompaniment (Day 1999a; Abaza 2001).

It is important that researchers fully acknowledge that different types of public space fulfill different primary functions and that fulfilling those functions may require restrictions. The specialization among types of public space actually benefits users in giving them choices. What some see as a key function of public space – as an arena for making claims public through political demonstrations – need not be fulfilled by all types of public space, just as the U.S. Supreme Court has recognized in its decisions, even though these spaces are indeed public in other ways by being accessible and providing a public good (Parkinson 2006).

A contrasting phenomenon to the privatization and control of public space is people's appropriation of public spaces for their own short- or long-term purposes for recreational, commercial, gardening, and other activities. Sites of appropriation include vacant, unused land and as well as streets, sidewalks, squares, parks, playgrounds, and parking lots. This phenomenon has been named: "found space" (Carr et al. 1992), "everyday urbanism" (Chase et al. 1999), "loose space" (Franck and Stevens 2007), "insurgent" or "guerilla" urbanism (Hou 2010) and "temporary use" or "temporary urbanism" (Overmeyer 2007; Oswalt et al. 2013). Grassroots initiatives by individuals and organizations to use and enliven public space are increasingly celebrated as evidence of the "temporary city" (Bishop and Williams 2012) and as acts of "tactical urbanism" (Lydon and Garcia 2015) and "DIY urbanism" (Talen 2015).

Now city government programs are inviting and supporting community and local business initiatives to create new public spaces in existing publicly owned spaces to function as public-private partnerships. In New York City, through the Department of Transportation's Public Plaza Program launched in 2009, 73 "pedestrian plazas" have been created on existing roadways next to remaining traffic lanes. At least 2,000 square feet in size, with movable tables and chairs and possibly plantings, bike racks, and lighting, these are funded by DOT and are proposed and maintained by local organizations (Department of Transportation 2016). And what started in 2005 as a single intervention by Rebar to convert one parking space into a small park in one on-street parking space in San Francisco for a single day has become an international event to create similar temporary installations on the annual PARK(ing) Day (Bela 2015). Following San Francisco's adoption of the Pavement to Parks program in 2011, many city governments now sponsor similar, permanent or seasonal installations in what are commonly termed "parklets" (San Francisco Planning Department et al. 2016). These examples demonstrate that public space, rather than shrinking, as some public space researchers claim, may well be increasing in number and variety – that is in variety of types.

Types as newly devised classification systems

In the section above we used commonly used names to frame and reflect upon past research. The names of traditional types and then slowly the newer types, capture both their primary functions and their commonly understood design features. Society-wide naming and shared understanding building and other place types is essential for guiding (and ordering) everyday life and for producing, modifying, and regulating the environment. So, when a new type appears, such as those resulting from PARK(ing) Day, they are quickly named ("parklets") in order to identify them, to define their use and often to direct and regulate that use and their design.

In other circumstances professionals and researchers invent new classifications to describe existing or newly envisioned kinds of places. Such a classifying of places (buildings or other spaces) into types is frequently used to aid in designing and regulating the built environment. For instance, to adopt a zoning ordinance that grants additional floor space to developers for buildings that incorporate public space, the New York City Department of City Planning developed a detailed set of types of POPS based on their location and their spatial or physical configuration (Kayden et al. 2000). Kayden et al. used this classification as a descriptive tool to present the frequency of these types and the number built each year and to compare their sizes and identify the particular type of all 503 privately owned public spaces in New York.

Researchers also invent their own classifications of public spaces composed of categories that are neither mutually exclusive nor exhaustive. For example, Flusty (1997) uses five adjectives to capture the "strategies" used "to intercept and repel or filter would-be users" in the spaces he observed in Los Angeles: stealthy, slippery, crusty, prickly, and jittery. He attaches each of these adjectives to the word "space" to identify "five species" (48). Crusty space, for example, is space that cannot be accessed due its enclosure while prickly space is uncomfortable to occupy. Flusty presents examples of the spaces he observed that exhibit the features he identifies. One space may exhibit more than one feature and the classification system is not used to describe sites in an exhaustive manner (or to count them).

Carmona (2010) also developed a set of possibly overlapping categories of public spaces as a means to review critiques of public space as being either "under managed" or "over managed" deriving these categories from a scrutiny of the relevant and expansive body of research. What is categorized is relevant literature rather than actual spaces. As with Flusty's list, the categories presented are not discrete kinds of spaces but rather features or characterizations of spaces.

For descriptive purposes researchers also compile mutually exclusive categories of actual public spaces by observing a group of them. Sometimes this involves making new distinctions between existing kinds of places, allowing readers to see such spaces from a new, possibly more fine-grained perspective. For instance, Cooper-Marcus, Francis and Russell (1990) composed a "typology of downtown plazas" in San Francisco: street plaza, urban oasis, corporate foyer, transit foyer, and grand public place.

Deriving types from historical analysis of archival material and actual places whose design and management have changed over time is another analytic method. This is the technique Cranz (1982) used to identify four types of urban parks in the US: the Pleasure Ground (1850–1900), the Reform Park (1900–1930), the Recreational Facility (1930–1965), and the Open Space System (1965–?). Through her research Cranz discovered how different features of urban parks – social goals, activities, key design features, promoters, and beneficiaries – changed in tandem with each other and thereby could serve to characterize the four types. In this approach, types become a means of discovery and analysis and then a way to describe the findings in an integrated historical narrative. The types then become useful in an additional way – to recognize and document the emergence of a new type, the Sustainable Park (1990s–present), which is the first type in her system to embrace human and ecological health (Cranz and Boland 2004). It is extremely informative, and valuable, that the features they use to characterize the five types remain distinct from each other and are clearly apparent in the presentation of the research. No information is lost in the compiling of the types.

Based on field observations of privately owned public spaces in New York City and interviews with users, Kayden et al. (2000) compiled a set of distinct contemporaneous categories that are, like Cranz's, mutually exclusive and exhaustive categories. The five types are: destination spaces, neighborhood spaces, hiatus spaces, circulation spaces, and marginal spaces and are distinguished from each other by a combination of physical and social features – who frequents the spaces, what activities occur, size, other design and physical features, and the presence of amenities. Using the legally defined types compiled by the Department of City Planning, Kayden et al. related one set of types to the other. Notably, however, Kayden et al.'s own five types are distinguished from each other by several different kinds of features such as aspects of both use and design. This prevents an exploration of possible relationships between features (e.g. whether frequency of use or type of users are correlated with being located indoors or the presence of amenities).

Subsequently, researchers studying bonus spaces in New York adopted a similar approach: that is combining different kinds of features of the spaces to define the types. For instance, Nemeth and his colleagues do not distinguish between bonus spaces located indoors from those located outdoors and combine characteristics of design and furnishing with characteristics of management and occupancy in their classification system (Németh 2009; Németh and Hollander 2010; Schmidt et al. 2011; Németh and Schmidt 2011).

The analysis Carmona and Wunderlich (2012) conducted is quite different. From their survey of new or substantially renovated squares in London they identify six, mutually exclusive types: piazzas, courtyards, incidental spaces, garden squares, forecourts, and other spaces. They distinguished these types from each other only according to their physical form and no other features and categorized the individual cases accordingly. While the squares vary widely with respect to several other features, the researchers noted some clear differences between types. For instance, piazzas and garden squares are generally larger than the other types and piazzas are more sanitized.

Based on their observations of use, Carmona and Wunderlich (2012) also present a set of types distinguished by function although these categories are not mutually exclusive. Coding spaces for function alone, independently of any other features, allowed them to determine key, shared characteristics among the types. For instance, consumption, civic, and transit spaces show the most "vibrant character;" seating is most common in community spaces; greenery is a feature of all types except civic ones; and corporate, civic, and consumption squares are the most likely to display public art. This research demonstrates that categorizing spaces according to a series of types that are distinct and independent of each other (e.g. types characterized by form, function, or management) allows for the discovery of correlations between features. Typing then becomes a fine-grained and powerful analytic tool.

To study 24 cases of interior privately owned public space in New York City (see Table 17.1), Huang (2014a) employed a similar method of classifying individual cases in order to discover similarities and differences between cases and possible relationships between the different features of design, management, and use. He classified the individual cases according to a combination of three features, all related to physical design: the relationship to adjacent streets and the presence and location of seating. Figure 17.1 shows the resulting five spatial types. Preliminary observations of all 24 cases revealed that two were inaccessible and two were used solely for short term waiting. Since the objective was to explore different uses of the spaces these cases were eliminated from further analysis.

Subsequent observations of the use and users of the 20 remaining spaces revealed dramatic differences in occupancy across the five spatial types. Namely, cross-block atria showed the highest levels of occupancy over the course of the day and evening, more than double the occupancy levels of the other four types. Further analysis of the cross-block atria showed that they are used for a great variety of individual and group activities such as table and card games and various self-improvement activities including learning foreign languages. By separating out the cross-block atria from the other four interior types of POPS, these aspects of their use could be identified (Huang 2014a, 2014b; Huang and Franck 2018).

Conclusion: typing as a research tool

Faced with a large amount of information, such as a wide range of existing studies, a series of changes over time in the design and use of a particular kind of environment, or a great many cases of one type of environment, a useful approach is to create a classification system. A series of categories, which are clearly distinguished from each other and are preferably mutually exclusive, is developed and research material or individual places are placed into those categories according to explicit criteria.

We call this approach "typing." We have used it ourselves in this chapter as a descriptive and comparative tool in reviewing existing research and we have shown how researchers have used typing as a descriptive tool in reviewing previous research and as an analytic tool in classifying actual places (Carmona 2010; Carmona and Wunderlich 2012; Cranz 1982; Cranz and Boland 2004; Kayden et al. 2000; Huang 2014a; Huang and Franck 2018).

Typing is effective for reducing a very large number of individual cases into a smaller and more manageable and more analyzable number of items (e.g. 10 instead of 100) without

Table 17.1 Twenty cases of interior privately owned public spaces in 19 buildings in Manhattan.

Twenty cases of interior privately owned public spaces in 19 buildings in Manhattan

Spatial type	Cross-block atrium	Atrium	Space with one side entrance	Linear space with designated seating areas separated from circulation route	Linear space sometimes with linear seating along the circulation route
Names of spaces observed	180 Maiden Lane 60 Wall Street Park Avenue Plaza Sony Plaza Former IBM Atrium CitiCorp Atrium Rubenstein Atrium	Former Altria Atrium 575 Fifth Avenue 805 3rd Ave Atrium 875 3rd Ave Atrium	650 Fifth Avenue Trump Tower Plaza 1991 Broadway	52 Broadway Olympic Tower Atrium Galleria 499 Park Avenue	Grand Central Plaza 875 3rd Ave Mezzanine
Street level	All except CitiCorp (one floor below)	All except 805 3rd Ave Atrium & 875 3rd Ave Atrium (one floor below)	All except 650 Fifth Avenue & Trump Tower Plaza (one floor below)	All except Galleria (one floor below)	All except 875 3rd Ave mezzanine (one floor above)
Cross-block connection	All	None	None	Only at 52 Broadway	Only Grand Central Plaza
Seating	Majority of area	Majority of area	Majority of area	Small area	Along circulation route except two spaces*
Food service	All	All except former Altria Atrium*	All	Only 52 Broadway*	Only Grand Central Plaza*
Drinking fountain	Only Park Avenue Plaza & Rubenstein Atrium*	Only 575 Fifth Avenue*	None	None	None
HVAC	All except former IBM atrium (heating only)	All	All except 1991 Broadway (heating only)	All	All

(Continued)

Table 17.1 (Cont).

Twenty cases of interior privately owned public spaces in 19 buildings in Manhattan

Spatial type		Cross-block atrium	Atrium	Space with one side entrance	Linear space with designated seating areas separated from circulation route	Linear space sometimes with linear seating along the circulation route
Restrooms		All except former IBM atrium	All but are locked in two spaces*	Only Trump Tower Plaza	Only Olympic Tower Atrium	None
Average number of occupants per hour	Weekday	48 (7 cases)	24 (4cases)	9 (3 cases)*	8 (4 cases)	8 (2 cases)*
	Weekend	29 (6 cases)**	12 (3 cases)**	17 (2 cases)**	4 (2 cases)**	2 (1 case)**

Notes: (1) Information from Kayden et al. (2000) or from this study (marked with *)
(2) ** Number of cases observed note in parentheses because some cases are closed on weekends.

(a) Five Physically Defined Types

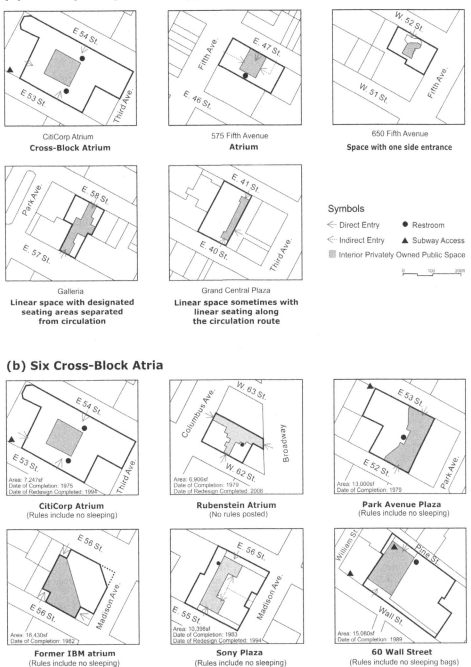

CitiCorp Atrium
Cross-Block Atrium

575 Fifth Avenue
Atrium

650 Fifth Avenue
Space with one side entrance

Galleria
**Linear space with designated
seating areas separated
from circulation**

Grand Central Plaza
**Linear space sometimes with
linear seating along
the circulation route**

Symbols

← Direct Entry   ● Restroom
← Indirect Entry   ▲ Subway Access
▨ Interior Privately Owned Public Space

0   100   200ft

(b) Six Cross-Block Atria

Area: 7,247sf
Date of Completion: 1975
Date of Redesign Completed: 1994
CitiCorp Atrium
(Rules include no sleeping)

Area: 6,906sf
Date of Completion: 1979
Date of Redesign Completed: 2008
Rubenstein Atrium
(No rules posted)

Area: 13,000sf
Date of Completion: 1979
Park Avenue Plaza
(Rules include no sleeping)

Area: 16,430sf
Date of Completion: 1982
Former IBM atrium
(Rules include no sleeping)

Area: 10,398sf
Date of Completion: 1983
Date of Redesign Completed: 1994
Sony Plaza
(Rules include no sleeping)

Area: 15,080sf
Date of Completion: 1989
60 Wall Street
(Rules include no sleeping bags)

Figure 17.1 Analysis of 24 Interior Privately Owned Public Spaces in New York City.
Image credit: Kayden et al. (2000)

sacrificing any of the information that is provided by the large number of individual cases (see Table 17.1). The different types can then be compared to each other in ways that reveal similarities and differences not previously known, particularly when the types are distinguished from each other according to singular criteria (such as physical design *or* use *or* management).

Typing however needs to be done with care and an eye to accuracy and to preserving information. Without basing the distinctions between types upon clearly articulated criteria, the researcher may group quite different kinds of spaces together in one category and will then be unable to discover important differences and important similarities, possibly drawing conclusions about the entire sample of cases that are inaccurate. Generalizations about the emptiness and strict management routines of all privately-owned public spaces in New York are one example: the generalizations overlook important differences between different kinds of POPS, such as those located indoors or outdoors and those with or without strict management practices.

At the same time, the cases classified as one type can nonetheless be significantly different from each other with respect to other features that are not captured by the criteria for defining that type. For instance, among Huang's cases of cross-block atria, which mostly showed high levels of occupancy, was one case located at 180 Maiden Lane with a very different management routine that, as a result, was not well used: its hours of operation were limited to weekdays during office hours (Huang 2014a, 2014b; Huang and Franck 2018).

Thoughtful, carefully defined and rigorously applied classifications of public spaces according to type can be used to describe and compare places and to analyze them. This makes "typing" both a good means for describing and comparing and also for discovering circumstances in the environment that have not been identified before. Typing as a means of analysis can be just as revealing if not more so. And, as importantly types can be used to present a well-organized and also cohesive narrative.

References

Abaza, M. (2001) Shopping Malls, Consumer Culture and the Reshaping of Public Space in Egypt. *Theory Culture & Society*, 18: 97–122.

Bela, J. (2015) User-Generated Urbanism and the Right to the City. In: *Now Urbanism: The Future City is Here*, Hou, J., Spencer, B., Way, T. and Yocom, K. (eds.). New York: Routledge, pp. 149–164.

Bishop, P. and Williams, L. (2012) *The Temporary City*. New York: Routledge.

Blomley, N. (2011) *Rights of Passage: Sidewalks and the Regulation of Public Flow*. New York: Routledge.

Borrell, L.N., Jacobs, D.R., Williams, D.R., Pletcher, M.J., Houston, T.K. and Kiefe, C.I. (2007) Self-Reported Racial Discrimination and Substance Use in the Coronary Artery Risk Development in Adults Study. *American Journal of Epidemiology*, 166: 1068–1079.

Campo, D. (2013) *The Accidental Playground: Brooklyn Waterfront Narratives of the Undesigned and Unplanned*. New York: Fordham University Press.

Carmona, M. (2010) Contemporary Public Space: Critique and Classification, Part One: Critique. *Journal of Urban Design*, 15: 123–148.

Carmona, M. and Wunderlich, F.M. (2012) *Capital Spaces: The Multiple Complex Public Spaces of a Global City*. New York: Routledge.

Carr, S., Francis, M., Rivlin, L.G. and Stone, A.M. (1992) *Public Space*. New York: Cambridge University Press.

Chase, J., Crawford, M. and Kaliski, J. (eds.). (1999) *Everyday Urbanism*. New York: Monacelli Press.

Cooper-Marcus, C., Francis, C. and Russell, R. (1990) Urban Plazas. In: *People Places: Design Guidelines for Urban Open Space*, Cooper-Marcus, C. and Francis, C. (eds.). New York: Van Nostrand Reinhold, pp. 9–68.

Cooper-Marcus, C. and Francis, C. (eds.). (1990) *People Places: Design Guidelines for Urban Open Space*. New York: Van Nostrand Reinhold.

Cranz, G. (1982) *The Politics of Park Design: A History of Urban Parks in America*. Cambridge: MIT Press.

Cranz, G. (1985) Skyways. *Design Quarterly*, 129: 1–31.

Cranz, G. and Boland, M. (2004) Defining the Sustainable Park: A Fifth Model for Urban Parks. *Landscape Journal*, 23(2): 102–120.

Davis, M. (1990) *City of Quartz: Excavating the Future in Los Angeles*. New York: Vintage Books.

Day, K. (1999a) Introducing Gender to the Critique of Privatized Public Space. *Journal of Urban Design*, 4(2): 155–178.

Day, K. (1999b) Strangers in the Night: Women's Fear of Sexual Assault on Urban College Campuses. *Journal of Architectural and Planning Research*, 16(4): 289–312.

Department of Transportation. (2016) "NYC Plaza Program." www.nyc.gov/html/dot/html/pedestrians/nyc-plaza-program.shtml.

Duneier, M., Hasan, H. and Carter, O. (2000) *Sidewalk*. New York: Farrar, Straus and Giroux.

Flusty, S. (1997) Building Paranoia. In: *Architecture of Fear*, Ellin, N. (ed.). New York: Princeton Architectural Press, pp. 47–59.

Francis, M., Cashdan, L. and Paxson, L. (1984) *Community Open Spaces: Greening Neighborhoods through Community Action and Land Preservation*. Washington, DC: Island Press.

Franck, K.A. and Schneekloth, L.H. (eds.). (1994) *Ordering Space: Types in Architecture and Design*. New York: Van Nostrand Reinhold.

Franck, K.A. and Stevens, Q. (eds.). (2007) *Loose Space: Possibility and Diversity in Urban Life*. New York: Routledge.

Gehl, J. (1987) *Life between Buildings: Using Public Space*. New York: Van Nostrand Reinhold.

Hou, J. (ed.). (2010) *Insurgent Public Space: Guerrilla Urbanism and the Remaking of Contemporary Cities*. New York: Routledge.

Hou, J., Johnson, J.M. and Lawson, L.J. (2009) *Greening Cities, Growing Communities*. Seattle, WA: University of Washington Press.

Huang, T.-S. (2014a) "Is the Public Invited? Design, Management and Use of Privately Owned Public Spaces in New York City." PhD diss., New Jersey Institute of Technology.

Huang, T.-S. (2014b) Not 'Fortress Los Angeles:' Design, and Management of Privately Owned Public Spaces in New York City. In: *Positive Criminology: Reflections on Care, Belonging and Security*, Schuilenburg, M., Steden, R.V. and Breuil, B.O. (eds.). The Hague, The Netherlands: Eleven International Publishing, pp. 117–127.

Huang, T.-S. and Franck, K.A. (2018) Let's Meet at Citicorp: Can Privately Owned Public Spaces Be Inclusive? *Journal of Urban Design*, 23(4): 499–517.

Jacobs, J. (1961) *The Death and Life of Great American Cities*. New York: Random House.

Kayden, J.S., Department of City Planning, and Municipal Art Society. (2000) *Privately Owned Public Space: The New York City Experience*. New York: John Wiley.

Kohn, M. (2004) *Brave New Neighborhoods: The Privatization of Public Space*. New York: Routledge.

Loukaitou-Sideris, A. and Ehrenfeucht, R. (2009) *Sidewalks: Conflict and Negotiation over Public Space*. Cambridge: MIT Press.

Low, S. (2004) *Behind the Gates: The New American Dream*. New York: Routledge.

Low, S., Taplin, D. and Scheld, S. (2005) *Rethinking Urban Parks: Public Space and Cultural Diversity*. Austin, TX: The University of Texas Press.

Low, S.M. (2000) *On the Plaza: The Politics of Public Space and Culture*. Austin, TX: University of Texas Press.

Lydon, M. and Garcia, A. (2015) *Tactical Urbanism: Short-Term Action for Long-Term Change*. Washington, DC: The Streets Plans Collaborative.

Mehta, V. (2014) *The Street: A Quintessential Social Public Space*. New York City: Routledge.

Miller, K.F. (2007) *Designs on the Public: The Private Lives of New York's Public Spaces*. Minneapolis, MN: University of Minnesota Press.

Németh, J. (2007) "Security and the Production of Privately Owned Public Space." PhD diss., Rutgers-New Brunswick.

Németh, J. (2009) Defining a Public: The Management of Privately Owned Public Space. *Urban Studies*, 46(11): 2463–2490.

Németh, J. and Hollander, J. (2010) Security Zones and New York City's Shrinking Public Space. *International Journal of Urban and Regional Research*, 34(1): 20–34.

Németh, J. and Schmidt, S. (2011) The Privatization of Public Space: Modeling and Measuring Publicness. *Environment and Planning B: Planning and Design*, 38: 5–23.

Oldenburg, R. (1989) *The Great Good Place*. New York: Paragon House.

Oswalt, P., Overmeyer, K. and Misselwitz, P. (2013) *Urban Catalyst: The Power of Temporary Use*. Berlin: Dom Publishers.

Overmeyer, K. (2007) *Urban Pioneers: Temporary Use and Urban Development in Berlin*. Berlin: Jovis Verlag.

Parkinson, J. (2006) Holistic Democracy and Physical Public Space. *British Journal of Political Science Conference*.

Project for Public Spaces. (2016) *Making Your Market a Dynamic Community Space*. New York: Project for Public Spaces.

San Francisco Planning Department, Department of Public Works, and Municipal Transportation Agency. (2016) "Pavement to Parks." http://pavementtoparks.org/.

Schmidt, S., Nemeth, J. and Botsford, E. (2011) The Evolution of Privately Owned Public Spaces in New York City. *Urban Design International*, 16(4): 270–284.

Schneekloth, L.H. and Franck, K.A. (1994) Types: Prison or Promise?. In: *Ordering Space: Place Types in Architecture and Design*, Franck, K.A. and Schneekloth, L.H. eds.). New York: Van Nostrand Reinhold, pp. 15–38.

Smithsimon, G. (2006) "The Shared City: Using and Controlling Public Space in New York City." PhD diss., Columbia University.

Smithsimon, G. (2008) Dispersing the Crowd: Bonus Plazas and the Creation of Public Space. *Urban Affairs Review*, 43: 325–351.

Sorkin, M. (1992) *Variations on a Theme Park: The New American City and the End of Public Space*. New York: Hill and Wang.

Stevens, Q. and Franck, K.A. (2015) *Memorials as Spaces of Engagement: Design, Use and Meaning*. New York: Routledge.

Talen, E. (2015) Do-It-Yourself Urbanism: A History. *Journal of Planning History*, 14(2): 135–148.

Watson, S. (2009) The Magic of the Marketplace: Sociality in a Neglected Public Space. *Urban Studies*, 46: 1577–1591.

Whyte, W.H. (1980) *The Social Life of Small Urban Spaces*. Washington, DC: Conservation Foundation.

18

PUBLIC SPACE USE

A classification

Sverre Bjerkeset and Jonny Aspen

Introduction

Can the great mix of activities that people carry out in cities be categorized in any meaning-ful way? In this chapter, we shall attempt to do so, focusing on how the city's open *public spaces*, i.e. outdoor spaces that, in principle, are accessible to all, are put to use.

Public space has been a central topic in urbanism, planning, and urban design for some time now. A physical manifestation of this is the extensive upgrading and construction of new public spaces in many cities worldwide. Justifications vary from issues of inclusion, local democracy, qual-ity of life, and public health, to ones of marketing and urban branding. This renewed interest in public space is also reflected in the United Nations' *New Urban Agenda* (2016) as well as in an increasing amount of scholarly work.[1]

Given such a heightened concern for the city's public spaces, there is, we claim, a need to develop a more nuanced and comprehensive vocabulary for how public spaces actually work in terms of use. Although a range of public space classifications exists, those that specifically address issues of use tend to be either too partial or too general. This impedes finer understandings of how public spaces function. Our proposal is a more comprehensive classification system. As a recording tool, this classification system can facilitate the identification and documentation of the full range of activities taking place in specific public spaces. As an analytical tool, it can be employed to compare public space use across various settings, thus contributing to a more informed, empirically based analysis of shifting features of urban public spaces, as well as to help bridge discussions across differ-ent scholarly disciplines and fields of practice. Along these lines, the classification system can also aid in identifying whether specific categories of public space use are in deficit or suppressed. Although not necessarily the case, the more activities and varied use of public space is often a sign of a more inclusive and social place.

In brief, what follows is an outline of a proposed comprehensive classification system in which the main aim is to facilitate the identification and recording of all major types of activities that take place in urban public spaces.

Existing classifications

Making sense of the world by way of classification is a prime concern within the social sciences and humanities. This extends to disciplines and fields in which public space either constitutes a research subject in itself or an object of more concrete planning, design, and governance. Thus, one can find a broad range of approaches for describing and categorizing features and aspects of public space. These mirror more practical concerns or disciplinary interests on contextual, morphological, or functional features of public space. Attention is also given to issues of ownership, management, and publicness, as well as topics of use more specifically. Given our particular purpose, the review focuses on the latter kind of literature.

Let it also be mentioned that, in writings on cities and urban cultural matters, one can find examples of more implicit ways of naming and classifying features of urban use. The carnivalesque (Bakhtin 2009 [1968]), the flâneur (Baudelaire 1970 [1964]; Benjamin 1999), and issues of play (Huizinga 1949) are but some examples. Much the same applies to research in urban studies, for instance on women's and ethnic minorities' public space use.

When reviewing existing ways of classifying public space use, one ought to look into the definition of 'use' upon which they are based. Two main approaches correspond to a basic distinction between action and behavior. The former refers to *activities*, which is our focus, and includes actions that are carried out with intent or a specific purpose. In contrast, behavior refers to *how* people go about. This might relate to social norms (e.g. 'x' behaved poorly or well), such as the range of low-level incivilities experienced in daily life that is documented and classified by Phillips and Smith (2006). One important strand of work focuses on ways of classifying issues of more normal social behaviors in public (e.g. Goffman 1966 [1963]; Lofland 1998). More recently, Mehta (2014), has attempted to classify street sociability (passive, fleeting, and enduring). Although efforts like these have limited relevance in our context, one should be aware that behavior and action often are ascribed much the same meaning.

Other classifications focus more specifically on activities taking place in public space. Topics cover public space use seen in relation to specific urban settings and contexts (Jacobs 1992 [1961]); types of public space and corresponding use (Carr et al. 1992; Franck and Stevens 2007; Carmona and Wunderlich 2012); work-related use (Whyte 1988); pedestrian use (Whyte); and playful use (Stevens 2007). Carmona et al. (2008) present a list of 17 public space uses, which in fact constitutes a mix of single activities, classes of activities, and public space-related sites or institutions. One can also find more specialized forms of use-related classifications. This goes, among other things, for categorizations of economic activities (e.g. formal versus informal) and transport activities (e.g. human-powered versus motorized), including a range of sub-categories. What use-centered classifications like these have in common is that they are partial; what is highlighted is just one or a few types of activities.

Spanning five decades and bridging academia and practice, the work of Danish architect Jan Gehl has been particularly influential. Through research and publications as well as public life surveys and urban design projects around the globe, Gehl and colleagues have had an especially strong bearing on how many planning and urban development practitioners perceive city life and public space. Gehl (1987) has classified public space use in various ways. One has been to group activities into basic categories such as 'walking, standing, and sitting,' and 'seeing, hearing, and talking' (133–172). A related approach is reflected in what is termed 'stationary activities.' These are subdivided into categories – which are neither exhaustive nor mutually exclusive – such as commercial activities, cultural activities, standing, secondary seating, café seating, and bench seating (Gehl and Gemsøe 1996).

The categorization Gehl most widely applies, though, and for which he is best known, is based on a distinction between necessary and optional activities. These two categories are presumed to capture all activities that transpire in public space. They roughly correspond to utilitarian and recreational activities, occasionally thought of as representing a continuum, but mostly treated as fixed categories. The more significant and less controversial argument is that 'the use of public space has gradually evolved from activities primarily motivated by necessity to those more optional in nature' (Gehl and Svarre 2013, 17). Although the two categories attribute significance to some essential dimensions of public space use, they are, we argue, too general to capture the sheer diversity of, or the internal variation between, such uses.

Issues of public space also tend to engage a range of community-oriented actors. Central among these is the global placemaking movement. The movement's hub, Project for Public Spaces (PPS), was founded in 1975 and inspired by the work of American urbanist and 'people-watcher' William H. Whyte. Based on an evaluation of 'thousands of public spaces around the world,' PPS has proposed a way of categorizing uses and activities: fun, active, vital, special, real, useful, indigenous, celebratory, and sustainable. These are categories that represent more 'intuitive or qualitative aspects' of activity and use (Project for Public Spaces 2018); hence, we find them of little help for our purpose.

Assumingly, a more comprehensive classification of public space use could be of value not only for research but also for local agencies involved in the planning and management of public spaces. A study of eleven cities worldwide found that public space classifications are widely used in public space management, and this sometimes extends to long-term planning. Most such classifications are based on size and function, though some also highlight issues of actual and potential uses (Carmona et al. 2008, 122). In formal land-use planning, four general classes tend to recur (retail, commercial/industrial, residential, and institutional) (Kropf 2017, 24). However, referring to land-use as such, they are not transferable to public space.

To conclude, we have not been able to find any examples upon which our efforts to build a more comprehensive classification system could be grounded. As the review shows, existing classifications are few and seldom are they comprehensive. They are often not more than a series of listings, and a common feature is that emphasis is placed upon a small selection of activity types. One exception is Gehl's broad distinction between necessary and optional activities; however, for our purpose, this represents an overly general approach. It is upon this background that we aim to develop a more complete and fine-grained way of classifying public space use. Before going into the details of this classification system, we shall provide a brief account of key concepts and of the type of classification system towards which we are heading.

Clarifications

By public space we refer to outdoor spaces in cities that in principle (but not always in practice) are open and accessible to all: squares, streets, parks, and promenades, but also more mundane spaces like parking lots, walkways, and bus stops. In terms of locational characteristics, the classification is loosely confined to dense, mixed-use urban areas.

By activity and use, we refer to the individual or collective action of using public spaces for various purposes (which, in most cases, can be observed). Thus defined, activities or use are not meant to cover behavioral features, which, to remind ourselves, are more related to *how* people go about. Neither do we take into account aspects of subjective experience, meaning, and imagination.

In its simplest form, a classification implies an ordering of cases by similarity. Principally, there are two ways of going about this: to make either a typology or a taxonomy (Bailey 1994; Lofland et al. 2006). The former is primarily conceptual, based on

Weberian ideal types; the latter is empirical. A typology is generally multidimensional, the topics under study possessing some complex but systematic interrelations. In contrast, a taxonomy – which is the appropriate approach here – is an elaborated list of all possible types into which a meaningful, empirically observable cultural phenomenon can be subdivided.

In the process of creating a classification system, two basic rules apply: the categories should be both exhaustive and mutually exclusive. That is, the categories developed should make it possible to classify all (or almost all) of the relevant cases (the rule of exhaustiveness). The contents of the classification should also be so defined that each case only can be placed within one category (the rule of mutual exclusiveness). Occasionally cases defy categorization. Residual cases should be as few as possible and explicable in terms of the setting or context in which they are embedded (Bailey 1994; Lofland et al. 2006). For our purpose, in dealing with a fairly complex and ambiguous phenomenon such as public space use, we interpret the principles of mutually exclusive and exhaustive categories more as an ideal than an absolute rule.

Our proposed classification system is primarily grounded on knowledge of Nordic settings. More specifically, it is based on extensive field research on issues of public space use in mixed-use areas in central parts of Oslo, Norway, mainly in the form of observation. This research has been carried out over the last seven years. A principle of data saturation guided the fieldwork; we ceased gathering data when it no longer provided new information (i.e. added anything new to the classification). While the classification partly rests on established classes of human activities, the naming, definition, and compilation of the categories are ours. We have earlier published a simplified version of the classification (Bjerkeset and Aspen 2018; in Norwegian), which generated some useful feedback.

In sum, what we propose is the following: a descriptive and tentatively exhaustive classification of individual and collective activities taking place in centrally located outdoor public spaces. Even though technically a taxonomy, for reasons of accessibility, we stick to the more common term classification.

An alternative classification

All together 16 categories make up the classification, each representing distinct activities. The categories are listed in Table 18.1 with appurtenant definitions and examples.

Some initial remarks: We differentiate between recreational activities that individuals typically carry out alone or with acquaintances, personal recreation activities (e.g. going for a walk, reading, hanging out), and recreational-like activities where some kind of interaction with strangers is more typical (i.e. selling and buying, civic, culture and entertainment, and ceremony and celebration activities). It is useful to note that users of public space can be divided into two groups: There are those who have specific roles to play or are committed to specific tasks (often related to income-generating work or voluntary engagements), and then there are those using the city for their own practical and recreation purposes, that is, regular users, of which the city mostly consists.[2]

Case study

We illustrate the potential usefulness of this classification system by examining two neighborhoods that constitute Oslo's western waterfront: Aker Brygge, and its more recent extension, Tjuvholmen. Gehl has followed the Aker Brygge urban redevelopment

Table 18.1 A classification of public space use

Category	Definition	Examples of public space use
Mundane activities	Activities that are of a daily, practical character, i.e. activities that are more or less imperative for the individual.	Passing through (e.g. walking or biking to/from home, work, school, kindergarten, supermarket). Walking the dog. Accompanying children to leisure activities. Waiting (e.g. for transport to arrive, for green lights, when queuing). Fixing and maintaining personal belongings (e.g. bikes, cars).
Personal recreation activities	Activities that are of a more optional character; which are often, but not always, related to leisure time and which are typically performed alone or with acquaintances.	Going for a walk. Sightseeing. Enjoying peace and quiet in a park. Pausing. Smoking. Reading and using the internet. Lying down. Sunbathing. Hanging out. Socializing. Flirting. People-watching. Working out. Playing. Window shopping. Eating and drinking (e.g. at outdoor restaurants, in parks). Partying.
Transportation activities	Activities that are about transporting people, goods, and products from one location to another (as well as pick-up and delivery, where relevant).	Private transportation (driving one's car).[3] Running public transport. Taxi-driving. Delivery of goods and services (mail, packages, food, etc.). Cash-in-transit. Ambulance transportation.
Selling and buying activities	Activities that are about marketing and selling goods and services (primarily economically motivated), as well as acts of buying such goods and services (motivated by varying degrees of necessity and choice).	Formal (selling): Outdoor serving. Marketing and solicitation. Sales from street and market stands (of goods, services, tickets, etc.). Informal (selling): Prostitution. Ambulatory vending. Shoe-shining. Buying: Acts of browsing, bargaining, paying, etc.
Civic activities	Activities by citizens, activists, (non-governmental) interest groups, etc. that are about expressing and representing opinions and will.	Political and religious activism. Marches. Demonstrations. Strikes. Information campaigns and petition signing. Recruitment for clubs and organizations. Non-profit fundraising.
Culture and entertainment activities	Activities that are about organizing, staging and performing events addressed to the general public – in order to entertain, enliven, enlighten, or disquiet – as well as acts of attending such events.	Organizing, staging, performing or attending: outdoor exhibitions, concerts, theatre, shows, fairs, and sports events, etc. Street performances. Sightseeing tours. Guided tours.
Ceremony and celebration activities	Activities that are about marking or celebrating important historical and contemporary events and phenomena.	Marking or celebrating: Religious and spiritual events. Historical victories and disasters. National days. Anniversaries. State visits. Newly elected office-holders. Carnivals. Parades. Graduations. Marriages. Funerals. Sports victories.

(Continued)

Table 18.1 (Cont.)

Category	Definition	Examples of public space use
'Production activities'	Activities of making goods and contents, mostly for later use, i.e. for sale, distribution or consumption.	Cooking (e.g. street food). Crafts-making. Urban farming. Media and film production related to movie shooting, news coverage, reporting, advertisements, information purposes, etc.
'Management activities'	Activities that are about maintaining law and order, safety and security, as well as providing general physical maintenance and attractiveness.	Street cleaning. Maintenance work. Minor repairing. Garbage disposal. Planting and gardening. Decorating. Parking enforcement. Traffic patrolling. Neighborhood watching. Security guarding. Policing. Military patrolling. Firefighting.
'Construction and renovation activities'	Activities that are about constructing, transforming, improving, renovating, re-modelling, dismantling, and demolishing buildings and other physical structures and features (for example, infrastructures).	Road works. Construction work. Earth-works. Foundation engineering. Façade renovation and repairs. Construction site inspection. Clean-up work. Setting up and dismantling provisional edifices.
'Teaching and learning activities'	Activities that are about organized teaching, learning, training, and investigation.	Open air classes and colloquiums. Excursions. Kindergarten outings. Research and training-related fieldwork. Practical outdoor training (e.g. driving lessons, apprenticeships, law and order enforcement training). Archaeological excavations.
'Work-related activities'	Activities that are about office-related work, tasks, or obligations carried out in public space settings.	Working from a café terrace or a park bench. Working 'on the go' (phoning, reading, texting). Work-related meetings, meals, etc. and social events (e.g. team building, after-work drinking) outdoors.
'Public aid activities'	Activities that represent "instances of helping behavior among the unacquainted that are the right of citizens to expect and the duty of citizens to provide" (Gardner 1986: 37).	Requesting minor favours (e.g. a match, correct change, help to cross the street, help to retrieve lost objects) and information (e.g. time, directions), and acts of complying with such requests. Acts of begging and giving money. Helping out in cases of emergency.
'Homeless activities'	Activities that are imperative for some individuals to undertake in public space due to their life situation, such as homeless people.	Taking care of basic bodily and hygienic needs (e.g. preparing food, eating and drinking, body washing, washing and drying clothes, sleeping). Bottle and trash collecting.

(Continued)

Table 18.1 (Cont.)

Category	Definition	Examples of public space use
'Deviant activities'	Activities or behavior that break with social norms, be they formal rules and laws, or more informal norms and conventions.[4]	Eccentric behavior and action. Addictive use (e.g. taking drugs, drinking). Uncivil acts and remarks. Stealing. Vandalism. Physical and sexual harassment. Violence. Terrorism.
'Other activities'	Activities that cannot be accounted for by any of the main categories.	

project closely and sees it as particularly well-working when compared with most international counterparts. The area's popularity, argues Gehl, is related to a combination of physical density, a mix of functions, and attractive public spaces (e.g. Gehl 2010, 69). That being said, Gehl does not specify what public space uses makes for this attractiveness, nor does he comment on what type of uses are absent. We suspect that Gehl's classification schemes are too general, or too partial, to capture the more specific use patterns of this area.

A more comprehensive system is needed if one is to be able to identify activity types that are present in specific urban areas, and ones that are partially or wholly missing. We

Figure 18.1 Aerial photo of the Tjuvholmen and Aker Brygge neighborhoods, 2014
Image credit: Agency for Planning and Building Services, City of Oslo/Mapaid

undertook an initial testing of our classification system at Aker Brygge and Tjuvholmen (over two days in June 2019). We walked through the selected spaces in the neighborhoods every second hour and recorded every activity we encountered (i.e. what each person present was doing) according to the categories of the classification (for Gehl's application of this method to map stationary activities, see Gehl and Svarre 2013). In doing so, we discovered that it was often difficult to distinguish between two dominant activity types, mundane and personal recreation activities. This was especially so when a person's visual appearance gave no clear indication of the specific activity taking place.

An activity recording of this kind primarily aims to reveal the breadth of public space use, including variations throughout the day, week, or year. Applying a more fine-tuned classification system allows for a fairly precise reading of the nature and shifting character of public spaces. Such a nuanced understanding is needed in order to compare and discuss public space qualities. Importantly, by recording activities taking place, one also uncovers activities *not* taking place. Our trial run revealed that informal buying and selling activities (e.g. street vending), improvised culture and entertainment activities (e.g. street performances), civic activities (e.g. political and religious ventures, demonstrations and so on), and certain public aid activities (e.g. begging) and homeless activities (e.g. sleeping) were more or less non-existent in this part of central Oslo. This points to the little acknowledged fact (in a Norwegian context) that in privately owned and managed neighborhoods like these, nearly all such activities are either forbidden or strongly curbed, testifying to restricted publicness (Bjerkeset and Aspen 2017).

Figure 18.2 Bryggetorget ('Harbour Square'), Aker Brygge neighborhood
Image credit: Sverre Bjerkeset

We argue that the proposed classification system can be useful beyond research. For instance, it can provide urban planning and design with an overview of activity types that should be further stimulated in order to achieve the highly valued aim of creating inclusive environments. It may also inform urban design by raising awareness of activities that can (and possibly cannot) be directly designed into the city fabric. In this way, and more generally, the classification system could also be used to build more systematic knowledge on how public spaces actually work, both understood as specific places and as a more comprehensive urban whole. Thus, it could guide policy development. For example, it may remind policymakers that a general regulation for public purposes might not be sufficient to enable certain variations in use and users, as the Oslo case shows. Some activity types may need to be specified in zoning plans to avoid being subjugated to prohibition.

Having introduced the classification and provided a brief case study, we turn our attention to some more overarching issues.

Challenges and potential

A comprehensive classification system for a complex phenomenon such as public space use does not come without weaknesses and limitations. We will now look into some of these before further considering what strengths and practical impacts the system can have.

Urban public cultures are not as plain and schematic as the draft of activity types may signal. By definition, urbanity is dynamic and shifting – and no less so in a time of expansive

Figure 18.3 Holmens gate ('Holmen's Street'), Aker Brygge neighborhood
Image credit: Sverre Bjerkeset

globalization, increasing demographic diversity, and massive digitalization. Above all, this last phenomenon complicates the picture: digital technology has become an integral part of everyday life. New layers of digital communication are interwoven with most contemporary social and cultural activities (Crang et al. 2007; Del Signore and Riether 2018), and perhaps especially so in urban contexts. Thus, people routinely engage in both online and offline activities when using public spaces. For this reason, we have chosen to treat people's everyday involvement with digital technology as an integral aspect of other kinds of public space use rather than as a separate category.

Many forms of contemporary public space uses are, as already indicated, mixed and multi-purposed. People often do several things at once or are engaged in activities of a shifting character. A woman hurrying through the streets on her way from work (mundane), might stop shortly to text a colleague (work-related), then by impulse join a crowd on a square watching a busker (culture and entertainment), before resuming her walk home (mundane) listening to music with headphones (personal recreation). A freelancer working on a laptop from a café terrace may continuously switch between practical, work-related, and recreational activities. A child on his way from school might run into some friends, get distracted in play for a while, simultaneously chat with other friends on the phone, then recall his promise to buy some items for dinner on the way home and therefore rush on. Instances such as these, where different activities quickly succeed each other or take place more or less at the same time, are many and diverse.

Moreover, the context of the situation in question can radically change the meaning or character of an activity. Activities that are recreational for some, like having a drink at an outdoor café, might be addictive for others. The same type of activity may also change character during the day due to alterations in purpose or pace. Take a mother walking her child in a stroller between home and kindergarten, for instance. Leaving home in the morning, she moves quickly and determinedly to her destination. On the way back in the afternoon, she might find herself walking at a much slower pace, having plenty of time to let herself and the child distract and amuse themselves by things they encounter on their way.

What all such nuances, ambivalences, and complexities add up to is that many activities might be challenging to place under one specific category or to place at all. In such instances, it can be helpful to pay careful attention to more behavioral aspects, such as walking pace or purposefulness of movement. Further, what someone wears or carries with them may provide useful clues, as may contextual factors like time of day. Sometimes it could, however, be difficult to identify what the activities are without having some knowledge about the individual's more subjective experiences or motivations. Whether walking the street is done for transit or recreational purposes (or both), it might be hard to decipher based upon external appearances alone. On such occasions, one would have to inquire with the person in question directly (even then, one may not necessarily get an answer, let alone an accurate or truthful one).

Despite all such limitations and precautions, we claim that the proposed classification system can contribute to a better understanding of what goes on in public spaces. Most importantly, it can ease the task of identifying, recording, and describing types of use in given public spaces. Consequently, it can also enable more finely tuned comparisons between public spaces in different urban settings. This may, of course, be combined with other types of investigations, such as on more structural features or issues of experience.

Urban life is constituted by a broad range of human activities, many of which are characterized by necessity, others by freedom of choice, spontaneity, and coincidence. The issue of necessity seems, however, often to be downplayed in many of today's discourses on public space and public life, in which recreation-oriented perspectives tend to dominate. Thus, what happens to be overlooked are everyday activities of a more prosaic kind that contribute to the 'ballet of the street,' to paraphrase Jane Jacobs (1992 [1961]). Assisted by a classification of the sort here suggested, it should be possible to break down this intriguing though often poorly understood ballet into some of its constituent parts and to identify how and why it plays itself out in different ways from one place to another.

More specifically, the proposed classification system can shed light on some core challenges of contemporary public space development, such as management and questions of over- or under-management (Carmona 2010a, 2010b). Whereas the latter might result in decay and lower use, over-management commonly reduces the public character of public spaces. This is shown in the case study from Oslo, which also demonstrates how the classification can be useful in policy, planning, and design.

More generally, the patterns of use that we have identified here, and the mainly non-technical terms applied to describe them, should make the classification simple to use as an analytical instrument for scholars and non-scholars alike. Hence, we hope this contribution also can help bridge discussions both within and between different scholarly disciplines and fields of practice on a crucial urban issue – the uses of the city's open public spaces.

Concluding remarks

In this chapter, we have proposed a comprehensive classification system to identify and record urban public space use. In doing so, we hope to meet a growing need for analytical and methodological tools that can make for a better understanding of, and dialogue about, how urban public spaces work in terms of activities and use. Living in a time of heightened public space interest and concern, it is somewhat surprising that no similar tools exist.

The classification presented here is based on a Northern European urban setting. Even though people's use varies and plays out differently across geographical and cultural contexts, we hold that the categories can have a heuristic value of a more general kind. In our view, the categories correspond to basic features and common functions in a great many cities and public spaces, especially so for post-industrial and neoliberal Western cities, but hopefully (with some adjustments) for cities in other parts of the world as well.

Although we have aimed at making a comprehensive classification tool, what has been presented in this chapter should be considered a proposal. Further adjustments may be needed to create a classification system of public space use that is as clear and robust as possible, and that easily facilitates spatial observation and documentation.

Acknowledgements

We would like to thank Setha M. Low and Karen A. Franck for valuable comments on earlier versions of this chapter.

Notes

1 The increase in the number of articles in *Urban Studies* that deal with aspects of public space is symptomatic: while only six articles were found for the period 1964 to 1990, close to 300 were published in the period 1990–2015 (Bodnar 2015, 2090).

2 The classification system might of course be further developed. One type of activity that could deserve a category of its own (to distinguish it from other kinds of management activities) has to do with handling merchandise, organizing goods on display, decorating facades, arranging chairs and tables at café terraces, etc. The same goes for activities of free services to the general public, e.g. information campaigns, health controls, serviced (mini) recycling stations, etc. It may also prove useful to develop subcategories, especially for some of the larger categories, such as personal recreation and mundane activities; e.g. for purposes of distinguishing between stationary activities and those involving motion, or between pedestrians and individuals moving around by mechanical or motorized means (cars apart). When it comes to the category personal recreation activities, a division into types such as play, romance, consumption, relaxation might result helpful. It should here be reiterated that, due to the complexity and ambiguity of the phenomena in question, some categories will partly overlap.

3 Notes: Although car driving, unlike other 'transportation activities,' to a large extent has to do with personal mobility, we have chosen to define it as a transportation activity, since it, like most other such activities, normally requires separate traffic lanes.

4 Like 'public aid activities.' it is often difficult to distinguish between activity and behavior when it comes to 'deviant activities.. While some types of deviance are proper activities (e.g. stealing), others (e.g. verbal harassment) can be seen as behavioral aspects of other kinds of use. What is considered 'deviant' is further subject to historical and cultural variation – values, norms, and laws often vary and change across time, place, and cultures. This might also shift from one urban context to another, depending on management regimes. On the other hand, deviance may also be considered to be a positive feature, depending on circumstance and people involved, as in cases of more eccentric behavior and action.

References

Bailey, K.D. (1994) *Typologies and Taxonomies: An Introduction to Classification Techniques.* Thousand Oaks, CA: Sage.

Bakhtin, M.M. (1968) *Rabelais and His World.* Cambridge, MA: MIT Press.

Baudelaire, C. (1970 [1964]) *The Painter of Modern Life and Other Essays.* London: Phaidon.

Benjamin, W. (1999) *The Arcades Project.* Cambridge, MA: Harvard University Press.

Bjerkeset, S. and Aspen, J. (2017) Private-Public Space in a Nordic Context: The Tjuvholmen Waterfront Development in Oslo. *Journal of Urban Design,* 22(1): 116–132.

Bjerkeset, S. and Aspen, J. (2018) Byromsbruk – et utkast til klassifikasjon [Urban Space Use: A Draft Classification]. *Plan,* 49(2): 12–19.

Bodnar, J. (2015) Reclaiming Public Space. *Urban Studies,* 52(12): 2090–2104.

Carmona, M. (2010a) Contemporary Public Space: Critique and Classification. Part One: Critique. *Journal of Urban Design,* 15(1): 123–148.

Carmona, M. (2010b) Contemporary Public Space. Part Two: Classification. *Journal of Urban Design,* 15 (2): 157–173.

Carmona, M., de Magalhães, C., and Hammond, L. (2008) *Public Space: The Management Dimension.* London: Routledge.

Carmona, M. and Wunderlich, F.M. (2012) *Capital Spaces: The Multiple Complex Public Spaces of a Global City.* London: Routledge.

Carr, S., Francis, M., Rivlin, L.G., et al. (1992) *Public Space.* Cambridge: Cambridge University Press.

Crang, M., Crosbie, T., and Graham, S.D. (2007) Technology, Timespace and the Remediation of Neighbourhood Life. *Environment and Planning A,* 39(10): 2405–2422.

Del Signore, M. and Riether, G. (2018) *Urban Machines: Public Space in a Digital Culture.* Woodbridge: LIStLab.

Franck, K.A. and Stevens, Q. (2007) Tying Down Loose Space. In: *Loose Space: Possibility and Diversity in Urban Life,* Franck, K.A. and Stevens, Q. (eds.). London: Routledge, pp. 1–33.

Gardner, C.B. (1986) Public Aid. *Urban Life,* 15(1): 37–69.

Gehl, J. (1987) *Life between Buildings.* New York: Van Nostrand Reinhold.

Gehl, J. (2010) *Cities for People.* Washington, DC: Island Press.

Gehl, J. and Gemsøe, L. (1996) *Public Spaces – Public Life.* Copenhagen: Danish Architectural Press.

Gehl, J. and Svarre, B. (2013) *How to Study Public Life.* Washington, DC: Island Press.

Goffman, E. (1966 [1963]) *Behavior in Public Places: Notes on the Social Organization of Gatherings.* New York: The Free Press.

Huizinga, J. (1949) *Homo Ludens: A Study of the Play-Element in Culture.* London: Routledge & Kegan Paul.

Jacobs, J. (1992 [1961]) *The Death and Life of Great American Cities.* New York: Random House.

Kropf, K. (2017) *The Handbook of Urban Morphology.* Hoboken, NJ: John Wiley.

Lofland, J., Anderson, L., Lofland, L.H., et al. (2006) *Analyzing Social Settings: A Guide to Qualitative Observation and Analysis*, 4th ed. Belmont, CA: Wadsworth.

Lofland, L.H. (1998) *The Public Realm: Exploring the City's Quintessential Social Territory.* New York: Aldine de Gruyter.

Mehta, V. (2014) *The Street: A Quintessential Social Public Space.* London: Routledge.

Phillips, T. and Smith, P. (2006) Rethinking Incivility Research: Strangers, Bodies and Circulation. *Urban Studies*, 43(5–6): 879–901.

Project for Public Spaces. (2018) "What Makes a Successful Place?" Accessed 7 July 2018. www.pps.org/article/grplacefeat.

Stevens, Q. (2007) *The Ludic City: Exploring the Potential of Public Spaces.* London: Routledge.

United Nations. (2016) "New Urban Agenda." Accessed 15 March 2019. http://habitat3.org/the-new-urban-agenda

Whyte, W.H. (1988) *City: Rediscovering the Center.* New York: Doubleday.

19

MAPPING THE PUBLICNESS OF PUBLIC SPACE

An access/control typology

Kim Dovey and Elek Pafka

Introduction

Let's go on an imaginary journey through the contemporary city. We begin in our private apartment, take the shared elevator to the car, and exit into the street – are we now in public space or is the car a private space? We drive to a publicly owned but privately managed carpark, exit to the public street, walk to the nearby train station, and pay for public transport into the city. As we exit the main station, we thread our way through a mix of buskers and hawkers on the sidewalk. Some shopping errands take us in and out of a string of shops and deep into a labyrinthine shopping mall. We meet our friend at a café that spills onto a small corporate park where private security guards ensure that the homeless do not encroach. Our friend invites us back to a party at her place in a large gated community with strict entry protocols and armed security. How might we best understand "publicness" in this journey through "public" space? How might rights to access the city be understood by those who lack the mobilities, resources, and invitations?

While there has been sustained debate over the privatization of public space, there remains a good deal of confusion about how the "publicness" of public space might be defined. The distinction between public and private space has long been framed as a continuum from the fully public to the fully private with various kinds of semi-private or semi-public space between. Questions of privatization, however, engage with the Lefebvrian notion of the "right to the city," incorporating rights of access along with differing kinds of action, appropriation, and transformation. In this chapter we suggest a typology of public spaces that locates differing conditions of privatization and publicness according to the criteria of accessibility on the one hand and ownership or control on the other. This typology generates six broad overlapping categories of publicness, which have been mapped at streetscape scales for three contrasting urban neighborhoods in Melbourne. This typology is neither a continuum nor a discrete set, rather it reveals a range of overlapping and intersecting types. The purpose of this mapping is not to fix our understanding of publicness into these categories but to expose some of the ambiguities and dilemmas facing urban designers; not to mount further arguments against privatization but to create the conditions for a more focused critique. The goal is to move beyond the general call for the "right to the city" in order to map – in a more literal sense – what this means for the design of public space.

Rights to the city

Henri Lefebvre's extended philosophical essay *The Right to the City* was written in 1960s Paris in the midst of student uprisings and Situationist calls for rethinking the potential of public space. At its heart is a claim for the right to appropriate urban space beyond any instrumental functional purpose, to utilize public space for drama, art, fun, play and politics in ways that have not been choreographed. This was both a critique of the modernist ideology to segregate and stabilize the city into functional zones and top-down order, and a broad-based claim of the right to occupy, appropriate, and transform public space. As Kafui Attoh (2011) puts it:

> At the heart of Lefebvre's conception of the right to the city is his notion of the city as... a work produced through the labor and the daily actions of those who live in the city. The right to the city... signifies the right to inhabit the city, the right to produce urban life on new terms (unfettered by the demands of exchange value), and the right of inhabitants to remain unalienated from urban life.
>
> *(674)*

Since Lefebvre's essay was translated into English, we have seen a burgeoning academic literature articulating the case for the "right to the city" and documenting political struggles for the use and occupation of public space (Lefebvre 1996; Mitchell 2003). David Harvey's (2008) essay with the same title involves a critique of capitalist accumulation through dispossession rather than the privatization of public space. For Harvey, neoliberalism involves an expropriation of the city in general rather than of public space in particular. His call for a democratization of public space is part of a wider chorus that is lacking in particulars when it comes to the practice of urban design and planning. If the right to the city is not to be an empty proclamation, we need a better basis for the critique of privatization. How might we map this notion of "publicness" in ways that might enable comparison between different cities, neighborhoods, urban design ideologies and privatization regimes?

Public/private

We begin with a brief account of some previous attempts to model and map different types of public space. In his seminal book *A Theory of Good City Form*, Kevin Lynch (1981) suggests five dimensions through which the performance of a city might be understood: vitality, sense, fit, access, and control. His discussion of "access" includes access to open space, friends, jobs, services, and information, while "control" is a critique of how various forms of public space are governed. Lynch's typology in this regard lists rights of presence (who gets access), action (what we can do there), appropriation (the right to exclude others), modification (the right to change the design or move the loose parts), and disposition (sale, demolition). The right of presence or access is the pre-condition for the following three: action, appropriation, and modification. Rights to action and appropriation are social and political freedoms of expression associated with the vitality and intensity of public space – while access does not ensure them, they can mean nothing without access. Rights of action, appropriation, and modification will differ from one place type to another and with different regimes of control. Lynch's final right of disposition and sale is of a different order again, linked to legal ownership; it identifies the principal connection to regimes of control. Lynch never sought to map these different forms of control over public space.

In a more recent series of papers Matthew Carmona (2010a; 2010b) analyses how new forms and types of public space and its management have proliferated under contemporary regimes of modernism, globalization, and neoliberal capitalism. He argues that critiques of contemporary public space can be divided into "a broad over/undermanagement dichotomy" (Carmona 2010a, 144). From this perspective the undermanagement of public spaces often leads to various types of "neglected," "invaded" (by cars), and "segregated" space while over management leads to "securitized" and "corporatized," "exclusive" spaces of consumption (Carmona 2010b, 157–9). He proposes a new typology of public space that incorporates 20 types in four groups: "positive" (good for public life, including streets and plazas), "negative" (bad for public life, including leftover spaces), "ambiguous" (including shopping malls), and "private" (including gated communities). The 20 categories are based on a mix of criteria including form, function, control, vitality, ownership, meaning, visibility, and user groups. Two neighborhoods in London are then mapped in accordance with these categories. While Carmona's analysis and critique is articulate, the typology incorporates too many categories for the maps to ever be readable. In our view, he also leaps prematurely to normative judgements about what is good or bad for public life.

Another means of measuring "publicness" is that developed by Jeremy Németh and Stephen Schmidt (2011) who suggest a triple axis model that seeks to gauge "ownership," "management," and "use" as measures of publicness. They argue that while ownership and control define the potential for publicness, the level and diversity of uses and users reveals the "actual publicness." From this view a place that is intensively used by a broad range of people for a broad range of activities has a higher degree of "publicness." Margaret Kohn (2004) has also suggested a critique of "publicness" that focuses on the particular forms of social and political encounter that are enabled and enacted. Drawing on the work of Guy Debord (1994) she distinguishes between sites geared to the consumption of spectacle and those that facilitate dialogue and interaction: "Movie theaters and sports stadiums do not feel like public places because they do not facilitate interaction between people. They aggregate individuals but they do so in a way that positions them as spectators rather than participants" (Kohn 2004, 10). Neither Kohn nor Németh and Schmidt engage in the mapping of publicness.

While volumes of use and the diversity of users of public space are important research questions, it does not follow that the concept of "publicness" might be established in this way. To equate "publicness" with popularity and diversity of use is to imply that a successful shopping mall becomes more public as it depletes the life of the street which then becomes less public. From such a view, the privatization of public space becomes self-legitimating and can escalate with no easy stopping rule. When people flock to private spaces (for whatever reason) this may simply produce an illusion of publicness. In a survey of new urban spaces in London, Carmona (2015) found that 45 per cent were privately owned/controlled yet publicly accessible and that the level and range of public use of these spaces compared favorably with those that are publicly owned/controlled. He concludes that ownership is relatively insignificant and that the critique of privatization has been overstated: "no evidence was found during the research of an unwritten agenda to subvert the experience of public space for any set of users, or to make it any less public" (21). Such agendas are of course hidden and generally market driven as quasi-public space is produced by commercial imperatives. One of his cases is Paternoster Square where Occupy protesters were evicted in 2011 (Köksal 2012).

In terms of the right to the city, we suggest that "publicness" is better construed as a space of possibility – a capacity for genuinely public life that may or may not be actualized. One model in this regard is the famous Giambattista Nolli map of 18th-century Rome. While often referred to as a figure/ground or building footprint map, in this case the white "ground" incorporates many publicly accessible interiors (such as churches and courtyards) that functioned as extensions to the street network (Bosselmann 1997). Figure 19.1 contrasts the morphology of a 16-hectare neighborhood of the Nolli map with a similar map of Naples by Giovanni Carafa from the same period (Thomas 2013). These two maps were produced for quite different purposes and audiences, revealing different morphologies, yet they share a seminal approach to mapping cities. The long-standing power of such maps in the urban imagination lies in the capacity to reveal the city as a network of accessible public space and to contrast such levels of access in different neighborhoods and cities. Like all good maps, the key lies in the selection of what to show and not show; these maps do not seek to show how public space is used. But how might we map the 21st-century city where a proliferation of corporate plazas, gated communities, and shopping malls has brought new levels and complex types of public access, ownership, and control of the public realm?

An access/control typology

Our approach then is to conceive of this conceptual space as a field of differences in degrees of publicness aligned along two axes, one representing degrees of public access and the other representing degrees of public control. Like any typology this conception reduces a broad field of differences and excludes some aspects of the public/private divide in urban space. We have excluded consideration of formal design as a criterion of publicness except as design mediates access. Many private spaces are designed to look public, but this does not make them public. We have also excluded consideration of the level of use or street life vitality as a criterion of publicness – a deserted street retains its publicness; vitality is crucial to urbanity but not to publicness.

Figure 19.2 is a diagram of this field of differences with increasing levels of access from bottom to top and increasing levels of public control from left to right. The vertical axis generates a distinction between spaces of open versus restricted public access. The category we might call "publicly accessible space" will include all of those places that a citizen can walk to without payment, invitation, or membership. The network revealed by the Nolli map of Rome is effectively "publicly accessible space" because it continues into publicly accessible buildings. While the Nolli map embodied some ambiguities of access (Bosselmann 1997, 14–15), 18th-century Rome was easy to map compared with 21st-century cities where so many new kinds of publicness have emerged. The horizontal axis distinguishes between private and public control with control largely identified with ownership. Various restrictions of access can apply to both publicly and privately owned/controlled spaces, whether internal or external.

These two axes generate a broad field of different types and degrees of publicness. Like any typology this is a simplification; a city is not a tree and public space does not submit neatly to any critique of publicness. Indeed, the ambiguities are crucial to understanding the ways that the publicness of urban space becomes blurred in everyday life; this diagram is a tactic for drawing such ambiguities into the light for critique. Our goal is to enter into such difficulties rather than to simply resolve them. We have therefore divided this field into a set of overlapping types as follows. "Open-public" space (top right) is the traditional

(a)

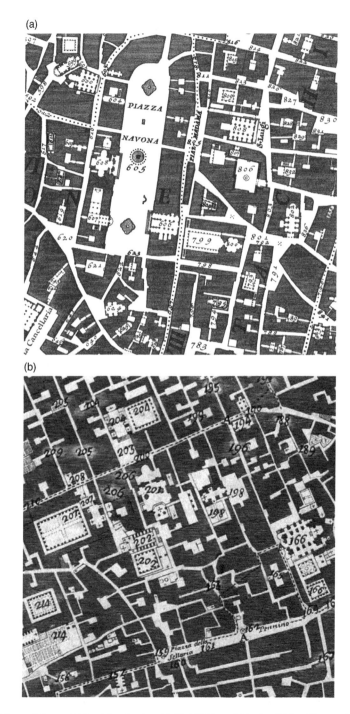

(b)

Figure 19.1 Maps of Rome (a) by Giambattista Nolli (1748) and Naples (b) by Giovanni Carafa (1775)
    Scale: 400 × 400 meters

Source: Wikimedia Commons

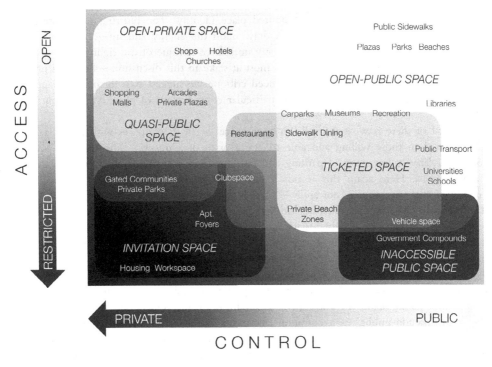

Figure 19.2 Six Types of Publicness – an Access/Control Typology
Image credit: Kim Dovey

pedestrian space of sidewalks, plazas, parks, and beaches but also includes internal spaces such as libraries. "Open-private" space (top left) incorporates privately owned and controlled space that is open for public access. Examples include churches, shops, restaurants, and hotels along with shopping malls, arcades, and corporate plazas/parks – whether internal or external. The category in the lower left quadrant we call "invitation space" comprising privately owned and controlled space where the public may or may not gain access by invitation. Examples here include all forms of housing and private workspace as well as gated communities, private parks, apartment foyers, and clubs. In the lower-right quadrant is a category we call "inaccessible public space" – all publicly owned and controlled spaces where access is restricted in some way. This includes public space that has been surrendered to vehicular movement (cars, rail, ports) as well as government compounds, security zones, and state facilities.

These four types are not mutually exclusive; we find overlaps and slippages between different kinds of publicness and privacy. Theatres, schools, and stadiums may be privately or publicly owned and controlled, or a mix of the two. The focus on access and control also produces some strange bedfellows: shopping malls, hotels, and churches are all forms of open-private space with full public access under private control. Sidewalk cafes are restricted on a daily and seasonal rhythm; access to shopping malls and arcades is closed at night; gated communities are subject to some forms of public control; ownership and control can be shared through public/private partnerships. The "Business Improvement District" is an example where local property owners are granted limited private control over public space for purposes of revitalization which may or may not include the removal of people and

239

practices that don't contribute to the desired place identity. The blurring of private and public control enables private interests to draw upon the energy and image of public life and channel it into profit. In this context any normative critique of the right to the city or the privatization of public space – what is most at stake in this discussion – becomes problematic. In order to move to a more nuanced critique we suggest two further overlapping categories that can be seen to encompass particular constellations of place types within this field.

The first of these is what we call "ticketed space" – all those spaces with public access but restricted to those willing and able to pay a price of admission. The price may vary from the relatively low costs of parking and public transport to those of exclusive theatres and restaurants. Here again this is not a discrete collection of place types so much as a condition that may or may not apply – schools and parking may or may not have a price of admission. This category includes sidewalk dining where there is often a clear public benefit in the private use of public space, but also private beach zones where there is not. In Figure 19.2, this category cuts horizontally across the field of publicness between those spaces that are open and closed to the public.

The second category of "quasi-public" space is defined as privately controlled spaces that operate "as if" they were public. The terminology here is designed to capture the quality of being almost public (from the Latin *quasi*: as if, almost) – apparently, but not really, public. The term "pseudo-public" is a common but more loaded term that carries overtones of deception. As the diagram in Figure 19.2 shows, this category cuts vertically across this field to incorporate shopping malls, arcades, corporate plazas, gated communities, and private parks, overlapping with open-private space to invitation space. This category marks a distinction between the mall or corporate plaza on the one hand and the church, shop, and hotel on the other – the former are designed, operated, and used "as if" they were public space while the latter are not. Shops, hotels, churches, and restaurants do not replicate public spaces as shopping malls, gated communities and corporate plazas do.

This model enables a distinction between forms of privatization that restrict access and those that do not (vertical axis); as well as a distinction between private and public control (horizontal axis). The model incorporates a large range of privately owned but publicly accessible attractions that are crucial to the urban life of any city – shops, theatres, libraries, museums, sporting arenas, and restaurants. While access may or may not be restricted, such place types are often in synergy with the open-public space network – particularly when they are structured as cul-de-sacs off the public street network. Many types of public and private space are interwoven in the most urban districts.

Mapping Melbourne

We now want to test this typology as a basis for mapping, applied to three sites in Melbourne: part of central Melbourne and two districts within the recent Docklands redevelopment (see Figure 19.3). Central Melbourne is a grid-based morphology developed from the 1830s where a series of laneways and arcades have emerged through successive layers of speculation and subdivision (Dovey et al. 2018a). Melbourne Docklands is an "instant" neighborhood developed since the 1990s under a neoliberal regime of largely private control over the development of public land (Dovey 2005).

Figure 19.4 shows a series of 16-hectare maps of a section of central Melbourne incorporating the main railway station (lower left), Federation Square (lower right), and the civic/retail district to the north. In these maps we have de-layered the city in order to understand

(a) Central Melbourne

(b) Victoria Harbour, Dockands

(c) Waterfront City/The District, Docklands

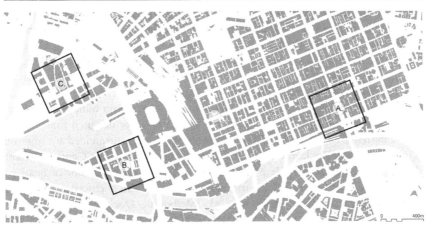

Figure 19.3 Three Melbourne Sites
Image credit: Elek Pafka

the categories separately in terms of ground floor functions (derived from City of Melbourne data). We have arranged the layers to correspond to their positions on the access/control field outlined in Figure 19.2. The "open-public" layer (top right) is essentially a network of sidewalks, plazas, and a library. Quasi-public space (center left) is a constellation of patches – mostly private arcades that serve as connections within the "open-public" network. The "open-private" layer (top left) is a patchwork of shops, hotels, and churches that plug into this network and show the western section as a vibrant shopping district. As we move down Figure 19.4, we find the less accessible layers of "ticketed," "invitation," and "inaccessible public" space. Ticketed space incorporates the railway platforms and museum, along with strips of sidewalk dining and parking. Invitation space is a constellation of patches, while inaccessible public space is mainly dedicated car and rail space and inaccessible public institutions.

Before we discuss the various relations between these layers, we want to present the identical mappings of two parts of Docklands. Victoria Harbour (see Figure 19.5) incorporates a slice of former docks between a river to the southwest and harbor to the north (for a brief history see Dovey 2005, 178–81). Here again open-public space forms the network, augmented with parks and waterfront promenades rather than plazas. Open-private space lines this network and there is minimal quasi-public space. Ticketed space is primarily parking and waterfront restaurants, while invitation space is a constellation of large island patches. Waterfront City/The District (see Figure 19.6) is a precinct on the opposite side of the harbor where the access network is based on the strategy of using a giant ferris wheel (at the top) as a "magnet" to anchor one end of a long shopping mall with the waterfront at the other end with housing on top, flanked by film studios and substantial parking (Dovey 2005).

In order to see the interconnections between layers more clearly, in Figure 19.7 we have superimposed two sets of layers that we call "accessible space" and "restricted space" while omitting inaccessible public space. "Accessible" space (left-side maps) is effectively the open pedestrian network of the city where open-public, quasi-public, and open-private space are juxtaposed. There are, of course, no island patches on these maps. The central Melbourne map shows a more extensive range of public access (less white space) partly due to its role as a retail center. "Restricted space" (right-side maps) is an overlay of ticketed and invitation space; these are largely constellations of island patches interspersed with strips of parking.

These maps reveal many of the effects of very different morphogenic processes on these sites. Central Melbourne has a more intricate grain with higher levels of public access and more complex interfaces between public space types; despite the rigid grid there is an emergent order of levels of access and control which differ across different parts of the grid. The Victoria Harbour morphology is more formularized with larger grain bundles of public/private mix that are repeated in multiple locations – typical of a city designed from top down. Waterfront City has a network of quasi-public space at its core, surrounded by large islands of restricted space. Both Docklands sites have more restricted space than Central Melbourne, mostly linked to car-dependency. It is notable that there is a large amount of quasi-public space (mostly arcades) in central Melbourne where they perform a crucial role as connections within the intricate public-space network. Melbourne Docklands was all originally public space that has been privately designed with minimal public oversight as part of a neoliberal planning agenda; Victoria Harbour is by far the best-designed part of a mostly poor outcome that has been documented elsewhere (Dovey 2005). The privatized parts of Docklands are more substantial than in the central city, but they remain as patches that are plugged into the public space network.

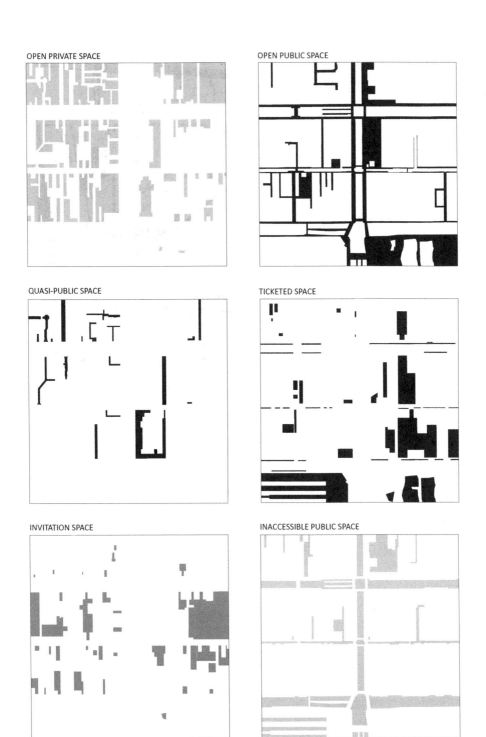

Figure 19.4 Central Melbourne

Image credit: Elek Pafka

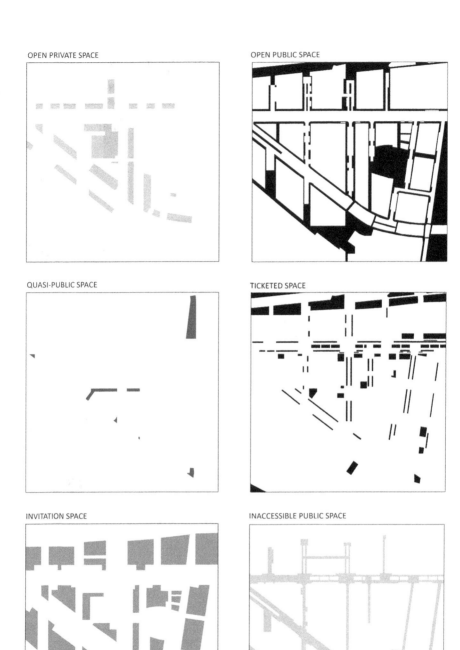

OPEN PRIVATE SPACE

OPEN PUBLIC SPACE

QUASI-PUBLIC SPACE

TICKETED SPACE

INVITATION SPACE

INACCESSIBLE PUBLIC SPACE

Figure 19.5 Victoria Harbour

Image credit: Elek Pafka

OPEN PRIVATE SPACE

OPEN PUBLIC SPACE

QUASI-PUBLIC SPACE

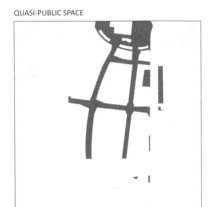

TICKETED SPACE

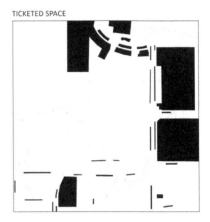

INVITATION SPACE

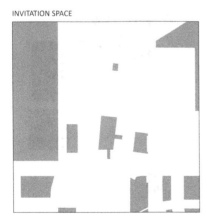

INACCESSIBLE PUBLIC SPACE

Figure 19.6 Waterfront City/The District
Image credit: Elek Pafka

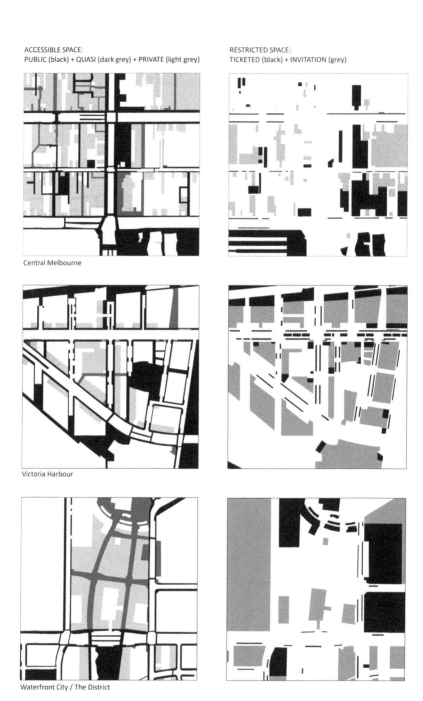

ACCESSIBLE SPACE:
PUBLIC (black) + QUASI (dark grey) + PRIVATE (light grey)

RESTRICTED SPACE:
TICKETED (black) + INVITATION (grey)

Central Melbourne

Victoria Harbour

Waterfront City / The District

Figure 19.7 Accessible versus Restricted Space
Image credit: Elek Pafka

The slippery commons

Rather than conclude here we want to open up questions that have not been sufficiently explored. This mapping is a small part of a very large undertaking to better understand and map the right to the city. If there is one clear lesson in this attempt to model and map the concept of publicness in public space, it is the challenge posed by the extraordinary complexity and ambiguity of the public/private framework. After excluding issues of formal expression and use, and then reducing issues of control and access to six types (which is close to the limit for effective mapping) we remain confronted by the ambiguous overlaps as shown in Figure 19.2. This overlapping is not a failure to get the typology correct but rather emerges from the attempt to derive a typology that is rigorously geared to the range of different access and control conditions. Ticketed space is both open and yet restricted to those who can pay the price; vehicle space is often accessible yet unwalkable. It is not that these categories are blurred, indeed the boundaries of accessible and controlled territories are generally specific to a meter or less. We have necessarily dealt with the ambiguity for mapping purposes by allocating specific spatial types into one category or the other. Vehicle space is mapped as inaccessible even though we all cross streets without crossings.

Rather than see these overlaps and slippages as problems to be erased we suggest they are where the critique of privatization needs to be focused, indeed the privatization of public space proceeds by exploiting slippages between public/private categories. While it has long been argued that clear boundaries between public and private space are characteristic of good city form (e.g. Jacobs 1961), for Lynch (1981) it is not about the clarity of public/private boundaries but clarity of rights. We would suggest that ambiguities of publicness can also be key to urban intensity. Well-accepted urban design principles such as "active edges" and "eyes on the street" suggest the value of interpenetrations between public and private space, and the public/private interface has been the subject of sustained mapping (Gehl and Gemzoe 1996; Dovey and Wood 2015). The value of a functional mix has long been understood, along with ideals of a social mix and small grain mix of building styles and types (Jacobs 1961). We have argued elsewhere that this "mix of mixes" is vital to public life and would extend this argument here to include a rich mix of different types of public and private space (Dovey and Pafka 2017). Any reduction of public–private relations to a clear separation can damage the deeper qualities of urbanity.

It is not so much that particular spatial categories are good or bad – the normative critique of privatization will need to focus on the relational assemblages of different types. The value of sidewalk dining in public space for instance is geared to the flows of pedestrian traffic. The open private space of shops, churches, and hotels works better when it is connected to open-public rather than quasi-public space. The quasi-public space of arcades works better as shortcuts between open-public spaces than when it constructs a separate network as in the shopping mall. We note the key role of quasi-public arcades in adding access through central Melbourne, compared to the relative paucity of such space in the privately designed Victoria Harbour. Any normative critique must engage with the complex synergies between different types. In this regard the overlaps between categories, far from being the problem, become the focus of attention. Particular interest should focus on the overlap between "invitation space" and "quasi-public" space – the "common interest developments" of gated communities and private parks that operate as if they were public. The most significant threat from privatization emerges where spaces of private access and control are no longer discrete patches that plug into a public space network, rather they become that network, as in the Waterfront City/The District example.

Another obvious limit of our mapping here is its two-dimensionality, we have only mapped access and control at ground level and without considering temporal change. Mapping is a task of producing spatial knowledge and is inevitably a selection of those layers of data that best work as a lens onto how the city works (Dovey et al. 2018b). Mapping the city as a dynamic three-dimensional assemblage remains a great challenge. Our goals for this chapter are modest and have been focused on the task of mapping the categories of private/public space, implicitly reducing the "right to the city" to questions of legal access and control. Yet our understanding of the "right to the city" cannot ultimately be limited in this way because it will also rest upon a better understanding of the urban "commons" – the rights produced by everyday collective use of shared space regardless of legal ownership (Ostrom 1990; Blomley 2008). Whether and how the urban commons might be better modelled and mapped remains a significant question. This brings us back to the ways that different kinds of shared space are used but it will surely entail more than a measure of the volume or diversity of uses and users.

Acknowledgement

This chapter is expanded and adapted from Dovey, K. (2016) *Urban Design Thinking*. London: Bloomsbury, pp. 153–158.

References

Attoh, K.A. (2011) What Kind of Right is the Right to the City? *Progress in Human Geography*, 35(5): 669–685.

Blomley, N. (2008) Enclosure, Common Right and the Property of the Poor. *Social and Legal Studies*, 17 (3): 311–331.

Bosselmann, P. (1997) *Representation of Places*. Berkeley, CA: University of California Press.

Carmona, M. (2010a) Contemporary Public Space, Part One: Critique. *Journal of Urban Design*, 15(1): 123–148.

Carmona, M. (2010b) Contemporary Public Space, Part Two: Classification. *Journal of Urban Design*, 15 (2): 157–173.

Carmona, M. (2015) Re-theorizing Contemporary Public Space. *Journal of Urbanism*, 8(4): 373–405.

Debord, G. (1994) *The Society of the Spectacle*. New York: Zone Books.

Dovey, K. (2005) *Fluid City*. London: Routledge.

Dovey, K, Adams, R. and Jones, R. (eds.). (2018a) *Urban Choreography: Central Melbourne 1985–*. Melbourne: Melbourne University Press.

Dovey, K. and Pafka, E. (2017) What is Functional Mix? *Planning Theory and Practice*, 18(2): 249–267.

Dovey, K., Ristic, M. and Pafka, E. (2018b) Mapping as Spatial Knowledge. In: *Mapping Urbanities*, Dovey, K., Pafka, E. and Ristic, M. (eds.). New York: Routledge, pp. 1–16.

Dovey, K. and Wood, S. (2015) Public/Private Urban Interfaces. *Journal of Urbanism*, 8(1): 1–16.

Gehl, J. and Gemzoe, L. (1996) *Public Spaces and Public Life*. Copenhagen: Danish Architectural Press.

Harvey, D. (2008) The Right to the City. *New Left Review*, 53: 23–40.

Jacobs, J. (1961) *The Death and Life of Great American Cities*. New York: Random House.

Kohn, M. (2004) *Brave New Neighbourhoods*. New York: Routledge.

Köksal, I. (2012) Activist Intervention. *Social Movement Studies*, 11(3–4): 446–453.

Lefebvre, H. (1996) In *Writings on Cities*, Kofman, E. and Lebas, E. (eds.). Oxford: Blackwell.

Lynch, K. (1981) *A Theory of Good City Form*. Cambridge: MIT Press.

Mitchell, D. (2003) *The Right to the City*. New York: Guilford Press.

Németh, J. and Schmidt, S. (2011) The Privatization of Public Space. *Environment and Planning B*, 38(1): 5–23.

Ostrom, E. (1990) *Governing the Commons*. Cambridge: Cambridge University Press.

Thomas, R. (2013) Propagandizing Nolli in Naples. In *Giambattista Nolli and Rome*, Verstegen, I. and Ceen, A. (eds.). Rome: Studium Urbis, pp. 149–159.

20

PRIVATE, HYBRID, AND PUBLIC SPACES

Urban design assessment, comparisons, and recommendations

Els Leclercq and Dorina Pojani

Introduction

At the beginning of the 21st century, the distinction between public and private space is blurring. The production of public space is no longer solely a state affair. Nor is public space universally accessible, collective, or inclusive. A wide variety of private actors ranging from large developers to small not-for-profit community groups, now take responsibility for the design, implementation, and maintenance of public spaces. These multiple stakeholders program spaces and set aesthetic and access rules according to their own agendas. Consequently, corporate, parochial, and/or individual interests often override public values such as accessibility, diversity, and inclusivity.

Progressive academic commentators have condemned this state of affairs. A major concern has centered on the evaporation of a physical public sphere, with knock on effects on democracy (Sennett 1977; Németh and Schmidt 2011). Also, private or hybrid spaces have been criticized as over-designed, sterile, inauthentic, and formulaic – in sum, simulacra of 'real' public spaces (Carmona 2015; Kohn 2004). Theme parks are cited as the prime example of 'invented' space, which stands in stark contrast to the great medieval and baroque squares of Europe, the vibrant marketplaces of Asia, or even the grand parade grounds produced by communist regimes (Sorkin 1992). However, between the extremes of Disneyland and Piazza del Campo, a number of everyday private or semi-private public spaces now exist in cities everywhere.

Hence a broader and multifaceted definition of the concept of publicness may be needed at this point. A new definition needs to reflect contemporary economic, societal, and cultural narratives (Carmona 2015; Carmona et al. 2008; De Magalhães and Freire Tigo 2017). Of course, notions of ownership, management, and access will continue to form an integral part of the publicness of space.

However, the essence of publicness also depends on how users are permitted to experience, evaluate, negotiate, and appropriate space to meet their own needs (Crawford 2008). Gehl (2011) argues that although the physical framework does not have a direct influence on the quality, content, and intensity of social contacts, architects, and planners can affect the possibilities for meeting, seeing, and hearing people – possibilities that both take on a quality of their own and become important as background and starting point for other forms of contact (13). Based on an understanding of urban space as an enabler rather than a determinant of human behavior, a number of commentators have advanced ideas on what a quality space entails and which criteria designers should follow.

But, does 'high quality space' equate 'public space'? Can spaces produced through private finance and planning embody both the design and 'public' qualities of traditional public space? Can designers create a physical setting that fosters publicness, as well as sociability and aesthetic value? A handful of scholars have conducted evaluations of public space in the privatization era – both in terms of design and 'publicness' (Kohn 2004; Varna and Tiesdell 2010; Németh and Smith 2011; Langstraat and Van Melik 2013; Mehta 2014). The purpose of this chapter is to add to this body of literature by undertaking a systematic visual assessment of three different projects in Liverpool (UK), which are ostensibly public. The case studies include an entirely private development, a public–private partnership, and a public project taken over by a community organization. The three projects are evaluated, compared, and contrasted using an existing tool developed by Ewing and Clemente (2013), which has been adapted to measure publicness in addition to urban design quality.

The analysis reveals that, while lacking in 'publicness,' privatized spaces may be quite popular with users because they possess high urban design quality and offer engaging activities. Hence, we argue in favor of expanding the notion of public space to include spaces produced through private finance and planning. Designers can enhance the public nature of these spaces by allowing for greater flexibility in use and more opportunities for spontaneous and temporary appropriation. This is crucial in order to safeguard space that is open to the public as an arena for democratic participation, deliberation, and action. Another issue is what the scope of surveillance – including public or private CCTV cameras, police/security officers, and facial recognition devices – should be. This issue must be the subject of serious public debate and engagement, which designers could facilitate. The consequences of surveillance on the public sphere are too serious to be left in the hands of a single group of professionals, however technically capable.

Methodology

Case studies

The case studies all involve Liverpool, England, which was once a flourishing city with a developed maritime economy. When port activities left for more profitable locations in the post-war period, the city entered a spiral of economic and spatial decline. To combat decay, in the 1990s the local council adopted a series of 'urban renaissance' strategies, which were advocated by the national government at the time (Towards an Urban Renaissance 1999). These strategies were based on the notion that the private sector must become a partner in urban affairs in order to reduce public costs – privatization and private involvement having been cornerstones of Thatcherite politics and Tony Blair's Third Way approach, respectively. Three urban districts in Liverpool, which are briefly described below, went through regeneration processes during that period (see Figure 20.1). Different production and management regimes applied in each case provide an opportunity to assess and compare the urban design quality of spaces, which were developed with differing levels of public and private control (see Figure 20.2a–c).

Figure 20.1 Case study locations in Liverpool

Image credit: Els Leclercq and Dorina Pojani; Google Maps

Liverpool ONE is an inner-city redevelopment project, encompassing 16 hectares. In 2000, the council appointed a private development firm to proceed with the design and construction of a predominantly retail based development, which it still manages. The land was practically sold on a 250-year land lease. The masterplan designers made a conscious effort to mimic a traditional city center, and visually and physically link the existing core to the waterfront. The first of its kind at this scale in the United Kingdom, Liverpool ONE earned much praise in design circles, winning a number of architectural awards.

Ropewalks was created in 1998, with a public–private partnership formed to run this project. It included a mix of stakeholders. The masterplan envisioned a mixed-use district, which was set to become Liverpool's new 'creative quarter.' The public sector funded the initial public space improvements, including pocket parks and pedestrian connections, in an effort to attract private investment in the area. This concept ultimately paid off. However, with investment depending on a large number of private actors, the area's regeneration has been slow and organic.

Granby4Streets started in 1995 as a publicly-led urban renewal scheme in an impoverished area which had experienced turmoil and riots. Instead of refurbishing Victorian houses in the area, the council opted for demolition and rebuilding. This approach met with strong local resistance as residents preferred to keep their homes and refurbish them. In 2007, residents took ownership of the plan, appropriating public spaces, painting boarded windows,

Figure 20.2a Case studies after redevelopment: Liverpool ONE
Image credit: Els Leclercq and Dorina Pojani

planting front gardens, and organizing a very popular street market. Hence a top-down urban renewal plan was transformed into a bottom-up community initiative, eventually supported by the council. The first Community Land Trust in the UK was formed here, as a result of the Granby4Streets project.

Analytical framework

The urban design literature is rife with frameworks that highlight the relationship between urban design features and the level of human activity in public spaces. Hassen and Kaufman (2016), Ewing et al. (2006), Gehl and Svarre (2013), Marshall (2012), and Hooi and Pojani (2020) have compiled lists and critical commentary regarding some of the classic frameworks that have been proposed in more than half a century. These authors' reviews show that few of the existing frameworks have been validated in real world settings (for an exception, see Mehta 2007; 2014). Most are derived from personal observations and theory. As such, they reiterate many of the same elements, in a circular referencing pattern. Also, not all existing frameworks are meant to be used for post-hoc evaluations. Most are prescriptive, i.e., intended to guide new projects (see Hooi and Pojani 2020).

Among the frameworks which are intended for post-hoc evaluations of urban design quality, the framework set forth by Ewing and Clemente (2013) is the most systematic one available; in it, urban design quality is linked to walkability. It builds on a strong theoretical foundation to discern why certain variables are more meaningful than others. Its other strengths include

Figure 20.2b Case studies after redevelopment: the Ropewalks

Image credit: Els Leclercq and Dorina Pojani

thorough testing and validation, and detailed step-by-step instructions for use (including a scoring sheet). For these reasons, this framework was selected and adapted for this study. A limitation is that, while the framework allows for the comparison and ranking of spaces in relation to one another, it does not provide absolute maximum and/or minimum scores for each variable. Therefore, it is impossible to determine whether an assessed space is 'good' in absolute terms.

Ewing and Clemente's framework encompasses five urban design aspects, which can be measured *quantitatively* by following a prescribed stepwise procedure. These are: (1) imageability, (2) enclosure, (3) human scale, (4) transparency, and (5) complexity. All five have passed the test of validity and reliability conducted by the original authors of the framework. In addition, the framework includes four urban design aspects that did not pass the test of quantitative validity and reliability, and hence can only be discussed *qualitatively*. These are: (6) coherence, (7) legibility, (8) linkage, and (9) tidiness. In the present study, we assessed these aspects too.

The aforementioned qualities primarily address the aesthetic value, physical setup, and functionality of public spaces. As such, the framework constitutes a holistic tool to evaluate design quality. However, it is of limited value when it comes to measuring the level of

Figure 20.2c Case studies after redevelopment: Granby4Streets
Image credit: Els Leclercq and Dorina Pojani

publicness in urban spaces (public or privatized). By contrast, other researchers consider dimensions of publicness as the core measures, including:

- Accessibility, inclusion, and tolerance of difference (Young 2000);
- Access, agency, and interest (Madanipour 2003);
- Ownership, accessibility, and intersubjectivity (Kohn 2004);
- Right of access, right of use, and ownership and control (De Magalhães 2010);
- Ownership, control, civility, physical configuration, and animation (Varna and Tiesdell 2010);
- Ownership, management, and use/users of a space (Németh and Schmidt 2011); and
- Ownership, management, accessibility, and inclusiveness (Langstraat and Van Melik 2013).

Accessibility, ownership, and a concern with the users of public space are recurring themes here. Therefore, based on this portion of the literature and our study objectives, two other *qualitative* aspects were deemed essential, including: (10) safety, and (11) appropriation. The case study areas were assessed for presence of physical elements that make up 'safety' (i.e. CCTV cameras, security guards, blank facades, etc.) and 'appropriation' (visual signs such as outdoor seating and user-contributed planters, artwork, buskers and the like). Although these elements imply quantitative results, they were assessed qualitatively because (same as coherence, legibility, linkage, and tidiness) they could not pass the test for quantitative validity and reliability. However, a qualitative evaluation does provide insights into the degree of publicness of the case study spaces. The adapted framework, which includes a total of eleven elements, is presented in Table 20.1.

Table 20.1 Analytical framework. Adapted from Ewing and Clemente (2013)

QUALITIES	ELEMENTS
Quantitative assessment **Imageability** Relates to sense of place and reinforces the innate human ability to see and remember patterns. It refers to the quality of a place that makes it distinct, recognizable, and memorable. It is determined by physical elements such as buildings and landmarks and their ability to capture, create, and evoke emotions.	street furniture, historic buildings, identifiers, number of people, events, noise level, landscape feature
Enclosure is the degree to which streets and public spaces are defined by vertical elements. It helps users gain a sense of position in space and identity with their surroundings. To determine enclosure, buildings, trees, artwork, light posts, and other vertical elements are considered as outdoor 'walls,' while streets, footpaths, and public space become outdoor 'floors'	sight lines, proportion street wall and sky, presence of physical elements
Human scale is important to create an intimate environment. It is the proportion of physical elements' size, texture, and articulation to the human form and function. To determine it, human height is compared to the height and length of buildings and their components – such as doorways, windows, awnings, lampposts, and planters. Human scale is also a function of human speed (as opposed to the speed and bulk of vehicles)	sight lines, active frontages, street furniture, trees and planters, signs of appropriation
Transparency pertains to the level of human activity that can be seen or perceived beyond the edge of the street or public space. See-through physical elements such as non-reflective windows, walls, doors, fences and landscape influence the transparency of the street	active uses, bay windows, porches, balconies, proportion of street walls and windows
Complexity refers to the visual nature and richness of a street. Its value lies in providing sensory pleasure and avoiding sensory deprivation among users. The variety, number, and type of physical elements, such as buildings, ornaments, landscape features, street furniture, and human activity, shapes the complexity of a street	variety in appearance buildings outdoor dining, presence of street furniture and art, people, variety in uses

Ewing Clemente (2013)

(Continued)

Table 20.1 (Cont).

Qualitative assessment

Coherence

refers to a sense of visual order. It ties spaces together through built form and helps with orientation. The degree of coherence is influenced by consistency and complementarity in the scale, character, and arrangement of buildings, landscaping, street furniture, paving materials, and other physical element

variety of building ages, proportion of windows, trees, street furniture, symbolic meaning

Legibility

refers to the ease with which the spatial structure of a place can be understood and navigated as a whole. The legibility of a place is improved by a street or pedestrian network that provides travelers with a sense of orientation and relative location and by physical elements that serve as reference points

vistas, urban grid, signage, edges, memorable architecture, landmarks, trees

Linkage

refers to physical and visual connections from building to street, building to building, space to space, or one side of the street to the other which tend to unify disparate elements and help with wayfinding. Tree lines, building projections, marked crossings all create linkage

ease of movement, positioning of trees, building lines, physical connection to adjacent areas

Tidiness

refers to the condition and cleanliness of a place. A place that is untidy has visible signs of decay and disorder; it is in obvious need of cleaning and repair. A place that is tidy is well maintained and shows little sign of wear and tear.

loose litter, bins, street furniture, graffiti, pavement, landscaping

Safety

depends on the surveillance measures that are present within an area, such as CCTV, private security guards, and police, but also on the 'eyes on the street' effect, which hinge on the number of pedestrians, blank vs transparent façades, and active vs vacant properties

CCTV, security guards, policing, blank facades, vacant or underused properties

Appropriation

refers to bottom-up initiatives to alter and adjust public space in order to better suit users' needs. Initiatives include 'urban acupuncture' and 'guerrilla urbanism'

planters, outdoor seating, street performance, art

Ewing Clemente (2013)

Leclercq (2018)

Data and analysis

For each case study, two street sections of approximately 100 meters were assessed (see Figure 20.3), as recommended by Ewing and Clemente (2013). However, for aspects such as safety and appropriation the entire case study area was observed. A quantitative assessment was carried out for the first five urban qualities, following a checklist provided by Ewing and Clemente (2013). For instance, to assess the quality of 'human scale,' the following elements were considered: amount of long sight lines, proportion windows at street level, average building height, number of small planters, and pieces of street furniture. The data derived were recorded on a spreadsheet, and the scores were weighted based on coefficients provided with the tool (Ewing et al. 2006). To compute the final score, a constant was added to the weighted score. For example, in the case of Paradise St.: 27 [pieces of furniture] × 0.04 [weight coefficient] + 0.56 [weighted total of the other elements] + 2.62 [constant] = 4.25.

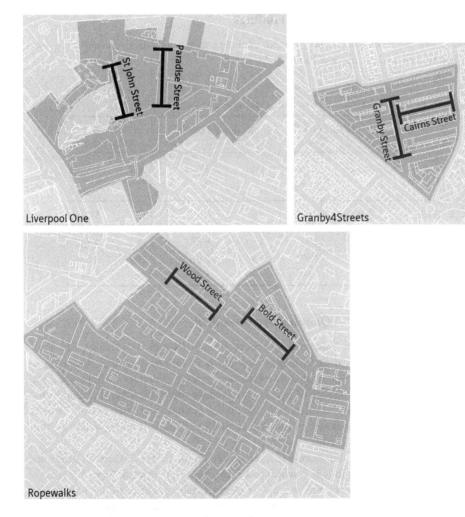

Figure 20.3 Assessed street sections in each case study area

Image credit: Els Leclercq and Dorina Pojani

The remaining six elements of the framework were assessed qualitatively, whereby each element was assigned a 1–5 score, with 1 meaning 'poor' or 'none' and 5 representing 'very good' or 'many.' For instance, 'presence of trees' was assigned 1 if there were no trees, 2 if there were a few small trees, and 5 if there were many mature trees present, and so forth. All qualitative scores were assigned equal weight and the scores were added up to a total number. For instance, to assess the quality of 'tidiness,' the researchers examined the condition of the pavement and the overall landscaping, noted the presence of any loose litter and graffiti and the evaluated the positioning of street furniture. The number of elements considered in the assessment varied by urban design quality; hence the total sums vary accordingly. The assessment took place during two two-day site visits in May 2015 and May 2016 in weather conditions varying from sunny to rainy. While the assessment was conducted by a researcher with substantial professional experience, some level of subjectivity is inevitable in qualitative research. The findings are set forth below. All accompanying photos are by the authors.

Findings

A detailed discussion of the eleven elements included in the analytical framework is presented below. For a summary, see Table 20.2.

Table 20.2 Quantitative assessment of five urban design aspects

	LIVERPOOL		ROPEWALKS		GRANBY 4 STREETS	
	Paradise St	*South John St*	*Bold St*	*Wood St*	*Cairns St*	*Granby St*
Imageability	7.26	6.83	9.51	5.59	4.67	5.10
Enclosure	3.62	4.04	3.61	3.19	3.17	2.64
Human Scale	4.25	3.74	3.79	2.01	3.66	2.15
Transparency	3.77	4.07	3.89	2.91	3.10	2.63
Complexity	7.26	5.68	7.38	4.24	4.94	4.54

Quantitative assessment

The highest score in terms of 'imageability' goes to Bold St (Ropewalks) followed by Paradise St (Liverpool ONE). Bold St is imagable owing to its historic buildings, including an identifiable landmark, buildings with non-rectangular shapes, buildings with identifiers at ground floor level, and the number of people and outdoor dining facilities. Paradise St too has identifiers at ground level, it offers outdoor dining options, and attracts large crowds. While the three streets at the lower end of the spectrum encompass historic buildings, they lack other identifiers and pedestrians.

South John St (Liverpool ONE) earns the highest score on the quality 'enclosure.' This is the result of its height-to-width ratio (1:0.4) which makes the street feel enclosed. Here, the sense of enclosure is further heightened by the presence of elevated walkways at first-floor level, which connect retail frontages on both sides of the street (see Figure 20.4). The sight lines of South John St are blocked on both ends: to the south, by a large warehouse (John Lewis) and to the north by a set of ascending stairs. These contribute to a perception of enclosure. Paradise St (also in Liverpool

Figure 20.3 Qualitative assessment of six urban design aspects

	Liverpool ONE			ROPEWALKS			GRANBY 4 Streets		
	Paradise St	South John St	Overall area	Bold St	Wood St	Overall area	Cairns St	Granby St	Overall area
Coherence	19	16	17	18	18	19	19	16	19
Legibility	26	23	27	29	25	27	29	26	30
Linkage	15	15	15	15	15	15	15	15	15
Tidiness	25	25	25	21	20	16	16	16	16
Safety	23	22	23	18	16	18	12	10	12
Appropriation	6	6	9	11	10	14	17	11	16

Els Leclercq and Dorina Pojani

Figure 20.4 Elevated walkways at South John St (Liverpool ONE) provide enclosure
Image credit: Els Leclercq and Dorina Pojani

ONE) is wide relative to the buildings that line it because, in the original plans, a lane was allocated for a tramline which never came to fruition. Here the height-to-width ratio is 1:1.1, which is still adequate to make the street feel enclosed. Bold St and Wood St (Ropewalks) score well on enclosure because, while both streets have excessively long sight lines, they are also narrow and continuously lined with medium-rise buildings (about four-stories). Uninterrupted façades contribute a sense of clarity of purpose to these spaces, while the streets at Granby4Streets are much more open. Paradise St and South John St (Liverpool ONE), Bold St (Ropewalks), and Cairns St (Granby4Streets) score high in terms of 'human scale.' All incorporate many design details at eye level. In the first three streets, the presence of active windows and pedestrians during the daytime help boost the scores. At the same time, these streets score lower on the levels of individual expressions in public space. Cairns St, on the other hand, does well as here the local residents have transformed the public space in piecemeal fashion by adding plants, seats, and paintworks.

Wood St (Ropewalks) and especially Granby St (Granby4Streets) score particularly poorly on the quality of 'human scale.' At Granby St low scores can be explained by a lack of active windows and doors as many shop frontages are boarded up even when in (infrequent) use. Moreover, Granby St incorporates few details at eye level, such as trees and greenery, and has long sight lines. There is little sign of residents or users taking ownership of the space here. Within the Ropewalks area, the difference in rating between Bold St and Wood St is due to the latter having fewer active windows and pedestrians on the street and much longer sight lines.

South John St (Liverpool ONE) scores high in terms of 'transparency' because its walls are continuous, without any intersections or gaps, and 95% of the façades consist of glass. The entire stretch comprises active shopping venues. Stores in Paradise St (also in Liverpool ONE) are larger than in South John Street, resulting in proportionally fewer active uses. Similar to South John St, Bold St (Ropewalks) is lined with continuous walls and scores 100% on active uses. However, the building façades here are built in brick (instead of glass) and thus have a lower proportion of window space. Wood St scores poorly on transparency for the same reason: few visible windows and part of the street frontage covered by shutters. Nonetheless, active use was rather high. In Granby4Streets, both surveyed streets have continuous walls but little active use and a small proportion of transparent windows (see Figure 20.5).

Paradise St (Liverpool ONE) and Bold St (Ropewalks) are the highest scoring streets in terms of 'complexity,' followed by South John St (again Liverpool ONE), owing to their façade variety. The other three streets sections have a similar, but rather low rating for complexity. Although the streets are endowed with a variety of colors, ornaments, and accents in the building façades, the low-scoring streets attract few passers-by.

Qualitative assessment

Both streets in Liverpool ONE have a high level of visual 'coherence,' owning to their well-positioned street furniture. Tree crowns of similar foliage, height, and diameter and continuous walls of active windows help to visually unify these streets. The entire Liverpool ONE area was built between 2004 and 2008; hence it has little variety in architectural styles, which contributes to coherence (but also to monotony). Similarly, the Ropewalks has stylistic coherence as the majority of its buildings were built during Liverpool's maritime heyday. But there is more variety here than in Liverpool ONE as a number of old façades have been renovated and some derelict ones have been replaced with modern versions. The narrow streets of the Ropewalks cannot host large shade trees (Figure 20.6); therefore, trees are predominantly located in pocket parks. Cairns St. (Granby4Streets) is lined with mature streets but has little street furniture, which could unify its appearance.

Figure 20.5 Boarded up residential properties on Cairns St (Granby4Streets)
Image credit: Els Leclercq and Dorina Pojani

Figure 20.6 Wood St is too narrow to accommodate any trees (the Ropewalks)
Image credit: Els Leclercq and Dorina Pojani

All three areas are highly 'legible': their internal routing, wayfinding, and connections with adjacent urban quarters are well thought out. Liverpool ONE transitions seamlessly to the existing city center and the riverfront. However, the connections to neighboring Ropewalks and a social housing estate are very poorly designed. The boundaries consist of buildings with blank, inactive facades and marginal uses, such as car parking. They rupture the city, thus unnecessarily highlighting the visual differences between privately- and publicly-produced space, or space created though public–private partnerships. All three areas have well-designed street connections, within and without. Walking is facilitated, especially in Liverpool ONE which is a fully pedestrianized zone. In the Ropewalks and Granby4Streets too – where cars have access – pedestrian movement is prioritized though the provision of sidewalks of adequate width.

In terms of 'tidiness,' Liverpool ONE looks meticulously clean and manicured, whereas the Ropewalks is less well maintained – although built of similarly durable, high-quality materials. In both the Ropewalks and Granby4Streets there is loose litter lying around, and private refuse collection bins stand in the public domain, as camouflaged storage for those has not been provided. Granby4Streets's boarded windows produce a disquieting feel, although they have been painted over by residents, and will eventually be fully redeveloped.

As for 'safety' the assessment suggests that Liverpool ONE is perceived as very safe compared to the Ropewalks and Granby4Streets. It is endowed with numerous CCTV cameras throughout, and private security guards (wearing distinctive red coats) patrol the area – although there is no public police presence (see Figure 20.7). The Ropewalks is also equipped with CCTV, but cameras do not cover the entire area. Only incident-prone spots, such as Concert Square, a very popular night-time destination, are surveilled. Some clubs or bars here have small private cameras facing their entrance. No police presence was noted, although environmental enforcement officers were observed patrolling the area once. In Granby4Streets there is single CCTV camera (in Granby St) but a police car was seen patrolling the area. A number of properties are still vacant here, with boarded windows.

Low levels of 'appropriation' are observed in Liverpool ONE. In a few cases, a low ledge below an escalator was customized as a seating spot. Otherwise, there are no signs of planters (provided by, or tended to, by users), graffiti, street artists, or buskers. Any musicians are subjected to a selection process; outdoor music is limited to certain times at fixed locations. In the Ropewalks there are many more signs of users taking ownership of public space. Users have contributed to the greening of the area by providing planters and hanging baskets. Graffiti are also visible at various locations. Similarly, individual interventions in public space are present throughout Granby4Streets (see Figure 20.8). Residents have planted and landscaped front gardens (including those of vacant properties) and decorated the area with planters (handcrafted from miscellaneous pieces of leftover furniture). They have also painted in lively colors the window boards previously installed by the Council to cover the windows of vacant properties. Finally, residents self-organize a monthly street market. On the downside, there are no outdoor dining options or public benches in Granby4Streets. Local residents use low walls, ledges, or bollards as seating spaces or gathering points. Residents in Cairns St. have placed a picnic table and benches on the sidewalk to provide seating opportunities.

Figure 20.7 Private security guards ('red coats') at Liverpool ONE
Image credit: Els Leclercq and Dorina Pojani

Conclusion

Here we return to the research question set forth earlier: Can spaces produced through private finance and planning live up to both the design qualities and the level of publicness which are typical in traditional public space? Our systematic assessment of three spaces in Liverpool, which were produced through different approaches, shows that privatized or hybrid spaces may be quite successful from an urban design standpoint. They are pleasant and full of human activity (being ostensibly accessible at all times to most members of the public). A full calendar of events, with plenty of opportunities for entertainment, is often programmed here in order to attract customers (Langstraat and Van Melik 2013).

However, these types of spaces may be overly controlled by their private owners – who employ security guards, over-sanitize the environment, install cameras, and adopt exhaustive lists of corporate rules. Private spaces may thus lack opportunities for appropriation, while individual reproduction of space for daily use may be limited – thus undermining the very essence of publicness (Lefebvre 1991). By contrast, public spaces which are created or restored through grassroots initiatives provide more opportunities for individuals to intervene

Figure 20.8 Flower pots and picnic tables have been contributed by residents along a sidewalk at
      Granby4Streets

Image credit: Els Leclercq and Dorina Pojani

in, and take ownership of, their surroundings. The difficulty is that communities may have
limited financial resources while public space improvements are quite costly. In addition,
local authorities, which were traditionally in charge of public space provision and oversight,
may find it difficult to trust other actors with these tasks – and accompanying budgets.
While private developers have been successful in forging relationships with the public sector
and taking over many planning duties, local residents and small-scale non-profit organizations are rarely given executive power.

These findings suggest that proclaiming the death of public space as it has been known
since antiquity is unwarranted at this point (see Sennett 1977; Sorkin 1992; Mitchell
1995). The inclusion of private interests in public space production does not necessarily
lead to poor quality urban spaces (see Pojani 2008). Urban designers can play a key role in
enhancing the publicness of the urban spaces they are in charge of – including those
spaces which are privately funded or produced. To achieve this goal, designers need to
treat urban design plans as a starting point rather than as an end product. They should
ensure 'soft' transitions between 'public' and 'private' urban quarters and break rigid borders. Spatial design, programming, and management are important in all cases, especially
where designers are from out-of-town and not involved in day-to-day operations. Finally,

design tools and elements should allow for appropriation of urban spaces by citizens rather than customers. Importantly, the term 'citizen' does not apply only to a select group from the white middle-class populace, but it also encompasses the poor, homeless, immigrants, people of color, and women.

The following are some recommendations:

Design as a starting point rather than an end product. The image of place should be seen as an evolving process rather than a fixed outcome (Van Assche et al. 2013). Based on this understanding, a site design should consist of a flexible framework rather than a detailed masterplan. This framework may form the basis of communication among different stakeholders and help coordinate interests, resources, and problems. Design could accompany decision-making at every step of the process rather than being merely added on as a 'finish' at the end (Van Assche et al. 2013). A menu of concepts can open the door to a variety of spatial solutions and visual appearances. Flexible and adaptable designs are more likely to remain resilient for decades to come and cater to the needs of future generations in addition to present users. The responsibility falls on urban designers to educate corporate clients – who typically demand finished visions and total certainty – on the benefits of this approach.

Designing smooth transitions rather than rigid borders. By focusing on transition or liminal zones, designers can ensure that 'public' spaces produced by the government, private developers, or PPPs, blend together seamlessly. In this manner, the city can be perceived as a unit rather than an archipelago of segregated islands – some public and possibly undermanaged and others private and potentially overmanaged (Carmona 2015). To achieve a visual and functional blend, the design should consider both the appearances of a transition zone, and the uses of the buildings facing it. Any leftover spaces should be treated as an opportunity to provide linkages.

Design tools to allow appropriation. Successful urban spaces, including privatized ones, allow appropriation by users for the purpose of meeting physical and psychological needs. Opportunities for citizens to appropriate space, enrich public life and multiply the public sphere should be embedded in the site designs. Public space should be envisioned not as a finished product but as a *bozzetto*, which users are invited to customize in a collaborative and evolutionary process. This can produce more creative design patterns, as users may be more attuned to the socio-cultural identities that define a place than professional designers (especially if the latter are out-of-town imports). This approach deliberately places people (i.e. the users rather than legal owners of public space) at the heart of public space production.

References

Assche, K.V., Beunen, R., Duineveld, M., and Jong, H.De. (2013) Co-Evolutions of Planning and Design: Risks and Benefits of Design Perspectives in Planning Systems. *Planning Theory*, 12(2): 177–198.

Carmona, M. (2015) Re-Theorising Contemporary Public Space: A New Narrative and a New Normative. *Journal of Urbanism*, 8(4): 373–405.

Carmona, M., Magalhães, C., and De Hammond, L. (2008) *Public Space: The Management Dimension*. New York: Routledge.

Crawford, M. (2008) The Current State of Everyday Urbanism. In: *Everyday Urbanism*, Chase, L., Crawford, M., and Kaliski, J. (eds). New York: Monacelli Press, pp. 25–29.

De Magalhães, C. and Freire Trigo, S. (2017) Contracting Out Publicness: The Private Management of the Urban Public Realm and Its Implications. *Progress in Planning*, 115(1): 1–28.

De Magalhães, C. (2010) Public Space and the Contracting-out of Publicness: A Framework of Analysis. *Journal of Urban Design*, 15(4): 559–774.

Ewing, R. and Clemente, O. (2013) *Measuring Urban Design: Metrics for Livable Spaces*. Washington, DC: Island Press.

Ewing, R., Handy, S., Brownson, R., Clemente, O., and Winston, E. (2006) Identifying and Measuring Urban Design Qualities Related to Walkability. *Journal of Physical Activity and Health*, 3(1): 223–240.

Gehl, J. (2011) *Life between Buildings: Using Public Space*. Washington, DC: Island Press.

Gehl, J. and Svarre, B. (2013) *How to Study Public Life*. Washington, DC: Island Press.

Hajer, M. and Reijndorp, A. (2001) *In Search of New Public Domain*. Rotterdam: NAI Publishers.

Hassen, N. and Kaufman, P. (2016) Examining the Role of Urban Street Design in Enhancing Community Engagement: A Literature Review. *Health & Place*, 41: 119–132.

Hooi, E. and Pojani, D. (2020) Urban Design Quality and Walkability: An Audit of Suburban High Streets in an Australian City. *Journal of Urban Design*, 25(1): 155–179.

Kohn, M. (2004) *Brave New Neighborhoods: The Privatization of Public Space*. New York: Routledge.

Langstraat, F. and Van Melik, R. (2013) Challenging the 'End of Public Space': A Comparative Analysis of Publicness in British and Dutch Urban Spaces. *Journal of Urban Design*, 18(3): 429–448.

Leclercq, E. (2018) "Privatization of the Production of Public Space." PhD Dissertation, Delft University of Technology, Delft, The Netherlands.

Lefebvre, H. (1991) *The Production of Space*. Oxford: Blackwell Publishing.

Madanipour, A. (2003) *Public and Private Spaces of the City*. London: Routledge.

Marshall, S. (2012) Science, Pseudoscience and Urban Design. *Urban Design International*, 17(4): 257–271.

Mehta, V. (2007) Lively Streets: Determining Environmental Characteristics to Support Social Behavior. *Journal of Planning Education and Research*, 27(2): 165–187.

Mehta, V. (2013) *The Street: A Quintessential Social Public Space*. New York: Routledge.

Mehta, V. (2014) Evaluating Public Space. *Journal of Urban Design*, 19(1): 53–88.

Mitchell, D. (1995) The End of Public Space? People's Park, Definitions of the Public, and Democracy. *Annals of the Association of American Geographers*, 85(1): 108–133.

Németh, J. and Schmidt, S. (2011) The Privatization of Public Space: Modelling and Measuring Publicness. *Environment and Planning B*, 38(1): 5–23.

Pojani, D. (2008) Santa Monica's Third Street Promenade: The Failure and Resurgence of a Pedestrian Mall. *Urban Design International*, 13(3): 141–155.

Sennett, R. (1977) *The Fall of Public Man*. New York: Knopf.

Sorkin, M.. (ed.). (1992) *Variations on a Theme Park*. New York: Hill and Wang.

Urban Task Force. (1999) *Towards an Urban Renaissance. Chaired by Lord Rogers of Riverside*. London: E& FN Spon.

Varna, G. and Tiesdell, S. (2010) Assessing the Publicness of Public Space: The Star Model of Publicness. *Journal of Urban Design*, 15(4): 575–598.

Young, I.M. (2000) *Inclusion and Democracy*. Oxford: Oxford University Press.

21

THROWNTOGETHER SPACES

Disassembling 'urban beaches'

Quentin Stevens

Introduction

In recent decades, urban design practice and research have given increasing attention to open spaces that are very temporary and malleable. Alongside permanent open spaces that are funded and managed by governments and private landowners, there have recently emerged a range of impermanent, more-or-less-public spaces that are facilitated by other actors. This chapter, inspired by Latour's (2005) Actor-Network Theory, considers these 'actors' in the widest possible sense, including not just persons with financial, material, and legal control, but the many non-humans that also have agency in transforming urban spaces. The chapter examines the diverse plurality of materials, people, and intangible forces that dynamically come together to shape open spaces today. It illustrates these through one particular new temporary space type that is widespread in Germany, and also common elsewhere: artificial 'city beaches' (Stevens and Ambler 2010; Stevens 2011; 2015).

This account emphasizes the constant actions that are necessary to create, maintain, or transform the relationships between the various actors that constitute these spaces. City beaches' dynamic, opportunistic, throwntogether nature reflects wider trends in open space design, planning and management that have arisen in the context of de-industrialization and growing local participation. These approaches are messy, and constantly evolving to take tactical advantage of shifting, localized resources, and opportunities (Hou 2010; Lydon and Garcia 2015). I draw here on Doreen Massey's (2005) conception of places as 'throwntogether' to emphasize that non-human objects and forces, like people, have their own trajectories (or 'aims'), and that city beaches highlight the nature of all places as the result of continuing acts of conflict, negotiation and alignment among these actors.

The archetypal German city beach is a low-key, temporary, independent hospitality venture on the ex–industrial riverfront of a large city, with sand, palm trees, thatched huts where drinks are sold, deck chairs, and a pool (see Figure 21.1). The city beach concept originated in St. Quentin, France in 1995, although it first became widely-known through *Paris Plages* in 2002, a big-budget project developed by Paris's city government on a temporarily-closed section of a northern-riverbank freeway.

(a)

(b)

Figure 21.1 Typical city beaches in Germany, with portable pools, decking, deck chairs, potted palm trees, and surfboard props. Strandsalon, Lübeck (a) and La Playa, Leipzig (b)

Image credit: Quentin Stevens

This chapter examines in turn six distinct categories of actors that have come together to constitute various city beach settings in Germany, and the various ways they are constellated, and explores the ongoing efforts required to bring and hold them together as places.

Vacant land

The city beach concept presented in *Paris Plage*, with its riverside view and southern exposure, has undergone translation to a variety of empty urban sites. Many city beaches temporarily occupy spaces on Germany's extensive ex–industrial urban riverfronts. These spaces have presented themselves as available through the departure of former uses. Vacancy requires that rapid large-scale redevelopment is somehow precluded, whether by economic conditions or inadequate infrastructure. In Hamburg, a string of city beaches opened on the north bank of the Elbe River in an historic port area long made redundant by larger-scale shipping. Seasonal flooding kept this area free from urban redevelopment. A long, narrow strip of land between Berlin's East Side Gallery (the remnant section of the Berlin Wall) and the Spree River was for many years a prime city beach location, because it was unviable for major development. Such former port spaces are often disconnected from their cities, and action is typically required to improve pedestrian access, through the agency of new bridges and access stairs; by demolishing the walls of ports and factories; and through new signs, maps, and place identities that reconnect these sites into people's image of the city.

One vivid illustration of a newly-created alignment of under-used urban spaces with this new spatial concept is Stuttgart's *Skybeach*. Its creator had seen *Paris Plage* in 2003, but no river runs through Stuttgart's center. After seeing an aerial photograph of a downtown department store, he connected the beach concept and the popularity of rooftop bars to the previously unused, elevator-accessible top level of its parking garage, with its excellent inner-city views, quietness, and sunshine (see Figure 21.2). This new association of city beaches with parking garages was itself subsequently translated to the rooftops of 19 other garages in 14 other German cities. Sometimes assembling these new pathways from the pedestrian shopping street below to the rooftop beach required installing temporary scaffolded staircases. Large banners and flags improve city' beaches local visibility. Maps on many beaches' websites help visitors find their way down minor alleys and truck routes to access their ex–industrial waterfront sites.

Only half of Germany's city beaches occupy waterfronts (Stevens 2011). Several other kinds of vacant sites attract beach projects: underused public parks and plazas, railway easements, and inland industrial areas. All these sites offer plentiful low-rent space, flexibility regarding construction and operation, and few conflicts with neighbors. City beaches are often stimulated by other low-rent leisure and creative activities nearby, including nightclubs, performance venues, and artists' studios, 'plugging into' these emerging creative milieu. New city beaches sometimes form larger precincts around pioneering projects, including 'Costa Hamburgo' on the port authority's carpark and Dresden's 'Elbiza'. A wider physical, representational, functional and historic context produces a sense of place that facilitates further new, escapist leisure settings.

Landscaping and furnishings

The development of city beaches has also been influenced by a palette of landscape elements that are readily-available, cheap, light, collapsible, and relocatable, yet welcoming and comfortable.

Figure 21.2 Skybeach, Stuttgart, on the roof of the parking garage of the Galeria Kaufhof department store, adjacent to the city's main pedestrian axis, Königstraße

Image credit: Quentin Stevens

In Berlin, sandy ground was already apparent before the first city beaches, revealed as the city's geological substratum by building demolitions (Schulz and Abele 2011). In pre-existing outdoor beach-volleyball venues, underused sandy areas were reimagined by the city beach concept. For some city beaches, construction-grade sand is donated or rented cheaply by building supplies businesses, who collect it after the project ends and sell it for concreting.

Sand is a disorderly, unfixed material that invites people to adjust it to make themselves comfortable, and which encourages initiative, playfulness and creativity in its use (Stevens 2009). Yet maintaining sand's role as a leisurely beach environment demands work of beach managers, who must constantly remove cigarette butts and smooth and replenish the sand. Specialist suppliers have developed proprietary mixes of two grain sizes that combine appearance and comfort without clinging to shoes and street clothes (David 2010). Because sand is annoying for prams, high heels, keeping tables and chairs level, and waiters, it is often confined to small patches surrounded by decking, for looking rather than touching.

'City beaches' are also defined by loose, shifting assemblages of small, easily-transportable items, especially potted palm trees, beach umbrellas, collapsible swimming pools, and portable cabins. Palm trees are a key scenographic element, invariably placed framing the beach's entrance. Transportability, cost, sub-zero-winters, and a need to minimize permanent shade dictate that palms are typically small and rented. Some hire firms provide low-maintenance synthetic ones. Cabins include thatched huts, market kiosks, and trailers, shipping containers, and portable toilets, all easily rented, transported, and repositioned to form visual and acoustic barriers to surroundings and screen back-of-house functions. Sail roofs, tents, buildings with openable facades and awnings, and large umbrellas allow rapid reorganization, relocation, and disassembly (see Figure 21.3) (Stevens 2011).

Figure 21.3 Temporary and relocatable buildings, play equipment, awnings, south sea umbrellas, palm trees, and a ship create a beach atmosphere. Bundespressestrand, Berlin

Image credit: Quentin Stevens

Deck chairs are ubiquitous on city beaches, even though they are rare on real beaches, having been imported from cruise ships. They thus connect to Europeans' preconceptions and memories of the Mediterranean. While at real beaches people often sit or lie on towels, deck chairs on city beaches enable their many short-term visitors in street clothes to relax, put just their feet in the sand, and sit comfortably so they can see and be seen. This furniture combines city comforts with informal, escapist appearances. Deckchairs also forge relationships between beach operators and drink, food, and cigarette vendors, who provide free, pre-branded ones.

A beach atmosphere is also created by 'tropical' cocktails (sold from thatched huts), music, lighting, heat-lamps, and props including surfboards, anchors, rowboats, paddles, life preservers, lighthouses, cargo chests, pirate flags, fishing nets, and shells. Equipment that supports activities that attract beach patrons also force other elements into city beach landscapes: volleyball pitches, children's playgrounds, sandcastle-building tools, dance floors, and stages. Children's sand play with buckets and spades demonstrate city beaches as unfixed, 'do-it-yourself' landscapes. Beaches' managers and their users are constantly reorganizing, removing, or adding deck-chairs and umbrellas, to optimize escapist atmosphere, patron capacity, social arrangements, activity needs, views, sun angle, and weather. This is a post-Fordist 'co-production' of a comfortable leisure environment and experience (Richards and Wilson 2006; Stevens and Ambler 2010).

Human actors

Despite strong consistencies in city beaches' formats, these spaces involve diverse human actors. Several early German city beaches were developed by artists, stimulated by government funding for new ideas for open spaces (Stevens 2015). These artists introduced creativity to seemingly-prosaic issues and processes around the design and management of open spaces, challenging expectations and sometimes circumventing regulations (Stevens 2015). Most German beaches are commercial ventures. A shortage of summer income at indoor bars and restaurants, vacant sites, and the city beach concept have combined to attract hospitality entrepreneurs, who then gather an idea, a site, funding, permits, and equipment. They must transfer their existing relationships with public bureaucracy, neighbors, and patrons, and develop new ones. Their interests inevitably influence their beaches' clientele and openness. The innovation and challenges of both artistic and commercial city beach projects bring together a wide range of actors from the creative industries; 'culturepreneurs' with creative, technical, and organizational expertise, who are nonetheless new to the materials and processes of urban placemaking (Lange 2011). They are compelled to develop new practices, discover new spatial potentials, and also to creatively navigate bureaucratic, regulatory, and financial relationships (SenStadt Berlin 2007).

Other German city beach projects have attracted involvement from non-profit organizations and local community groups, highlighting that innovative, risk-taking open space projects are not always driven by profits. These citizen initiatives generally emerge in cities small enough for community actors to maintain strong contacts in local government. Their protagonists emphasize the importance of existing organizations and strong engagement and trust, to draw in the many other necessary actors. They also note that city beach projects help to build these relationships (Stevens 2015). Similarly, the novel requirements of city beach projects help to establish new cross-department working relationships within local governments (Stevens 2011). Several of these beaches were consciously developed by local

Figure 21.4 Demonstration by local gymnastics club. Strandleben, Vaihingen an der Enz
Image credit: Quentin Stevens

organizations, as mechanisms for drawing their communities together and facilitating social interaction in public spaces (see Figure 21.4). *Bristol Urban Beach* in England engaged local service providers and sought to attract performances and activities led by diverse local community members, with the aims of encouraging broad community participation and interaction and strengthening community institutions (Mean et al. 2008). Notably, community beach spaces recognize children and teenagers as members of the public, who are in some cases negotiators, builders, and performers, not just passive consumers of adult-determined environments (Stevens 2015).

German city governments do not just regulate privately-instigated beach proposals; many cities proactively explore, market, and underwrite opportunities (SenStadt (Senatsverwaltung für Stadtentwicklung) Berlin 2007). Whether city beaches are public- or private-led, their novel spaces, experiences, and use programs bring together very diverse user groups. These users all have an important role in continually performing the beach space and its atmosphere, with their sunglasses, bare feet, sunbathing bodies, swimwear, and shovels (Stevens and Ambler 2010). The continual production of beaches draws together the desires and skills of their employees, artists and designers, sponsors and suppliers, legislators and bureaucrats, and changing financial conditions, weather, and surrounding events. Beach environments are continually being re-arranged, packed away, extended or compressed, and being redeployed to new locations. City beaches' intangible atmosphere relies on the labor of entrepreneurs with backgrounds in theatre, music, media, and advertising, and constant reinforcement from local citizens passionate about sports, yoga, and dancing. Participatory events and changing themes maintain novelty and keep visitors engaged (Gale 2009).

While city beaches all involve temporarily throwing similar spaces, furnishings, aesthetics, and creative energy together, they can thus involve quite different groups of human actors, linking to different constellations of personal and collective aspirations, responsibilities, and risks. These in turn are linked to different financing arrangements and management approaches (Stevens and Ambler 2010). Temporary city beach projects also involve dynamic and varied flows and combinations of initiative, creativity, authority and risk-taking among human actors. These factors all affect the publics and publicness of the spaces that emerge. Yet beaches' human actors are still somewhat interchangeable. Sometimes entrepreneurs operate beaches under contract for local governments. One short-term artist-initiated and city-financed beach was subsequently revived by a local residents' association. The atmosphere and activities it had stimulated provided a durable nexus for continuing it with quite different actors, funding, and labor. These shifts in motivations and responsibilities thus reflect a wider development dynamic of the city beach concept: ideas first created and explored by artists with public funding were adopted and translated by numerous entrepreneurs who tested the profitability of various formats, sites, services, and clienteles, and were then adapted by local residents, who optimized their social benefits (Stevens 2015).

Constellations of human actors and beach sites also evolve at a larger scale. As the city beach concept quickly became stable, replicable, and transferrable, professionalization and systematization began developing in the operation of private-sector city beaches. This networking and stabilization is demonstrated by the emergence of eleven chain operators. Five companies in Berlin run fourteen beaches between them. Two chains created or acquired new beaches annually, using past profits and contacts and existing equipment and staffing. In ten cities, sites already developed into beaches were taken over by different operators in later years, applying new branding and management to the same scenery. The founder of Stuttgart's *Skybeach* copyrighted his concept, sold franchises, and licensed the name to others. *Skybeach* and another chain specialize in parking garage rooftops, where site and management factors are highly standardized. These sites remain dependably vacant each summer; eight are owned by one department store chain. Other independent operators copied the *Skybeach* model in 16 other cities. City beach chains can also share brand names, advertising, websites, furnishings, suppliers, performers, and capital. A franchise is a durable set of such relationships that new entrepreneurs can rent access to. Independently-run city beaches are also linked in various ways, coordinating resources, interests, and processes. The large number of city beaches in Germany has subsequently shaped other industries, which rent potted palms and sand, print deck chair logos, and structure beverage consignment contracts. Some entertainers tour city beaches. These open spaces may be temporary and thrown together from loose materials, but their locations and operations are often very durable.

Energy

Alongside the various constellations of individual people, objects, and spaces that give life and form to city beaches, there are also several kinds of broader intangible actors that stimulate these projects, and influence how, when, and where they arise, and which other actors acquire agency in shaping them. Prime among these intangible actors are the various flows of energy that erratically circulate through cities, including creative energy, excitement, and weather. Attention to these energies highlights the dynamism of the actor relationships that usher projects like city beaches into existence, and the forces that can drive them apart.

City beaches first emerged where creative individuals were immersed in urban milieus, like Berlin's, that were latent with creative energy (SenStadt (Senatsverwaltung für Stadtentwicklung) Berlin 2007; Lange 2011). These milieus enabled rapid transfers and cross-fertilizations of new ideas, skills, customers, and further contacts, bringing new knowledge and unconventional approaches to the development of urban spaces. Under such conditions, the enthusiasm, imagination, and initiative for creating city beaches could spread, from artists to individuals in the hospitality and entertainment industries, and to city planners and untrained residents. New ideas also keep transforming the relationships between the various actors that produce city beaches. This is evident from the diverse ways that city beaches have acquired sites, materials, equipment, and funding; the shifting roles of local governments in relation to temporary uses like city beaches (from regulation to toleration to facilitation); and the emergence of mediating agencies that link other actors together (Stevens 2018).

The spread, form, and success of city beaches has also been precipitated by short bursts of energy and excitement, in particular Germany's hosting of the Football World Cup in 2006, when large crowds visited city beaches to watch games on outdoor screens (see Figure 21.5). Several avant-garde artist- and community-led city beach projects were stimulated by local, regional, or European events and funding schemes with unrelated purposes, including campaigns to clean up Germany's rivers, catholic pilgrimages, and horticulture competitions (Stevens 2015).

Weather is also a fluctuating form of energy. Increasingly-warm European summers, especially 2006's '100-year-summer,' had helped prompt the growth of city beaches, but Germany's unpredictable, uncontrollable, often-rainy summers cause major operational difficulties, and influence profitability, often causing closures. Weather forces beach operators to provide deckchairs and weather protection for guests that can quickly be adapted to the quotidian passage of the sun, clouds, and rain. Operators consider rain protection a more essential element of a German city beach than waterfront access. Fickle weather also requires operators to provide weather forecasts and daily updates about their opening hours on

Figure 21.5 Public Viewing of World Cup football matches. Strandbar, Magdeburg
Image credit: Quentin Stevens

webpages and social media. This energy flow also conditions the hiring conditions of service staff and entertainers, who typically have flexible or no contracts, so that they can be called to work or declined work according to changing weather. When the sun comes out, staff, deck chairs, umbrellas, and music must all spontaneously materialize (Stevens 2011).

Concepts

The city beach as a concept is more enduring than any specific project, site, or operator. The *Paris Plage* exemplar was rapidly translated in turn by artists, entrepreneurs, and community organizations, which have each had to adapt it to a diversity of German sites, and work with different materials, funding, and legislation (Stevens 2015). Many projects have been inspired by people seeing a city beach, and then seeking sites, funding, and clientele to create one. One month after *Paris Plage* appeared, the first German commercial beach, *Strandbar Mitte*, found a south-facing waterfront site within a run-down, former-East-Berlin park, a prohibition (as with Paris's site) on permanent construction, an operator who already had a theatre and clientele there, and sand exposed by building demolitions (Stevens and Ambler 2010; Schulz and Abele 2011). The pre-existing concept connected these. Its 2002 appearance in Berlin linked it to an abundance of other vacant sites, underemployed creative entrepreneurs, tourists, and a government receptive to new ideas. The idea spread quickly to other major German cities and then smaller towns (Stevens 2011). Some entrepreneurs purchased franchises to *Skybeach's* successful concept; other merely copied elements.

The materials, design, and programming of individual beaches have been shaped around several distinct themes that connect them to specific desired clienteles: white-plank cruise-

Figure 21.6 South Pacific atmosphere through extensive use of thatching. Strand Pauli, Hamburg

Image credit: Quentin Stevens

ship chic; Mediterranean resort town; laid-back 'castaway' driftwood; and 'South Pacific native' bamboo and thatch (see Figure 21.6). Beaches' names help anchor these themes. Six beaches on parking garages play on the word 'deck.' Eight sites are called 'Island,' although none actually are. Many names evoke the Mediterranean (del mar, sol, plage, playa), tropics (cabana, coco, Copa, 'Baykiki'), or deserts (oasis, dune, Sahara, Casablanca, Zanzibar). The surrounding industrial waterfront aesthetic also influences city beach décor, introducing shipping containers, 44-gallon drums, dock cranes, and graffiti.

Use programs are also important in enlivening and differentiating city beaches' atmospheres. Yoga classes and children's playgrounds attract early-morning female patrons. Free wireless internet attracts students and creative workers. Dance floors and free lessons attract couples and singles. Community-run beaches host social and educational activities, and even religious services. Entrepreneurs with backgrounds in event management and the arts bring live music, poetry readings, debates, and exhibitions to their beaches. Germany's hosting of Football's World Cup prompted beaches to host 'public viewing' – watching sports and films on outdoor screens; a new, very profitable concept which financed a major expansion and inspired entrepreneurial imitators. Many of these activities have very tenuous links to sandy beaches. Playing sport is another major use: predominantly volleyball, but also beach soccer, boules, minigolf and canoeing.

The broader concept of temporary use (*Zwischennutzung*) has both facilitated and shaped opportunities for various people and landscape materials to appropriate land for city beaches, and other projects. It configures time as both a constraining and an enabling factor. It establishes terms for new actor agreements on rents and responsibilities, reducing financial overheads and risks, valorizing new uses as experiments with planning regulations, and quashing local resident and commercial opposition to such projects by circumscribing their impacts (Stevens 2018). Such reimagining allows *Strandbar Mitte* to still claim temporariness in its seventeenth year of operation.

The flexibility of the city beach concept has allowed beaches to move to new sites. Düsseldorf's first beach operator relocated onto a moored boat in nearby Cologne when his original harbor-front location was developed for an international hotel. Munich's *Kulturstrand* has occupied six different locations; they eventually negotiated with the city to rotate annually between four sites (David 2010).

Numerous administrative concepts have also shaped the possibilities of city beaches. Beverage companies minimize operation costs by supplying beach operators with drinks on consignment: they deliver, take no up-front payment, and earn only from drink sales. Some beaches have a *Minimumverzehr*, a minimum-purchase scheme, where patrons must pre-purchase a voucher for food and drinks on entry. Existing German building code provisions for temporary 'flying buildings' allow beaches exemptions from many time-consuming and expensive requirements and approval processes that would otherwise thwart such low-budget initiatives. Instead, performance-based regulations – related to levels of inebriation, noise, light, and waste – shape city beaches as events, optimizing the usability of their sites. Two Berlin projects, Peter Arlt's *Bad Ly* (1999) and Susanne Lorenz's *Badeschiff* (2004), were able to place swimming pools and beaches on urban sites without obtaining planning permits or meeting normal health and safety requirements, by defining them as temporary art projects, and not charging entry fees (Stevens 2009; 2015).

Austerity

The lure of money has encouraged some entrepreneurs to pursue profits by assembling whimsical beach atmospheres out of degraded urban spaces, cheaply-rented furnishings, and casually-employed staff, and by circumventing expensive regulatory requirements. But city beaches are not necessarily profitable, and investment finance is often very limited. Money also has great influence on city beaches through its absence. An absence of market demand and financing to redeveloped Germany's extensive urban brownfields has created spatial opportunities for new users and uses with small budgets and time-horizons. Land owners often provide sites for little or no rent, or even pursue tenants, to help advertise their site, to facilitate repairs, or just to cover holding costs (SenStadt (Senatsverwaltung für Stadtentwicklung) Berlin 2007). Operators' lack of capital shapes beaches by encouraging new practices of renting, donations and making-do, inputs of sponsorship, and sweat equity and creativity from operators and from local residents. Market weakness also means that city beaches' operations (their admissions policies, use programs, noise levels, employment) remain strongly conditioned by their financial dependence on local governments, through grants, loans, rent-free land, free services, and tax reductions.

Emphasizing operators' power to shape city beaches also ignores operators' and staff's dissatisfaction with conditions and outcomes they cannot control. A lack of existing profitable business opportunities and stable employment require individuals with media or hospitality backgrounds to work in a quite different sector, designing and managing landscapes. This low-budget entrepreneurship is in part an effect of the social circumstances of German cities' economic crisis and the atmosphere and practices of austerity that follow from it (Färber 2014). City beaches are more numerous in German cities with high unemployment. To focus on operators' control also ignores the temporary and fragile nature of the relationships constituting these beaches. Premises must remain rapidly vacatable on occasion of flooding or property sale. The vicissitudes of weather easily bring financial ruin. A shortage of conventional urban development funds and customer receipts has also drawn arts and social funding into these open-space projects, especially in smaller, poorer cities (Stevens 2015).

What brings the beach together

This analysis of how city beaches are thrown together highlights four general insights into the contemporary production of public spaces, which are in marked contrast to much literature about urban redevelopment that foregrounds the roles of large multinational corporations and architecture firms, long-term planning, partnership contracts and standardized financing, land-forming, and construction approaches.

First, successful spaces can be thrown together by extremely diverse and distributed sets of agents and mechanisms, including materials, sites, and contextual conditions (see Figure 21.7). City beaches and other public spaces can be stimulated, created, managed, and supported through entrepreneurship from the private sector, the public sector, and/or diverse community actors.

Second, city beaches exemplify an exploratory, experimental attitude toward placemaking. City beaches are both tangibly and metaphorically 'sandboxes': physical and social milieus that are open and unstructured, gathering together a diversity of elements that can easily be manipulated to serve different roles, and which thus encourage playful, collaborative exploration of possibilities and learning, and foster new approaches (Stevens 2015). Although their appearance verges on stereotype, they innovate through new kinds of actors, sites, timeframes, new funding and management approaches and programming (Lydon and Garcia 2015). They re-imagine and re-purpose previously under-utilized urban spaces to discover varied new uses and benefits. These outcomes in turn produce other changes, by financially underwriting culturepreneurs' other ventures, enhancing German social cohesion during the World Cup, and stimulating the cleaning up of waterways and the gentrification of former harbors.

The third distinctive characteristic of city beaches is their variability. They illustrate the dynamic relationships and processes through which actors are brought together to constitute spaces; and the variable power relations which produce stability or change in these relationships and these spaces. These projects embrace impermanence. They demonstrate the potential flexibility of public spaces that can transform from hour to hour (with sun, rain, and human visitors), from day to day (with seasons, programs, new local ideas, and events) and from year to year (with changes in equipment, budget, site, landlord, and operators). They show how even loose furnishings can significantly modify a space's atmosphere and functionality. They encourage various actors to modify and extend the setting, its uses and even its location. The combination of the city beach's robust and transferable central concepts and components and its extremely variable details is surely fundamental to its success. The flexible and dynamic approaches of city beaches, and other temporary uses, as post-Fordist, 'just-in-time' open spaces, seem well-suited to recent economic uncertainties (Tonkiss 2013). Their short cost-recovery timeframes limit their need for fixed relationships to sites, robust landscape design, popular enthusiasm, political will, regulatory certainty, or capital. They harness the fluid, short-term availability of all these factors. Nothing is tied down, and every element can be rapidly exchanged or re-purposed in response to changing needs.

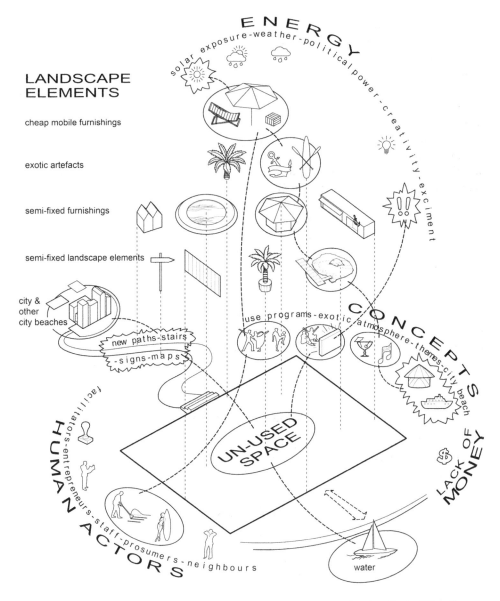

ENERGY

solar exposure-weather-political power-creativity-exciment

LANDSCAPE
ELEMENTS

cheap mobile furnishings

exotic artefacts

semi-fixed furnishings

semi-fixed landscape elements

city &
other
city beaches

new paths-stairs
-signs-maps

use programs-exotic atmosphere-themes-city beach

CONCEPTS

facilitators-entrepreneurs-staff-prosumers-neighbours

HUMAN ACTORS

UN-USED SPACE

LACK OF MONEY

water

Figure 21.7 The six sets of actors that are thrown together to create city beaches. Dashed lines illustrate four ways different actors temporarily assemble other actors, clockwise from top

- sunshine draws together staff, prosumers, and activities such as sunbathing that perform the beach
- specific city beach themes draw together exotic atmosphere, landscape elements, exotic artefacts and mobile furnishings
- the excitement of the World Cup draws together un-used spaces and the new activity of public viewing
- new paths, stairs, signs, and maps draw together cities, un-unused spaces and waterfronts

Image credit: Quentin Stevens and Ha Thai

Fourth, the conditions outlined above raise an open question about the diverse, exploratory, and varying publicness of these and other throwntogether urban spaces. The human relationships that produce city beaches may seem particularly transient. But the success of German city beaches as public spaces rests on the ongoing, active involvement of a wide range of actors. These include beach users, who are typically 'prosumers' actively engaged in the performative co-production of the space and its uses and meanings (Richards and Wilson 2006). They include community organizations representing residents, youth, churches, and arts organizations; creative industry synergies; and cross-departmental working within government. These projects illustrate the important role of existing social networks, and the role of the beach projects in initiating and strengthening such connections. The community organizers of Vaihingen's *Strandleben* highlight that the assemblage that produced it did not exist beforehand. Trust developed through the process of making the space, through detailed discussion and negotiation over rights and responsibilities between the council, the community association, and individual event organizers (Gassner et al. 2011).

The power to act and to shape the formation of a city beach, whether for people, or for sites, furniture, concepts, excitement, or specific activities, comes from responsibilities that are dynamically created, assigned, and transferred. Munich's *Kulturstrand* highlights the negotiated nature of these responsibilities, as each of its new host sites and neighborhood contexts inspired new spatial forms, programs, and operational constraints (David 2010). These negotiated processes also help to explain regional variations in the location, number and form of city beaches (Stevens 2011). As spaces that are temporarily thrown together and negotiated, city beaches illustrate a potential future for public spaces that remain highly contingent, participatory, and open to change.

Acknowledgements

This research was supported by a grant from the Australian Research Council (Project DP180102964), funded by the Australian Government, and by a Senior Research Fellowship from the Alexander von Humboldt Foundation, Germany.

References

David, B. (2010) Manager of *Kulturstrand*, Munich. Personal interview.

Färber, A. (2014) Low-Budget Berlin: Towards an Understanding of Low-Budget Urbanity as Assemblage. *Journal of Regions, Economy and Society*, 7: 119–136.

Gale, T. (2009) Urban Beaches, Virtual Worlds and 'The End of Tourism'. *Mobilities*, 4(1): 119–138.

Gassner, S., Schmidt-Hitschler, U. and Hitschler, T. (2011) Organisers of *Strandleben*, Vaihingen an der Enz. Personal interview.

Hou, J. (ed.). (2010) *Insurgent Public Space: Guerrilla Urbanism and the Remaking of Contemporary Cities*. New York: Routledge.

Lange, B. (2011) Professionalization in Space: Social-spatial Strategies of Culturepreneurs in Berlin. *Entrepreneurship and Regional Development*, 23(3–4): 259–279.

Latour, B. (2005) *Reassembling the Social: An Introduction to Actor-Network Theory*. Oxford: Oxford University Press.

Lydon, M. and Garcia, A. (2015) *Tactical Urbanism: Short-term Action for Long-term Change*. Washington, DC: Island Press.

Massey, D. (2005) *For Space*. Los Angeles, CA: Sage.

Mean, M., Johar, I. and Gale, T. (2008) *Bristol Beach: An Experiment in Place-Making*. London: DEMOS.

Richards, G. and Wilson, J. (2006) Developing Creativity in Tourist Experiences: A Solution to the Serial Reproduction of Culture? *Tourism Management*, 27: 1209–1223.

Schulz, C. and Abele, M. (2011) Managers of *Strandbar Mitte* and *Oststrand*, Berlin. Personal interview.

SenStadt (Senatsverwaltung für Stadtentwicklung) Berlin. (ed.). (2007) *Urban Pioneers: Temporary Use and Urban Development in Berlin*. Berlin: Jovis.

Stevens, Q. (2009) Artificial Waterfronts. *Urban Design International*, 14(1): 3–21.

Stevens, Q. (2011) "Characterising Germany's Artificial 'City Beaches:' Distribution, Type and Design." *3rd World Planning Schools Congress*, Perth.

Stevens, Q. (2015) "Sandpit Urbanism." In: *Enterprising Initiatives in the Experience Economy: Transforming Social Worlds*, Knudsen, B., Christensen, D. and Blenker, P. (eds.). New York: Routledge, pp. 60–80.

Stevens, Q. (2018) Temporary Uses of Urban Spaces: How are They Understood as "Creative?". *Archnet-IJAR: International Journal of Architectural Research*, 12(3): 90–107.

Stevens, Q. and Ambler, M. (2010) Europe's City Beaches as Post-Fordist Placemaking. *Journal of Urban Design*, 15(4): 515–537.

Tonkiss, F. (2013) Austerity Urbanism and the Makeshift City. *City*, 17(3): 313–324.

22

DESIGNING PARKS FOR OLDER ADULTS

Anastasia Loukaitou-Sideris

Introduction

Today, older adults represent the fastest growing demographic segment in the U.S. But while advances in medicine and public health have helped this group live longer and healthier lives, planning and urban design are lagging behind in their efforts to create age-friendly cities. Nowhere is this more evident than in neighborhood parks, where older adults are largely missing (Kemperman and Timmermans 2006). Indeed, elders remain a highly underserved group in regard to parks, even though studies have discovered a positive relationship between their use of parks and their physical health and emotional well-being. Parks provide an opportunity to socialize with others, spend time in a natural setting, and undertake activities including exercising, gardening, walking, or simply meditating in nature (Rodiek 2002). However, American parks have not been designed with seniors in mind: only a few parks offer senior-friendly programming or have senior centers with public recreational or educational programs. It is not then surprising that people over 60 comprise only 4% of neighborhood park users (Cohen et al. 2016).

The World Health Organization (WHO) and the American Planning Association (APA) have initiated efforts to accommodate "aging in place" and create age-friendly cities (WHO 2007; Winick and Jaffe 2015). These efforts include open space as an important element of the urban form (see Figure 22.1) but provide little detail on how to design it to better fit the needs of older adults. Little knowledge exists about the preferences of older adults in regard to neighborhood open space or the influence of objective and subjective features of the neighborhood built environment on their physical activity patterns (e.g. walkable streets, street lighting, parks and recreational facilities, park exercise equipment, safe crosswalks, and pleasant scenery).

This chapter draws information from the literature and empirical research to discuss the open space preferences of older urban-living adults and translate them into design guidelines for senior-friendly neighborhood parks. The chapter is composed of three parts: a brief literature review about the benefits of parks for older adults; a discussion of the findings of the empirical research; and a compilation of design guidelines.

Figure 22.1 Age-Friendly City Elements, according to the World Health Organization

Image credit: World Health Organization (2007)

Benefits of parks for older adults

Neighborhood parks (usually up to 5 acres in size) represent valuable assets for cities because they provide recreational opportunities, serve as places for social interaction, and offer a natural respite to urban dwellers. Studies have shown that neighborhood parks are beneficial for older adults, offering them physiological and psychological benefits, which contribute to their quality of life, self-reported physical and mental health, and longevity (Takano et al. 2002; Rappe et al. 2006; Gardner 2008; Sugiyama, Thompson, and Alves 2009).

Older adults represent the most inactive population segment, and physical activity can benefit even the oldest and most frail elders (Pahor et al. 2006), and even slow the aging process (Sun et al. 2010). Walking is the most common physical activity for older adults, and public health experts have observed a positive relationship between walking and physical health. Older adults who have good access to neighborhood open space are more prone to walking and physical activity (Kaczynski, Potwarka, and Saelens 2008), and the total park acreage in a neighborhood is positively associated with walking by older adults (Li et al. 2005). A study in Helsinki, Finland found that the self-reported health of elders—a major predictor of physical health outcomes— related positively to more frequent visits to outdoor green spaces (Rappe et al. 2006). Indeed, researchers observe that physical health benefits from outdoor space pertain to even the frailest of older adults (Aspinall 2010), especially if they raise among them feelings of comfort, safety/security, and aesthetic pleasure (Tinsley, Tinsley, and Croskeys 2002). Therefore, providing parks as a safe, welcoming outlet for elders to walk and exercise is of vital importance for their physical health and quality of life (Gibson 2018).

Researchers have also found that time spent in neighborhood parks and outdoor gardens can positively influence their mental health, reducing stress and improving feelings of well-being (Rodiek 2002; Hansmann, Hug, and Seeland, 2007). People get emotional attachment to places other than their home, sometimes referred to as "third spaces"

(Garvin et al. 2013). Parks represent third spaces for elders, as they can help them develop a positive connection to their surrounding environment and feel part of their community. At the same time, the connection between people and nature has been associated with healing processes, stress reduction, and relaxation (Ulrich et al. 1991). Stress increases cortisol release by the brain. High cortisol levels in elders may cause decrease in memory and increase the risk for dementia or other cognitive impairment (Kiraly 2011). Exposure to nature can reduce stress, thus improving cognitive functions (Pappas 2009).

Parks can also help reduce social isolation that is often prevalent among older adults, provide opportunities for intergenerational interaction, and create a sense of place and attachment. A study of older people in the U.S. shows that the strength of ties among neighbors in an inner-city neighborhood was related to the availability and proximity of parks and green common spaces (Kweon, Sullivan, and Wiley 1998). Opportunities for socializing in parks may be more important to some elders than a park's facilities (Gardner 2008; Cohen et al. 2009; Parra et al. 2010). Studies find that seniors perceive open spaces as gathering spaces (Garvin et al. 2013).

Open space preferences of urban-living older adults

To get the perspectives about parks of low-income elders living in central city U.S. neighborhoods, we conducted five focus groups with members of Saint Barnabas Senior Services, the oldest and largest senior services center in Los Angeles. Each focus group included six to nine participants older than 65 and lasted 1.5 to 2 hours. Focus groups were diverse in terms of gender and race/ethnicity, and some were conducted in Spanish and Korean. Discussions revolved around issues of park access and use (including challenges and impediments) and preferred park designs, amenities, and programs.

The focus groups showed that older adults were very concerned about issues of safety and security at parks (or on their way to parks) and mentioned both human and environmental sources of danger. They were fearful of crime but also of falling or being hit accidentally by unsupervised youth at the parks, especially those skateboarding, bouncing balls, or running around. They were also concerned that they might fall on the way to the park because of broken sidewalks.

In summary, participants emphasized six desirable characteristics of parks: 1) security from human threats and environmental hazards; 2) good accessibility; 3) presence of natural elements (greenery, wildlife); 4) age-friendly park design and programming; 5) opportunities for walking and physical activity; and 6) settings and programs that encourage social interaction. They also gave a number of suggestions on how to achieve these park features. Figure 22.2 illustrates the essential features of a senior-friendly park from the participants' perspective. A detailed discussion of this focus group research can be found in Loukaitou-Sideris et al. (2016).

Design guidelines for senior-friendly parks

How should we design neighborhood parks that include these attributes deemed desirable by older adults? This section synthesizes information to suggest design guidelines for senior-friendly parks. Primary sources of information included the responses of focus group participants; secondary sources came from a detailed examination of the scholarly and professional literature on the open space needs of seniors, including design guidelines on healing gardens

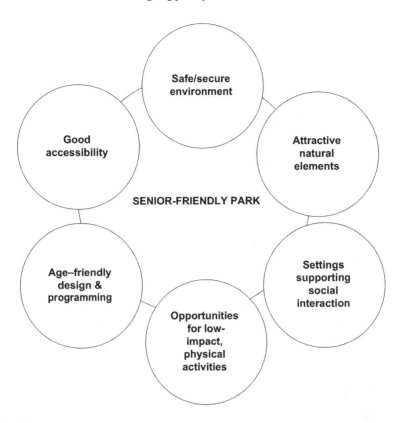

Figure 22.2 Elements for a senior-friendly park

Image credit: Anastasia Loukaitou-Sideris

and therapeutic landscapes, as well as guidelines, toolkits, and manuals about the design of age-friendly cities (Cooper-Marcus and Barnes 1999; World Health Organization 2007; Stafford 2009; Cooper-Marcus and Sachs 2014).

The information gleaned from these different sources is used to propose ten purposes that senior-friendly parks should strive to satisfy. These may be valued by most park users but are particularly significant for older adults and are as follows (in no particular order): 1) control; 2) choice; 3) safety and security; 4) accessibility; 5) social support; 6) physical activity; 7) privacy; 8) contact with nature; 9) comfort; and 10) aesthetic and sensory delight.

Control

Control refers to persons' "real or perceived ability to determine what they do, to affect their situation, and to determine what others do to them." (Ulrich 1999, 37). A sense of control is of particular importance to elders, who often see their physical or cognitive abilities lessening. Control is achieved when park users have a good sense of orientation and understanding of the park's layout and different offerings. Orientation is particularly important for elders suffering from cognitive impairments. For a park to provide a sense of orientation, people must know that it exists and be able to see it from some distance, can visit its different parts without getting lost, and understand and be able to take advantage of its offerings. The following elements can promote a better orientation and way-finding:

- A visible sign that marks a park's entrance and indicates its name.
- Way-finding signage in the parts of the neighborhood from where the park is not visible.
- A legible park layout that can be comprehended easily from the main entry.
- A map with the park's layout posted at the entrance and at other areas inside the park.
- A clear layout of park paths and avoidance of dead-end paths.
- Distinctive and highly visible features (e.g. a kiosk, a clock tower, a fountain, a distinctive gate [see Figure 22.3]) that can serve for orientation. This is particularly important for larger parks.
- Educational and informational signs at the park (e.g. directions on how to use equipment and facilities, plant labels, etc.).
- Signs in Braille and also visual graphics on signs.
- Utilization of sensory cues (sound patterns, pennants, flower fragrances) for way-finding of significant destinations.
- Boxes or newsstands near the park entrance holding flyers with information in different languages about park programs and activities.

Control is not only about acquiring a sense of orientation, or what Kevin Lynch (1960) called "legibility" of the urban form, but is also related to having a sense of choice, safety/security at the park, and easy accessibility that allows older adults to access the park easily on their own.

Figure 22.3 A distinctive gate and sign makes the presence of this park visible from afar
Image credit: Liz Devietti

Choice

Park users, both young and old, value choice. Visitors to a park should have a variety of places to wander and look at and a mix of activities and programs for passive and active recreation and enjoyment. This can be achieved by encouraging flexibility in park design and offering different choices in the ways that a park can be enjoyed. Some of the following elements can promote choice.

- Park areas that offer different qualities and opportunities (e.g. walking paths, exercise stations, areas with lush greenery and vegetation, meditation garden, barbeque areas etc.).
- Different options for people to sit—spaces in the sun and shade; spaces that are more open and public; and spaces that are more enclosed and private.
- Different views and vistas.
- Opportunities for passive recreation (places to sit, read, people-watch, play cards or board games, and socialize with friends) and active recreation (walking paths, exercise activities, gardening).
- Walking paths that offer different visual and sensory experiences, have different lengths, and various levels of difficulty.
- Equipment for active recreation that can accommodate different levels of physical activity (from low-impact exercise to more strenuous activities).
- Flat, grassy, multi-purpose areas that can accommodate a variety of different activities (e.g. tai-chi, yoga, picnicking).
- A variety of seating options for a person alone, for small and larger groups: movable chairs, right-angled seating that allows conversation between a few people, circular inward-facing seating that accommodates larger groups, and more secluded individual seating for those who seek privacy.
- Some features and materials that can be moved, manipulated, and changed.

Safety

The focus group discussions revealed that feelings of safety at the park are very important for older park users. Frequently, a perceived lack of safety leads them to avoid visiting neighborhood parks. The fear of victimization from crime, of tripping and falling, and of crossing high-traffic streets to reach the park represented sources of stress for elders. As a result, the design guidelines that follow are arranged under these three themes.

Designing against crime

- Clear lines of walking paths; avoidance of dense foliage and shrubs that obstruct views.
- Park enclosure via transparent decorative fencing that does not obstruct visibility from the street.
- Locking the park after sunset (unless there are organized activities).
- Good lighting throughout the park; avoidance of dark, concealed areas.
- Emergency buttons/phone boxes.
- Technological innovations that increase security (motion-activated lights, surveillance cameras, emergency report systems, etc.).
- Keeping the park clean and well-maintained.
- Organizing community volunteers to oversee safety.

Designing against falls

- Non-slippery paving materials that provide appropriate traction on walking surfaces.
- Even sidewalks and smooth interfaces between paved and unpaved surfaces.
- Narrow joints on paving units and concrete to prevent canes, wheelchairs, or high heels from becoming trapped (Cooper-Marcus and Sachs 2014).
- Paths that are flat or have a gentle slope; avoidance of steps.
- Keeping all paths clean from trash to avoid tripping hazards.
- Curbs along paths to make movement safer for those with walkers or on wheelchairs.
- Handrails on stairs, ramps, and paths.
- Good lighting along all paths.
- Allowing skateboarding or cycling inside the park only in designated areas.

Designing against traffic

- Location of parks or at least park entrances along low-traffic streets.
- Provision of safe crossings at intersections surrounding the park with clearly visible and audible crosswalk signals.
- Signalized intersections near the park that allow more time to cross the street before the light turns red.
- Traffic calming measures on streets bordering the park such as "bulb-outs" (sidewalk extensions that decrease the street crossing distance) and medians.

Accessibility

The ability to access a park quickly, safely, and with ease influences individuals' decisions to visit it. Indeed, the proximity of parkland to the place of residence is an important determinant of park visitation (Borst et al. 2008). This is particularly true for elders, who typically have less physical stamina than younger adults. For them, the ease of the journey to and from the park and the ease of movement and orientation while at the park become particularly important. Accessibility has both physical and psychological dimensions. Certain locational and park design characteristics can make a park more welcoming to elders and also enhance its physical accessibility. Additionally, the provision of supportive programs and activities at the park promotes its psychological accessibility. The following design elements are, thus, appropriate:

- Location of parks near other facilities used by seniors (e.g. senior citizen centers, churches, community centers).
- Connection of the park to the larger community through community events and activities that appeal to seniors (e.g. outdoor health clinic; yoga for seniors, farmers markets).
- Location of public transit stops near the park.
- Ample handicapped parking close to the park.
- Provision of adequate and barrier-free sidewalks around the park.
- ADA (Americans with Disabilities Act) accessible and universal design for all sidewalks leading to the park and all paths at the park.
- Limited grade changes and provision of gently-sloped ramps.

Figure 22.4 Multiple benches along pathways make it easier for older adults to rest while walking
Image credit: Waltearrrr (Flickr account)

- Wide park pathways, easy to use by people on wheelchairs.
- Multiple sitting opportunities along common routes to the park and along park paths (see Figure 22.4).
- Adequate signage for easy way-finding, with large-font signs that are easily visible by people on wheelchairs.

Social support

Social support refers to the human need of wanting to be connected with other human beings and be cared for and supported by them. Parks and park activities can encourage interaction and socializing among elders and between elders and other groups. As discussed in the focus groups, particular activities taking place at the park can connect older adults to their larger community and promote intergenerational exchanges. Appropriate design can create spaces that enable people to see and interact with one another, while appropriate programming can involve older and younger adults in common recreational or educational activities. The following design and programmatic elements can help promote social support.

- Seating that facilitates social interaction and allows people to watch human activity, such as seats at right angles, on a circle facing each other, and movable seats.
- "Props" for social interaction and small-group activities such as barbeque pits and tables for chess, board, and card games.

- Areas for informal gatherings and large-group activities and outdoor classes.
- Architectural, natural or landscape elements (kiosks, water fountains, flowers, pigeons, etc.) that bring people together around a common focus.
- Neighborhood events at the park such as community picnics and parties, annual cultural celebrations, such as 4th of July and Chinese New Year, and more regular events such as farmers markets, food banks, music concerts, and movie nights at the lawn.
- Community exchanges such as food banks, community-supported agriculture, and exhibits of neighborhood photos taken by elders.
- Organized events that bring youth and seniors together such youth orchestra performances.
- Nodes for local information exchange such as a bulletin board or newsstands.
- An outdoor reading room, such as a "little free library," where community members can donate and read books.

Physical activity

Parks can encourage physical activity by providing appropriate settings for active recreation and walking. Elders are more likely than other groups to live sedentary lifestyles and become intimidated by the prospect of exercise. The reluctance of many older adults to get involved in physical activity may be a result of fear because of declining capacities and limited stamina but also lack of appropriate spaces and social support for exercise. Some elders in the focus groups expressed embarrassment of exercising alone at the park. Walking is the easiest and most common type of physical activity for seniors. Park designers can encourage walking by creating ability-appropriate and attractive walking paths that reach interesting destinations (e.g. fountains, gazebos, vistas, outdoor coffee carts). Incentives and opportunities for walking and exercise should be provided at different lengths and levels of difficulty to address varying levels of ability. Additionally, exercise can be achieved through activities and games such as yoga, tai chi, bocce ball, lawn bowling. Gardening is yet another activity that involves movement and exercise. Elements able to promote physical activity include:

- Walking loops and paths out of non-slippery and rubbery materials that are easy on the feet.
- Destination points at the end of paths that encourage/attract people to reach them.
- Markers along long trails that measure walking progress.
- Low-impact exercise equipment.
- Spaces for exercise away from heavy-traffic areas, under shade and with interesting views.
- A spacious area with grass and trees for group classes or games.
- Physical activity classes for seniors, such as dance, tai-chi, or yoga.
- Gardening.

Privacy

Even in public spaces, individuals often yearn for some level of privacy and tranquillity. Parks can offer a break from the high-pace urban setting and serve as small urban oases within the hustle and bustle of busy city life. Many elders in the focus groups, who lived in small urban apartments, emphasized their need for a tranquil environment at the park. Thus, design and landscaping should offer sufficient levels of privacy in some subareas of the park, allowing people to avoid social interaction, if they so wish. The following guidelines seem appropriate:

- Park location on a quiet street.
- Screening of outside noises with natural sounds (water, breeze moving through tree leaves, etc.).
- Placement of seating areas away from street noise.
- Use of buffer planting to minimize street noise and create a sense of enclosure around the quieter and more private park areas.
- Park subareas that enable visitors a level of physical and visual privacy.
- Some seating where park visitors can sit alone.
- Allocation of certain park areas for more individual uses such as personal garden plots or bird feeders.

Contact with nature

Parks bring nature into the city and can offer positive natural distractions, that foster relaxation and decrease stress and worrisome thoughts (Cohen et al. 2007). Elders in the focus group discussed their desire to walk in natural surroundings, "see squirrels," "hear birds," and "smell flowers." The following elements can help emphasize parks as natural settings.

- Abundance of "softscapes" (flowers, verdant plans, vegetation).
- Mature trees that offer adequate shade to park visitors.
- Water elements such as ponds, creeks, small waterfalls (see Figure 22.5).

Figure 22.5 A small water pond brings the natural element of water into the park
Image credit: Lené Levy Storms

- Unthreatening wildlife such as birds, butterflies, squirrels, ducks.
- Instillation of nature sounds (water, birds, breezes, wind chimes).

Comfort

Provision of physical and psychological comfort should be an explicit goal of park design, and this is important for older adults. Psychological comfort relates to feelings of safety, which was discussed previously. This section proposes elements that aim to provide elders and other park visitors with physical comfort.

Seating

- Rest stops and multiple opportunities for seating.
- Comfortable seating with ergonomic designs, backs, and arms (see Figure 22.6).
- Natural seating materials such as wood or stone that are more resistant to high temperatures; avoidance of materials that retain excessive heat, such as metal.
- Since elders are particularly sensitive to glare, avoidance of materials that have bright and reflective surfaces (e.g. aluminium, white surfaces).
- Light movable seats that can be easily moved.
- Sufficient space around benches and tables for people on wheelchairs and walkers.
- Placement of seating and tables under large trees, kiosks, gazebos, trellises, or canopies for shade.

Figure 22.6 Seating without a backrest and surrounding traffic make this park setting uncomfortable for older adults

Image credit: Liz Devietti

Protection from the elements

- Park settings with adequate exposure to sunlight (especially for cold climates).
- Trees and vegetation strategically placed to protect from direct sunlight and minimize glare from the sun.
- Overhead protection from sun (umbrellas, kiosks, or arbors).
- Orientation of park or park elements to protect from prevailing winds but allowing cool breezes in hot climates.

Amenities

- Universally accessible water fountains.
- Multiple restrooms with handicapped access and hooks for canes.
- Multiple trash receptacles.
- Low or intermediate lighting levels to avoid glare.
- Fixed and movable tables where people can have a picnic or eat lunch.
- Water fountains and electric outlets near seating areas.
- Food stands.

Comfortable movement

- Paving that does not inhibit movement.
- Curvilinear and flat paths that are more easily navigable for people on wheelchairs.
- Contrasting colors on pavement materials and seating to respond to some seniors' compromised depth of field.
- Gently-sloped ramps instead of steps.

Aesthetic and sensory delight

Parks should offer an aesthetic respite in the city and opportunities for sensory enjoyment. Sensory stimulation should not only be visual but also auditory, tactile, and olfactory. Park designers may employ aesthetically pleasing design features and landscaping, art pieces, enchanting sounds, and fragrant flowers. Parks designed with seniors in mind should also consider that older adults walk at slower paces and cover shorter distances than younger adults. For this reason, sensory-interesting features should be placed at shorter intervals than it would be necessary for spaces designed for the general public. However, the amount of visual variety should not result in visual clutter. The following design elements seem appropriate:

- Placement of paths and benches in ways that take advantage of interesting views.
- Screening unpleasant views (e.g. blank walls, parking lots) through vegetation, mural painting, or art placement.
- Gently curving paths.
- Plants and flowers of varying sizes, contrasting colors, pleasing fragrances and textures, and seasonal variety.
- Plants that have special meaning for particular cultural or ethnic groups living in the neighborhood.
- Outdoor art pieces and fun and whimsical features.
- Water elements and mature trees generating soothing natural sounds.

Conclusion

Older adults represent a fast-growing segment of the population but cities are often not prepared to accommodate their needs. US cities lack senior-friendly parks, despite the fact that research has shown that parks can offer many benefits to older adults. This chapter compiled primary and secondary information to generate design guidelines for senior-friendly neighborhood parks. We should note that these guidelines intend to give general directions; park planners and designers should also consider the physical characteristics of their context (topography, size, microclimate, surrounding land uses, street network, etc.), and the social characteristics and stated preferences of the intended users, particularly since older adults are a heterogeneous group in terms of age, physical and cognitive capacities, and socio-cultural characteristics.

Older adults in focus groups indicated that they were reluctant and fearful to mingle in the park with other age groups, particularly teenagers and young adults. But in inner-city neighborhoods that lack the space and resources for new parks, the demand for "seniors-only" parks is hard to satisfy. It is, therefore, important to identify acceptable ways to create neighborhood parks that are truly intergenerational and can be shared by different age groups. Cooper-Marcus and Francis (1990) talk about "layering and separation"—the formation of activity zones at the park that allow different groups to coexist—and argue for a park design that "permits regular groups of users to lay claim on certain areas" (73–74). Planners and designers should consider these ideas and designate certain areas within a park that seniors can call their own, away from the hustle and bustle and danger for them generated by more active uses and settings, such as soccer fields. It is time to design age-friendly cities, and this should start with the provision of age-friendly public spaces.

References

Aspinall, P.A. (2010) Preference and Relative Importance for Environmental Attributes of Neighbourhood Open Space in Older People. *Environment and Planning B: Planning & Design*, 37(6): 1022–1039.

Borst, H.C., Miedema, H.M.E., De Vries, S.I., Graham, J.M.A., and Van Dongen, J.E.F. (2008) Relationships between Street Characteristics and Perceived Attractiveness for Walking Reported by Elderly People. *Journal of Environmental Psychology*, 28(4): 353–361.

Cohen, D.A., Han, B, Nagel, CJ, et al. (2016) The First National Study of Neighborhood Parks: Implications for Physical Activity. *American Journal of Preventive Medicine*, 2016; 51(4): 419–426.

Cohen, D.A., Mckenzie, T.L., Sehgal, A., Williamson, S., Golinelli, D., and Lurie, N. (2007) Contribution of Public Parks to Physical Activity. *American Journal of Public Health*, 97(3): 509–514.

Cohen, D.A., Sehgal, A., Williamson, S., Marsh, T., Golinelli, D., and Mckenzie, T.L. (2009) New Recreational Facilities for the Young and the Old in Los Angeles: Policy and Programming Implications. *Journal of Public Health Policy*, 30: S248–S263.

Cooper-Marcus, C. and Barnes, M. (1999) *Healing Gardens: Therapeutic Benefits and Design Recommendations*. New York: John Wiley and Sons.

Cooper-Marcus, C. and Francis, C. (1990) *People Places: Design Guidelines for Urban Open Space*. New York: John Wiley and Sons.

Cooper-Marcus, C. and Sachs, N. (2014) *Therapeutic Landscapes*. New York: John Wiley and Sons.

Gardner, P.J. (2008) "The Public Life of Older People Neighbourhoods and Networks." PhD Diss., University of Toronto.

Garvin, E., Branas, C., Keddem, S., Sellman, J., and Cannuscio, C. (2013) More than Just an Eyesore: Local Insights and Solutions on Vacant Land and Urban Health. *Journal of Urban Health-Bulletin of The New York Academy of Medicine*, 90(3): 412–426.

Gibson, S. (2018) "Let's Go to the Park:" An Investigation of Older Adults in Australia and Their Motivations for Park Visitation. *Landscape and Urban Planning*, 180: 234–246.

Hansmann, R., Hug, S., and Seeland, K. (2007) Restoration and Stress Relief through Physical Activities in Forests and Parks. *Urban Forestry & Urban Greening*, 6(4): 213–225.

Kaczynski, A.T., Potwarka, L.R., and Saelens, B.E. (2008) Association of Park Size, Distance, and Features with Physical Activity in Neighborhood Parks. *American Journal of Public Health*, 98(8): 1451.

Kemperman, A. and Timmermans, H.J.P. (2006) Heterogeneity in Urban Park Use of Aging Visitors: A Latent Class Analysis. *Leisure Sciences*, 28(1): 57–71.

Kiraly, S. (2011) Mental Health Promotion for Seniors. *British Columbia Medical Journal*, 53(7): 339–340.

Kweon, B.S., Sullivan, W.C., and Wiley, A.R. (1998) Green Common Spaces and the Social Integration of Inner-City Older Adults. *Environment and Behavior*, 30(6): 832–858.

Li, F., Fisher, K., Brownson, R., and Bosworth, M. (2005) Multilevel Modelling of Built Environment Characteristics Related to Walking Activity among Older Adults. *Journal of Epidemiology and Community Health*, 59(7): 558–564.

Loukaitou-Sideris, A. Levy-Storms, L., Chen, L., and Brozen, M. (2016) Parks for an Aging Population: Needs and Preferences of Low-Income Seniors in Los Angeles. *Journal of the American Planning Association*, 82(3): 236–251.

Lynch, K. (1960) *The Image of the City*. Cambridge: MIT Press.

Pahor, M., Blair, S.N., Espeland, M., Fielding, R., Gill, T.M., Guralnik, J.M., Hadley, E.C. et al. (2006) Effects of Physical Activity Intervention on Measures of Physical Performance: Results of Lifestyle Interventions and Interdependence for Elders Pilot LIFE-P Study. *Journal of Gerontology and Biological and Medical Sciences*, 61: 1157–1165.

Pappas, A. (2009) Nature-Related Contact for Healthy Communities. In: *Re-Creating Neighborhoods for Successful Aging*, Abbott, Carman, Carman, and Scarfo (ed). Baltimore: Health Professionals Press, pp. 53–84.

Parra, D.C., Gomez, L., Sarmiento, O.L., Buchner, D., Brownson, R., Schmidt, T., and Lobelo, F. (2010) Perceived and Objective Neighborhood Environment Attributes and Health Related Quality of Life among the Elderly in Bogotá, Colombia. *Social Science & Medicine*, 70(7): 1070–1076.

Rappe, E., Kivela, S.L., and Rita, H. (2006) Visiting Outdoor Green Environments Positively Impacts Self-Rated Health among Older People in Long-Term Care. *Horttechnology*, 16(1): 55–59.

Rodiek, S. (2002) Influence of an Outdoor Garden on Mood and Stress in Older Persons. *Journal of Therapeutic Horticulture*, 13: 13–21.

Stafford, P.B. (2009) *Elderburbia: Aging with a Sense of Place in America*. Santa Barbara: Praeger.

Sugiyama, T., Thompson, C.W., and Alves, S. (2009) Associations between Neighborhood Open Space Attributes and Quality of Life for Older People in Britain. *Environment and Behavior*, 41(1): 3–21.

Sun, Q., Townsend, M.K., Okereke, O.I., Franco, O.H., Hu, F.B., and Grodstein, F. (2010) Physical Activity at Midlife in Relation to Successful Survival in Women at Age 70 Years or Older. *Archives of Internal Medicine*, 170: 194–201.

Takano, T., Nakamura, K., and Watanabe, M. (2002) Urban Residential Environments and Senior Citizens' Longevity in Megacity Areas: The Importance of Walkable Green Spaces. *Journal of Epidemiology and Community Health*, 56(12): 913–918.

Tinsley, H.E.A., Tinsley, D.J., and Croskeys, C.E. (2002) Park Usage, Social Milieu, and Psychosocial Benefits of Park Use Reported by Older Urban Park Users from Four Ethnic Groups. *Leisure Sciences*, 24(2): 199–218.

Ulrich, R. (1999) Effects of Gardens on Health Outcomes: Theory and Research. In: *Healing Gardens: Therapeutic Benefits and Design Recommendations*, Cooper Marcus, C. and Barnes, M. (eds). New York: John Wiley and Sons, pp. 27–86.

Ulrich, R., Simons, R.F., Losito, B.D., Fiorito, E., Miles, M.A., and Zelson, M. (1991) Stress Recovery During Exposure to Natural and Urban Environments. *Journal of Environmental Psychology*, 11: 201–230.

Winick, B.H., and Jaffe, M. (2015) *Planning Aging-Supporting Communities* (PAS Report 579). Chicago: American Planning Association.

World Health Organization. (2007) "Global Age-Healthy Cities: A Guide." www.who.int/entity/ageing/agefriendly_cities_guide/en/index.html.

23

EXPANDING COMMON GROUND

Ken Greenberg
(ADAPTED FROM GREENBERG, K. (2019) *TORONTO REBORN: DESIGN SUCCESSES AND CHALLENGES.* TORONTO: DUNDURN.)

Introduction

Public spaces are as varied and specific to places as the urban cultures that have given birth to them and the circumstances of their creation. They come in all sizes and shapes. The best ones have a living, ongoing symbiotic relationship with their host cities, from the traditional *campi* of Italian cities, the *pleins* of the Netherlands, the *places* of France, the *plazas* and *paseos* in the Spanish speaking world, to the traditional urban parks in the heart of cities like Central Park in New York City (and all the great Frederick Law Olmsted parks for that matter), Chapultepec Park in Mexico City, and the Bois de Boulogne in Paris. Chicago's Millennium Park and Melbourne's repurposed pedestrian laneways, newly carved out urban public spaces, are recent and modern variations on this relationship.

City parks and squares have emerged as a potent new focus for community allegiance, civic pride, and neighborhood identification. Even where government is severely challenged in providing and maintaining such spaces, civil society is providing a renewed groundswell of support for the commons. These efforts range from the high-profile, New York's Central Park Conservancy and Toronto's Bentway Conservancy, for example, to hundreds of more modest 'Friends of' organizations. Public–Private Partnerships have become the way to accomplish what government can no longer manage alone.

There is another, even more basic, argument for the presence and quality of public space, one where social equity and enlightened self-interest come together. It is the economic counter to the ultimately self-defeating neo-conservative legacy of neglect and withdrawal, an ideal which leads to private wealth inside gated communities and public squalor beyond. There is now a broadly shared understanding among economists that cities are the principal generators of wealth, subject to their ability to attract investment in a diverse and creative economy where quality of life and quality public (and publicly accessible private) space are key attractors.

There is substantial evidence that the presence of attractive and well-maintained parks and open spaces—the commons—contribute to that quality, create economic value in surrounding areas and enable its long-term retention. We often get to know and experience other cities through their great parks and public spaces; they are magnets which attract investment. In a world of intense

Figure 23.1 Derives—a site specific performance—June 2019
Image credit: SeeWhatIcy Media 7

competition, sense of place becomes a key consideration in location choices for businesses, institutions, and individuals.

Generosity in the public realm is also a major contributor to a public's physical and mental health. Exacerbated by sedentary lifestyles, caused by an overreliance on the automobile and a tendency to spend long hours in front of screens, our current public health crisis has produced an obesity epidemic with corresponding increases in diabetes and heart disease rates. This growing prevalence, especially among children, is alarming. The inevitable (and perhaps desirable) shrinking of private living space, as a reflection of economic pressures, puts a premium on public spaces. Driving to the gym or health club is no substitute for public spaces near people's home and work, where all ages and abilities can participate in an array of active pastimes, from walking to cycling to the whole range of year-round sports and athletic activities. Proximity and quality of experience can embed health promoting activities into part of everyone's daily routines.

Paradoxically, as parks and open space, key components of the public realm, have been better appreciated of late, they have been equally threatened and undermined by funding challenges. This is a public crisis, and if crisis is the mother of invention, non-conventional and creative strategies are necessary to create a new vision for the public realm. One avenue to success resides in exploiting the arteries and veins of the city and stitching them together in new and remarkable ways. By focusing on linkages between existing and new green spaces, we can expand and improve public open space, turning disjointed, disconnected places into a continuous, interconnected web of regional and neighborhood linear greenways comprised of not only traditional parks, but also trails, bridges, and green streets.

In June 2016, an extraordinary gathering at the High Line's offices in New York City brought together a diverse group of creators, including designers, philanthropists, and community leaders, of a new generation of public space projects from across North American cities to share experiences in creating these new forms of public space. Despite each project's particularities, they shared a common language around repurposing post-industrial spaces found throughout their cities—the neglected bridges, elevated highways and transit lines, rail lines, viaducts, neglected river banks and bayous, transit stations, and flood ways. By way of this sharing, new kinds of public–private partnerships were forged including a variety non-profit conservancies., enabling new resources to be brought to the table; in the process new models managing and programming were invented combining public funding, private philanthropy and sponsorships Out of that initial encounter came the idea of forming The Highline Network, a group that would share experiences and best practices, engage in advocacy, and communicate the power of these ideas and approaches to a broad range of audiences.

While it is clear that these initiatives have the potential to bring enormous positive impacts for their cities, local communities, and for the spaces themselves, they raise important questions about roles and responsibilities. What are the overarching goals and priorities? Who is being served? What are the points of intersection? How do we meet the changing needs of cities, residents, and communities, and where do these projects fit in the discussion of larger social issues?

In that meeting, we wrestled with the meaning of democratic, accessible, and inclusive public space, acknowledging that in many situations our initiatives have significant impacts on existing low-income communities and the subsequent potential for displacement through gentrification. We talked about the obligation to respect the places and people that are and were present, as well as the need to work with a broad range of other actors and build the

Figure 23.2 Singing Out choir performs on The Bentway Skate Trail opening day
Image credit: Andrew Williamson

capacity and civic muscle of those affected in order for them to secure a stake in the outcomes. We stressed that this commitment, addressing the full spectrum of society, should be reflected in the diverse makeup of boards and staff, as well as in the nature of programming.

We acknowledged that the ultimate role of the city is to defend the public's interests and that there is a need to work in close partnership with city bureaucracies—public agencies, parks departments, and transportation departments—to identify our place in the ecosystem of public spaces. These collaborative efforts are not a panacea, intended to supplant or diminish commitment of governments to the public realm, but rather present an opportunity to strengthen our commitment of adding to the resource base. By pioneering new models of shared community stewardship and engagement, we are able to take on projects which otherwise might not happen.

The Bentway

The opening of the Bentway in Toronto took place on January 6–8, 2018, one of the coldest weekends of the winter with temperatures around -30 Celsius. I had been working on this project for years and for many months I couldn't wait to see how it would actually work. In the event some 20,000 Torontonians came out to brave the cold that weekend. It was remarkable to see them bundled up enjoying a choir, watching an amazing demonstration of break dancing on ice and discovering artwork enlivening the space. Mayor John Tory, Councilors Cressy and Layton officiated at the ribbon cutting along with Judy and Wil Matthews, whose extraordinary donation had made all this happen. Intense curiosity quickly turned to a sense of wonder and pleasure as the skaters took to the ice for the first time. The reaction by Torontonians as reflected in media attention and all forms of social media was overwhelmingly positive and enthusiastic.

It felt odd lacing up my skates for the first time while sitting on a bench under the Gardiner Expressway. And a little wobbly (after years of not skating) once I joined the procession of skaters of all ages and skill levels on the ribbon of ice, weaving through the 'bents,' the combined columns and beams holding up the elevated highway. The experience of being there on skates was exhilarating and surprising as I looked up at the enormous height of the road deck some five stories above or at the familiar earthen ramparts of Fort York on one side and the neighboring towers across Fort York Boulevard on the other. High above me the unseen traffic above sped by oblivious to what was happening below.

There were newly planted trees in the islands formed by the 220-meter-long figure eight of the skate trail; containers had been arranged form an improvised cluster in the middle of what was still a construction site with sheltered skate rental, lockers, a vending stand with hot chocolate and cider, benches and fire pits. It felt both entirely natural as if it had always been there and totally unexpected at the same time, almost like a pilfered pleasure in a former cavernous no man's land. Something so familiar had turned into something completely different. The highway above had all but disappeared and in its place another reality had materialized in the void on the ground.

The Bentway project was launched in 2015 to exploit a great untapped resource previously ignored in a highly strategic location. No longer a symbol of division, this almost two-kilometer stretch of the elevated Gardiner Expressway from west of Strachan Avenue to Spadina has been re-appropriated to frame and shelter a great new civic living room under a five-story high 'roof' shared by the seven neighborhoods it touches. It has transformed this space

Figure 23.3 Skaters skate past Janine Miedzik's installation Pro Tem
Image credit: Andrew Williamson

Figure 23.4 A Fall Day at the Bentway
Image credit: Nic Lehoux

from a psychological and physical barrier into a connection, from a back to a front, from hostile and off-putting to welcoming. It is a new kind of hybrid public space that doesn't exactly fit into existing category; not a park, a square or a trail although it has elements of all of those.

It was a treasure hidden in plain sight. In an article I wrote in 2011, I included a rough sketch showing where all the new development was occurring in this highly strategic central location was taking the form of dense, high-rise buildings with little open space. These immediate neighborhoods were eventually to house some 77,000 residents in about 15 years including young families with several thousand children and a high percentage of new arrivals; approximately 39% were born outside Canada. With radical growth and the need for public space, necessity became the mother of invention.

The stewardship model

It took an extraordinary act of civic generosity and willingness to take a risk. Judy Matthews and I were friends and colleagues from our days in the Planning and Development Department at the City of Toronto. She and her husband Wil, an investment banker, are dedicated urbanites and long-time Torontonians. Their profound love for Toronto had led them to a unique form of philanthropy with a focus on public space. In April of 2015 I got a call from Judy. She and Wil were searching for a legacy public-realm initiative and were very particular in their requirements. They were seeking to do something that would be more than a discrete public space, a park, or square that might in itself be quite delightful, but rather were looking for an opportunity that would be catalytic and have a significant impact stimulating change beyond itself. I immediately thought of the 'under Gardiner' potential and described it to Judy who was intrigued.

This was the beginning of a creative journey to reimagine this forgotten stretch of land beneath the Gardiner expressway as great civic space. With the highly creative and talented team of young landscape architects Marc Ryan and Adam Nicklin, who had come together to form the firm Public Work, we landed on a simple way of showing what was a very complex project by producing a three-dimensional concept plan that lifted the 'cover' off the Gardiner. It revealed the space below, highlighting its shifting and unique relationships to context along its two-kilometer length and how these could be exploited to provide a highly diverse set of experiences as a public space. Although there were no perfect parallels, we gathered a number of the most relevant precedent images we could find to convey the potential. We also prepared a high-level estimate of the capital cost of the project which was $25 million.

A critical part of the invention of the Bentway was the exploration of stewardship models. How would it be managed? What would its relationship be to City Hall and to the many constituencies it would serve?

Once again, the Bentway was to be a pioneer in Toronto, providing a new model of stewardship and funding of public space by setting up the non-profit conservancy. In the early days of planning for the Bentway Judy and Wil Matthews joined my wife and me in a visit to New York where in addition to visiting the High Line and various other innovative projects, we met with Betsy Barlow Rogers, the founder of the Central Park Conservancy wanted to get advice based on the Central Park Experience.

We learned about the Conservancy's origins and ways in which it has been structured to combine the resources of the City of New York and philanthropy including its Education and Service programs providing groups and individual youth with hands-on opportunities to learn about Central Park and help preserve and maintain its landscapes. We were suitably impressed and encouraged to come up with a made-in-Toronto model for a Bentway Conservancy, which became the first urban version in Canada. This was a significant departure for Toronto, however, and many of the implications had to be worked out and built into the agreements with the city.

The phenomenon of taking something built for one purpose in the life of a city and by challenging preconceptions, turning it into something else is not without precedent. It is a matter of seeing what is familiar with fresh eyes, often spurred by pressures demanding radical innovation and re-appropriation. A familiar historic example of such a re-purposing is the removal of fortifying walls surrounding medieval cities, Paris being a prime example. These city walls were walls originally designed for defense but as armaments evolved and the city expanded, the redundant fortifications were demolished and their vacated 'footprints' converted into 'Grands Boulevards,' which became defining features of the modernizing city. Or a more contemporary Parisian example the banks of the Seine transformed seasonally with truckloads of sand into 'Paris Plage,' making the city its own resort.

The elevated Gardiner was an object we loved to hate and for many good reasons, as a brutal heavy-handed intrusion in the urban fabric, trampling on its surroundings. Yet seen from below with fresh eyes its supporting structure is a thing of strange beauty, forming a magnificent space. It was this quality that my colleagues at Public Work and I wanted to exploit. The plan we developed transforms this portion of the Gardiner Expressway's neglected but grand under belly defined by the heavy-duty rib cage of concrete post-and-beam structural elements ('bents') supporting the highway deck. The ever-so-slightly curved free-standing colonnade they form bends sinuously around Fort York, opening up perspectives that are only fully revealed as you move through the space. The approach was not to radically change this, but to embrace it and reveal its intrinsic qualities and rugged character.

Network of public spaces

A great 'civic living room' for the underserved neighborhoods surrounding it, the Bentway hosts a sharing and overlapping of activities in a dynamic and vital public space for all citizens that links physical and cultural communities, neighborhoods, and people. The space has been outfitted like an immense stage set with power and lighting making the monumental frame of the Bentway 'adhesive' with rigging to hang things off the bents using specially designed friction clamps that don't penetrate the concrete surface. Ground level treatments vary from hardscape to softscape creating many configurations of places where things can readily happen but are not prescribed leaving maximum opportunity for improvisation and discovery.

The Bentway does not stand in isolation. The complementarity and interpenetration of the large green space of the Fort and the space of the Bentway offers rich program opportunities for combined use. The seamless overlap with the National Historic site creates a fluid connected landscape allowing for vegetation to also thrive within the bents. The Bentway functions like an immense linear shade canopy, providing more welcome shade when the sun is high in mid-summer but because of its height there is considerable sun penetration in the shoulder and winter months. The sound of the highway traffic above is also attenuated within the Bentway shielded by the deck above.

Figure 23.5 Visitors watch a streetdance battle at The Bentway Block Party
Image credit: Andrew Williamson

Across the pedestrian cycle bridge the multi-use trail of the Bentway will weave through and connect a series of already publicly accessible spaces under the Gardiner, sandwiched between adjoining residential developments and leading across several north–south streets to Canoe Landing Park where it rises up a slope to connect with the new schools and community center in City Place and then gently descends again to Spadina Avenue where there are plans for another pedestrian/cycle bridge.

The Bentway is responding to a powerful trend for innovative programming in public spaces as artists and creators seek direct dialogue with the city with greater opportunity to reach broader, more diverse, and unexpected audiences. The working mandate of the Bentway adopted by the Conservancy Board expresses this relationship to the evolving city as a cultural project:

> The Bentway is a new public space and programming platform deeply rooted in Toronto's urban fabric. It is a project both *of the city* and *about the city working* to deliver accessible, participatory and responsive programming at the intersection of arts and urbanism. Through the lens of culture and recreation the Bentway explores Toronto's changing urban landscape as well as the opportunities and issues that unite cities across the globe.

The Bentway is special; it grew out of a particular response to need and opportunity in Toronto. But this is a case of simultaneous discovery. The combination of all of these factors is leading to an intensive re-imagining of the city—seeing the city through fresh eyes including a major new role for arts and culture in public spaces, a participatory, interactive, democratization of institutions moving out into the public sphere. In the process we are changing our mental maps of the city as we move from the perception of public space as isolated green spaces to webs or matrices, growing, organically as networks. The Bentway clearly embodies all of the above.

While there are clearly significant differences among the projects that were presented in the initial meeting of the High Line Network—distinctions between organizations/projects and communities—they have much in common in the ways in which they respond to these drivers and there was real interest expressed in continuing and expanding on this conversation. The Network members have spent considerable time discussing the broader implications of our initiatives that come with the stewardship models they are developing. Public/private partnerships and philanthropy bring many powerful advantages—the opportunity to grow the pie bringing more resources, to address the urgent for public space nimbly and flexibly in innovative ways allowing the unique and exceptional to emerge, to 'prototype' exploring new forms. They foster and enable hybrids that do not neatly fall within the traditional categories and to incubate new and exciting ways of engaging communities in new models of stewardship that also bring major responsibilities.

These include the need for transparency, accessibility and keeping in close touch before, during and after the creation phase including programming that would precede full buildout of projects. Most significantly perhaps was the concept of achieving balance, maintaining an unwavering commitment to publicness while tapping a range of new funding resources and inventing new organizational models. We recognized the dual responsibility for these projects—for the space that is in our control (delivering equitable, inclusive public space within the project limits) and the space that is out of our direct control (being strong advocates in the surrounding neighborhoods).

Equity and health

Toronto has a particular public realm challenge which has to do with overcoming a real and perceived pre-war city and post-war urban/suburban divide. Central Toronto lives on its historic capital of relatively dense and interconnected pre-war neighborhoods with shared public space, but David Hulchanski, Director of the Centre for Urban and Community Studies at the University of Toronto, painted an alarming picture of the erosion of city-wide common ground as income polarization is reflected geographically in Toronto's accelerating city-wide division into three very separate cities: a growing area of poverty (housing a large population of recent immigrants) in the post-war suburban fringe, a growing high-income enclave in the city center, and a shrinking middle class in between. The questions this raises include equity in terms of access to public space and public facilities, and overcoming barriers that divide us.

Increasingly, regional health authorities are recognizing this relationship between urban form that enables walking and cycling and public health. Patterns of development that discourage parents from allowing their children to walk to school, or which require the use of a car for every journey are now proven to be problematic. In Toronto, former Medical Officer of Health, David McEwen actively drew attention to the relationship between civic design that promotes active lifestyles, and reduced health care costs. While government policy lags behind the research more and more members of the design community are acting on the evidence.

A similar correlation holds true for mental health. In October of 2017 I was invited to speak a conference at the Royal College of Physicians in London entitled 'Unleashing

Figure 23.6 Sunday Social Yoga with Muse Movement
Image credit: Nicole Pacampara

Figure 23.7 Bocce at the opening of CITE
Image credit: Denise Militzer

Health by Design' bringing together a mix of design professionals and health practitioners. The issues of mental health related to social isolation and the importance of public space for social connection were seen as significant as those related to physical well-being.

Great parks and open spaces are the hallmark of a great city. Given the lack of availability and cost of land conventional methods are not up to the task. This begs the question of what our next public spaces will be as we evolve into a great and densely populated city. We have outgrown the limited amount of dedicated public space that is the inheritance from a much smaller city of Toronto. Space for new conventional parks is scarce and with current land values extremely expensive. Even with 'carve-outs' and land takings from development, and cash put into a park reserve fund traditional means of buying discrete parcels of land to create new parks are strained.

Linking the public realm

By tapping all of these latent resources simultaneously and in combination, putting the pieces together—laneways, street redesign, ravines, hydro corridors, rail lines, stormwater management, floodproofing and transportation initiatives along with development-related park contributions and consolidation of major public sites, a vastly expanded public realm can emerge addressing many of the current deficiencies. It will be different both in scale and kind. Rather than discrete and bounded public spaces carved out of a grid of streets blocks—parks and squares—it has the potential to become the fully continuous connective tissue of the urban fabric itself.

One of the greatest challenges to thinking and operating holistically this way is getting beyond the 'siloed' mandates of the many different agencies tasked with providing these functional pieces of the public realm within prescribed boundaries—be it streets or parks or hydro corridors or protected natural areas. It requires getting those responsible to pool their efforts and bring these components together to form the physical places we experience in the city.

This is easier said than done; jurisdictional, bureaucratic inertia and risk aversion are powerful impediments. But breaking through these administrative and legal barriers is being driven by an increased sense of the public's perceived need and right to overcome these barriers, and progress is being made as agencies redefine their mandates and develop collaborative strategies. What the growing number of examples is demonstrating is that the entire city can become more park-like, green and connected for people on foot and on bicycle. The ability to move around relatively freely and experience the city this way breaks through perceived barriers between neighborhoods and districts. Discrete places are joined; relationships and flows become more continuous.

Key linking public-realm elements play a vital role in overcoming previous divides like the Bentway or the High Line in New York, the Simone de Beauvoir Bridge over the Seine in Paris linking the new Tolbiac and Bercy Neighborhoods, and the George C. King Bridge across the Bow River in Calgary connecting the Bridges Neighborhood and the new East Village. Local Toronto examples include our own Humber Cycle and Pedestrian Bridge, a wonderfully well-used landmark which performs a similar vital role linking South Etobicoke and the Western Beaches across the Humber River and tying together the river valley and waterfront trail systems

Unlike the more formal public spaces of a planned city in previous eras, this new approach to public space conveys a very different feeling. It is often opportunistic based on 'found spaces,' improvised and episodic. The new sensibilities in play for the public realm, like the city itself, tend to be multivalent and overlapping, linear and organic, weaving through the city and intertwined with natural and infrastructural systems. The new green sinews—links as opposed to self-contained 'nodes' in planning jargon—are penetrating and open up the city in areas that were formerly terra incognita. Unlike the grand and formal front-of-the-house spaces, they provide a unique vantage point and back-of-house experience, which is in tune with the complex, overlapping, flash back and forward, multi-layered condition of twenty-first-century life.

This desire to meander through the city in public space is at the same time ancient—witness the promenade, the passeggiata, the paseo. And Frederick Law Olmsted's green ribbons through cities (Boston's Emerald Necklace, for example) and other similar nineteenth-century-visions such as Horace Cleveland's Grand Rounds and Chain of Lakes in Minneapolis are antecedents to be sure. But there is also something new here in the immense popularity of walking, jogging, cycling, in-line skating (often with kids in tow), or otherwise exercising while exploring new territories.

In some ways this is the twenty-first-century version of the mind-and-body sensorial experience of the nineteenth-century boulevard flâneur, with the open-ended allure of the 'road' but in another form. Venturing into uncharted territory can be deeply satisfying psychologically, a quest combining discovery and self-discovery. Encountering others enroute, people-watching on the fly with occasional acknowledgement, greeting, and communication of the shared experience and the possibility of gregarious encounters—all of this reinforces a sense that 'I am not alone.' For me, especially from the raised vantage point of a bicycle seat, the feeling is reminiscent of snorkeling; like an exotic seascape, the city reveals itself in new ways.

Figure 23.8 Playing in The Bentway water feature
Image credit: Nic Lehoux

For many the ostensible reason or pretext for getting out is that we also stretch our bodies and enjoy the health benefits of exercise, increased cardiovascular stimulation, and movement of our limbs outside the gym or fitness center. But it is much more than exercise. Moving self-propelled at relatively low speeds restores the intuitive geographic understanding of spatial relationships that the car weakened—a feel for the real distances between things, a sense that it's all connected. The 'kinetic' feeling in navigating these networks can be exhilarating. The city reveals itself in new ways. More than exercise it offers a chance to commune with our neighbors.

This phenomenon is still in the early stages, and it will be fascinating to see where it goes. It is not too difficult to imagine cities traversed by extensive green and sociable networks as the legacy of a post-automobile age. This increased physical and psychological access to the city may portend a new form of democratic public space and a powerful reversal of the postwar retreat from life in public spaces.

Conclusions

With this shift to a larger more expansive sense of the public realm, a new reading of the city is emerging as our individual mental maps evolve and change. By no longer primarily orienting ourselves by highways or major arterials, we can increasingly think of green networks as guideways and landmarks throughout the city. This fluid idea of the landscape, where flows are more organic and seamless, following natural features like watercourses and topography, returns us, in some interesting ways, to a pre-colonial sense of the land we inhabit; the shared commons becomes less hard-edged and bounded by surveyed property divisions and functional categories.

Not only are we witnessing a new way of expanding the commons, by reconceiving the public realm, we are also inventing new forms of co-management and bringing new partners to the table. As we have seen, shared public space or common ground is the life-blood of a great city; it is far too important to be neglected or shortchanged. The redefining of public and private roles has many benefits, but just like government, civil society needs to be held accountable in clearly defined ways. Philanthropy should be seen as a complement to and stimulant of, not as a substitute for government involvement. The Bentway was intended to be a model demonstrating what is possible, in both design and programming, but also an example of how to partner with the city and inspire others to come forward with innovative projects. Critical to this type of endeavor is that the essential publicness of the space not be compromised in any way, and that accountability to the public be preserved.

24

THE SKYSCRAPER AND PUBLIC SPACE

An uneasy history and the capacity for radical reinvention

Vuk Radović

Introduction

This essay focuses on the uneasy relationship between public space and contemporary skyscrapers. Developers and owners of contemporary skyscrapers have often promised the equivalent quality of truly publicly owned spaces – to be accessible and open to all –through various forms such as gardens, plazas, lookouts and so on. Yet the pseudo-public reality offered by the much-lauded Privately Owned Public Spaces (POPS) falls short of expectations. This essay presents an analysis of the idea of territoriality within the edifice of the skyscraper in several paradigmatic examples, which illustrate the origins of concept of POPS in skyscraper design, current trends, and the ideas about future. The emphasis is on the positioning of Privately Owned Public Spaces within skyscrapers themselves, where they are usually relegated to the periphery of the building. This periphery gets conceptualized as an externality to the building itself, as internality to the site and externality to the structures in the form of public central courtyards or even internal to the structures (as in the case of the Rockefeller Center, Manhattan), yet critically segregated by peripheral positioning at the ground floor (the Leadenhall Building, London) and the top floor (20 Fenchurch Street Tower, London). Common to all of these externalities is hard control over highly regulated, pseudo-, and semi-public space, which questions the use of the term public in describing them.

For the discussion that follows it is important to emphasize the physical nature of public space, which is always an arena for diverse manifestations of the public discourse, the realm of free thought, expression, and association. From the foundational thought of the Greeks to Hannah Arendt, Jane Jacobs, Jürgen Habermas, Bill Hillier, and Judith Butler (to name a few), urban and political theorists of various orientations stress the constructed aspects of public space – as streets, squares, parks, spaces between buildings, spaces of buildings and functions etc. – in which the full richness of public life can unfold. Being simultaneously socially defined and defining, urban public spaces inevitably conform to concrete rules and regulations that keep the interests of the people firmly above and beyond any interests, private or corporate. Central to public life and public space remains the quality which Lefebvre et al. (1996) postulated as *the right to the city* but, as Mitchell (2003) argued, "just what public space is – and who has the right to it – is rarely clear, and certainly cannot be established in the abstract … " (5).

Public space enshrines the fundamental, ancient notion of public good. While that concept is an ideal, the experience of public quality has to be real, constantly re-defined, and defended. Since the 1970s, with rising awareness of an environmental crisis, definitions of public good started to include demands for environmental and cultural sustainability, which in 2015 became sanctioned as official United Nations Sustainable Development Goals.

What matters here are the physicality, vulnerability, and that need for regular redefinition of the fundamental and intertwining values that underpin the concepts of what true public space is, and what it ultimately is not. The definition and expressions of those concepts have to be as alive as those concepts are, to account for fundamental continuity and perpetual change within forces shaping them. New forms of urban life seek new expressions, which need to be placed in positive relationship to the time-honored values of urbanity. That is why, rather than seeking a blueprint or some prescriptive advice about future of public space within already present and emerging urban realities, this essay attempts to elucidate the nexus of physical public space and the conceptual framework of publicness by focusing on what is, perhaps, the most rapidly evolving architectural type of contemporary era, the skyscraper.

The idea of the skyscraper

In order to understand the historically antithetical relationship between the ideal of public space and the skyscraper as development, we need to examine the role which public space has had in formation of that architectural type.

An early visualization of the concept of a skyscraper, as famously illustrated by A.B. White in the 1909 Real Estate issue of LIFE magazine (Figure 24.1) explains the origins of the idea of modern vertical settlement (Koolhaas 1978). White presented a structure of columns and platforms, repeating "land" subdivision, building types and form which were common at the time. That structure barely relates to what was to become a skyscraper.

Such conceptualization of life above the ground was a direct response to the non-urban reality and the firm framework of values within which the developer strived to innovate. *The Commissioners' Plan* of 1811 was, simply, a subdivision of the island of Manhattan above Houston Street up to what is now the 155th Street. The grid here expressed colonization of the New World through urban means, the appropriation by an imposed "neutrality" of the Cartesian urban form, making land reducible to financial transactions, to money. In order to prevent chaotic expansion of the new wealth into the hinterland, a homogenous and rigid pattern of allotments was laid out over the island (even beyond the geographical limits of what has been known and mapped at that time). The subsequent Plan of 1818 was, again, the physical manifestation of values of European Enlightenment, thus relating to the ideals of democracy, private property, and equality. Through its understanding of native land as *tabula rasa*, that plan is also the most abhorrent tool for colonial expansion (Highmore 2006). In this concept of space, from the outset there was no place for the notion of "public good," and by extension that of "public space." Those values reserved for a select "public" only. One could argue that the exterritorial character of the skyscraper was an unavoidable and truly American projection of society, aimed exclusively at the entitled citizens only.

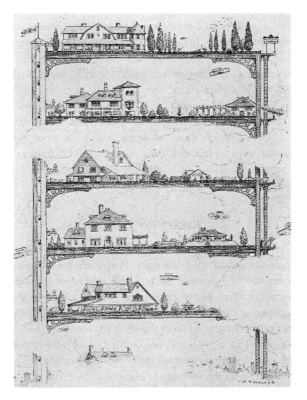

Figure 24.1 The open-space skyscraper concept

Image credit: A.B. White, 1909 Real Estate issue of LIFE magazine

A.B. White's idea of how to circumvent the orthodoxy of the ground plane, for profit, of course, was a vertical subdivision of the air above the grid, establishment of a rectilinear, shelf like structure which only replicated the idealized vision of the New World of his era. Rather than imaging the new type as an enclosed object more akin to traditional European towers, this new skyscraper was a multiplication of rural open space, with it all the trappings of an idealized rural life, such as meadows, orchards, and homesteads. Although, predictably never constructed, this utopian concept provides an intriguing starting point for the analysis of tenuous relationship between what would be a thoroughly private object and the idea of open space, open yet not public domain in nature.

The reality of the skyscraper

The physical reality of skyscrapers as we know them today is one of territorial integrity of the object, with potential for added externalities. The examples of interactions between the two that follow will help open discussion about the dialectics between public and private interests in contemporary urbanism, within the limitations of our theme. They illustrate the evolving meaning associated with the syntagma "public space" in the context of skyscrapers and provide a starting point for discussion of the idea of public inside a decidedly architectural space. Not surprisingly, the examples which best cater for that purpose are of Anglo-American provenance, as both the architectural type (skyscraper) and the ideological framework which introduced it and imposes it globally (neoliberal capitalism) have also originated there.

Historically, skyscraper developments have externalized public space. Such, externalized spaces, for which developers claim the public quality, tend to be privately owned. Although conceptually created much earlier, since the 2000s the acronym POPS, also coined in New York, was promoted, and imposed as a "solution" to total commercialization, portrayed as an inevitable new urban condition as with a lack of publicness and a limit to public access (Kayden 2000). While in social terms, with some good will and positive spin, one could argue that these spaces still seek their socio-cultural definition, their physical positioning within the project challenges such attitude. POPS tend to be located at the periphery of architectural objects.

Rockefeller center

The Rockefeller Center in Manhattan (designed by Raymond Hood in 1939) provides a paradigmatic example needed to open this story. Rather than developing the entirety of the site, the developer "donated" a portion of the buildable land towards public use and skirted the perimeter with skyscrapers. The whole of the site, which pans several of Manhattan's city blocks, has two main open spaces; a) a pedestrian thoroughfare which dissects the site east/west; and b) an open plaza underneath the 30 Rock Building. At the time, this type of urban addition within a purely commercial site was unique, unlike what could be seen elsewhere within Manhattan. However, this pedestrianized zone was firstly and foremostly intended as an extension of the shopping environment, with the promenade originally conceived as an artery to funnel pedestrians towards the interior of the site. The sunken plaza, originally intended to house the Metropolitan Opera Company building, was never constructed. The remaining gap between the buildings was retained primarily due to commercial considerations, and as a publicly accessible open space only later (Adams 1985). The Rockefeller Center provided a much-copied model where public space is only a means of bringing pedestrians in, towards shopfronts and offering resting areas, spaces for exercise and seating. The space labeled "public" is peripheral to the architectural objects. It is conceived to invite the public to the interiors of the ground plane, whilst acting as a barrier to the shield private structures.

Walkie-talkie and Cheesegrater

A recent trend in newly constructed skyscrapers has been to place "semi-public" spaces and POPS within the physical boundaries of the architectural objects themselves, conforming to local public space laws whilst retaining as much of the possible building footprint for commercial area. Two exemplary cases can be found in London, only a few hundred meters apart. 20 Fenchurch Street Building and 122 Leadenhall Street Tower – colloquially, because of their respective silhouettes, referred to as the Walkie-Talkie and the Cheesegrater. The designers and developers of those skyscrapers approached the integration of public amenity within their physical boundaries, in two diametrically opposite ways.

The Leadenhall (Cheesegrater) building, designed by Roger Stirk Harbour + Partners, had a large portion of the ground floor footprint entirely open, devoid of enclosed spaces with only several sets of escalators and large columns interfering with public thoroughfare. The darkened area (see Figure 24.2) in the underbelly of the tower represents the forecourt area. The building itself is designed as a horizontally extruded right-angle triangle, with the top tapering so as to allow privileged (and, in public interest legally protected) views of

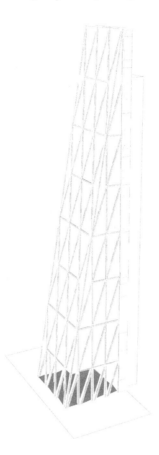

Figure 24.2 Leadenhall Cheesgrater building by Roger Stirk Harbour + Partners (2013)
Image credit: Vuk Radović

St. Paul's cathedral (Wood 2015). The frontage facing Leadenhall St. is minimal, with a cavernous overhang providing shelter to the forecourt.

Ultimately, this intervention is precisely that is what that external demarcation of a forecourt as public space is all about. The space itself contains an elongated stretch of concrete public seating to the perimeters of the site, with several trees planted around those benches, with a hard-surfaced entry and bollards preventing vehicular access. Although seemingly generous by giving whole ground-floor away to public, this type of insertion of public space within a private structure still adheres to the principle of externality. However, morphing into an addition-by-subtraction by "sacrificing" an internal space provides open external space. The separation is still placed at the decidedly hard edge, segregating the outside from the inside and, consequently the private from the public. The formal authoritative structures of private interest dominate the public. The rules and regulations defined under the authority of the private and imposed over the public in the literal and functional sense.

The Walkie-Talkie in 20 Fenchurch Street was opened a few months prior to the Leandehall tower. In this case, the designer, Rafael Viñoly approaches the partial integration of public space within the objective space of the private by allowing a "public" viewing gallery, referred to as the Sky Garden in promotional advertising of the skyscraper to the general public, at the summit of the commercial building (see Figure 24.3 – darkened area). Due to the shape of the skyscraper, that of an unusual floorplate expanding from bottom to top, the volume also nominally allows the largest floorspace to be designated for the viewing gallery. While the platform is an enclosed space within the edifice of the building, it operates through its connection via a lift available to public. Thus, similarly to the Leadenhall Building, the publicly accessible space sits within the planar confines of the private space. It remains segregated with the private owner providing the rules and regulations governing its function, whatever they may be (Shenker 2017). The architects explain this inversion of public space from its rootedness to the urban fabric of the city as an innate desire for humans to ascend, and thus to provide the public a free of charge experience of otherwise privileged views or "otherworldliness." The interior of the "Sky Garden" is an amalgamation of public seating areas, green planted zones, and publicly accessible spaces, such as restaurants. However, security reasons are cited for the need to pre-book place in the restaurant or the "public" space itself. While the justification for design concept came from a noble idea to challenge the socioeconomic stratification of cities, particularly the likes of London, by literally raising the public to the highest strata of the building, the actual quality has been reduced, predictably at the cost of truly public space (Viñoly et al. 2015).

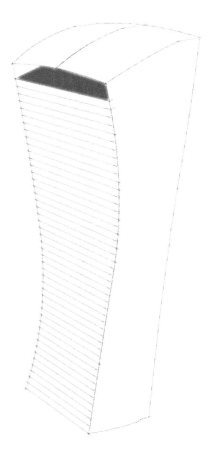

Figure 24.3 20 Fenchurch Street (Walkie-Talkie) building by Rafael Viñoly (2014)
Image credit: Vuk Radović

As usually, the final realization of the public space has differed drastically from the grand architectural vision. The Sky Garden upon completion was not free to the public, but rather required an entrance fee. That already happened to 30 Street Mary Axe (designed 2003 by Foster and Partners), affectionately known as the Gherkin, which now hosts a restaurant at its previously labelled public space (Reike 2015). These ongoing controversies associated with Privately Owned "Public" Space show a clear pattern, where the idea of public accessibility dominates the concept and design, then gets pushed to periphery during the design development phase, to eventually become completely compromised through management based on restrictions. Those practices are by no means limited to London, although the examples listed above have a secondary concerning attribute. The approval of these towers is contingent on the developers providing public space "back" to the city, and thus by ultimately requalifying public space to POPS, to semi-public, to potentially even private, should not be viewed as an act of kindness rescinded, but rather as a subtraction of required public space. A cynical view at reoccurrence of such practice suggests that it could well be the part of comprehensive business planning.

The Vessel

The most recent of example of privately owned public space within the context of skyscraper developments to be mentioned here can, yet again, be found in Manhattan. The newly constructed Hudson Yards complex (2019) has been described as the most expensive site in the world, houses a unique structure within its sprawling outdoors space. Named *the Vessel*, British designer Thomas Heatherwick designed a tower and lookout which is part sculpture, part public viewing gallery and, supposedly, part public space. The initial concept was for a set of intertwined staircases leading upwards, to nowhere, particular with the act of climbing providing the general public a unique view of the Hudson River. Heatherwick described the structure as a "three-dimensional public space, like a park, but taller" in an attempt to infuse an architectural object with a public good (Schwab 2019). Initially the structure was to be completely public, in the sense that although owned by the developers could be scaled by all.

As has been the trend with most public offerings within private structures the end result is very different to that initial idea. Although free, an appointment must be made with the agency operating the structure prior to visiting and is subject to a lengthy terms of service agreement which stipulates that any photographs, videos, or media collected of the project remains the property of ERY Vessel LLC (Hudson Yards 2019). This clause, a thoroughly contemporary incarnation reflective of the social media era where images are much more likely to be seen through mobile apps rather than printed media, is yet another form of soft power being exerted over the general public visiting a place that has been marketed as public, yet has a purely commercial role.

Likely futures and desirable futures

As frequently argued by David Harvey (2005; 2006) the above examples and indeed most global examples featuring a public/private interface predictably favor the private over public interest, not only in terms of the disproportionate share of the risks and profit, but in terms of the essence of the idea of public, Lefebvre's right to the city, the right to appropriation, which should not be confused with ownership.

Within our narrow focus on a specific type and its constraints, we have explained how all attempts towards integrating public space within the confines of the private object remain tokenistic at best, not only in terms of its spatial relegation to the periphery. That periphery can be the external ground plane, the undercroft, or the crown, yet nonetheless – the periphery. That limits the potential impacts of "public" as the set of values on the object itself to an absolute minimum. As long as thinking continues to be within the existing paradigm, the primacy of private over public and banal tokenism in design and definitely in reality are the most likely outcomes.

But, future directions in the integration of public space within skyscrapers or any suitably sized residential buildings can challenge this false dichotomy, and the perceived inevitability of separation of the private object from the public realm. Several contemporary theorists and practitioners have focused specifically on this lack of integration, also powering their critique on the imperatives of environmental and cultural sustainability. These conceptual projects tend to refer to eco- or green-skyscrapers, or an integration of these buildings into the broader, renewed garden-city movement (Bingham-Hall and WOHA Architects 2016).

The theoretical basis for claims that skyscrapers need to be sustainable are based on the well-grounded proposition that in the contemporary world sustainability is more than simply ecological, but rather it requires the careful balancing of economic, cultural, and ecological requirements (Elkington 1998). Only when combined in a coordinated and locally adequate manner, those measures provide for whatever a "truly" sustainable outcome might be. The economic viability of skyscrapers is rarely questioned. The result of economic non-viability, particularly in the private, commercial sense, would result in the building simply not being constructed. Recently the secondary, ecological requirement has been successfully adopted. That was often in the form of technological and material advancements in the building fabric, under the moniker smart-buildings and smart-cities. However, the aspect of culturally inclusive design has been much harder to attain, as can be observed in the general "sameness" of skyscrapers, irrespective of their geographical location. In commercial skyscraper districts, be they in Canary Wharf in London, Lower Manhattan, Dubai or even tropical Asia, configurations and aesthetics have more in common with one another than their cultural surroundings.

If we assume that the quality public realm is inherently interwoven with the specific "public" of the urban, then the culturally sustainable buildings require an integration of the public realm within the skyscrapers themselves. If the right to the city is the right to that "public," then it includes the right to the complex of issues that frame the broadly understood sustainability. This is exactly the point Ken Yeang (2002) approaches when calling for the new typology of skyscrapers to be linked to urban design, rather than object-based architecture, although his book could easily be described as suggesting insertion of public space as the solution to the cultural sustainability of skyscrapers. Yeang's major critique of the prevalent architectural design of modern skyscrapers is that the architects tend to design a floor-plate which yields the largest financial returns to the developer as a primary consideration, and they simply multiply this "successful" floor-plate indefinitely with little to no variance. He suggests that this type of design is not responsive to the requirements needed for the buildings inhabitants and therefore provides a suboptimal, inhuman solution. His radical questioning of the current trends in skyscraper design focus on several key observations related to public space; a) a need for continuity between and distinguishability of private and public space within skyscrapers, b) a legibility, both of identity and character of public space within skyscrapers, and c) adaptability over time embedded in design (Yeang 2002).

Yeang's approach demands redefinition, from architectural singularities towards urban multiplicities, and it is not limited to the insertion of public space within the towers. The separation of the inhabitants of high-rise residential skyscrapers from the ground plane, and thus the public realm of the city, is of major concern, as their lives are also segregated into discrete public and private components more so than their compatriots living at lower levels, and thus more immersed in their surroundings. Sky-courts, plazas, and streets-in-the-sky are among the proposed solutions to mitigate the alienation of the occupants from public life and each other. Integration of the city and skyscraper is conceived in the way that public spaces within these eco-skyscrapers are not intended as accessible only to the residents (and, thus only pseudo public), but rather as required components of a larger urban strategy of their home cities. One of Yeang's (2002) most radical ideas is the insertion of readapted quality public space in major metropolises, such as New York's Central Park, as green "lungs" of the towers, providing air circulation, pollution reduction, and so on. His premise is that in vertical urbanism "urban design concepts meant for the horizontal plane will be flipped to a high-rise condition and reinterpreted as vertical propositions" (32). Similarities between Yeang and A.B. White's imagining of a future skyscraper are inescapable. Both of them advocate a re-insertion of already seen possibilities within their newly conceived constructs.

Yeang, while providing some critical steps in the right direction, is not radical enough in this specific interpretation of what quality public space in skyscrapers could and should be. His theory remains fundamentally limited to an architectural top down vision, where an enlightened expert or singular designer manifests his/her own vision of an ideal public space, which materializes as a "solution" to vertical spaces. This does not take into consideration that true public space is never limited to grand gestures within cities but demands multiplicity of actions by active citizens. Central Park in New York does function as a reprieve for New Yorkers from the concrete jungle that surrounds it, but cities also need fine-grained, spontaneous, and irregular pockets of public space. Precisely those acts are never to be seen in skyscraper spaces designated as "public."

Conclusions

For true small-scale public space to be achieved a radical rethinking of skyscrapers is required. In order to achieve this spontaneity of the public realm, an antithesis to traditional and conceptually conservative, unimaginative skyscrapers needs to be established. Under the rubric of vertical urbanism, the *urbanism* component must be promoted to equal footing as the *vertical* – which assumes an object. In quantitative terms skyscrapers, especially very large ones, have a potential which their smaller siblings do not possess. Long ago, architect Rem Koolhaas referred to this as the attribute of *bigness*, something that can be achieved only with scale (Koolhaas and Mau 1995). This bigness provides skyscrapers beyond certain size to become conceptually "more" than just larger version of their smaller selves. The research that underpins this essay suggests that bigness itself may also provide the ground for reinterpreting public realm, as the potential for true *urbanism* in the context of skyscrapers.

As suggested at the opening of this essay, in the way urban life itself does so the ideals of public space and sustainable development demand regular redefinition of underpinning values, balancing their fundamental continuity and perpetual change. Public space should not be considered as a welcome addition, or an excuse for otherwise unsustainable buildings, but rather an integral component and the new force that challenges their very being. Supertall residential skyscraper buildings of immense height (officially defined as taller than 300 meters), that sometime house thousands of occupants, have the potential to transition from architectural singularity into public multiplicity. However, this can only be achieved through their conceptual re-interpretation as cities, towns, or communities. Successful cities are seldom top-down designed, never by a single designer. They have morphed over time to become what they are and what they might become. Self-governance, in the form of city councils and city ordinances, seems to hold be key aspect of this transformation, along with the thorny relationship between ownership and citizens' rights.

Further investigations into possibilities of such transition aims at re-evaluating skyscrapers above certain heights and numbers of occupants into discrete urban wholes within broader urban conglomerations. If we understand the scalar jumps of urban morphology within contemporary cities as being; a) building, b) village, c) city and d) region, then we can see how the *bigness* of skyscrapers places them between the scale of building and village (Cataldi 2018).

Perhaps a qualitative jump from understanding these skyscrapers as very large buildings towards treating them as parts of the *village* subtype hold the potential for the emergence of true urbanism, including new manifestations of public space. That is where the right to the city could be established, precisely as "the right to the *oeuvre* (participation) and appropriation (not to be confused with property but use value) was implied in the right to the city" (Lefebvre, Kofman, and Lebas 1996).

References

Adams, J. (1985) *Rockefeller Center Designation Report*. In: M. Pearson (ed). The City of New York: New York City Landmarks Preservation Commission.

Bingham-Hall, P. and WOHA Architects. (2016) *Garden City Mega City: Rethinking Cities for the Age of Global Warming*. Singapore: Pesaro Publishing.

Cataldi, G. (2018) Towards a General Theory of Urban Morphology: The Type-Morphological Theory. In: *Teaching Urban Morphology*, V.T. Oliveira. (ed). New York: Springer, pp. 65–79.

Elkington, J. (1998) *Cannibals with Forks: The Triple Bottom Line of 21st Century Business*. Oxford: Capstone.

Harvey, D. (2005) *A Brief History of Neoliberalism*. Oxford: Oxford University Press.

Harvey, D. (2006) Right to the City. In: *Divided Cities*, R. Scholar. (ed). Oxford: Oxford University Press, pp. 83–104.

Highmore, B. (2006) *Michel De Certeau: Analyzing Culture*. New York: Continuum.

Hudson Yards (2019) "Vessel Terms and Conditions." Hudson Yards New York. Accessed 1 January 2019. www.hudsonyardsnewyork.com/discover/vessel/terms-conditions.

Kayden, J. S. (2000) The New York City Department of City Planning and the Municipal Art Society of New York. *Privately Owned Public Space: The New York City Experience*. New York: John Wiley & Sons, Inc, p. 3.

Koolhaas, R. (1978) *Delirious New York: A Retroactive Manifesto for Manhattan*. Oxford: Oxford University Press.

Koolhaas, R., B. Mau and Office for Metropolitan Architecture. (1995) *S, M, L, Xl*. New York: Monacelli Press.

Lefebvre, H., E. Kofman and E. Lebas. (1996) *Writings on Cities*. Cambridge: Blackwell.

Mitchell, D. (2003) *The Right to the City: Social Justice and the Fight for Public Space*. New York: Guilford Press.

Reike, S. (2015) The Promise of Public Realm: Urban Space in the Skyscraper City. In: *Global Interchanges: Resurgence of the Skyscraper City*, Chicago: The Council on Tall Buildings and Urban Habitat, pp. 246–251. https://static1.squarespace.com/static/599422ff2994cae8c28857b0/t/5a1bfd2af9619afa6a6ce376/1511783727577/2466-the-promise-of-public-realm-urban-spaces-in-the-skyscraper-city.pdf

Schwab, K. (2019). "Why Everyone Hates the Vessel." *Fast Company*. 29 March. www.fastcompany.com/90326416/why-everyone-hates-the-vessel 2019.

Shenker, J. (2017) "Revealed: The Insidious Creep of Pseudo-Public Space in London." *The Guardian*. 24 July. www.theguardian.com/cities/2017/jul/24/revealed-pseudo-public-space-pops-london-investigation-map.

Viñoly, R., C. Blomberg and M. Blanes. (2015) Challenges and Benefits of Integrating Public Space into Tall Buildings. In: *Global Interchanges: Resurgence of the Skyscraper*, Chicago: The Council on Tall Buildings and Urban Habitat, pp. 282–287. https://global.ctbuh.org/resources/papers/download/2471-challenges-and-benefits-of-integrating-public-space-into-tall-buildings.pdf

Wood, A. (2015) Rethinking the Skyscraper in the Ecological Age: Design Principles for a New High-Rise Vernacular. *International Journal of High-Rise Buildings*, 4(2): 91–101.

Yeang, K. (2002) *Reinventing the Skyscraper: A Vertical Theory of Urban Design*. West Sussex: Wiley-Academy.

25

IN PURSUIT OF INCLUSIVE SPACES?

Memory, monuments, and the politics of public story telling

Renia Ehrenfeucht

Introduction

On June 17, 2015, a young white American man entered the Emanuel African Methodist Episcopal Church in Charleston. He opened fire on the African American pastor and parishioners, murdering nine people and injuring others. This racially motivated violence renewed efforts to remove the monuments commemorating men who fought for the Confederacy, the Southern US States, which were fighting to uphold slavery during the American Civil War (1861–1865). Although located in public spaces, most of these monuments were placed by community groups a generation or more after the Civil War when white Americans were fighting to reassert white social and political power. To contemporary activists, the public monuments perpetuated white supremacy which led to the 2015 massacre. Others however argued they were historical symbols, representing complex events that should be remembered if not celebrated.

Public spaces are an intricate interplay of diverse activities, practices, and ways of being. Debates over monuments speak to the ongoing, and important, role that public spaces play in conveying societal values, histories, and collective memories. Monuments materialize societal stories through commemorating a person or event. While a community's history and future aspirations are inscribed in space in varied ways—for instance, when and how communities were built and how they have changed by daily practices and adaptions—monuments are unique in the way they intentionally tell public stories and tell stories publically. They, along with the names of streets, parks, and public buildings, enact shared histories. Unlike everyday practices and adaptations, monuments and memorials are symbolic and tangible representations of these histories. The monuments simultaneously draw attention to and produce specific narratives about them.

In diverse and inequitable societies, however, residents reach no consensus about historical meaning, and instead, different people advocate for representations that reflect their values and experiences. A critical challenge to the future of public memorialization is how to represent conflicting stories. In the case of the debates over removing the

Confederate monuments, Dell Upton (2017) has argued that "This is not ultimately a conflict over monuments. It is a conflict over the values that we wish to endorse in the contemporary public realm." Memorialization continues to be important because it recognizes shared experiences, which is a necessary step towards building inclusive societies. This includes acknowledging different groups' contributions and reparations for past wrongs. Monuments as public acknowledgement can be a form of symbolic reparations (Marschall 2010).

To capture the experiences of increasingly diverse societies, public spaces need a cacophony, the multiplicity of voices and experiences, with dissonance as well as harmony. Yet not all perspectives have equal validity because to work towards more just societies, such public actions must further the project of dismantling structures that perpetuate injustice. In doing this, no societal myth or produced historical truth can be sacred or unable to be revised or rejected. Instead, the questions might be why this monument in this place at this time? For what purpose? In whose interests?

The following sections explore the critical fissures in public memorialization in order to envision a politics of memory for just societies. Inclusive public spaces are paramount to this project. Changes in monuments, their removal, their content and form, reflect social change (Stevens and Sumartojo 2018; Croegaert forthcoming). Yet the monuments also make tensions visible, highlighting how communities and societies have attempted reconcile deeply irreconcilable perspectives about their histories, people, and events—whether reflected in extant monuments or under discussion for new ones—with little resolve. These debates continuously produce and remake the public realm.

Public unities

Monuments and memorials have numerous purposes, to assert power, commemorate loss, recognize the contributions of particular groups or individuals, and draw attention to a phenomenon. Governments, grassroots groups, artists, family, friends, and activists install monuments. Most memorials are installed with permission but some are unauthorized. Regardless, when enacted in or visible to public spaces, monuments take on public significance and their meanings are interpreted and reinterpreted by the people who experience them.

National governments have traditionally created memorial destinations in prominent public spaces to commemorate political leaders and political histories represented by events such as war. In capital cities, these memorial landscapes affirm state power and can help uphold the government as an institution. The National Mall and Memorial Parks in Washington, for example, is a grand open space stretching from the U.S. Capitol to the Potomac River, with the presidential residence, the White House, to the north. The monuments memorialize presidents George Washington, Abraham Lincoln, and Franklin Delano Roosevelt, the veterans of the Korean War, the Vietnam War, and World War II as well as civil rights icon Dr. Martin Luther King, Jr. Millions visit each year "to commemorate presidential legacies, to honor our nation's veterans, to make their voices heard, and to celebrate our nation's commitment to freedom and equality" in the words of the National Park Service (National Park Service (NPS) 2019).

Even within this traditional form and location, the narrative that upholds the country's myths change. Monuments to Japanese American veterans in World War II and African American veterans in the Civil War take their places alongside the founding fathers who upheld slavery and the President who allowed Japanese Americans to be relocated to internment camps. Since the National Mall and Memorial Parks is a site of nation building, the monuments that implicitly recognize injustice do so within a narrative of progress rather than accountability.

Monuments to individuals embody contemporary stories and values that reflect the time they are enacted. The Franklin Delano Roosevelt monument in Washington, DC. exemplifies how representing the president's legacy came to embody politics that Roosevelt did not embrace. In January 2001, landscape architect Lawrence Halprin's monument to the president was unveiled. The memorial has two statues of the president, one in which he is seated in a wheelchair. Throughout his presidency, President Roosevelt could not walk without assistance because he had contracted poliomyelitis as an adult. Despite his reliance on canes, walkers, and wheelchairs, his public persona showed him standing or seated in ways to hide his disability and only once, six weeks before his death, did he appear before the U.S. Congress in a wheelchair. The monument, the largest in the memorial park, took decades to build, and after two unsuccessful attempts, Halprin designed as series of spaces that depicted different dimensions of Roosevelt's presidency including for example people representing the defining period of the Great Depression. Halprin depicted the president seated. When the designs became public, however, disability rights activists argued that the monument obscured the president's disability. Their activism led to the second statue of the president in a wheelchair (Stein 2004). While representing a historical reality, the monument also indirectly commemorates the influence of disability rights movement.

While national monuments can add elements to commemorate a wider range of experiences, can they contribute to reconciling internal social and political divisions associated with traumatic histories in unequal societies? To this end, national governments have attempted to build monuments as spaces of reflection in order to pay tribute and heal. After the 1994 elections that ended South Africa's apartheid regime, numerous new memory sites and monuments were installed to preserve the memory of both resistance and repression, to honor the victims, and to attempt to foster healing and reconciliation (Marschall 2010). Established in response to the findings of the Truth and Reconciliation Committee, Freedom Park in Pretoria symbolizes the transition to a democratic government from the apartheid regime. It intended to be an official site for reconciliation and unity. Its usable public space offers event areas like the amphitheater and places for reflection. Its explicit narrative is inclusive, democratic, and venerates centuries of fighting for what is right, which is embodied in a wall of names. Attempting to represent everyone who was sacrificed for South Africa, it includes the colonial wars, World War I and World War II, and the struggle against apartheid. As a heritage destination, it interprets the country's history as a path to the future (Marschall 2010).

Yet, unified visions do not exist. South Africa's Freedom Park's wall names anti-apartheid activists, recognizing that black South Africans actively worked towards the contemporary democracy rather being passive victims in an unjust regime. At the same time, a general wall that commemorates everyone who fought for freedom is an attempt at unity that avoids blame or responsibility. At least in the immediate period after its creation, many were unsatisfied for different reasons. Some of the white minority disparaged the memorials or consider them black Southern African monuments. Some black South Africans disagreed with the official interpretations of history that attempted to neutralize the actors that perpetuated the discriminatory and violent regime (Marschall 2010).

In the 1990s, after the Pinochet dictatorship (1973–1990), Chile's newly elected democratic government used memorials to mark a transition to democracy. These sites were associated with the dictatorship's execution, torture, and disappearance of thousands of people.

During Pinochet's regime there were over 82,000 documented political arrests and hundreds of thousands of people were forced into or chose exile (Lerer 2016). Two important monuments are the Santiago General National Cemetery and the "Peace Park," built in 1995 on the site of Villa Grimaldi, a former torture center on the outskirts of Santiago (Marschall 2010; Lerer 2016). The sites include monuments that list names of those who were murdered and disappeared. No national consensus has developed about Pinochet's regime, however, and the monuments acknowledge the victims, but do not call out perpetrators. As Lerer (2016) has shown, national monuments recognizing human rights atrocities while seeking domestic reconciliation use design techniques such as minimalist and abstract style to name the victims. While recognizing the harm, the design, and framing hold no one accountable. In the perspective of a national state, this might be necessary to promote reconciliation. At the same time, it denies justice to the survivors, families, and those that were killed and tortured, and it fails to acknowledge that in these cases, ordinary people, not only the leaders, perpetuated injustice. These memory sites intend to both reconcile and say never again, but do little to show how such atrocities came into being.

No monument can be a neutral representation of the present or past. Michael Warner's definition of a public can be useful to move beyond the notion that any one story can represent a society. He defines a public as a self-organized entity based in relations among strangers (Warner 2002). Publics are constituted by attention as people articulate commonalities and differences and understand themselves to be part of a whole. Members of given publics can invoke the idea of the public, a generalized social totality of people, as a way to normalize specific visions and further particular objectives. However, no singular public could develop in cities with people who are differently situated and have varied priorities and values. No singular interpretation can encapsulate the complex responses in diverse and unequal societies. Governmental perspectives whether national pride or unity are also the perspective of a public, one perspective on the nation, rather than a multiplicity of experiences. National perspectives can be both aspirational and concealing, and the monuments produce history by framing stories for future generations after those who were directly affected stop speaking.

Public multiplicities

Diverse perspectives and complex public processes make it increasingly difficult for local groups to install public memorials. In the last decades, countermemorials and grassroots monuments, many of which are placed without authorization or allowed as temporary installations, have been one response. Vernacular monuments often respond to painful pasts in places associated with the memorial subject and in ways that people cannot avoid confronting them.

Hundreds of such memorials have recognized car accidents, epidemics such as the heroine crisis, and other situations that elicit mourning and anger (Margry and Sánchez-Carretero 2011; Stevens and Ristic 2015). Peter Margry and Christina Sánchez-Carretero (Margry and Sánchez-Carretero 2011) define "grassroots memorialization … as the process by which groups of people, imagined communities, or specific individuals bring grievances into action by creating an improvised and temporary memorial with the aim of changing or ameliorating a particular situation" (2). Grassroots memorials uplift events that might otherwise be overlooked or ignored. Countermemorials interpret a situation or event in a way that differs from the official public or governmental perspective. Cecily Harris (2010) defines the countermemorial as "characterized by a non- or anti-presentational character" that establishes a "non-hierarchical, anti-authoritative" relationship with those who interact with it (35). Vernacular monuments can be ephemeral but others remain indefinitely, and they can be—but are not always—oppositional.

Vernacular memorials differ from monumental landscapes in scale and location. As Quentin Stevens and Mijana Ristic (Stevens and Ristic 2015) have shown, pavement or sidewalk memorials in urban areas engage the passing public. Gunter Demnig's *Stumbling Blocks* is possibly the world's largest memorial with 43,000 sites in 16 countries and 1,000 German towns. Remembering the Holocaust atrocities, 10×10 cm pavement or sidewalk insets indicate the last place where deportees resided or worked, as shown in Figure 25.1. The brass plaques show the deportee's name, date of birth, arrest, deportation, and murder. Germany has accepted complicity in Nazi atrocities through numerous state sanctioned memorials such as *The Memorial to the Murdered Jews of Europe*, designed by architect Peter Eisenman and engineer Buro Happold, which is part of Berlin's state-sanctioned monumental landscape. Figure 25.2 shows the monumental scale of the 2711 concrete slabs that are reminiscent of coffins. The design has nevertheless been criticized as abstract, never directly speaking to the Holocaust or holding anyone accountable for the murders (Brody 2012). An underground space holds the names of millions of the Jews who were murdered and other information about the Holocaust, but it is separate from the monument itself. The thousands who make the pilgrimage to Berlin have a fundamentally different experience than those who stumble upon plaques as they walk through a neighborhood.

In some cases, monuments can offer a place to mourn for people who are directly affected, and symbolic memorialization also creates tangible space for remembrance.

Figure 25.1 Stumbling blocks in Chemnitz, Germany, artist Gunter Demnig
Image credit: Harry Härtel/dpa picture alliance

Figure 25.2 The Memorial to the Murdered Jews of Europe
Image credit: Alf Simon, The University of New Mexico

For decades, in Canada and the United States, First Nations and Indigenous women and girls have been murdered and gone missing. In 2014, a report published by the Royal Canadian Mounted Police acknowledged that over 1000 women had been murdered or were reported missing (Mas 2014). Public officials from the police to local governments refused to recognize that this was happening and put off the women's family and friends who demanded that their disappearances be investigated and stopped. The movement to demand attention to Missing and Murdered Indigenous Women and Girls took many forms, including temporary and permanent memorialization. One transitory act was to tie red ribbons to the Selkirk Bridge in Canada's Manitoba Province (Crabb 2017). While such recognition might appear to be non-confrontational, the power these actions hold become clear in the responses. In 2017, unknown people removed the ribbons, leading to others offering assistance in retying them. In 2014, the first permanent monument in Canada was placed, designed to represent the beauty and roughness of life, and the passing between two worlds (Canada Broadcasting Corporation (CBC) 2014). The memorial was permanent and state-sanctioned recognition of a situation that the women's family and friends had been experiencing. Other actions preceded and followed the official memorial because the situation continued.

Ghost bikes have become a widespread grassroots memorial. The cycling advocacy community recognized that calling out premature death was a way to assert injustice. The first ghost bike appeared in St. Louis in 2003, after a local bike shop owner saw a cyclist struck and killed by a car (Dobler 2011). Family, friends, or cycling organizations place the bikes.

Figure 25.3 Ghost bike in Albuquerque, 2018
Image credit: Renia Ehrenfeucht

Although each is placed in response to a specific death, the white bikes on street corners, attached to utility poles or in the medians have become a reminder of the risks that bicyclists face as much or more than a memorial to the individual who was killed. Figure 25.3 shows a ghost bike memorializing a rider who was killed by a car in Albuquerque. New York's Street Memorial Project is an organized initiative that has installed 164 ghost bikes, 54 for people for whom they had no information (Ghost Bikes, n.d.).

Ghost bikes have become recognized as memorials. When New York City's Sanitation Department proposed to include ghost bikes in its plans to remove abandoned bicycles from city streets, hundreds of people opposed the idea. In response, the city removed ghost bikes from the list, providing a form of official sanction. More than 603 ghost bikes have been documented, spanning at least 28 countries and 210 cities and communities (Ghost Bikes, n.d.).

Public memorialization is an ongoing negotiation to define priorities and construct belonging. People take space when they need it and alter it to fit their visions. Thousands of different groups take action to assert their remembrances in public. Even governmental commemorative planning is often decentralized, with the authority granted to different agencies and levels of government (Stevens and Sumartojo 2018), and this way it also parallels other dimensions of public life. Subsequent management—or the decision to leave unauthorized memorials—can also be decentralized, and falling under the jurisdiction of the agency that manages the space.

Blurred public boundaries

All actions to establish monuments produce public stories, and the lines between individual or community action and official sanction become blurred. Few who come in contact with public monuments know the context in which they are enacted, and few monuments include this information. These boundaries are also fluid because local organizations may place or advocate for a sanctioned monument that is located in public space, or change its form thereby changing its meaning. Sabine Marschall (2010) has astutely observed that:

> The ways in which memories of pain, suffering and resistance are conserved, produced, displayed, consumed and built into the collective historical consciousness reveal much about a society's process of political and societal transformation [as well as] patterns of emancipation, persistent marginalization and new hegemonies.
>
> *(362)*

An action can be directly transformed from that representing the perspective of an artist or activist to one that receives public endorsement. A case of this was another piece, the *Trail of Remembrance*, created by artist Gunter Demnig. It was a 16 km painted trail of text in Cologne ("Mai 1940–1000 Roma and Sinti") memorializing the route where 1,000 Sinti and Roma had traveled to a train station that took them to extermination camps in Poland during the Holocaust. As the paint faded, the city museum installed short brass strips with the same wording at 22 sites along the route, thereby formally claiming it as a public memorial (Stevens and Ristic 2015). The monument recognizing missing and murdered Indigenous women was also an example of community action that led to official action.

Many sculptures are located in public spaces, along the streets or in parks where people come in contact with them during other activities. Most are placed in public space as a singular art object with minimal interpretative information. When people come across monuments, their presence signifies the importance of the person or event, and through this, the memorials indicate who is welcome and whose interests are upheld, even if only minimal information can be be learned from the site itself. Community groups need official permission or a permit to site these sculptures, a public act that sanctions the community action. Either allowing or removing monuments therefore reflects an official position. The Confederate monuments referenced in the introduction were established by community groups in past eras. In 2017, the New Orleans mayor and city council removed four Confederate monuments after the City Council declared them a public nuisance. Figures 25.4 and 25.5 show the statue of Jefferson Davis, the only President of the Confederacy, and the site after removal. There was no immediate agreement how to repurpose the space. The monument removal was a controversial decision that elicited organizing on both sides, even resulting in death threats to the first contractor hired to remove the statues (Croegaert forthcoming). In 2016, the Columbus city council authorized a plaque that memorialized a site where a black man was lynched or murdered in 1923. This was requested and funded by the members of the Association of Graduate and Professional Students, but located in a public plaza, thereby becoming a public memorial (Tolbert 2017).

Installing monuments as well as allowing them to be placed or remain is an explicit process of decision making that chooses which stories become public and will be repeated and reinterpreted. Prior to the movement to take down the Confederate monuments, cities throughout the U.S. South started to install Civil Rights Movement monuments through which African Americans and their allies fought to end systemic racial injustice. Most recognizably depicted by the leader Martin Luther King, Jr., Civil Rights memorials show an alternative view of injustice by showing the struggle against it. King has been interpreted as a leader who advocated for building equality among groups rather than power for African Americans. In constrast, even though he was highly influential, Malcom X is less commonly represented because he was a controversial leader. Civil Rights leaders are commemorated in street names, statues and commemorative plaques, and public actions.

Figure 25.4 Jefferson Davis, New Orleans, 2016
Image credit: Ana Croegaert, The University of New Orleans

Ana Croegaert (forthcoming) has shown in the case of the Confederate monument removal in New Orleans, the debates were not only about monuments but about the legacies of racial injustice that resulted in poor quality schools, unequal access to jobs, and neighborhood disinvestment. In 2018, when the National Memorial for Peace and

Figure 25.5 The site where Jefferson Davis stood

Image credit: Ana Croegaert, The University of New Orleans

Justice opened in Montgomery, it was the first memorial dedicated to the legacy of enslaved black people and deliberate acts of terror and humiliation against African Americans and other people of color. According to the Equal Justice Initiative, it is a memorial to the practice and legacy of lynching to create a place to reflect on this brutal history of racial inequality. At the same time, it brings the legacies forward by linking them to contemporary racial inequity in the U.S. criminal justice system.

Memorials or removing monuments do not substitute for other forms of reparations or changes to the structural systems that uphold inequality. The failed promises of memorials can be both symbolic and material. In South Africa, a memorial and museum were built in Sharpeville to honor victims of the 1960 Sharpeville Massacre. Anticipated development surrounding these sites that would have benefited local communities never materialized, and subsequently residents protested the lack of service delivery and conditions, including protests and threats to boycott the commemoration ceremonies (Marschall 2010). The memorial came to represent the failed promises and black South Africans' ongoing marginalization.

Just cacophony

Enacting as well as challenging, defending, or taking down monuments is a debate over broader societal values in which publics are differently situated and public narratives are told and retold. The challenge occurs because the legacies of past oppression continue, and dismantling oppression includes holding members of privileged and repressive regimes accountable. When and why monuments are placed, their design, and when they are removed reflect the values of given people exercising power at particular times. Inclusive public memorials are varied and conflicting, and societies repeatedly make intentional decisions about which stories to tell and retell. Yet recognizing this does little to answer the questions about whose stories to tell, how to tell them, or what to do with extant memorials when perspectives and power shifts.

The memorials in this chapter speak to both efforts to seek justice and assert governmental and racial power, and they include both state-initiated and community based monuments. Monuments vary, as do their intents, from destination-oriented official interpretations to countermemorials in the pavement that become visible unexpectedly. Nevertheless, all monuments produce history and possible futures by shaping which stories are told and interpreted. They embody contemporary values that are enacted and continue to reproduce meaning. Since people have different perspectives on events or people, a remembrace might be educational or a political demand for one person but bring solace or pain to another. How to recognize, commemorate, or memorialize histories in diverse societies remains a critical question for future of public space.

References

Brody, R. (2012) "The Inadequacy of Berlin's 'Memorial to the Murdered Jews of Europe'." *The New Yorker*, 12 July. www.newyorker.com/culture/richard-brody/the-inadequacy-of-berlins-memorial-to-the-murdered-jews-of-europe

Canada Broadcasting Corporation (CBC). (2014) "Winnipeg Monument Honours Missing, Murdered Aboriginal Women." *CBC News*, 12 August. www.cbc.ca/news/canada/manitoba/winnipeg-monument-honours-missing-murdered-aboriginal-women-1.2734302

Crabb, J. (2017) "Memorial Honouring Missing and Murdered Indigenous Women Removed from Selkirk Bridge." *CBC News*, 18 May. https://winnipeg.ctvnews.ca/memorial-honouring-missing-and-murdered-indigenous-women-removed-from-selkirk-bridge-1.3420353

Croegaert, A. (forthcoming) "Architectures of Pain: Confronting Racism through Monuments Removal in 'New' New Orleans." *City & Society*.

Dobler, R.T. (2011) Ghost Bikes: Memorialization and Protest on City Streets. In: *Grassroots Memorials: The Politics of Memorializing Traumatic Death*, Margry, P.J. and Sánchez-Carretero, C. (eds). New York: Berghahn Books, pp. 169–187.

Ghost Bikes. (n.d.) "Home Page." *Ghostbikes.org*. http://ghostbikes.org/

Harris, C. (2010) German Memory of the Holocaust: The Emergence of Counter-Memorials. *Penn. History Review*, 17(2): 34–59.

Lerer, M. (2016) Chilean Memorials to the Disappeared: Symbolic Reparations and Strategies of Resistance. In: *A Companion to Public Art*, pp. 51–74.

Margry, P.J. and Sánchez-Carretero, C. (2011) *Grassroots Memorials: The Politics of Memorializing Traumatic Death*. New York: Berghahn Books.

Marschall, S. (2010) The Memory of Trauma and Resistance: Public Memorialization and Democracy in Post-Apartheid South Africa and Beyond. *Safundi: The Journal of South African and American Studies*, 11 (4): 361–381.

Mas, S. (2014) "Number of Murdered, Missing Aboriginal Women Surprises Top Mountie." *CBC News*. 17 May. www.cbc.ca/news/politics/number-of-murdered-missing-aboriginal-women-surprises-top-mountie-1.2645674

National Park Service (NPS). (2019) "Icons of the Nation's Capitol." *NPS*, 17 July. www.nps.gov/nama/index.htm

Stein, S. (2004) The President's Two Bodies: Stagings and Restagings of FDR and the New Deal Body Politic. *American Art*, 18(1): 32–57.

Stevens, Q. and Ristic, M. (2015) Memories Come to the Surface: Pavement Memorials in Urban Public Spaces. *Journal of Urban Design*, 20(2): 273–290.

Stevens, Q. and Sumartojo, S. (2018) Shaping Seoul's Memories: The Co-evolution of Memorials, National Identity, Democracy and Urban Space in South Korea's Capital City. *Journal of Urban Design*, 24(5): 757–777.

Tolbert, D. (2017) "Addressing Our Tortured History, One Monument at a Time." *The Huffington Post*, 22 June. www.huffingtonpost.com/entry/addressing-our-tortured-history-one-monument-at-a_us_594add15e4b062254f3a5b31

Upton, D. (2017) "Confederate Monuments and Civic Values in the Wake of Charlottesville." *Society of Architectural Historians Blog*, 13 September. www.sah.org/community/sah-blog/sah-blog/2017/09/13/confederate-monuments-and-civic-values-in-the-wake-of-charlottesville

Warner, M. (2002) *Publics and Counterpublics.* New York: Zone Books.

PART 4

Actions

Public space is not just a common good described by physical attributes and by a position in relation to other urban components. The significance of public space has to do with time too. The meaning and the role of a public space can change during the time of its activity and because of contingency. Neglected or ordinary spaces can all of a sudden become the spaces of resistance when social and cultural conditions determine the need to react and to resist (Zuccotti Park in New York, Gezi Park in Istanbul, Azadi Square in Tehran, the Hong Kong International Airport). The uses and activities in a public space can change during time due to the emergence of new phenomena or population. The internal spatial organization of public space and its relations with other spaces may form a permanent and almost immutable cluster or network (as the sequence of plazas, and galleries, and promenades in historical city centers.) Although, spaces—and relations among them—can be temporaneous, being activated by different groups of citizens during different times of the day and night, or designed by intentional actions (a pop-up intervention, an art performance), or by the demand of a group or a community to establish a space of presence and visibility.

The space in public space becomes truly public through action. In the context of the increasingly heterogeneous contemporary city, public space is no longer a product of limited and predetermined identities and meanings. As a space for myriad functions, interests, and institutions, the city is a container of multiple "publics." The actors that occupy, maintain, and control the space, and also ones that identify with the public space vary, as does their claim and power over the space. As a result, the issues related to these public spaces are varied, as are the conflicts and struggles of people who claim a right to the space and ones that are excluded. Through their actions and new expressions, numerous groups identify and make existing and new locations public. But these actions also result in conflicts with authorities and among user-groups about the claim, appropriate behavior, and use of space. The twenty-first century has seen a plethora of actions toward the making of public space. These actions have been as wide-ranging as the influences on the thinking on public space; the chapters that follow explore the different meanings of actions in public space.

In the first chapter, Jeffrey Hou explores the multiple dimensions of public space as a space of resistance and resilience, and how public space functions as a space of assemblies and amalgamation. The author argues that understanding these dimensions is critical to clarifying the significance of public space as a space of resistance and resilience in the struggle for an active and meaningful democracy.

Three key concepts of urban studies: public space, urban resistance, and urban emancipation, are revisited by Sabine Knierbein. The chapter, through consideration of practices of resistance and emancipation in public space, argues for a dialectical study of emancipatory spatial praxis in public space and changing aspects of everyday life, since it is at their interface where "the political" emerges.

The third chapter investigates the changes in symbolic attributes of public spaces and their associated meanings. Tali Hatuka suggests that the change of meanings of a public space could be modified and challenged if its representation and associated practices are not agreed by the public. In particular, during protests activists correspond with or challenge the public space narration as a means to develop a new idea or paradigm.

Rike Sitas explores the roles of pavements and public-facing art as ways to make public space more appealing, hence capable of generating more democratic engagement in cities. The examples explored in this chapter, in Ghana and South Africa, challenge normative notions of public space providing opportunity for decolonial encounter.

The role of public spaces in creating positive context of reception for immigrants is investigated by Jeremy Németh using the case of *el Raval*, a long-time immigrant arrival neighborhood in Barcelona. Support to immigrant integration by planners and designers should, according to the author, focus on creating diverse space of encounter and places where affinity groups can share concerns, claims, and interests.

Latin American public spaces that support and embody socio-cultural and political dimensions, are explored by Andrei M. Z. Cristani and Clara Irazábal using three everyday life examples on how publicness is claimed by different groups, often against disciplinarization attempts.

The concept of the Temporary City unfolds out of an understanding that the city is rooted in a four-dimensional, highly dynamic scenography. Mona El-Khafif describe typologies of temporary public spaces where citizens and public agencies cooperate in the construction of temporary projects that reshape the public realm until the "permanent" design will be implemented.

The final chapter in this part, by Troy Glover, focuses on the animation of public space, an active effort to engage in transformative placemaking initiatives that draw people to socialize, linger, and be together in public spaces. To encourage animation efforts, this chapter identifies a variety of tactics evident in current practice that consider the value and potentialities of inclusivity and belonging.

26

PUBLIC SPACE AS A SPACE OF RESISTANCE AND DEMOCRATIC RESILIENCE

Jeffrey Hou

Introduction

Public space has returned to the center stage of political struggles in recent years. From the Arab Spring and Occupy Wall Street to the Umbrella Movement, squares, streets, parks, plazas, and even privately owned public spaces have emerged with fresh identities and meanings through both organized and spontaneous citizen actions. In one instance after another, everyday places and even infrastructure sites have been transformed into highly visible spaces through mass gatherings (see Figure 26.1). With diminishing democracy in the face of neoliberalization and resurging totalitarianism, these instances highlight the important role of public space as a vehicle for resistance and democratic resilience.

As a space of resistance and actions, public space provides a stage especially for those without a voice to be seen, heard, and recognized. Through large assemblies and greater visibility, their voices and presence could accrete and gain support of others to become significant political forces. However, as the outcomes of recent events suggest, public space is also a site of continued struggles, as groups negotiate and compete for their claims, and as movements seek transformative outcomes through protests and occupation of space but often with limited success.

This chapter explores the multiple dimensions of public space as a space of resistance and resilience. Specifically, based on a review of relevant literature and cases, it explores how public space functions as a space of *assemblies and amalgamation, visibility and meanings, encounters and negotiations*, and *everyday resistance*. Through a discussion of these processes, I argue that public space as a vehicle for debates, negotiations, struggles, and critical reflections is critical to the function and resilience of democracy.

Public space and urban resistance

Cities and urban spaces have long been sites of political control as well as resistance. In *The City and the Grassroots*, Castells (1983) traces a history of urban resistance to the revolt of Castilian cities against the royal authority of Carlos V that challenged the feudal order five centuries ago, and later the Paris Commune of 1871, which inspired uprisings elsewhere. Resistance in urban

Figure 26.1 During Occupy Wall Street, Zuccotti Park was transformed from an ordinary plaza in a financial district into a site of occupation and protest

Image credit: Jeffrey Hou

spaces continued to define the history of the twentieth century, through the examples of Mahatma Gandhi's movement of non-violence in India, the Civil Rights marches in the 1950s and 1960s in the United States, the anti-globalization protests in Seattle, and countless examples elsewhere. Most, if not all, these events took place in prominent public spaces of the city, namely streets, squares, and other forms of civic spaces.

The vast literature of resistance and social movements have explored factors and processes that shape revolutions and social movements, including the framework of political opportunity structure, mobilization, and framing (Mcadam, Mccarty, and Zald 1996). However, the role of space, and particularly public space, has arguably been missing or remaining marginal. The recent rise of public space literature focusing on the agency and subjectivity of citizens and collectives represents a refreshing departure from predominant concerns over surveillance, security, and militarization, as well as gentrification and privatization (Davis 1992; Sorkin 1992; Smith 1992; Boyer 1992; Kohn 2004). By examining the everyday, lived aspect of urban spaces, insurgent appropriation and adaptations, and other short-term transformation under the rubrics of DIY, tactical, and temporary urbanisms, these discussions point to the role of individuals and collectives in shaping and reinventing public space, although not all have addressed the dimension of political resistance (Crawford 1999; Franck and Stevens 2007; Hou 2010; Bishop and Williams 2012; Iveson 2013; Douglas 2014).

The events of 2011, namely the Arab Spring, Occupy Wall Street, and numerous anti-austerity protests around the world, represent a recent historical watershed not only in the respective geographical and political contexts, but also in discourses regarding the relationships between public space and political resistance. Building on the earlier work of Hardt and Negri (2000), Mitchell (2003), Mayor (2009), and many others, as well as a renewed interest in Lefebvre's concept of the right to the city, this growing body of thoughts and observations has greatly expanded what we know about the specific relationships between resistance and public space. This chapter explores this current literature to identify specific ways public space functions as a space of resistance and how these functions contribute to the resilience of liberal democracy as a political system and a lived practice.

Public space and democracy

Democracy as a political system and practice commonly refers to a system of governance by members of a society. Through representation and direct participation, a democracy becomes accountable to the public. But as members of society can hold different values, identities, and beliefs, democracy also requires deliberation and debates. Public space, at least in the Western tradition, is historically where such processes take place and how the "public" as members of the society holds the state or the government accountable. In her seminal work, *The Democratic Paradox*, political theorist Chantal Mouffe (2000) brings such dynamic further by arguing that "[a] well-functioning democracy calls for a vibrant clash of democratic political positions," and that fortifying and defending liberal-democratic institutions requires understanding that such tension can never be completely overcome (104). With the notion of "agonistic struggles," she further states, "the public space is the battleground where different hegemonic projects are confronted, without any possibilities of final reconciliation" (Mouffe 2007, 3).

Mouffe's conception of public space and democratic agonism serves as a useful point of departure for understanding the role of public space in a liberal democracy, which other theorists have picked up upon. Working toward a notion of anarchistic, radical democracy, Springer (2010) argues that the relationship between radical democracy and space is crucial, because democracy "requires spaces where ideas can be contested," and that "public space can be understood as the very practice of radical democracy including agonism [...]" (537). Similarly, Shepard (2009) argues, "Access to space for dialogue is fundamental for democracy to thrive" (273). In contemporary democracies, such public space has been represented by parliaments, councils, other forms of representative body, and even press and media, where deliberation and debates occur. However, for those without proper access to these institutions, and in instances where such institutions fail to represent the interest of the people, protests and demonstrations can serve as an important vehicle for contestations, debates, and resistance. In these instances, public space can play a critical role.

The history of political struggles in societies around the world serves as evidence for this important role of public space. Specifically, while public space can indeed be used to demonstrate power and domination, they can also be appropriated or occupied by the disenfranchised and dispossessed members of the society to express their grievance and forge solidarity and alliance. But how does such a dynamic actually work? How do public spaces function as a space of political resistance? Through what features and processes do public spaces serve as a space of resistance? How do those functions contribute to the resilience of democracy as a political system and practice? In the following, four specific functions of public space are examined to provide a holistic, though not exhaustive, understanding of the dynamics of public space, resistance, and democracy.

Space of assemblies and amalgamation

Let's begin with public space as a space of gathering. In most protests and demonstrations, crowd size, location, and access have been associated with success of gathering. Central and/or symbolic locations matter to the effectiveness of the gathering and its ability to attract more participation and project strength. This is where public space's primary function in resistance resides. In *Rebel Cities*, Harvey (2012) acknowledges that "certain environmental characteristics are more conducive to rebellious protests than others—such as the centrality of squares like Tahrir, Tiananmen, and Syntagma, and the more easily barricaded streets of Paris" (117). In examining four cases of protests in Cairo, Manama, Barcelona and New York City, Franck and Huang (2012) find creative use of public space as a tool for political action, including quick adaptations in response to changing conditions. Examining the encampment at Plaza Catalunya in Barcelona, De la Llata (2015) identifies innovative forms of self-organizations in the use of public space during the protest that echoes the commitment to openness by the movement.

As a space of assemblies, public space protests can lead to other intended and unintended consequences. One of such consequences has been the spiralling of events that leads to greater and potentially more diverse mass gatherings. For example, the 2013 Gezi Park protest in Istanbul was initially started by a group of environmental activists protesting against the loss of the park to commercial development (Gül, Dee, and Cünük 2014). The protest was soon joined by local residents, football fan groups, workers, students, the unemployed, and a number of left-wing organizations, and spiralled in size and scope (Aytekin 2017). Similarly, Marom (2013) noted that the 2011 summer protest movement in Israel had initially focused on a limited agenda, namely the lack of affordable housing, a concern among the young, "creative" middle class in Tel Aviv. As the crowd expanded on the streets, the movement became more inclusive, "addressing long-standing socio-spatial inequalities between Israel's 'center' and 'periphery' [...]" (2826).

Observers of recent resistance movements have focused on the role of social media in mobilizing people from diverse backgrounds. For instance, Juris (2012) has commented on how social media have contributed to "an emerging logic of aggregation in the recent #Occupy movements—one that involves assembling of masses of individuals from diverse backgrounds within physical spaces" (260). The feedback loop of virtual and physical spaces has also become a common theme in recent movements. Besides well-known cases such as the Tahrir Square protest in Cairo, Padawangi (2013) notes that activists in Jakarta have utilized social media to raise awareness of protest events. The internet accelerates mobilization processes, paving the way for "smaller and more varied groups to assemble, and informs journalists of impending actions" (856). As sites of gathering and enabling diverse publics to participate and aggregate, public space can function as an important vehicle for mobilization, diverse engagement, and amalgamation in a movement.

Space of visibility and meanings

Through mass gatherings, public space enables a movement to be visible to a wider audience. With such visibility, public space plays an important role in conveying messages and meanings whose formation also represents a critical function of public space. Citing Arendt's (1958) notion of the "space of appearance," Springer (2010) argues that such function is vital to democratic representation, "Action and speech require visibility because for democratic politics to occur, it is not enough for a group of private individuals to vote anonymously [...]" (537). Furthermore, visibility can translate into political power or at least attention. In their study of four protest movements from

Cairo to New York City, Franck and Huang (2012) note that the occupation of physical spaces "gave these political movements international visibility through transmission of detailed and evocative images in the media" (17).

As a space of visibility, public space helps communicate messages and forge meanings. Observing the everyday struggles of people in the Middle East, Bayat (2013) notes that streets "are not only where people express grievances, but also where they forge identities, enlarge solidarities, and extend their protests beyond their immediate circles to include the unknown, the strangers" (13). These identities and meanings operate also at a symbolic level. Commenting on the Occupy Wall Street movement, Marcuse (2011) notes the role of the occupation in "giving expression to and concretizing an inchoate but widely shared and deeply felt unhappiness … " Movements can also imbue existing sites with new meanings. In Madrid, the temporary occupation of Puerta del Sol gave the location a new symbolic meaning which the city administration and corporation have tried to erase by transforming it into a space of consumption (Beltran and Garcia-Hipola 2014; Kränzle 2017). In the Occupy movement, Murphy and O'Driscoll (2015) argue that the movement "shifted the meaning of public spaces that formerly represented the power and authority of the state and corporations […] to one in which the people present felt themselves to be powerful participants in democracy" (330).

The function of public space as a space of visibility is particularly important to those lacking a voice in the political process. In her commentary on the events of 2011, Sassen (2011) notes, "The question of public space is central to giving the powerless rhetoric and operational openings" (573). In Jakarta, Padawangi (2013) argues that the protest site helps the urban poor in broadcasting their messages to the middle and upper classes. They function as a "megaphone" (857). Besides the voice of the marginalized, large-scale gatherings can also communicate the vulnerability of the establishment and reveal crises of political legitimacy. Similar to protests elsewhere, Merrifield (2014) notes that "encounters between dissatisfied people in Taksim Square created a democratic moment at the bottom of society, while heralding a crisis of legitimacy at the top" (xiv). Sassen (2011) also notes that " … the city makes visible the limits of superior military power" (573). As a space of visibility and meanings, public space can shape the public discourse and communicate political messages that are important to a movement.

Space of encounters and negotiations

As a space for mass gatherings and agonistic contestation, public space is also an important space of encounter among a variety of actors, including activists, protest participants, strangers, and passersby who may share or diverge in their worldviews, values, politics, and identities. These encounters may forge a sense of solidarity or entail contestations and negotiation. They present opportunities for learning and understanding that would not have happened if such mass assemblies did not take place (see Figure 26.2).

Murphy and O'Driscoll (2015) have identified "affect" as the key to transformation through physical participation; images, sounds, and activities produce a sense of political engagement. Similarly, Bayat (2013) thinks of streets as a medium through which latent communication can be established between strangers who recognize their shared interests and sentiments, writing "This is how a small demonstration may grow into a massive exhibition of solidarity" (13). Marcuse (2011) made a similar observation on Occupy Wall Street that public space provides a "glue function, creating a community of trust and commitment to the pursuit of common goals."

Though mass gatherings may produce comraderies and solidarity, they can also be sieged by conflicts and contentions among movement activists and supporters. At Athens' Syntagma

Figure 26.2 Spontaneous conversation between protestors and passersby during Occupy Wall Street
Image credit: Jeffrey Hou

Square, the discourses and repertoires of the anti-austerity protests were starkly different between "the higher square" and "the lower square," with the former focusing on a nationalistic narrative, and the latter on equality and justice and drawing a larger and more diverse crowd. The organizational and decision-making structure, through the General Assembly, was also highly unstable, and full of disagreement and conflicts among and with the groups themselves (Kaika and Karaliotas 2017). Similarly, during the Sunflower Movement in Taipei, protestors also formed "cliques and tribes" (Chen 2017). While the groups have come together during the movement, "their motivations of their participation were dissimilar, leading to tensions and disagreements between the groups and amongst movement members on protest tactics and strategies" (141).

While divisions and difference have complicated the function and operation of a movement, contestation and negotiation over the differences can also be opportunities for learning and understanding. Marcuse (2011) calls this an educational function, one that involves "questioning, exploration, juxtaposition of differing viewpoints and issues, seeking clarification and sources of commonality within difference." Observing the protests in Tahrir Square and Yemen's Sana'a, Sassen (2011) notes that "the overcoming of conflicts has become the source of an expanded civicness" (675). In the case of Syntagma Square, Kaika and Karaliotas (2017) find the necessity of dealing with competing imaginaries, strategies, and logics to "open up a fertile terrain of strategic experimentation and revision" (128). In Taipei, Chen (2017) argues that the cliques and tribes have "illuminated the Sunflower Movement as a conglomerate of multiple social movements and civil society groups" (141). As a space of encounters, public space provides opportunities negotiations of diversity and differences in a movement resulting in learning and understanding.

Space of everyday resistance

Besides large-scale gatherings that may last for days, weeks, or months and may ultimately cease or diminish, everyday resistance in public space takes on a distinctly different form and duration. As a practice that extends beyond ephemeral or short-lived occurrences, these everyday acts of resistance are particularly relevant in examining how public space contributes to democratic resilience. In examining the everyday forms of resistance that have led to large-scale events, Bayat (2013) argues that "ordinary people can change their societies through opportunities other than mass protests or revolutions" (x). He calls these "nonmovements"—"Precisely because they are part and parcel of everyday life, nonmovements assume far more resiliency against repression than the conventional activisms" (21).

A number of seminal works have focused on the function and operation of everyday resistance as a distinct form of social, political, and economic resistance. Using the practices of indigenous people under Spanish colonization as an example, Michel de Certeau (1984) argues that the indigenous subjects subvert the colonial rituals, representation, and laws "not by rejecting or altering them, but by using them with respects to ends and references foreign to the system … " (xii). He further distinguishes between strategies and tactics, with the former representing the work of institutions and structures of power, and the latter representing creative resistance to these structures as enacted by ordinary people. In *Weapons of the Weak*, James C. Scott (1985) conceptualizes the minor but persistent struggles of the subordinate class against the exploitation of the state or the dominant class as "everyday forms of resistance."

Despite their quotidian nature and even tendency to avoid direct confrontation, everyday resistance has the ability to scale up when the conditions exist. In his observation of Arab Spring protests, Bayat (2013) argues that the nonmovements may turn into organized social movements when the opportunity arises. In my own work, I have identified three specific processes—*rupturing, accreting*, and *bridging*—through which everyday resistance can be elevated into organized actions. Through *rupturing*, everyday acts create openings in the institutional and spatial structure of the city and set the stage for transformative actions. Through *accreting*, actions may aggregate, multiply, reproduce, and eventually become a force to be reckoned with. Through *bridging*, activists, non-profits, and community organizers can play an important role in transforming everyday acts of individuals and small groups into collective forces and influences and connecting them with institutional processes (Hou 2018).

Power (and limits) of public space as a space of resistance

As recent events around the world suggest, public space is critical to the functioning of democracy by serving as a vehicle for mass assemblies, by making issues and subjects visible, by forging and communicating identities and meanings, by supporting debates and dialogues, and by enabling the public to hold the state accountable. As Sassen (2011) has argued, instead of "a space for enacting ritualized routines," public space, or "the street" can be conceived as a space where "new forms of the social and the political can be made" (574). Harvey (2012) further suggests that "the collective power of bodies in public space is still the most effective instrument of opposition when all other means of access are blocked" (161–162). He further argues, "The actual site characteristics are important, and the physical and social re-engineering and territorial organization of these sites is a weapon in political struggles" (117–118).

As potent as public space can be in confronting and resisting political hegemony and oppression, it's worth noting that there are important limits or caveats. First, public space, especially the formalized and ritualized ones, should not be the only site of agonistic struggles. Other locations, such as workplaces as well as domestic and liminal spaces, are also potential sites of struggles, even as conventional workplaces have been disappearing in many parts of the world (Harvey 2012). Commenting on Tiananmen protests in 1989, for example, Emerton (2017) argues that the subsequent "unstaged resistance" by Chinese workers has proven to be more enduring by operating according to a logic of "rupture" and "surprise" (209). On Occupy Wall Street, Marcuse (2011) makes a similar argument by pointing out the danger in focusing too much on a specific space while losing the big picture in the movement, "The spaces sought for occupancy are not the prize for which the battle is being fought, but rather a terrain on which the battle takes place."

Indeed, cases have demonstrated that while public space plays a critical role in supporting and galvanizing a movement, the real political battle is often won elsewhere (see Figure 26.3). In Madrid, for instance, following the eviction from Puerta del Sol, movement activists continued their political activism at the neighborhood level, taking advantage of a strong grassroots political culture in many Spanish cities (Kränzle 2017). Such grassroots organizing eventually led to the victory of the leftist party in Madrid's 2015 Mayoral election after two decades of Conservative rule. However, not all attempts have been equally successful. In Israel, Marom (2013) observes that while some movement activists have entered formal politics or set up political initiatives, "with the loss of its widespread and prolonged presence in public space, [the movement] has also lost a significant part of its visibility, amplification and impact" (2839). On Occupy Wall Street, Kreiss and Tufekci (2013) take note of the movement's difficulty in engaging institutional politics—a process which they argue is key to broad and durable societal transformation.

Lastly, it is also important to note that protests in public space can also lead to backlash and further repression. Frer and Meier (2017) point out that protest may lead to outcomes that are contradictory to the protest's aim, including more restrictions and control through "new surveillance systems, increased police or military presence, more effective means of crowd control and domination or more effective weapons" (128). At the same time, reflecting upon the post-revolution politics in Egypt, Abaza (2017) points out that street activism has become "hazardous and highly costly in terms of human life" with the re-emergence of the army in civil life, after the ousting of President Morsi (170).

Despite these complex dynamics, uncertainties, challenges, and possible setbacks, it is precisely in this light that public space serves as a space of continued struggle. It is this struggle that allows for mobilization and assembly, that sheds light on important issues and questions facing society, that enables debates, exchanges, and contestations of ideas, values, and worldviews, particularly for the marginalized bodies and voices. It is through these struggles and engagements that democracy can become lived and resilient. Finally, it is the realization of public space as vehicle for social and political processes that we must view our role as scholars, activists, practitioners, and/or policymakers in uncovering the potentials as well as limits of public space as a space of resistance and collective actions.

Figure 26.3 Women's March 2017, USA: The battle was won not on the street but in the ballot box during the 2018 midterm election in which minority and woman candidates made significant gains in the Congress

Image credit: Jeffrey Hou

References

Abaza, M. (2017) Cairo: Restoration? and the Limits of Street Politics. *Space and Culture*, 20(2): 170–190.

Arendt, H. (1958) *The Human Condition*. Chicago: University of Chicago Press.

Aytekin, E.A. (2017) A "Magic and Poetic" Moment of Dissensus: Aesthetics and Politics in the June 2013 (Gezi Park) Protests in Turkey. *Space and Culture*, 20(2): 191–208.

Bayat, A. (2013) *Life as Politics: How Ordinary People Change the Middle East*. Stanford: Stanford UP.

Beltran, M. and Garcia-Hipola, M. (2014) The Democratization of Public Space: Insurgent or Institutional Occupation of Puerta del Sol in Madrid? *The International Journal of Interdisciplinary Cultural Studies*, 10(1): 1–14.

Bishop, P. and Williams, L. (2012) *The Temporary City*. New York: Routledge.

Boyer, M.C. (1992) Cities for Sale: Merchandising History at South Street Seaport. In: *Variations on a Theme Park: The New American City and the End of Public Space*, Sorkin, M. (ed). New York: Hill and Wang, pp. 181–204.

Castells, M. (1983) *The City and the Grassroots*. Berkeley: University of California Press.

Chen, K.W. (2017) Democracy, Occupy Legislature, and Taiwan's Sunflower Movement. In: *City Unsilenced: Urban Resistance and Public Space in the Age of Shrinking Democracy*, Hou, J. and Knierbein, S. (eds). New York: Routledge, pp. 133–144.

Crawford, M. (1999) Introduction. In: *Everyday Urbanism*, Chase, J., Crawford, M., and Kaliski, J. (eds). New York: Monacelli Press, pp. 8–15.

Davis, M. (1992) Fortress Los Angeles: The Militarization of Urban Space. In: *Variations on a Theme Park: The New American City and the End of Public Space*, Sorkin, M. (ed). New York: Hill and Wang, pp. 154–180.

De Certeau, M. (1984) *The Practice of Everyday Life*. Berkeley: University of California Press.

De La Llata, S. (2015) Open-Ended Urbanisms: Space-Making Processes in the Protest Encampment of the Indignados Movement in Barcelona. *Urban Design International*, 21(2): 113–130.

Douglas, G.C.C. (2014) Do-It-Yourself Urban Design: The Social Practice of Informal "Improvement" through Unauthorized Alteration. *City & Community*, 13(1): 5–25.

Emerton, R.H. (2017) Beijing '89: The Duration of the Event. *Space and Culture*, 20(2): 209–220.

Franck, K. and Huang, T.S. (2012) Occupying Public Space, 2011: From Tahrir Square to Zuccotti Park. In: *Beyond Zuccotti Park: Freedom of Assembly and the Occupation of Public Space*, Shiffman, R., Bell, R., Brown, L.J., Elizabeth, L. (eds). Oakland: New Village Press, pp. 3–20.

Franck, K. and Stevens, Q. (eds). (2007) *Loose Space: Possibility and Diversity in Urban Life*. London: Routledge.

Frer, L. and Meier, L. (2017) Resistance in Public Spaces: Questions of Distinction, Duration, and Expansion. *Space and Culture*, 20(2): 127–140.

Gül, M., Dee, J. and Cünük, C.N. (2014) Istanbul's Taksim Square and Gezi Park: The Place of Protest and the Ideology of Place. *Journal of Architecture and Urbanism*, 38(1): 63–72.

Hardt, M. and Negri, A. (2000) *Empire*. Cambridge: Harvard University Press.

Harvey, D. (2012) *Rebel Cities: From the Right to the City to the Urban Revolution*. New York: Verso.

Hou, J. (ed). (2010) *Insurgent Public Space: Guerrilla Urbanism and the Remaking of Contemporary Cities*. New York: Routledge.

Hou, J. (2018) Rupturing, Accreting and Bridging: Everyday Insurgencies and Emancipatory City-Making in East Asia. In: *Public Space Unbound: Urban Emancipation and the Post-Political Condition*, Knierbein, S. and Viderman, T. (eds). New York: Routledge, pp. 85–98.

Iveson, K. (2013) Cities with the City: Do-It-Yourself Urbanism and the Right to the City. *International Journal of Urban and Regional Research*, 37(3): 941–956.

Juris, J.S. (2012) Reflections On #Occupy Everywhere: Social Media, Public Space, and Emerging Logics of Aggregation. *American Ethnologist*, 39(2): 259–279.

Kaika, M. and Karaliotas, L. (2017) Athens' Syntagma Square Reloaded: From Staging Disagreement towards Instituting Democratic Spaces. In: *City Unsilenced: Urban Resistance and Public Space in the Age of Shrinking Democracy*, Hou, J. and Knierbein, S. (eds). New York: Routledge, pp. 121–132.

Kohn, M. (2004) *Brave New Neighborhoods: The Privatization of Public Space*. New York: Routledge.

Kohn, M. (2013) Privatization and Protest: Occupy Wall Street, Occupy Toronto, and the Occupation of Public Space in Democracy. *Perspectives on Politics*, 11(1): 99–110.

Kränzle, E. (2017) Public Space in a Parallel Universe: Conflict, Coexistence, and Co-Optation between Alternative Urbanisms in the Neoliberalizing City. In: *City Unsilenced: Urban Resistance and Public Space in the Age of Shrinking Democracy*, Hou, J. and Knierbein, S. (eds). New York: Routledge, pp. 186–198.

Kreiss, D. and Tufekci, Z. (2013) Occupying the Political: Occupy Wall Street, Collective Action, and the Rediscovery of Pragmatic Politics. *Cultural Studies Critical Methodologies*, 13(3): 163–167.

Marcuse, P. (2011) "The Purpose of the Occupation Movement and the Danger of Fetishizing Space." Accessed 14 August 2018. https://pmarcuse.wordpress.com/2011/11/15/the-purpose-of-the-occupa tion-movement-and-the-danger-of-fetishizing-space/

Marom, N. (2013) Activising Space: The Spatial Politics of the 2011 Protest Movement in Israel. *Urban Studies*, 50(13): 2826–2841.

Mayor, M. (2009) The 'Right to the City in the Context of Shifting Mottos of Urban Social Movements. *City*, 13(2–3): 362–374.

Mcadam, D., Mccarthy, J.D. and Zald, M.N. (eds). (1996) *Comparative Perspectives on Social Movements: Political Opportunities, Mobilizing Structures, and Cultural Framings*. Cambridge: Cambridge University Press.

Merrifield, A. (2014) *The New Urban Question*. London: Pluto Press.

Mitchell, D. (2003) *The Right to the City: Social Justice and the Fight for Public Space*. New York: Gilford Press.

Mouffe, C. (2000) *The Democratic Paradox*. New York: Verso.

Mouffe, C. (2007) Artistic Activism and Agonistic Spaces. *Art & Research*, 1(2), Accessed 25 August 2018. www.artandresearch.org.uk/v1n2/mouffe.html

Murphy, K.D. and O'Driscoll, S. (2015) The Art/History of Resistance: Visual Ephemera in Public Space. *Space and Culture*, 18(4): 328–357.

Padawangi, R. (2013) The Cosmopolitan Grassroots City as Megaphone: Reconstructing Public Spaces through Urban Activism in Jakarta. *International Journal of Urban and Regional Research*, 37(3): 849–863.

Sassen, S. (2011) The Global Street: Making the Political. *Globalizations*, 8(5): 573–579.

Scott, J.C. (1985) *Weapons of the Weak: Everyday Forms of Peasant Resistance*. New Haven: Yale University Press.

Shepard, B. (2009) Community Gardens, Convivial Spaces, and the Seeds of a Radical Democratic Counterpublic. In: *Democracy, States, and the Struggle for Social Justice*, Gautney, H.D., Smith, N., Dahbour, O., and Dawson, A. (eds). New York: Routledge, pp. 273–296.

Smith, N. (1992) New City, New Frontier: The Lower East Side as Wild, Wild West. In: *The New American City and the End of Public Space*, Sorkin, M. (ed). New York: Hill and Wang, pp. 61–93.

Sorkin, M. (ed). (1992) *Variations on a Theme Park: the New American City and the End of Public Space*. New York: Hill and Wang.

Springer, S. (2010) Public Space as Emancipation: Meditations on Anarchism, Radical Democracy, Neoliberalism and Violence. *Antipode*, 43(2): 525–562.

27

PUBLIC SPACE AND THE POLITICAL

Reconnecting urban resistance and urban emancipation[1]

Sabine Knierbein

Introduction

If we assume that democracy is a spatial praxis rather than an abstract political field, public space – understood as lived space of contemporary cities – needs to be reconsidered as place of doing democracy. Through an exploration of the linkages between the concepts of urban resistance and urban emancipation under post-political conditions, this chapter argues that publics need to be revisited as ever-changing and contingent foundations. Post-politics is characterized by consensual governance regimes that work to reduce political contradictions to policy problems (Wilson and Swyngedouw 2015). Under post-political conditions, these policy problems are managed by experts and legitimated through participatory processes in which the scope of possible outcomes of these processes, and their lines of argumentation are narrowly fixed in advance (cf. Wilson and Swyngedouw). Key thinkers detecting a post-political condition in urban development and wider politics share a post-foundational ontology, according to which there is no essential ground to any social order (Marchart 2010). In contrast to political philosophies that ground society in a state of nature, a primordial hierarchy, or an economic base, post-foundational theorists begin from the position that all social orders are profoundly contingent and structured to conceal their own absent ground (cf. Wilson and Swyngedouw, 10).

In public space, "the political" may eventually become enacted through the everyday spatial practices of publics producing space (e.g. recent and earlier waves of protest in public plazas, streets, airports, tube stations of Hong Kong) (Chen and Szeto 2017). Yet a fine line needs to be drawn between different types of face-to-face political acts of urban resistance, as resistance may move from anti-politics to alter politics, provoking real change in social structures and institutional governance arrangements (e.g. Taipei when protestors first occupied the legislative chamber of Parliament in 2014 to resist against a new trade agreement between Taiwan and China, and later became elected politicians and thus institutionalized) (Chen 2017). Different forms of resistance entail: occupations of parliaments, squares, streets, or factories; sit-ins, revolts, rage, human-chains, protest, appropriation, or passive bodily resistance against authoritarian regimes. An example for the latter has been Erdem Gunduz, the "Standing Man" in Turkey (Seymour 2013).

Wider public-space based urban struggles point to contingent structural imbalances between promises of political equality and increasing empirical evidence for enhanced patterns of social inequality (e.g. anti-austerity/*indignados* protests in Madrid's Plaza del Sol) (Kränzle 2017). Related to debates on public space and urban resistance are debates that ask for the emancipatory potential of public space. These debates render public spaces as places of public urban life in constant struggle between the utopian topoi of social/human emancipation which remained unfulfilled when political emancipation was achieved for wider groups of the society, leaving the more vulnerable and marginalized groups often behind in their struggle for full social emancipation.

By better understanding the processes and implications of the recent urban resistances and urban emancipations, this chapter seeks to contribute to the ongoing debates concerning the role and significance of public space in the practice of lived democracy.

Public space, urban resistance, and emancipation

According to Lefebvre (2003), urbanization processes are key vehicles that spatially catalyze growth, competition, and alienation. Simultaneously, they are characterised by continuous attempts of urban societies to self-organize through acts of resistance as part of their endeavour towards human emancipation, de-alienation, and meaningful lives.

This abstract analysis of the relation between capitalism, urbanization processes, and social change can be best analyzed at the level of lived space inquiry and everyday life research. Here, public space can be understood as a seismograph of social and political change. Public space research may unravel completely new qualitative aspects of the changing relations between capitalism and urbanization (see Bayat's 2013 innovative concept of social non-movements), and thus produces innovations in urban research, because public-space researchers evidence that global processes of capitalist urbanization accrete locally in manifold ways which might undermine one global narrative of capitalism (Madanipour 2010; 2019; Bayat 2013; Knierbein and Hou 2017). Thereby, researchers may analyze how specific, contingent social orders become instituted in specific places (e.g. military control of public space after protests as in the case of Mexico D.F. protests) (De la Llata 2017).

As public life and public space can, however, never become fully controlled, the process of instituting such an order is continuously overthrown, for instance when an order becomes contested by insurgent or counter publics who engage in resistance struggles for political possibilities along plural identity lines (Swyngedouw 2015). Protestors may be loud and collectively organised when democratic rights to protest in public spaces are granted and protected, or they may exercise these rights in a more silent and fragmented manner if democratic expression of opinion in public space is threatened e.g. by state violence against protestors (Bayat 2013; Knierbein and Hou 2017). On a theoretical level, public space can be scrutinized as a constant place of political struggle, a terrain "in which two heterogeneous processes collide: that one of government in an almost Foucauldian sense of governmentality [the police] and that one of emancipation [the political]" (Marchart 2010, cited in Mullis and Schipper 2013, 79).

Emancipatory thinking has increasingly been providing valuable impetus to both urban research and practice ever since linkages between emancipation and the city have been affirmed in the philosophical foundations of the social sciences (Marx 1844; Weber 1978). Throughout the 20th century, a series of emancipatory struggles and attending scientific debates realized the liberatory potential of urban spaces as grounds for opportunity and possibility, cosmopolitanism and freedom from a multitude of political, cultural, social, and economic constraints (Lees 2004). Emancipatory movements have involved almost all social identity constructs and social structures, including labor, gender, ethnicity, ecology, peace, freedom, and justice. In recent years, emancipatory struggles have predominantly taken place as a critique of different local forms of (neo)liberalization, and more recently, as an urgent call by the youngest generations to stop climate change. Struggles for rights and equality have shaped not only institutional politics at various scales, but also the symbolic order of our cities. Lefebvre's (1968) claim for the "right to the city" has found global resonance with professionals, cultural producers, and activists engaging in planning and designing urban spaces. His claims have also served as leitmotif of theoretical and practical endeavours against dispossession, displacement, and exclusion. Urban studies situate emancipatory resistance movements in the history of capitalist urbanization (Goonewardena 2011; Bayat 2013; Mayer 2013). Urban resistance movements, predominantly understood as constituted around a collective political ideal, have been studied to understand how certain symbolic and social orders have emerged, and how insurgent publics have influenced trajectories of urbanization. However, emancipation seems to preserve its ambiguous meaning for a broad portion of the social and political scientific communities, as well as the design and planning disciplines. Much of the contemporary debate in political theory tends to refrain from spatializing emancipatory praxis, while attempts at transferring post-political thought to the fields of urban studies and planning theory, with the exception of Eric Swyngedouw (2015), tend to conceptually circumvent emancipation (Metzger, Allmendinger, Oosterlynck 2014; Roskamm 2017). In addition, emancipation, if systematically reviewed, is often not used on its own terms, but rather serves as a mechanism, means, or bridge aligning with other concepts relevant for urban theory (cf. Knierbein and Viderman 2018, 4):

- forms of innovative *self-organization* and *self-management*;
- struggles *for equality and equity* and *against structural patterns of inequality*;
- articulations to *renew democracy* through *utopian praxis and action*;
- *attempts to overcome gridlocked ways of thinking* when conceiving relations between space, society, and urbanization;
- calls for *liberation from oppressive constraints* pointing to the fact that power relations are immanent in all types of social relations.

Karl Marx (1844) distinguished between *political emancipation* and *social or human emancipation*. *Political emancipation* concerns the relation between the individual and the state: the pursuit of equal access to political decision making in a modern state against the conditions of oppressive social relations. It is achieved when everyone is treated equally under the law of the state. Marx admits that "in the existing world order," i.e. capitalism, "political emancipation is […] a big step forward." Yet, he also expresses that "real practical emancipation" might go beyond a reduced version of political emancipation towards full social emancipation. Criticism was directed against bourgeois aspects of the emancipatory project, which separate political and social power(s), as the wider social striving for emancipation came to a halt when many people had been granted rights to vote, even though the material living conditions of all members of the society had not changed for better. For Marx, social emancipation could only be accomplished through individual de-alienation. This is achieved when "[hu]man[s] re-absorb in … [themselves] the abstract citizen" and turn again into "a *species-being* in [their] everyday life." Thereby, people recognize and organize their "'own powers' as social powers" including the separation between social powers and themselves, a separation previously resulting from political emancipation (Marx 1844, n.p.).

Marx's argument has been taken up in more recent thought when claiming the need for emancipation in processes of global urbanization. Here, social emancipation is a guarantor for political emancipation, whereas political emancipation does not automatically provide for social emancipation (cf. Merrifield 2006, 114). Differentiations between political and social emancipation can be transferred to current debates in planning theory that make similar distinctions. As Purcell (2009) outlines, present liberal democracies have been based on an unsolved tension between political equality and social inequality. These have been explained by frequent shortcomings of liberal democracies to overstress freedom and underemphasize equality (Mouffe 2000). In this sense, emancipatory struggle is unavailing for social suffering associated with lived experiences of exclusion, marginalization, or inequality (Bourdieu 1984). This contribution does not situate emancipation in the static space of the ideal(ized) city, but it takes on the challenge of revisiting the relation between emancipation and urbanization.

De Sousa Santos (2006) agrees that modernist means to achieve emancipation are in fact anachronistic colonizing forces, whereas he insists that the original aims of emancipation are still, if not even more, globally relevant. A tension between the everyday *experience* of people and their *expectations* is central to the understanding of emancipatory potential (cf. 13–14). Throughout modernity, capitalism has maintained the narrative that social improvement is possible for (nearly) everyone based on the rights resulting from capitalistic political emancipation. Despite the maintained discrepancy between political equality and social inequality in many of the Western democracies, during the 20th century the modern emancipatory project preserved social peace and order by coupling emancipatory struggle with social regulation effectively working to reduce tensions arising from discrepancies between the regulatory forces and emancipatory drivers of urbanization. Yet regulation did not eliminate the existing disjuncture between political equality and social inequality, leaving the struggle for social emancipation incomplete. The early 21st century witnessed a break away from these fragile regulatory routines, as for a great deal of the world population, the *expectations* have become less positive than the current *experience* (De Sousa Santos 2006 [author emphasis]). This is because the balance between regulation and emancipation has been distorted towards regulation, which meant control and order, whereas emancipatory action was delineated as chaotic, and thus to be regulated and ordered, particularly through means of planning and urban design. In fact, De Sousa Santos (cf. 2006, 14) argues that real emancipatory potential has shrunk. This shrinkage takes place through processes which Guerrero Antequera (2008) coins as *neoliberal democratic disciplining* as "the actual democracy which presents itself like our truth, is nothing but the interruption of the unfolding acts of democratization, of the liberating practices that escaped and put the controls and codifications" of previous forms of oppression in crisis (276).

Linking urban emancipation, urban resistance and "the political"

A contemporary critique of lived space and thus of the changing everyday life under capitalism remains a largely uncharted realm which carries a great potential for contributing to a needed pluralisation of emancipatory thoughts. Against this background, there is an urgent need – through a focus on emancipation – to shift attention again on the innovative and centrifugal powers of the critique of everyday life (Lefebvre 2014). In this sense, emancipation needs to be addressed in relation with Lefebvre's production of space as inherently characterized not only by control, domination, and colonization, but by insurgencies from everyday life and acts of resistance in the lived space of urbanized areas. The self-activating affects, passions, and powers that stem from ordinary life can be considered as the very pre-condition of emancipatory praxis embedded in local space and socio-historic context. This requires an understanding of politics as a practice, rather than as an abstract political field (Marchart 2007). Thereby, a conceptualization of emancipatory politics as "all-encompassing permanent dimension of all social life" is reinforced (55).

The question then is, why are emancipations predominantly scrutinized as emerging in various forms of articulated political insurgencies and spaces of social movements? Does an immanent Eurocentric focus on social urban movements as a main vehicle for a (modernist utopian version) of social transformation mainly distract scholarly attention away from the everyday unsettling of routines, and its genuinely emancipatory character (Bayat 2013)? Amin and Thrift (2004) have related emancipation to the politics of the lived city, involving politics of embodiment and politics of turf. They acknowledge a faltering, but vocal potenti-ality of cities performing the "constant hum of the everyday and prosaic web of practices" presupposing the "existence of the politics of the minor register full of small gains and losses, which never quite add up" (Amin and Thrift 2004, 233). Thereby, they recognize that emancipatory politics evolve out of the numerous forms of ordinary urban sociality.

While most of the literature linking spatial praxis to emancipation in public space tends to celebrate large-scale (revolutionary) acts, this chapter draws attention to ordinary city publics and everyday places of change that sprout in cracks of structural power systems, often emerging from the messy minutiae of everyday life (cf. Knierbein and Viderman 2018, 269). But how and why do these "cracks" come into being, transforming an ordinary public realm into a lived and relational (counter) space?

Publics as everchanging and contingent foundations

One of the core arguments of post-foundational thought is that the political re-enactment of equality can only emerge because of the inevitable contradictions of a social order which presupposes equality but simultaneously disavows it (cf. Rancière 2010, 9). A dual notion of foundation is central to post-foundational thought; this duality assumes that while grounding society in a solid foundation is impossible – it is possible to form 'contingent foundations' that operate as a plurality of competing foundational attempts "[seeking] to ground society without ever being entirely able to do so" (Marchart 2007, 7). The impossibility of found-ing a social order, as in structuralism, thus "serves as a condition of possibility of always only gradual, multiple and relatively autonomous acts of grounding" (155). Rancière (2010), for instance, is provoked by endeavours to implement emancipation, which will always overturn into a form of societal management by 'enlightened' experts. The ground can then only ever be ripe for forms of disappointment that interpret the dream of emancipation as the root cause of the injustices perpetrated by those same experts (Corcoran 2010, 3). This find-ing shows that there is an urgent need to revisit and reactivate the concept of emancipation with caution. Rancière's (2010) political subjects, for instance, are continually driven by endeavours to ground an unconditional equality as "lived and effective" and not simply "represented" by a particular set of institutions in power (177).

Shaping the urban fabric reveals a tension between particular and universal interests, a tension which is best mirrored in the efforts of conceptualizing public space in capitalism. Equality in public space can hardly be achieved, as there is an inherent division already employed when defining public space (Lofland 2009). Following the reasoning of post-

Figure 27.1 Volunteers at the Hauptbahnhof (Vienna New Main Station) organized the care of thousands
of refugees

Image credit: Christopher Glanzl

foundational thinkers, the need for constant and never-ending definition of who is part of
a city and who is not, of who is heard in a city and who is not, of who is seen in a city and
who is not, is an essential ingredient of defining and redefining publics. Any static articulation of what a public is or any attempt to socially ground its "essence" therefore condemn it
to failure. Radical democracy claims can be achieved through lived spaces in their broadest
sense, and under constant negotiation of whose lived space is concerned (e.g. in the course
of the "refugee crisis" of 2015, train stations in Vienna, Austria, became spots of welcoming
arriving refugees (see Figure 27.1), a process which was built on different conceptions of
publics by distinct leadings partners in two train stations (Knierbein and Gabauer 2017).

This plea reflects the urgent need to raise a storm of conceptual critique against public
space. For some, this storm will end in a reconceptualization of public space by first de-
constructing the use of overall positive connotations, omnipresent co-optation into business-
friendly policy agendas and general concept stretching, by unravelling its key critical aspects
and by successively reconstructing public space conceptually (Knierbein and Viderman 2018;
Madanipour 2019). For others this storm will offer an invitation to dive beyond conceptions
of public space which have historically been initiated with the rise of capitalist spatial separation, and thus cannot be unbound from conflict-ridden pasts (Lofland 2009).

Affect, face-to-face politics, and alter politics

Urban and Planning Scholars have recommended expanding urban and spatial analysis
beyond rational discourse (Gabauer 2018). However, an effort to involve theories on the

body and affect into the conceptualization of emancipation is yet to be made, notwithstanding Lefebvre's (2014) and Rancière's (2009) bodies of work. There is a need to establish connections between (1) an *emancipatory capacity to know* and (2) an *emancipatory power to act*, which requires situating emancipatory praxis and thought within embodied space of everyday lives characterized by affective encounters (Viderman and Knierbein 2018). Or, as Watson (2006) has it, by the dialectics of (dis)enchantments of urban encounters. The recognition of the spatial dimension of the political practice reveals emancipatory struggles as having path-dependency: even though their universal demands might be articulated in the global realm, they have a place of origin in a specific, spatial, everyday-life setting. Uneven development and conditions of experienced social inequality, however, also mediate through bodies and embodied space (Low 2003).

The spatial dimension of emancipatory action cannot be separated from everyday life. By conceptualizing public space as a vision for and a practice of radical democracy, Simon Springer (2010) emphasized the relationship between emancipation and its spatial form. Public space is where agonism and dissent sediment as embodied and affective geographies: "It is in spaces of the public that the discovery of both power and demos is made, and it is in the contestation of public space that democracy lives" (554). Any attempt to conceptualize emancipation thus needs to involve its dimension as an endlessly open pursuit unfolding through everyday affective encounters and experiences (Gabauer 2018). Accordingly, emancipatory struggle is enacted through ethical engagement with others, whether in conflict or "getting along" (Shields 2018). It is also place-bound and place-specific (Vidosa and Rosa 2018). Such an ethical engagement with the "other" may create a meaningful visible change for concerned publics in the present, sometimes in opposition to unmaterialized utopian fantasies. These struggles render everyday life political. Everyday life includes both urban design utopias and past pragmatisms of planning to display what society has desired and how it has desired, while simultaneously serving as both a motivation and medium for negotiating urban futures in the urban present (cf. Knierbein and Viderman 2018, 272). Research needs to consider a conception of urban emancipation where even the smallest engagement or act of negotiating difference in everyday life spatially institutes a certain dimension of emancipation. Emancipatory capacities of planning, however, cannot be unlocked by merely involving a growing number of stakeholders in communicative and collaborative procedures, which exclude affect and passion (Gabauer 2018). Rational approaches to find an ideal speech situation in participatory and communicative planning often fall short to acknowledge political antagonisms present in public spaces, where different interest meet and sometimes collide on an everyday basis. Instead of retaining its domain to representations of space, planning must embrace lived space of a pluralist society of which dissent is part (e.g. the Maunula Democracy Project in Helsinki, Finland) (Kuokkanen and Palonen 2018). Pløger (2018) in this respect, has called for planning to overcome its institutionalism and to discover its connection with everyday life and the political.

> To make contest and strife productive, planning might need a 'wandering planner' (...); that is, a planner that listens to and knows the 'street voice.' It needs a planner that is allowed to work with agonism as a discussant within people's everyday lives and as an 'editorial' organizer of dialogues on everyday life questions, sense of place, aesthetics, design, art, feelings, and desires contesting planning.
>
> *(273)*

Set against the perseverance of modernist planning to insert abstract planning institutions and instruments and top-down urban design visions, critical design praxis involves multiple publics in constant flux, supports the production of knowledge about space in a collectively shared manner, and fosters a positive politics of difference and encounter. Critical designers maintain relationships with diverse groups, hear dissent, give hope, and provide means to publics to become active agents of their own project of desired urban transformation.

Another dimension of critical planning and design involves time because "insurgent democratic politics, [...] are radically anti-utopian; they are not about fighting for a utopian future, but are precisely about bringing into being, spatializing, what is already promised by the very principle upon which the political is constituted, i.e. equalitarian emancipation" (Swyngedouw 2015, 174). In this way, practical emancipation helps raise concerns, develops desires, articulates needs, and provides solutions to make the city itself a real political project for social emancipation (e.g. ongoing demonstrations in Hong Kong, where demonstrators seek to defend the relative independence of the place in central squares, airports and underground lines; or from Vienna, where the Thursday demonstrations take place weekly both in central and peripheral sites of the city since the takeover of the far-right government in 2017, and even continued when government collapsed over the Ibizagate scandal in 2019). A lack of egalitarian politics and social justice is thus part of the problem the concept of emancipation describes.

Practical emancipation combines both anti- and alter politics. Sparks for practical emancipation may be achieved through practices of spatial resistance in public space (e.g. protests on *Heldenplatz* against a far-right wing ball organized in Vienna's former imperial palace, the *Hofburg* (see Figure 27.2), which resulted in UNESCO's taking the Viennese ball culture from the list of protected cultural heritage (Knierbein and Gabauer 2017).

But alter politics may go far beyond resistance when collectives appropriate space to proactively perform a different version of democratic city-making, as in the case of occupying Viennese train stations to welcome refugees. As Ghassan Hage (2012) has confirmed: "the structure of the radical political imaginary at any given time is characterized by a certain balance between 'anti' politics and 'alter' politics" (292). The question then is whether (counter) publics develop acts, tactics, and strategies of resistance, not necessarily in the sense of anti-politics, but more aligned with the idea of alter-politics, including the affective striving for a politics of change. An effective politics of change, promoted by the Plataforma de Afectados por la Hipoteca in Spain's public space or by pro-refugee movements in Vienna, connects abstract and universal political goals with face-to-face politics and lived space (García-Lamarca 2017; Viderman and Knierbein 2019). That is, it promotes a spatial, design and planning praxis that reinstates multiple use value of public space thus acknowledging the deep connection between places of everyday life and places of political transformation.

Public space and the practice of lived democracy

Through an exploration of public spaces and resistance in post-political times, the grand modernist narrative of emancipation yet to come needs to be rejected. This rejection includes the recognition of the plurality of social foundations as always varied, contingent, and temporarily established (Marchart 2011). The promise of change can no longer be conceptualized from a fixed idea of public space and public life, rather it must be sought in a multiplicity of hope-filled political actions that range in scale from the small ordinary act to the politics of paradigmatic social change. Spatial dimensions of emancipation entail: meaningful spaces of everyday encounters, affective interactions in lived space, as well as

Figure 27.2 Protest against the WKR-Ball in Vienna in 2014; the building of the Austrian Parliament is visible in the background

Image credit: Christopher Glanzl

spatially practiced solidarity in struggles that promote a counter conduct to shortcomings related to unfettered capitalist urbanization, racist bias in urban development, and new nationalistic attempts that focus on public space. This intellectual move unfolds around an understanding of public space research and action characterized by (political) passion to engender basic transformation.

> Emancipation must accordingly be understood as an awakening, a (re)discovery of power that is deeply rooted in processes of mobilization and transformation.
>
> *(Springer 2010, 554, referring to Kothari 2005)*

Studying public space as lived space thus means studying active emancipations, keeping in mind Rancière's (2009) understanding of emancipation as "the blurring of the boundary between those who act and those who look; between individuals and members of a collective body" (19). Lived experiences of emancipatory resistance are, therefore, in the focus of this plea to renew the critique of everyday life in the study of processes of urbanization. This encompasses (1) the tracing of *everyday, practical, critical,* and *active emancipations* through spatial inquiry focusing on lived space. It also involves (2) a continuous critical reflection on the positionality of professionals in emancipatory struggles, with regard to both *emancipatory capacity to know* and *emancipatory power to act*. For examples of critical and emancipatory design practices, see Mady (2018), Van Wymeersch and Oosterlynck (2018), Yigit Turan (2018), and Viderman and Knierbein (2018).

Wilson and Swyngedouw (2015) consider the post-political approach an opportunity to rethink the relationship between "our critical theories" and "egalitarian-emancipatory" struggles, with the focus on the political subject who through such a struggle, "aims to take control again of life and its conditions of possibility" (309). As Guerrero Antequera (2008) has emphasized, these struggles can be nourished by a liberating imaginary of social transformation that does not work in a utopian nor an eschatological manner but promotes a concrete path of emancipatory praxis which simultaneously transforms people, circumstances, and conditions (279). Such an understanding starts from the finding that an analysis of the spatial dimension of emancipatory action cannot be separated from the study of changing patterns of everyday life, as it is at their interface where "the political" emerges.

Note

1 This text connects earlier findings published in Knierbein and Viderman (2018) and in Hou and Knierbein (2017).

References

Amin, A. and Thrift, N. (2004) The 'Emancipatory' City? In: *The Emancipatory City? Paradoxes and Possibilities*, Lees, L. (ed). London: Sage, pp. 231–235.

Bayat, A. (2013 [2010]) *Life as Politics: How Ordinary People Change the Middle East*. Stanford: Stanford UP.

Bourdieu, P. (1984) *Distinction: A Social Critique of the Judgement of Taste* (Translated by R. Nice). London: Routledge & Kegan Paul.

Chen, K.W. (2017) Democracy, Occupy Legislature and Taiwan's Sun Flower Movement. In: *City Unsilenced: Public Space and Urban Resistance in the Age of Shrinking Democracy*, Hou, J. and Knierbein, S. (eds). New York: Routledge, pp. 133–144.

Chen, Y.C. and Szeto, M.M. (2017) Reclaiming Public Space Movement in Hong Kong. From Occupy Queens Pier to the Umbrella Movement. In: *City Unsilenced: Public Space and Urban Resistance in the Age of Shrinking Democracy*, Hou, J. and Knierbein, S. (eds). New York: Routledge, pp. 69–82.

Corcoran, J. (2010) Editor's Introduction. In: *Dissensus: On Politics and Aesthetics*, Ranciere, J. (auth) and Corcoran, S. (trans. and ed). New York: Continuum, pp. 1–24.

De la Llata, S. (2017) Operation 1DMX and the Mexico City Commune. The Right to the City Beyond the Rule of Law in Public Spaces. In: *City Unsilenced. Public Space and Urban Resistance in the Age of Shrinking Democracy*, Hou, J. and Knierbein, S. (eds). New York: Routledge, pp. 173–185.

De Sousa Santos, B. (2006) *Renovar la teoría crítica y reinventar la emancipación social*. Buenos Aires: Clacso Libros.

Gabauer, A. (2018) Conflict vs. Consensus. An Emancipatory Understanding of Planning in a Pluralist Society. In: *Public Space Unbound. Urban Emancipation and the Post-Political Condition*, Knierbein, S. and Viderman, T. (eds). New York: Routledge, pp. 173–188.

García-Lamarca, M. (2017) Reconfiguring the Public through Housing Rights Struggles in Spain. In: *City Unsilenced. Public Space and Urban Resistance in the Age of Shrinking Democracy*, Hou, J. and Knierbein, S. (eds). New York: Routledge, pp. 44–55.

Goonewardena, K. (2011) Critical Urbanism: Space, Design, Revolution. In: *Companion to Urban Design*, Banerjee, T. and Loukaitou-Sideris, A. (eds). New York: Routledge, pp. 97–108.

Guerrero Antequera, M. (2008) *Tras el exceso de la sociedad: emancipación y disciplinamiento en el Chile actual.* In: *De los saberes de la emancipación y de la dominación*, A.E. Ceceña. (ed). Buenos Aires: Clacso Libros, pp. 261–282.

Hage, G. (2012) Critical Anthropological Thought and the Radical Political Imaginary Today. *Critique of Anthropology*, 32(3): 285–308.

Hou J. and Knierbein S. (eds) (2017) *City Unsilenced. Urban Resistance and Public Space in the Age of Shrinking Democracy.* New York: Routledge.

Knierbein, S. and Gabauer, A. (2017) Worlded Resistance as "Alter" Politics: Train of Hope and the Protest against the Akademikerball in Vienna. In: *City Unsilenced: Public Space and Urban Resistance in the Age of Shrinking Democracy*, Hou, J. and Knierbein, S. (eds). New York: Routledge, pp. 214–228.

Knierbein, S. and Hou, J. (2017) City Unsilenced: Spatial Grounds of Radical Democratization. In: *City Unsilenced. Public Space and Urban Resistance in the Age of Shrinking Democracy*, Hou, J. and Knierbein, S. (eds). New York: Routledge, pp. 231–242.

Knierbein, S. and Viderman, T. (2018) *Public Space Unbound: Urban Emancipation and the Post-Political Condition.* New York: Routledge.

Kothari, R. (2005) *Rethinking Democracy.* New Delhi: Orient Longman.

Kränzle, E. (2017) Public Space in a Parallel Universe: Conflict, Coexistance and Cooptation between Alternative Urbanisms and the Neoliberalizing City. In: *City Unsilenced: Public Space and Urban Resistance in the Age of Shrinking Democracy*, Hou, J. and Knierbein, S. (eds). New York: Routledge, pp. 186–198.

Kuokkanen, K. and Palonen, E. (2018) Post-Political Development and Emancipation. Urban Participatory Projects in Helsinki. In: *Public Space Unbound: Urban Emancipation and the Post-Political Condition*, Knierbein, S. and Viderman, T. (eds). New York: Routledge, pp. 99–112.

Lees, L. (ed). (2004) *The Emancipatory City? Paradoxes and Possibilities.* London: Sage.

Lefebvre, H. (1968) *Le droit a la ville.* Paris: Éditions Anthropos.

Lefebvre, H. (2003 [1970]) *The Urban Revolution.* Minneapolis: University of Minnesota Press.

Lefebvre, H. (2014) *The Critique of Everyday Life.* The One Volume Edition. London/New York: Verso.

Lofland, L.H. (2009 [1998]) *The Public Realm: Exploring the Cities Quintessential Social Territory.* Brunswick: Aldine Transaction.

Low, S.M. (2003) Embodied Space(s). Anthropological Theories of Body, Space and Culture. *Space and Culture*, 6/1: 9–18.

Madanipour, A. (2010) *Whose Public Space?* London: Routledge.

Madanipour, A. (2019) Rethinking Public Space. Between Rhetoric and Reality. *Urban Design International*, 24: 38–46.

Mady, C. (2018) Public Space Activism in Unstable Contexts: Emancipation from Beirut's Post-Memory. In: *Public Space Unbound: Urban Emancipation and the Post-Political Condition*, Knierbein, S. and Viderman, T. (eds). New York: Routledge, pp. 189–206.

Marchart, O. (2007) *Post-foundational Political Thought. Political Difference in Nancy, Lefort, Badiou and Laclau.* Edinburgh: Edinburgh University Press.

Marchart, O. (2010) *Die politische Differenz. Zum Denken des Politischen bei Nancy, Lefort, Badiou, Laclau und Agamben.* Berlin: Suhrkamp Verlag.

Marchart, O. (2011) Democracy and Minimal Politics: The Political Difference and Its Consequences. *The South Atlantic Quarterly*, 110(4): 965–973.

Marx, K. (1844) "On the Jewish Question." Accessed 11 August 2017. www.marxists.org/archive/marx/works/1844/jewish-question

Mayer, M. (2013) First World Urban Activism. *City*, 17(1): 5–19.

Merrifield, A. (2006) *Henri Lefebvre: A Critical Introduction.* New York: Routledge.

Metzger, J., Allmendinger, P. and Oosterlynck, S. (eds). (2014) *Planning against the Political: Democratic Deficits in European Territorial Governance.* London: Routledge.

Mouffe, C. (2000) *The Democratic Paradox.* New York: Verso.

Mullis, D. and Schipper, S. (2013) Die postdemokratische Stadt zwischen Politisierung und Kontinuität. Oder ist die Stadt jemals demokratisch gewesen? *Sub/Urban Zeitschrift für Kritische Stadtforschung*, 2(1): 79–100.

Pløger, J. (2018) Conflict and Agonism. In: *The Routledge Handbook of Planning Theory*, Gunder, M., Madanipour, A. and Watson, V. (eds). Routledge: New York, pp. 264–275.

Purcell, M. (2009) Resisting Neoliberalization: Communicative Planning or Counter Hegemonic Movements? *Planning Theory*, 8(2): 140–165.

Rancière, J. (2009) *The Emancipated Spectator*. London: Verso.

Rancière, J. (2010) *Dissensus: On Politics and Aesthetics*. Corcoran, S. (trans. and ed). New York: Continuum.

Roskamm, N. (2017) *Annäherungen an das Außen. Laclau, die Stadt und der Raum*. In: *Ordnungen des Politischen. Einsätze und Wirkungen der Hegemonietheorie Ernesto Laclaus*, Marchart, O. (ed). Wiesbaden: Springer, pp. 145–167.

Seymour, R. (2013) "Turkey's 'Standing Man' Shows How Passive Resistance Can Shake a State." *The Guardian*, 18 June. Accessed 4 July 2019. www.theguardian.com/commentisfree/2013/jun/18/turkey-standing-man.

Shields, R. (2018) Improvising an Urban Commons of the Street: Emancipation-from, Emancipation-To and Co-Emancipation. In: *Public Space Unbound. Urban Emancipation and the Post-Political Condition*, Knierbein, S. and Viderman, T. (eds). New York: Routledge, pp. 69–82.

Springer, S. (2010) Public Space as Emancipation: Meditations on Anarchism, Radical Democracy, Neoliberalism and Violence. *Antipode*, 43(2): 525–562.

Swyngedouw, E. (2015) Insurgent Architects, Radical Cities and the Promise of the Political. In: *The Post-Political and Its Discontents*, Wilson, J. and Swyngedouw, E. (eds). Edinburgh: Edinburgh University Press, pp. 169–188.

Van Wymeersch, E. and Oosterlynck, S. (2018) Applying a Relational Approach to Political Difference: Strategies of Particularization and Universalization in Contesting Urban Development. In: *City Unsilenced. Public Space and Urban Resistance in the Age of Shrinking Democracy*. Hou, J. and Knierbein, S. (eds) New York: Routledge, pp. 38–53.

Viderman, T. and Knierbein, S. (2018) Reconnecting Public Space and Housing Research through Affective Practice. *Journal of Urban Design*, 23(6): 843–858.

Viderman, T. and Knierbein, S. (2019) Affective Urbanism: Towards Inclusive Design Praxis. *Urban Design International* (accepted for publication, forthcoming).

Vidosa, R. and Rosa, P. (2018) Emancipatory Practices of Self-Organized Workers in the context of Neoliberal Policies: IMPA, the Case of a Recovered Factory in Buenos Aires. In: *Public Space Unbound: Urban Emancipation and the Post-Political Condition*, Knierbein, S. and Viderman, T. (eds). New York: Routledge, pp. 225–238.

Watson, S. (2006) *City Publics: The (Dis)enchantments of Urban Encounters*. London: Routledge.

Weber, M. (1978 [1924]) Ch. XVI: The City. In: *Economy and Society. An Outline of Interpretive Sociology*, Roth, G. and Wittich, C. (eds). Berkeley: University of California Press, pp. 1212–1374.

Wilson, J. and Swyngedouw, E. (2015) *The Post-Political and Its Discontents*. Edinburgh: Edinburgh University Press.

Yigit Turan, B. (2018) Revitalizing Yeldeğirmeni Neighbourhood in Istanbul towards an Emancipatory Urban Design in the Landscapes of Neoliberal Urbanism. In: *Public Space Unbound: Urban Emancipation and the Post-Political Condition*, Knierbein, S. and Viderman, T. (eds). New York: Routledge, pp. 158–172.

28

ALTERNATING NARRATIVES

The dynamic between public spaces, protests, and meanings

Tali Hatuka

The narration of public space

Public spaces are public when (and inasmuch as) they *are not only* "mapped" by sovereign powers (including supranational organizations) or imposed by economic forces (the domination of the market) "but also 'used' and 'instituted' (or constituted) by civic practices, debates, forms of representations, and social conflicts, hence ideological antagonisms over culture, religion, and secularism" (Balibar 2009, 201). Thus, public space represents the sociopolitical dynamics of a particular time and place. That is, why every public space developed or initiated by formal institutions (e.g. governments, municipalities, planning authorities) is a political space, but not every political space is a public space.

People and powers are aware of both the political and temporal dimensions of public space. This temporality is most evident during protests, when protestors use their power to interpret the symbolic attributes of place, by adding to or modifying its meaning. Protestors' underlying rationale is that symbolic signs have arbitrary relationships to specific objects and are constructed through social and cultural systems. As a result, when society changes, the significance of its symbols changes (Edelman 1964). Hence, while physical forms clearly have an impact on human behavior, human actions can also modify the form and meaning of places. Thus, urban form and symbolism are interlinked and evolve through an ongoing process of interpretation and negotiation. As Murray Edelman notes, "The conspicuousness of public structures, together with either emptiness of explicit meaning, enables them to serve as symbolic reaffirmations of many levels of perceptions and beliefs" (Edelman 1995, 90).

The discussion of the role and power of narration of public space is associated with the debate surrounding historical narration, the ways in which it is constructed, and those who construct it. The unsettled conventions of historical narration have led to the contemporary perception of narration as a function of social power, a social expression of contextual settings. Thus, for example, the French sociologist Maurice Halbwachs (1992) proposed that social groups—families, religious cults, political organizations, and other communities—develop strategies to maintain their images of the past and the present through places, monuments, and rituals of commemoration. Benedict Anderson argued that "imagined communities" are constructed as public memories to concretely affirm abstract ideals (Anderson 1991).

From this perspective, public spaces are important venues for exposing and celebrating social, cultural, and political narrations as reminders of a group's power.

Parallel to and in association with the debate surrounding historical narration, a shift has occurred in the citizen's role in the construction of places. Citizenship has been perceived not only as membership in a polity but also as a reminder of the right and power to participate in the public sphere (Hatuka 2012). With the turn of the 21st century, these features have been developed and enforced, with governments focusing on enhancing civil participation and civil engagement as a tool that reinforces democratic legitimacy and power. This approach has significantly changed the citizen's approach in the production of place—today, the citizen is viewed as an active agent who participates in the development of the built environment (Hatuka 2018).

Most importantly, contemporary conceptualization of historical narration and citizenship, are similar in terms of their perceptions of time. Both are rooted in presentism—a counterpoint to the historicist idea of "progress" (Huyssen 2003, 2). Adopting presentism implies that history is no longer conceived as a continuous grand narrative, ideas which have informed the understanding of historical time in the modern age. With the growing significance and influence of these discourses, public spaces became the concrete sphere of negotiation over narratives. The physicality of places and the ability to experience them in daily life became more significant than historical textbooks. Moreover, citizens had the opportunity to negate or challenge a place's symbols, memories, and images as conceived by professionals. This acknowledgment of the mutable nature of narration defined new (and complex) relationships between place, symbols, and spatial practices in cities worldwide.

Figure 28.1a Public spaces are public when they are not only "mapped" by sovereign powers but also "used" and "instituted" by civic practices, debates, forms of representations, and social conflicts. Washington Mall, Washington, DC

Image credit: Tali Hatuka

Figure 28.1b The discussion of the role and power of narration of public space is associated with the debate surrounding historical narration, the ways in which it is constructed, and those who construct it. Tiananmen Square, Beijing

Image credit: Tali Hatuka

Challenging narration during protests

Offering narration (either supporting or negating a regime), is fundamental to all protests. All human beings can participate in challenging public spaces' narrations and meanings, highlighting existing boundaries and limits of an existing paradigm, and such participation is one of the basic modes of human progress. Protestors produce the new values "around which institutions of society are transformed to represent new values by creating new norms to organize social life" (Castells 2012, 9).

Different strategies are designed to address the narration or memory of place during protests. Generally, three key approaches are identified: *continuity, reconstruction, and negation* (Hatuka 2018). Each of these strategies results in a different approach to the narration of place.

First, *continuity* works well with protests that accept the legacy of a place and communicate with that place by adding another layer to the story, which works well with its underlying premises. Using this approach will also affect the performance of protest, with a tendency toward agreed-upon or known protest narration. A well-known gathering is the November 4, 1995 rally in Rabin Square in Tel Aviv. On that day, Prime Minister Yitzhak Rabin was assassinated during the "Yes to Peace, No to Violence" Rally in support of the Oslo Accords. The event took place in the city's main public square. The rectangular geometry of the space with its raised terrace, establishes an elevated hierarchy between the crowd and the speaker/performer (Hatuka and Kallus 2008). Similar to many other assembly areas, the architectural characteristics of the space were used to shape the November 4th event, when thousands of youngsters in the square waved banners and called for peace in Hebrew and Arabic while leaders stood on the balcony giving their speeches. After the assembly, the Prime Minister walked down the service stairs and was shot in the back by a young religious Jewish person. Rabin's assassination exposed the deep fissures in Israeli society and triggered ongoing public debate about how to heal them. The square became the locus of memory of the murder and a constant reminder of the tensions that led to it. As a memorial ritual, the name of the square was changed and a monument was placed at the exact site of the assassination (Engler 1999; Hatuka 2009; Vinitzky-Seroussi 2009). These acts reinforced the political formalization of the space and its assemblies, adding further symbolic meaning and thus magnifying the importance of the square's ritual and theatrical dimensions (Hatuka and Kallus 2008).

Second, different from continuity, *reconstruction* is the approach being held in reserve from the existing narration; this approach aims to reconstruct the meaning of space with what has been lost, adding missing components or emphasizing neglected parts. An example of such an event is the innovative performance of the Asociación Madres de Plaza de Mayo, whose members marched in circles around the pyramid in Plaza de Mayo in Buenos Aires wearing embroidered bandanas that displayed the names of their "disappeared" children and relatives. The performance of the Madres shows how groups appropriate space by redefining its access, appearance and representation and reclaim the space by using some of its physical attributes and modifying its cultural origin (Taylor 1997; 2006; Arditti 1999; Torre 2000; Bosco 2006; Allmark 2008). This well-known example reveals how an innovative act emerges from both the space's design (the paved circle around the monument) and the legal limitations of protesting against the government. Thus, despite the incentive to abandon the plaza for a safer location, the mothers sustained a symbolic presence in the form of a silent march around the May Pyramid. That form, so loaded with cultural and sexual associations, became the symbolic focus of what started as a literal response to the police's demand that the women "circulate." Thus, the ritual of the Mothers of the Plaza De Mayo involved challenging political and social ideas by acting in proximity to key national symbols, acting in proximity to political powers, and creating intimacy in the form of a march in which they carry their message on their bodies (Taylor 1997). Their action dramatically challenged the narrative told by the regime, alternating, and reconstructing the history of the people and this central public space.

Third, *negation* is the most radical approach. This approach implies the total disruption of a place's narration, resulting in the destruction or replacement of existing symbols, renaming, or the creation of new elements. Negation is also about creating new symbols that convey meanings, feelings, perceptions, and beliefs that have not been thus attributed in the particular context that is associated with the action. At the same time, this "new" icon must be legible to the public at large. It must simplify meaning but not dilute it. Protestors play a key role in creating and spreading the symbol or new icon; thus, the number of actors that support the symbol is vital to its dissemination and influence on the public at large. However, the power of this type of approach is about not only creating the symbol but also displaying it. Furthermore, the relationships between the symbol and the place are critical to the meaning of the event as a whole. Spaces, especially monumental places, often include many symbols that carry particular contextual meanings; thus, the act of presenting something new is a powerful act of appropriation which can also be perceived as a violent act. Thus, in its radical form, this approach will take place at a key

Figure 28.2 Reconstruction is an approach aims to reconstruct the meaning of space with what has been lost, adding missing components or emphasizing neglected parts. Plaza De Mayo, Buenos Aires

Image Credit: Tali Hatuka

focal public space in the city, one that includes symbolic icons of the ruling powers and which might dismantle these icons or replace them with new ones. When successful, i.e., when it is adopted by the public, the icon is impossible to repress. If it leaves a mark on a space, it might irrevocably change the place's symbolic meaning, even if it is only displayed in that space temporarily. An example of this approach is the Tiananmen Square protests, which were led by labor activists, students, and intellectuals in the People's Republic of China (PRC) between April 15 and June 4, 1989. The participants were generally critical of the ruling Communist Party of China (CPC), and they demanded democracy and broader civic freedoms. The demonstrations were focused in Tiananmen Square in Beijing, but large-scale protests also occurred in cities throughout China (Dingxin 2001; Wu 2005). The action that was perceived as violating the political symbolic order was the placement of the Goddess of Democracy, a seven-meter-high stature that was placed in front of Mao's portrait on May 30. Although inspired by statues such as the Statue of Liberty (a young woman holding a torch), the goddess was not a replica; she projected an image of a young Chinese woman. The goddess functioned as the students' monumental symbol, which was placed in the square and added to the five monuments that would stand there permanently. These acts were all part of the place-making process—a means of externalizing and spreading protestors' intentions. After its placement, the Goddess of Democracy served as an icon that represented the movement's needs and desires. The PRC government's subsequent military crackdown on the protesters in Beijing left numerous civilians dead or injured. Protests in other cities throughout China, including Shanghai, remained peaceful. The Goddess of Democracy was destroyed when a tank drove into it at full speed during the military crackdown in the square (Dingxin 2001; Wu 2005). However, although the Goddess of Democracy was destroyed, her ghostly image continued to haunt the regime. The icon could not be erased from the minds of the Chinese people.

Many public spaces have been modified after protests. Such modifications frequently involve renaming places, redesigning a particular object to memorialize a protest, or destroying objects or placing new objects that negate the existing paradigm. Therefore, although narrations may be physically manifested in stone and concrete, they are all replaceable; this impermanence reminds us that symbols and narration carry a dimension of betrayal and forgetting.

Public space, narration, and civil action

In sum, public space is a social frame where different powers work to establish and maintain social and political order. This order is based on the relationships established among key actors, including governments, religious parties, political parties, capital owners, etc. This order is in constant flux, as it is based on repetitive actions by social groups and citizens to support order and the regime.

However, public space is not merely a place of political power representation but also an arena of confrontation over worldviews and narratives. Protests and collective actions are common methods of challenging narratives told by the regime and the symbols that represent it in public space. Thus, public space should be viewed as a dynamic category that is always under negotiation by the people who use it and define its meaning. In that sense, the physicality of space has no meaning in itself and could be perceived and understood in different ways throughout history. Hence, although the configurations of some public spaces are quite important, it matters more what social and political performances are conducted within them, and thus what

Public space, narration and civil action: key strategies

Continuing	Reconstructing	Negating
Accepting the legacy of place and communicating with it by adding an additional layer.	Adding to narration, adding what has been "lost", or emphasizing neglected parts.	Replacing existing narration, symbols, re-naming, constructing new elements.

Modifications frequently involve renaming places, redesigning a particular object to memorialize a protest, or destroying objects or placing new objects that negate the existing paradigm.

Figure 28.3 Alternating narratives of public spaces

symbolic associations are built up over time. In other words, the physical (i.e. scale, size, topography, architecture, and furniture) and symbolic attributes of a public space are only important in terms of the ways in which they are used and interpreted.

Finally, the process of alternating public spaces' meanings could be viewed as *Revisioning Moments*, or socio-temporal processes that change our perceptions of agreed meanings (Hatuka 2010). In their radical form, revisionist moments, whether spontaneous or planned, aim at re-establishing the social order. The collective memory is critical to the construction of the revisioning moment, which formulates new social meaning. In that sense, despite its dynamic and temporary features, the protestors' actions in public spaces are part of a war on the memory of truth, order, worldviews, and power. Also, the process of altering meanings should not be seen as autonomous points in time and space but as a socio-spatial chain of actions. The value of perceiving the processes of alternating meanings as revisioning moments assists in shifting the discussion from an analysis of public spaces as objects to an analysis of the socio-physical processes of the built environment. Furthermore, it is a tool for understanding the personal and collective dimensions of political actions that take place in public spaces, which further allows exploration of the unexpected influences on the fundamental values of a society (Hatuka 2010).

Indeed, over the last decades increased privatization, technological control, and personalization have changed public spaces and have been given various names, such as *The Fall of Public Man*, *Bowling Alone*, and *Alone Together*, metaphors that characterize the decline of civic engagement in the public sphere (Sennett 1976; Putnam 2000; Turkle 2011). Undeniably, the private and the personal have taken precedence over the public; private spaces have replaced public gathering spaces; and society has become generally less interested in public matters and more driven by private interests and personal desires. As Zygmunt Bauman (1999) writes, there is currently no easy and obvious way to translate private worries into public issues and, conversely, to discern and pinpoint public issues in private troubles. Nevertheless, the wave of worldwide protests in the twenty-first century has shown that people are able to challenge Bauman's argument and engage in modifying and challenging public spaces narrations, orders, and meanings.

References

Allmark, P. (2008) Framing Plaza de Mayo: Photographs of Protest. *Continuum: Journal of Media & Cultural Studies*, 22(6): 839–848.

Anderson, B.R. (1991) *Imagined Communities: Reflections on the Origin and Spread of Nationalism*. London: Verso.

Arditti, R. (1999) *Searching for Life: The Grandmothers of the Plaza de Mayo and the Disappeared Children of Argentina*. Berkeley: University of California Press.

Balibar, E. (2009) Europe as Borderland. *Environment and Planning D: Society and Space*, 27(2): 190–215.

Bauman, R. (1999) *In Search of Politics*. Cambridge: Polity.

Bosco, F.J. (2006) The Madres de Plaza de Mayo and Three Decades of Human Rights' Activism: Embeddedness, Emotions, and Social Movements. *Annals of the Association of American Geographers*, 96(2): 342–365.

Castells, M. (2012) *Networks of Outrage and Hope: Social Movements in the Internet Age*. Cambridge: Polity.

Dingxin, Z. (2001) *The Power of Tiananmen: State-Society Relations and the 1989 Beijing Student Movement*. Chicago: The University of Chicago Press.

Edelman, M. (1964) *The Symbolic Uses of Politics*. Chicago: University of Illinois Press.

Edelman, M. (1995) *From Art to Politics*. Chicago: The University of Chicago Press.

Engler, M. (1999) A Living Memorial: Commemorating Yitzhak Rabin in the Tel Aviv Square. *Places*, 12(2): 1–4.

Halbwachs, M. (1992) *On Collective Memory*. Coser, L.A. (trans. and ed). Chicago: University of Chicago Press.

Hatuka. (2009) Urban Absence: Everyday Practices versus Trauma Practices in Rabin Square, Tel Aviv. *Journal of Architecture and Planning Research*, 26(3): 198–212.

Hatuka, T. (2010) *Violent Acts and Urban Space in Contemporary Tel Aviv: Revisioning Moments*. Austin: University of Texas Press.

Hatuka, T. (2012) Civilian Consciousness of the Mutable Nature of Power: Dissent Practices Along a Fragmented Border in Israel/Palestine. *Political Geography*, 31(6): 347–357.

Hatuka, T. (2018) *The Design of Protest*. Austin: University of Texas Press.

Hatuka, T. and Kallus, R. (2008) The Architecture of Repeated Rituals. *Journal of Architectural Education*, 61(4): 85–94.

Huyssen, A. (2003) *Present Pasts: Urban Palimpsests and the Politics of Memory*. Stanford: Stanford University Press.

Putnam, D.R. (2000) *Bowling Alone: The Collapse and Revival of American Community*. New York: Simon and Schuster.

Sennett, R. (1976) *The Fall of the Public Man*. Cambridge: Cambridge University Press.

Taylor, D. (1997) Making a Spectacle: The Mothers of the Plaza de Mayo. In: *The Politics of Motherhood: Activist Voices from Left to Right*, Jetter, A., Orleck, A., and Taylor, D. (eds). Hanover: Dartmouth College, University Press of New England, pp. 182–196.

Taylor, D. (2006) Trauma and Performance: Lessons from Latin America. *PMLA: Publications of the Modern Language Association of America*, 121(5): 1674–1677.

Torre, S. (2000) Claiming the Public Space: The Mothers of Plaza de Mayo. In: *Gender, Space, Architecture*, Rendell, J., Penner, B., and Borden, I. (eds). London: Routledge, pp. 140–150.

Turkle, S. (2011) *Alone Together: Why We Expect More from Technology and Less from Each Other*. New York: Basic Books.

Vinitzky-Seroussi, V. (2009) *Yitzhak Rabin's Assassination and the Dilemmas of Commemoration*. Albany: State University of New York Press.

Wu, H. (2005) *Remaking Beijing*. Chicago: University of Chicago Press.

29

PUBLICS, PAVEMENTS, AND PUBLIC-FACING ART IN POST-COLONIAL URBAN AFRICA

Rike Sitas

Introduction

The atmosphere of a public space, its aesthetics and physical architecture, its historical status and reputation, its visual cultures, subtly define performances of social life in public and meanings and intentions of urban public culture... The projections – cast out from billboards, public art, the design of space, public gatherings, the shape of buildings, the cleanliness of streets, the sounds and smells that circulate, the flows of bodies – come with strong sensory, affective, and neurological effects. They shape public expectation, less so by forcing automatic compliance, than by tracing the boundaries of normality and aspiration in public life.

(Amin 2008, 15)

It has long been assumed that vibrant public spaces are crucial for democratic engagement in cities. These romantic aspirations of public space have captured the imagination of urban thinkers for centuries – from Hannah Arendt's (Arendt 1958) action, freedom, and plurality to Jurgen Habermas' (Habermas 1962) public sphere. Democratic notions of the social encounter within public space have been paramount in many of these deliberations, where according to Ash Amin (2015), "civitas and demos" prevail as crucial to the value of public space. There has also been a heightened exploration of the role of public space in the right to the city and an ongoing call to protect the commons – be they virtual or real (Hardin 1968; Maclellan and Talpalaru 2009; Harvey 2012; Jeffrey, McFarlane, and Vasudevan).

Dominating many discussions has been to look at how public space can foster particular kinds of human interaction. In an attempt to counter the increasing privatization of the commons and to foster an equitable right to the city, a "good" public space is often claimed, according to Amin (2006), to be one that is inclusionary and "a visible emblem of order and harmony" (1010). By implication, public space can run the risk of being perceived as all things to all people, which is an ambitious expectation. This ideal does not take into account the myriad power relations that shape people's everyday activity and freedom of movement in unequal cities – of which none are exempt.

Increasingly, the relational turn in urban studies has asserted the importance of seeing the human and material as entangled in inextricable ways (Lefebvre 1991; Amin 2008). Southern thinkers have been recognizing the public dimensions of Abdou-Maliq Simone's (2010) cityness, the quiet encroachment of the informal of Asef Bayat (1997), Faranak Miraftab's insurgent planning, and Aihwa Ong's (2006) as spaces of citizenship. Despite shifts in thinking, how this lands in the minds of architects and planners remains limited, particularly in African cities facing a polycrisis (Pieterse 2008). Cities in Africa are growing rapidly, and face challenges related to land, basic service delivery (water and electricity), food security, and employment. Informal and formal entanglements are the new normal in economic, housing, and service delivery systems. With rapid internal and foreign migration, inequality and complex gender politics, scarce access to livelihoods and resources – social conflict is common. It is unsurprising that power dynamics (linked to race, ethnicity, class, gender, sexuality) that play out in every facet of society, also impact on the perceptions and uses of public spaces.

The implementation of creating public spaces has largely resulted in parks and cultural/recreational facilities – many of which are under-used, derelict and unsafe for women and children (Sitas 2015; Fortuin and Sitas 2019). The normative uses of public open spaces are sometimes inverted – playgrounds being used for drug deals, while children play on the pavements. The public imagination of urban planners, designers, and thinkers still tend to identify public spaces as discreet entities. This can be seen in CAD designs of architects and in the urban fabric alike (Rose 2001; Watson 2014). The reality is often much more blended into everyday spaces – the notions of public and private are opaquer, especially in dense informal parts of the city. This does not mean that public spaces do not exist, and publicness is not being fostered in other spaces and in unrecognized ways.

This chapter explores the relationship between public-facing art and pavements as enabling in-between spaces where the social, economic, and political coalesce and collide. It starts by exploring pavements and publicness, before drawing on three examples of public-facing art. It concludes by arguing that these examples challenge normative notions of public space by troubling public–private binaries; providing spaces for decolonial encounter; and enabling dialogue between human and material worlds.

Pavements and publicness

… the ordinary practitioners of the city live "down below," below the threshold at which visibility begin. They walk – an elementary form of this experience of the city; they are walkers, *Wandersmanner*, whose bodies follow the thicks and thins of an urban "text" they write without being able to read it.

(de Certeau 1984, 93)

Walking has fascinated urbanists and artists for some time. Walter Benjamin's pedagogy of walking conjured politicized people as "botanizing" the pavement. Michel de Certeau's (1984) *flânerie* is regularly invoked to substantiate this impulse, although Benjamin and de Certeau have both been critiqued for their gendered interpretations of the role of the *flâneur*. For Ash Amin and Nigel Thrift (2002) who recognize the importance of the everyday and walking scale, aimless meandering is not enough. "The flâneur's poetic of knowing is not sufficient. The city's transitivity needs to be grasped through other means" (14). They argue that other forms of knowledge are also needed to supplement the "wandering/wondering": maps, novels, and films in addition to academic enquiry. Fundamental to this is overlaying the poetic and the political in the complex power relations that shape public life. The pavement therefore becomes an important place where these congregate in cities, and, as Annette Miae Kim (2015) asserts "the sidewalk space of the city is the city to most people" (2).

Pavements are essentially interstitial spaces – in-between here and there, public and private, local and global, legal and illegal – and are where the social, economic, political and spatial intersect in complex, contested, coercive and collective ways – "Jane Jacobs (1961) called sidewalks 'the main public places of the city' and 'its most vital organs' … active sites of socialization and pleasure" (Loukaitou-Sideris and Ehrenfeucht 2009, 3). They are places brimming with possibility, but they are also heavily regulated.

The pavement is often the site of struggle between the competing and often irreconcilable agendas of economic development and informality; property market logics and sociospatial justice. As Loukaitou-Sideris and Ehrenfeucht (2009) assert, "public spaces have never been just, access has never been universal, and systematic solutions for public spaces have never been meant to integrate the priorities of all users" (272). This can be seen in regular street clearances where those deemed undesirable – primarily street traders and the indigent – are, often forcibly, removed from pavements. Public art is also heavily regulated through often punitive by-laws, where particularly marginalized youth are criminalized through the illegality of graffiti. The nature of pavements are also distinctive (Loukaitou-Sideris and Ehrenfeucht) depending on the neighborhood. This is most evident in southern cities where informal or slum dwelling can be the norm for up to 70% in some African cities. In urban contexts where private spaces are small, leisure activities spill over onto the sidewalks. Board games, pool tables, and betting games abound, and children and drunken revelers often share pavements as their playgrounds. Pavements therefore produce a plurality of publicness and public encounter.

But pavements are not merely vessels for human activity. Their form shapes the experiences of public life, and function as a representational mechanism in their materiality. According to Andrew Hickey (2010) "[t]his is what streets do – they give us access to collective, contemporary culture, but in ways that seem ordinary, or everyday" (161). As a result, pavements are microcosms where local and global power dynamics play out. For Hickey, the street is not benign:

> The street is the teacher we don't realize is there, sending out imagery and signage at every turn, requiring mediated behaviors as we negotiate the people and places it leads to and draw on our accumulated knowledge (our "street smarts") to safely arrive at the destinations we set out for. The everyday-ness of the street masks its influence; the mundanity of the street as a product of urbanized landscapes sees us encountering these spaces regularly but unquestioningly.
>
> *(161)*

Hickey is primarily concerned with the way signs – such as advertising and graffiti – shape community identity, much like Sarah Nutall (2008) does in her assessment of style in Johannesburg. For Hickey and Nutall, these signs fundamentally shape youth and consumerist culture. It is not only these signs that shape culture, and as assemblage thinkers in urban studies suggest, to understand public space, it must be thought of less as a vessel for civic life, and more as a set of relations between people and things (McFarlane 2011, 206).

Although relational thinking can run the risk of being "domesticated [by the] *soi-disant* politics of postmodernism" according to Peter McLaren (2010), the possibilities Colin

McFarlane refers to are interesting to consider in the context of public-facing art (565). Art has become recognized as an important tactic for understanding how the poetics and politics of the city overlay, and as Maxine Greene (2000) maintains, "we also have our social imagination: the capacity to invent visions of what should be and what might be in our deficient society, in the streets where we live, in our schools" (5).

Thinking of the pavement as a place where the social imagination can be triggered opens up another way of using public-facing art in understanding urban dynamics. This has largely got to do with the ability of art to operate beyond the real and the rational in order to tap into the affective dimensions of city life (Sitas and Pieterse 2013).

Public-facing art

… everywhere, artists working right now may be onto more far reaching ways of communicating what contemporary city life and cities are about. The city is always suspended as a case of 'heres' and 'elsewheres,' connected yet – yet … and that is why artists may be doing a better job than southern, or northern, theorists in 'painting,' 'composing,' 'dancing' and 'writing' cities into being. It remains to scholarship to go further.

(Mabin 2014, 32)

There is a rich discourse around public art globally (Deutsche 1998; Kester 2004; Kwon 2004; Bishop 2012; Evans 2018; Zebracki and Palmer 2018), but the term 'public art' runs the risk of being an all-encompassing term – including sanctioned interventions such as public sculpture and monuments, art murals, public performances, as well as subversive practices such as graffiti and art-based protest performance. The discourse largely stems from Anglophone and northern centric perspectives. This section seeks to be more specific – identifying art practices that take the interaction with publics more seriously – facing publics, in public, in critical ways in southern contexts (Sitas 2016).

Drawing from three examples – The *Chale Wote Street Art Festival* in Accra, *pumflet* in various places in the Western Cape, and dala's *CityWalk* in Durban – this section explores the role of cultural and creative life in shaping spaces in sometimes surprising ways. Although there has been an increased focus on the potential of public political becoming (Harvey 2012; Amin 2015), urbanists may not be as adept as engaging the sensory, transient, and ephemeral affects of socio-spatial entanglements as artists (Miles 1997; Bishop 2012; Sitas and Pieterse 2013; Sitas 2016; Zebracki and Palmer 2018). This section argues that creative action in interstitial public spaces such as streets and pavements can shape new public imaginaries.

CHALE WOTE street art festival

It is often said that you can't know where you are going until you look at where you came from. That is to say that if you want to create or re-imagine your future consider your past and what could be changed for you to achieve the desired future … Sionne Neely, co-founder of ACCRA [dot] ALT, the festival's organizers, said that the idea behind the theme was to "recapture these narratives of trauma, pain and survival and creating a different kind of interpretation of our destiny." (Adwoa McTernan 2013)

ACCRA[dot]ALT is an Accra-based organization exploring the role of culture "as a critical motor of sustainable development for the continent" (ACCRA[dot]ALT 2017b).

The organization emerged to respond to limited support for emerging artists in West Africa and has four activities: *Chale Wote Street Art Festival*, Sabolai Radio, Talk Party Series, and Strolling Goats. "We articulate an alternative vision of Africa by a creative and eclectic corps of young people. These social engineers are also invested in transforming their communities by creating imaginative solutions to the problems of poverty, unemployment, and miseducation" (ACCRA[dot]ALT 2017b). The *Chale Wote Street Art Festival* started in 2011 as "an alternative platform that brings art, music, dance, and performance out of the galleries and onto the streets of James Town, Accra" (Figure 29.1) (ACCRA[dot]ALT 2017a). The idea stemmed from discussions as part of ACCRA[dot]ALT's Talk Party Series, and in its eighth iteration has grown from strength to strength.

Festivals, historically, "have been important forms of social and cultural participation, used to articulate and communicate shared values, ideologies and mythologies," but increasingly

Figure 29.1 Chale Wote Street Art Festival, Accra, Ghana
Image credit: Simbi Seam Nkula

festivals have become popular top-down tactics for place branding (Bennett, Taylor, and Woodward 2014, 1); often copying and pasting festival forms from elsewhere, that may unintentionally may erase political histories (Perry, Auger, and Sitas 2018). The *Chale Wote Street Art Festival* has taken an active stance to avoid this approach to the historic James Town. Each festival has a theme that interrogates the past in the present, while cultivating future imaginations for Accra specifically, but African cities more generally as well.

James Town is one of the oldest colonial city districts in Accra, Ghana. Although largely inhabited by fishing communities, the neighborhood is a popular tourist attraction for Ghana's colonial past. James Town, according to Adwoa McTernan (2013), reveals "the very present and very real stories of slavery, colonialism and independence." These histories are not often spoken about in public discourse or schools, and there is an increasingly vocal caution about colonial nostalgia in African cities. The *Chale Wote Street Art Festival* actively encourages artists to engage with the socio-spatial histories of the city and the continent, resulting in critical engagement of the political power of activating the pavement. The neighborhood is not merely a vessel for action as many attempts at festivalization are – the spaces in the city are constitutive in themselves. The pavement becomes a place of engagement with spaces and times beyond the here and now – the colonial architecture is not fetishized; it is questioned in creative and critical ways.

In addition, in a context of an unequal city, where many people live in precarious situations, the festival organizers have made concerted attempts at ensuring local participation, particularly of youth and children. The festival is an artist led phenomenon committed to critical creative practices, but also promoting both formal and informal creative economies – providing a space for those falling outside of the mainstream art market to present and promote their work. African cities are quintessentially young (average age is 19) with high levels of unemployment, and in contexts with minimal cultural funding, the festival is carving out a new urban lexicon that youth on the continent can relate to. This challenges normative assumptions about expressions in and of public spaces.

pumflet

pumflet: art, architecture, and stuff is a publication series exploring the social imagination, stories of neighborhoods and reflecting on histories of the present. *pumflet*'s aim is to publicize research-in-process and to conceive of interventions in space and public culture based on research. It is a collection of conceptual art interventions and a collection of correspondence art practices.

(Wolff and Wa Lehulere 2017)

Ilze Wolff is one of the founding architects of Wolff Architects and Kemang Wa Lehulere is a visual artist, and collectively they have spent the last three years exploring the intersection of art, architecture (and stuff). The purpose of their collaboration was designed to explore artful means of architectural research that moves beyond the mere physicality of the built environment. Each exploration has involved looking at the social and political histories related to particular buildings and sites.

The first *pumflet* – "Daar gaan die Alabama" – explored the history of the Alabama Cinema and its demolition in 1984. There used to be a plethora of cinemas dotted in neighborhoods across Cape Town. These were important social and political sites during apartheid. The Alabama Cinema was mid-screening when the bulldozer arrived for demolition. The project involved the re-screening of the interrupted film on the pavement where the old cinema once stood; the display of newspaper clippings in the corner shop that now occupies the site; and the publication of *pumflet*, a set of letter exchanges dwelling on the events surrounding the demolition. The second *pumflet* – "Gaiety" – is also an ode to a cinema in a neighborhood in Stellenbosch. The Gaiety Bioscope was a cinema for non-white people in Die Vlakte and the intervention became guided walk by Wilfred Damon, an ex-resident, "a publication of Wilfred's recollections of both events: the earthquake inter-rupted screening at Gaiety Bioscope and the non-screening of La Bohème" (Wolff and Wa Lehulere 2017).

"Gladiolus" was the third *pumflet* produced. This iteration documents a conversation between residents who were forcibly removed during apartheid, visual documentation from the archive, and the poetry of Gladys Thomas, a current resident of the apartheid con-structed neighborhood of Ocean View that still remains today. The most recent *pumflet* – Rondhuis – was developed in response to the empty 2018 South African Pavilion in Venice Architecture Biennale. Wolff and Wlehulere publicized an open call for submissions of photographs:

> We invite you to submit your photograph of an oppressive space. We invite you to join the conversation around space, freedom, and non-freedom. We invite you to challenge, with us, the oppressive bureaucracies, which through their indiffer-ence, stifle public conversations on the unfreedom experienced by architecture and doing so, the articulation of ways out. We invite you to search with us for more antidotes.
>
> *(Wolff and Wa Lehulere 2017)*

The exhibition elicited hundreds of entries and provided a space for collectively challenging the notion and role of architecture "in the choreography of daily life."

pumflet is an ongoing initiative and continues to explore how to research and tell stories about collective public life. It makes direct links between contemporary social histories and everyday lived realities in the present. It explores the relationality between the historical, the social and the material in experimental ways (Figure 29.2). In doing this, it blurs the bound-aries between public and private, personal, and political. Intimate everyday stories are over-laid and threaded through political histories that are intimately connected to place. Again, public spaces are not merely vessels for engagement between people, but are also seen as vital components of democratizing the mortal and material.

Dala's CityWalk

> The CityWalk is an investigative journey, an exuberant exploration as well as a humbling and cautionary tale, an allegory on the infinite complexities of space and timings in Durban (South Africa).
>
> *(dala 2008)*

The conceptual starting point of the *CityWalk* (Figure 29.3) stems from the informal settle-ment of uMkhumbane (Cato Manor) and follows the pedestrian route that thousands of people take every day, through middle-class suburbs, and along the side of the main freeway entering Durban, as a way to save public transport fare to get to Warwick Triangle where the central minibus taxi rank lies. The route traverses a microcosm of South African city life and is therefore a useful tool for unpacking bigger urban issues. The walk passes through

Figure 29.2 pumflet, Cape Town, South Africa
Image credit: Ilze Wolff and Kemang Wa Lehulere

quintessential South African spaces: a squatter settlement, low-cost housing, a shopping mall, a middle-class suburb, a working-class suburb, an inner city "no-go" zone, pavements with and without street traders, a range of formal and informal markets. It also moves in and alongside different types of mobility, following footpaths, public transport (minibus taxis, buses, and train routes), and private transport (cars traveling on the roads and the major freeway).

Six lanes of traffic and six paths carved from footfall converge in Warwick Triangle, which is the gateway to Durban. Around half a million people circulate through Warwick Triangle daily. Warwick Triangle also boasts the largest number of markets including the Early Morning Market (fresh produce), the Muthi Market (medicinal herbs), a fresh fish market, a meat market, the Bovine Head Cookers market (street café selling cow brains), The Victoria Street Market (spices and curios), and Brooke Street Market (clothes, shoes and apparel). Informal street trade lines the pavements and it is the central transport inter-change for minibus taxis, buses, and the train. Warwick Triangle, although the busiest transport node in Durban, is perceived as a "no-go" zone for the affluent. According to Simbao (2013), in many ways one can read jahangeer's CityWalk

373

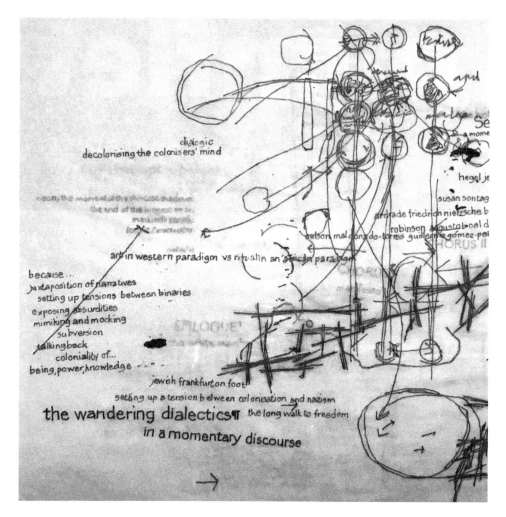

Figure 29.3 CityWalk, Durban, South Africa
Image credit: doung jahangeer

performances as a response to blindness, or at least as a way of challenging our blind spots with regard to the spaces around us, particularly spaces still marred by the laws of Apartheid. He encourages his participants to read between the lines of road markings, cracks in the tar, signposts, and a palimpsest of tar, dirt, and discarded objects that indicate the lurking presence of feet gone by. (411). Artist-architect doung jahangeer, spent almost 15 years of walking the same route which has meant building relationships with people along the way – from squatters in an informal section of uMkhumbane, to shop owners and security staff, the mall managers and the drug dealers, market traders and street traders, and the local municipal bureaucrats and NGOs. The *CityWalk* has become a network of relationships that cross between the grassroots and the gentry, but that is fundamentally located around the intersection of the social, economic, political, and spatial on the pavement. According to O'Toole (2014) sensory-rich, his purposeful meander was neither a tourist reverie nor a faux intellectual field trip. It was an open-ended and exploratory event, idealistic in its own way, but also fundamentally respectful of the travails of working class citizens navigating an increasingly sprawling and motorized post-apartheid city still marked by social divisions.

In order to explore the activation of urban narratives, and with the formation of dala, a series of *CityWalk* inspired projects were facilitated. Working with the KwaZulu-Natal Society for the Arts (KZNSA)'s *Young Artists' Project* (2009), dala hosted four artists from southern Africa. The purpose of this was to develop a cross-border public dialogue on cities, space, and art for social change in southern Africa. Three of the four artists who were selected became fascinated with the market-mall dichotomy and embarked on explorative artistic experiments to unpack the binary in a context where the market area of Warwick Triangle was being threatened of demolition in preparations for hosting the FIFA 2010 World Cup. *TIME_FRAME 2010* took up this challenge more explicitly during the World Cup, producing a series of digital art and performative installations. *Interface* (2011) brought anti-apartheid struggle artists (a painter, poet, photographer, and musician) living in uMkhumbane, together with contemporary new media artists to produce a series of video works installed on pavement spaces around a car wash along the route. In collaboration with local universities' urban design and performance departments, the *CityWalk* was explicitly used as a pedagogical tactic for urban learning – engaging students in developing site-specific creative interventions. The materiality of public spaces are as crucial as the encounters enabled through the creative initiatives.

Conclusion

[d]ialogue with the people is radically necessary to every authentic revolution … Sooner or later, a true revolution must initiate a courageous dialogue with the people. Its very legitimacy lies in that dialogue … The earlier dialogue begins, the more truly revolutionary will the movement be …

(Freire 1970, 98–9)

Chale Wote Street Art Festival, pumflet, and the *CityWalk* are all examples of public-facing art experiments aimed at engaging multiple publics, in public spaces. These examples challenge normative assumptions around public space and public life in three ways.

Firstly, unlike many public art interventions that are installed or happen on squares, in parks and in discreetly identified public spaces, all three examples spill out onto streets, and take place on pavements. The projects demonstrate that tapping into everyday practices that coalesce on pavements offer opportunities for re-thinking public–private binaries, and publicness. The quality of urban life is usually measured by the design of the public space and the kinds of human engagement it enables, but these projects demonstrate that seemingly banal interstitial spaces may offer more conducive opportunities for engaging public-political imaginaries.

Secondly, all three are also committed to exploring colonial pasts in the present through curating and choreographing intimate inter-generational encounters (Miles 1997) with diverse people, places, and histories. Whereas many creative interventions are critiqued for erasing political histories, bolstering place branding at the expense of socio-spatial justice, these projects enable spaces for critical conversations about place identities (Dovey 2014). They offer a range of readings of human-material entanglements, and particularly the past in the present. They also challenge the perception of African cities as "cesspools of disease and filth" and value cultural action as an important part of city life (Kim 2015, 6). These kinds of decolonial actions and imaginaries are crucial for realizing more just and equitable cities.

Thirdly, these spaces are essentially dialogic. Although Freire (1970) was referring to dialogue between people, these examples offer the opportunity to recast Freire's thinking through spatial and temporal lenses – the projects are underpinned by relational conversations, inspired by radically re-thinking relationships between people and places. Although these examples may not permanently alter the public spaces in which they inhabit, they create transitional moments (von Kotze and Wildemeersch 2014) where transformative thinking and learning is paramount: *Chale Wote Street Art Festival* transforms a part of the city steeped in the trauma of colonialism and slavery into a critical and creative engagement with histories, producing a new affective language to process the past, present, and future; *pumflet* creates a platform for conversations around social and political histories of place – where the architecture is as important a consideration as the socialities that it creates; and *dala* places political and cultural action through an explicit dialogic pedagogy of co-producing new urban knowledge creatively.

Finally, these three examples demonstrate that forms of resistance, of occupation, and political engagement need not happen in normative approaches to politics in public spaces – for example protest. In many African cities, political occupation through protest is common-place and often runs the risk of resulting in violent clashes with authorities. These projects engage an affective politics of enrolment that may capture the political imagination – particularly of youth – more than formal political forums and forms (Sitas and Pieterse 2013). The artful urban encounters between diverse publics and places allow an intimacy with social, spatial, and political histories that enrich democratic publicness and critical debate through less-threatening tactics than normative politics or scholarship may ordinarily allow.

This chapter started by challenging normative aspirations of public spaces as imagined by urban thinkers, before offering pavements as potential sites for public action. Drawing on three examples, the chapter argues that re-thinking publics and public spaces through public-facing art-architecture practices can offer new ways of exploring the perspectives of democratic and decolonizing potential of public space.

References

ACCRA[dot]ALT. (2017a) "Chale Wote Street Art Festival." Accessed July 13, 2018. http://accradotal tradio.com/chale-wote-street-art-festival/

ACCRA[dot]ALT. (2017b) "Vision and Mission." Accessed July 18, 2018. http://accradotaltradio.com/ vision-mission/

Adwoa McTernan, B. (2013) "Futuristic Folklore in James Town, Ghana." *Africaisacountry*, 16 September. Accessed 18 July 2018. https://africasacountry.com/2013/09/futuristic-folklore-in-james-town-ghana

Amin, A. (2006) The Good City. *Urban Studies*, 43(5/6): 1009–1023.

Amin, A. (2008) Collective Culture and Urban Public Space. *City*, 12(1): 5–24.

Amin, A. (2015) Animated Space. *Public Culture*, 27(2).

Amin, A. and Thrift, N. (2002) *Cities: Reimagining the Urban*. Cambridge: Polity Press.

Arendt, H. (1958) *The Human Condition*. Chicago: The University of Chicago Press.

Bayat, A. (1997) Un-civil Society: The Politics of the 'Informal People'. *Third World Quarterly*, 18(1): 53–72.

Bennett, A., Taylor, J., and Woodward, I. (2014) *The Festivalization of Culture*. Fanham: Ashgate.

Bishop, C. (2012) *Artificial Hells: Participatory Art and the Politics of Spectatorship*. London: Verso.

dala. (2008) "CityWalk." Accessed July 13, 2018. www.dala.org.za/citywalk2.html

de Certeau, M. (1984) *The Practice of Everyday Life*. Berkeley: University of California Press.

Deutsche, R. (1998) *Evictions: Art and Spatial Politics*. Cambridge: MIT Press.

Dovey, K. (2014) Planning and Place Identity. In: *The Ashgate Research Companion to Culture and Planning*, Young, G. and Stevenson, D. (eds). Farnham: Ashgate, pp. 257–272.

Evans, F. (2018) *Public Art and the Fragility of Democracy*. New York: Columbia University Press.

Fortuin, A. and Sitas, R. (2019) "Too Many Men." Perceptions of Public Spaces in Mitchel's Plein. Cape Town.

Freire, P. (1970) *Pedagogy of the Oppressed*. New York: Continuum.

Greene, M. (2000) *Releasing the Imagination: Essays on Education, the Arts, and Social Change*. Minneapolis: Jossey-Bass.

Habermas, J. (1962) *The Structural Transformation of the Public Sphere: An Inquiry into a Category of Bourgeois Society*. Cambridge: Polity Press.

Hardin, G. (1968) The Tragedy of the Commons. *Science*, 162(3859): 1243–1248.

Harvey, D. (2012) *Rebel Cities: From the Right to the City to the Urban Revolution*. London: Verso Books.

Hickey, A. (2010) When the Street Becomes and Pedagogue. In: *Handbook of Public Pedagogy: Education and Learning Beyond Schooling*, Sandlin, A., Schultz, B. and Burdick, J. (eds). New York: Routledge, pp. 161–170.

Jeffrey, A., McFarlane, C., and Vasudevan, A. (2012) Rethinking Enclosure: Space, Subjectivity and the Commons. *Antipode*, 44(4): 1247–1267.

Kester, G. (2004) *Conversation Pieces: Community and Communication in Modern Art*. Berkeley: University of California Press.

Kim, A.M. (2015) *Sidewalk City*. Chicago: University of Chicago Press.

Kwon, M. (2004) *One Place after Another: Site-Specific Art and Locational Identity*. Cambridge: MIT Press.

Lefebvre, H. (1991) *The Production of Space*. Oxford: Blackwell Publishers.

Loukaitou-Sideris, A. and Ehrenfeucht, R. (2009) *Sidewalks: Conflict and Negotiation over Public Space*. Cambridge: MIT Press.

Mabin, A. (2014) Grounding Southern City Theory. In: *The Routledge Handbook on Cities of the Global South*, New York: Routledge.

Maclellan, M. and Talpalaru, M. (2009) Editor's Introduction: Remaking the Commons. *Reviews in Cultural Theory*, 2(3): 1–5.

Mbembe, A. and Nutall, S. (2008) *Johannesburg: The Elusive Metropolis*. Durham: Duke University Press.

McFarlane, C. (2011) Assemblage and Critical Urbanism. *City*, 15(2): 204–224.

McLaren, P. (2010) This First Called My Heart: Public Pedagogy in the Belly of the Beast. In: *Handbook of Public Pedagogy: Education and Learning Beyond Schooling*, Sandlin, J.A., Schultz, B.D., and Burdick, J. (eds). London: Routledge, pp. 564–572.

Miles, M. (1997) *Art, Space, and the City: Public Art and Urban Futures*. London: Routledge.

Nutall, S. (2008) Styling the Self. In: *Johannesburg: The Elusive Metropolis'*, Mbembe, A. and Nutall, S. (eds). Johannesburg: Wits University Press.

O'Toole, S. (2014) "Commuters: Rush Hour Rest Stop." *Goethe Institute*, July. www.goethe.de/ins/za/en/kul/fok/auu/20807516.html

Ong, A. (2006) Mutations in Citizenship. *Theory, Culture & Society*, 23(2–3): 499–505.

Perry, B., Auger, L., and Sitas, R. (2018) Cultural Heritage Entanglements: Festivals as Integrative Sites for Sustainable Urban Development. *International Journal of Heritage Studies*, [*forthcoming*] DOI: 10.1080/13527258.2019.1578987.

Pieterse, E. (2008) *City Futures: Confronting the Crisis of Urban Development*. Cape Town: Zed Books.

Rose, G. (2001) *Visual Methodologies*. London: SAGE.

Simbao, R. (2013) Walking the Other Side: doung jahangeer. *Third Text*, 27(3): 407–414.

Simone, A. (2010) *City Life from Jakarta to Dakar*. London: Routledge.

Sitas, R. (2015) Community Centres in Crisis: The Story of the Tsoga Environmental Resource Centre. In: *State/Society Synergy*, Brown-Luthango, M. (ed). Cape Town: African Centre for Cities, pp. 178–199.

Sitas, R. (2016) Public-Facing Art and African Cityness. In: *Public Art in Africa: Art and Urban Transformations in Douala* Online edition, Pensa, I. (ed). Geneva: MetisPresses.

Sitas, R. and Pieterse, E. (2013) Democratic Renovations and Affective Political Imaginaries. *Third Text*, 27(3): 327–342.

von Kotze, A. and Wildemeersch, D. (2014) Creative Encounters in Public Art and Public Pedagogy. *Studies in Art Education*, 55(4): 313–327.

Watson, V. (2014) African Urban Fantasies: Dreams or Nightmares. *Environment and Urbanization*, 26(1): 215–231.

Wolff, I. and Wa Lehulere, K. (2017) "Pumflet: Art, Architecture and Stuff." *Oharchitecture*. http://oharchitecture.blogspot.com/

Zebracki, M. and Palmer, J. (2018) *Public Encounter: Art, Space and Identity*. New York: Routledge.

30

BRIDGING AND BONDING

Public space and immigrant integration in Barcelona's *el Raval*

Jeremy Németh

Introduction

Immigration has become one of the most important political and social issues of the 21st century. Politicians in the U.S., Brazil, Hungary, Italy, Sweden, and elsewhere have campaigned successfully on nationalist platforms, and in many countries anti-immigrant discourse, racially discriminatory incidents, and hate crimes are at an all-time high (Kishi 2017; Lopez, Gonzalez-Barrera, and Krogstad 2018). Such sentiments are particularly pronounced in countries that have struggled to create positive environments into which immigrants can integrate, and leaders of these countries frequently blame immigrants for their inability or unwillingness to adapt to the norms of the receiving society (Saunders 2010; Bittner 2018; Nakamura and Witte 2018).

As the vast majority of immigrants to the U.S. and Europe settle in cities and suburbs, planners and designers must play a larger role in shaping the everyday experiences of immigrants and, in turn, shaping the broader landscape of immigration worldwide. In this chapter, I examine the role that the built environment, and public spaces more specifically, play in creating positive "contexts of reception," or the combination of factors that shape how welcome immigrants feel in receiving communities (Portes and Rumbaut 2014). Scholars have long understood that neighborhoods play a powerful role in determining quality of life and social mobility, but few have unpacked the ways in which the immigrant experience is structured by the physical environment, in particular the streets, parks, squares, and other "sites of social infrastructure" that represent the quintessential stages for sociocultural activities (Mehta 2013; Chetty , Hendren, and Katz 2014; 2016; Klinenberg 2018).

Using the case of *el Raval*, a long-time immigrant arrival neighborhood in Barcelona's *Ciutat Vella* (Old City), I conduct interviews, field research, and a document review to explore why immigrant integration tends to thrive in certain contexts over others, and what lessons planners and designers can learn from this rich and complex neighborhood. I ask not whether *el Raval*'s public spaces *create* positive contexts of reception on their own, but whether certain types and combinations of spaces might create the *potential* for the types of exchange and encounter that support immigrant integration (Talen 2006). I find that the neighborhood's combination of both bridging and bonding spaces may contribute to its identity as a thriving arrival neighborhood (Putnam 2000; Németh, 2019). To that end, planners and designers seeking to support immigrant integration should focus not only on creating diverse spaces of encounter that bridge difference but also more socially homogenous places where affinity groups can share concerns, claims, and interests.

Living together in difference

Immigrant integration is a dynamic, two-way process wherein both immigrants and the receiving society have responsibility to work together to build a cohesive community. Integration goes beyond border policies to focus on how immigrants are being incorporated within cities, suburbs, and regions, and specifically how societies foster improved economic mobility for, enhanced civic participation by, and openness to immigrants (Pastor et al. 2012).

Successful immigrant integration – one mutually developed and practiced – can enrich daily life and infuse receiving communities with new creativity and prosperity, increase empathy and understanding of the "Other," and can prepare immigrants and host communities for life in a globalized future (National Academies of Sciences, Engineering, and Medicine 2015). Given that immigrant integration is a two-way street, the onus falls not only on immigrants to engage civically but also on the receiving society to create positive contexts of reception. In the context of significant socio-economic, cultural, racial, and ethnic diversification urban scholars and policymakers are once again debating whether to adopt either *assimilationist* or *multiculturalist* approaches to immigrant policy.

An assimilationist or "melting pot" approach asks newcomers to adopt the social, political, and cultural customs of the receiving society (Bloemraad, Korteweg, and Yurdakul 2008). Proponents believe that absorption into host communities can increase access to resources, capabilities, and rights that ultimately lead to health and quality of life benefits for immigrants and receiving societies alike (Popay et al. 2008). Nevertheless, critics claim that assimilationist approaches can force immigrants to abandon deeply held cultural practices and identities in order to conform to mainstream societal norms (Young 1990). Putnam (2007) showed that residents of the most diverse cities in fact tend to "hunker down" and interact less with their neighbors than those in more homogeneous settings. In addition, fears of racism, discrimination, and persecution can discourage immigrants from living in diverse areas in favor of co-ethnic enclaves that might "normalize the presence of [immigrants] and allow safe movement through the area" (Wessendorf 2019; Staeheli, Mitchell, and Nagel 2009, 638). Valentine (2008) challenges the idealization of diverse spaces of encounter, arguing instead that close proximity and forced interactions can in fact aggravate hostilities and increase stereotyping between diverse groups.

Multiculturalist approaches to immigration policy tend to support natural affinities toward group-based differentiation and clustering and the retention of immigrant cultural heritage, the deepening of in-group social ties, and the strengthening of political identities (Bader 2016; Young 2000; Bloemraad 2006). Tight-knit, homogeneous communities have proven more resilient when confronting disasters, shocks, and economic downturns (Klinenberg 2002; Rumbach, Makarewicz, and Németh 2016). May (2001) argues that group differentiation provides an opportunity for marginal groups like immigrants to clarify internal positions and dynamics and prepare themselves for the public realm with less fear of retribution or persecution. Nonetheless, critics argue that multiculturalism can breed isolation and discourage loyalty by immigrants to the receiving society (Bissoondath 1994). Isolation can foment extremism by those who feel distanced from mainstream society, such as in the Parisian *banlieues* (Packer 2015). The well-documented ills of racial and ethnic segregation include the development of what some call a spiral of deprivation or a "culture of poverty," the continual decline in quality schools in economically-distressed areas, and a lack of social mobility for residents of low-income neighborhoods (Massey and Denton 1993; Coates 2014).

I argue that successful immigrant integration policy finds a middle ground between assimilation and isolation. Such policies should prioritize *bridging* social capital, or networks between socially heterogeneous groups, and *bonding* social capital, or connections within a more homogenous group or community (Putnam 2000). Instead, successful immigrant integration should support immigrants in retaining much of their cultural customs and ethnic heritage but also contributing to the societal whole. Iris Young (2000) calls this notion *differentiated solidarity*, a political and social ideal that describes, quite simply, how we might live together in difference. Differentiated solidarity is unique in affirming respectful engagement and mutual identification with diverse others *and* celebrating and encouraging group-based differentiation and residential clustering. Since all bridging and bonding actions require a stage, we now make the simple leap from the conceptual to the material public realm.

This is important because, to date, sociologists have almost entirely ignored the role the built environment plays in shaping the immigrant experience, and planners and designers have long overlooked how spatial structure might specifically play a role in immigrant integration, perhaps for fear of being considered environmental determinists (Talen 2006). Nevertheless, if we agree that poorly designed environments can "become a dumping ground for migrants, cut off from everything," and that "the very shape of a neighborhood" can keep it out of contact with a receiving society, then it follows that that thoughtfully-designed neighborhoods can provide *positive* contexts of reception for immigrant integration (Saunders 2010, 290–292). But just what does such a spatial arrangement look like and how might it function? I turn to Barcelona's *el Raval* to explore how public spaces might shape the immigrant experience.

El Raval

El Raval represents a site of exploration that, taken together with emerging scholarship, helps me to develop a conceptual understanding of how public spaces can impact immigrant integration. That the neighborhood has been able to grow and sustain its foreign-born population implies that certain aspects of the neighborhood might be welcoming to newcomers. It does not, however, imply that all immigrants there feel a sense of *belonging* or that all racial and ethnic groups in *el Raval* live in some paradisiacal solidarity. To make such claims, one must conduct in-depth surveys with old and new residents; in this study I rely solely on repeated personal observations, secondary data from published reports, and a handful of interviews with neighborhood residents.

In order to map key public spaces in the neighborhood, I conducted field observations over 13 months from 2015 to 2018. During these visits, which focused on observing street life and spatial patterns in *el Raval*, I took detailed field notes and hundreds of photographs and conducted informal interviews with residents and employees of local establishments, focusing particularly on how these people experienced changes to the social and physical fabric of the neighborhood since the early 1980s. I reviewed dozens of documents, plans, and peer-reviewed articles on the history of the neighborhood and conducted a spatial and socioeconomic analysis using data obtained from the municipality's statistical clearinghouse (Ajuntament de Barcelona Department d'Estadística 2018).

El Raval is a dense warren of jumbled streets to which immigrants have flocked for centuries (Degen 2017). Since its beginnings as a site of agricultural production to its later role as the city's industrial zone, the neighborhood has always been considered a marginal space (the Catalan word *raval* translates most closely to "suburb"). In the late 19th century, this working-class neighborhood was coined *Barri Xino* (Chinatown) by a journalist, a derogatory term referring to its renown for prostitution, drugs, and crime. Since the 1980s, the neighborhood has seen the influx of middle-class residents from around Spain and abroad, which some attribute to the municipality's commitment to supporting new tourist infrastructure, such as cultural facilities and museums. Along with this commitment came the destruction of traditional hotspots for drug dealing and prostitution (Degen 2003) with parallel incentives provided to organic grocers and craft beer establishments, among others (Arbaci and Tapada-Berteli 2012).

Despite its prime location in one of the most popular travel destinations in the world, residents of *el Raval* remain "marginal" in a number of ways. The neighborhood's RFD index, a measure of family income, is 75.8, compared to 100 for the rest of the city and to 108.5 for the adjacent *Barri Gotíc* (Ajuntament de Barcelona Department d'Estadística 2018). The cadastral value is $70/ square foot in *el Raval* compared to $84/square foot for the rest of the city, and only 23% of residents have a university education compared to 38% of *Barri Gotíc* residents. At 174 residents/acre, *el Raval* may just be the densest neighborhood in Europe and approaches levels found only in Dhaka, the world's densest city at 180 residents/acre (United Nations Human Settlement Programme 2017; Ajuntament de Barcelona Department d'Estadística 2018). A number of *secciós censals* (similar to US Census tracts) have densities in the 500+ residents/acre range.

Notably, 49% of the neighborhood's 47,605 residents are foreign-born compared to only 17% of Barcelona residents (Ajuntament de Barcelona Department d'Estadística 2018). This number has increased dramatically since 1991, when only 4.8% of *el Raval* residents were born outside Spain. Now, residents of 125 nationalities live within an area just slightly larger than New York City's Greenwich Village, a neighborhood that also houses four Christian churches, four mosques, three Hindu temples, and a Sikh *gurdwara* (Ajuntament de Barcelona Department d'Estadística 2018; Németh 2019).

The physical fabric of the neighborhood has changed remarkably in the past several decades as a result of plans that have coincided with major events in the city's history, including the transition from Francoism to democracy and, most notably, the city's hosting of the 1992 Olympics. Preparation for the Olympics helped catalyze significant improvements around the city, including connecting the seafront to the city, improving hundreds of public spaces, creating an Olympic Village, and introducing dozens of museums, stadia, and other public works on Montjuic. This set of interventions led by Heads of Planning Oriol Bohigas and Joan Busquets has been called "the Barcelona Model" and is the subject of both praise and scorn from urban scholars (Marshall 2004).

Some of the key projects include attempts in the 1990s and 2000s to increase light and air in the remarkably dense *el Raval* through a philosophy championed by Bohigas called *esponjamiento*, or a "hollowing out." Figure 30.1 shows how these projects altered the urban fabric of *el Raval*. Fernández (2014) estimates that the *esponjamiento* displaced around 10,000 long-time residents of the neighborhood. Degen (2003) argues that the intent of these public space projects was to "civilize" the neighborhood by attracting well-heeled newcomers and tourists. But others argue that the *esponjamiento*'s goal of creating new bridging spaces in *el Raval* was somewhat successful in creating new places of encounter (Ortiz, Garcia-Ramon, and Prats 2004; Subirats and Rius 2006). Indeed, these decades saw the development of three new bridging spaces in which, although not without conflict, people from all walks of life tend to congregate. Figure 30.2 includes all public spaces mentioned in the subsequent sections.

Figure 30.1 Figure ground showing changes to urban fabric between 1956 (left) and 2012 (right)
Image credit: Uta Gelbke (2014)

Bridging spaces

The first new space created was the *Rambla del Raval*, a large linear pedestrian space in the center of the neighborhood that required the controversial demolition of five blocks of residential properties in the 1990s (see Figure 30.3). Besides breaking up the dense labyrinth of streets in this part of the neighborhood, planners intended for this space to become a new focal point for *el Raval* similar to the city's iconic Las Ramblas that bisects the *Ciutat Vella*. The space now hosts a number of annual festivals led by community organizations and cultural associations including the *Festa Major del Raval* (Horta 2010). During most days, the *Rambla del Raval* is populated by groups Moroccan and Pakistani men, the presence of which some claim tends to exclude non-immigrants and women; by night, however, the space consists of a mix of male and female tourists and groups of Catalan youth (Ortiz, Garcia-Ramon, and Prats 2004).

The second bridging space is the set of grand public squares around the *Casa de Caritat* complex, which includes the famous *Museu d'Art Contemporani de Barcelona* (MACBA) and its *Plaça dels Àngels*, which attracts skateboarders from all over the world (see Figure 30.4). The space is populated throughout the day by a cacophony of regulars and tourists, from people experiencing homelessness to chess players to dog walkers to cycling kids (Degen 2017).

The third new bridging space is the *Plaça de Salvador Seguí*, a space just blocks from the *Rambla del Raval* bounded on one side by the *Filmoteca de Catalunya*, an independent movie theater constructed in 2012, and on the other by *Carrer d'en Robador*, an infamous street long known as the center of prostitution and drug dealing in the city. In the center of the space, however, sits a well-used playground filled with neighborhood children and fronted by a bustling tapas bar. By day, the space often hosts flea markets and small festivals, and crowds come from all over the city. Police occupy the space both day and night, more as a means of protecting sex workers and keeping the peace than enforcing drug and prostitution laws (Fernández 2014).

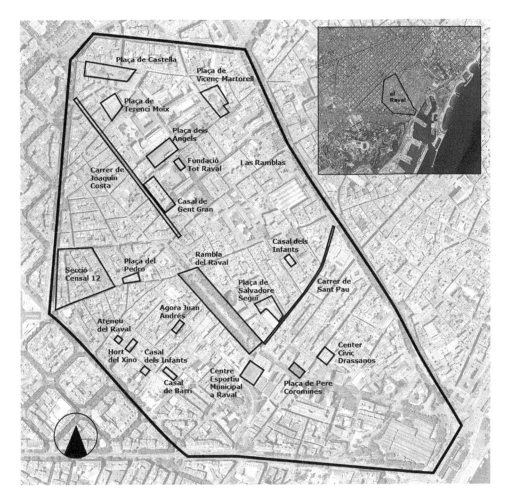

Figure 30.2 Map showing key public spaces in *el Raval*
Image credit: Jeremy Németh; Google Earth

In addition to these three newer bridging spaces, it is worth drawing attention to *Carrer de Joaquín Costa*, a long, vibrant, pedestrianized street filled day and night with activity (see Figure 30.5). The street is replete with hip bars and "third wave" coffee shops along with Italian pizza parlors, Turkish kebab houses, a Syrian tailor, Moroccan produce shops and *carnisseries*, Pakistani spice shops, Filipino salons, and *Torrons Licors*, one of the oldest continuously-operated, Catalan-run liquor stores in the city.

Bonding spaces

Complementing these spaces that intend to mix old and new residents are the dozens of community centers, places of worship, commercial clusters, community gardens, and other bonding spaces found throughout *el Raval*.

Plaça del Pedro and its surrounding streets house a cluster of Pakistani- and Moroccan-run fresh food shops. As of 2017, the small *secció censal* in which the space is located contains no

Figure 30.3 Rambla del Raval
Image credit: D. Mills

fewer than 35 stores selling "meat and pork," "fruits and vegetables," and "eggs and poultry" (Ajuntament de Barcelona Department d'Estadística 2018). Across the neighborhood, daytime beer drinkers – many long-time Raval residents – cluster in *Plaça de Pere Coromines*. Tourists and immigrants from Europe and the U.S. are the most frequent users of *Plaça de Vicens Martorell*, which contains a popular playground near Las Ramblas. In the university-adjacent part of the neighborhood, the *Plaça de Castella* is often full of boisterous *Universitat de Barcelona* students, and nearby *Plaça de Terenci Moix* is popular with Filipino teens playing full-court basketball games day and night. Within the larger bridging space of *Plaça dels Àngels* discussed earlier, clusters of skateboarders from around the world congregate, and while not skateboarding share cans of beer and *porros* (tobacco mixed with hashish) in the open air *plaça*.

In addition, a number of key bonding spaces are not outdoor *plaças* but are in fact what Amin (2002) calls "sites of banal transgression" such as recreation centers, libraries, shops, community centers, food markets, and day care centers. Berta Güell (2016) notes that 244 Pakistani-run businesses operate in the neighborhood, including 15 Pakistani-operated mobile phone stores along *Carrer de Sant Pau* alone. Commercial and residential density creates concentrations of activity and represents a key factor attracting immigrants to arrival neighborhoods as it creates positive contexts of reception for economic integration (Portes and Manning 1986). Business concentration is a distinctive feature of ethnic enclaves: a recent study of Pakistani workers in *el Raval* revealed that business clustering has created higher than average levels of social capital among this community due to the sharing of a co-ethnic workforce, among other factors (Ortiz, Garcia-Ramon, and Prats 2004).

Figure 30.4 Plaça dels Àngels
Image credit: F. Schiami

Related, *el Raval* houses hundreds of community groups and neighborhood organizations. The Department d'Estadística (2018) now counts 95 officially-recognized organizations in the areas of "health and assistance," "social services," and "associations." Subirats and Rius (2006) identified 571 cultural and arts-related businesses and over a dozen health centers, a number of *casals de barri* (neighborhood-based community centers), as well as three nursery schools, five primary schools, two secondary schools, and four libraries. The non-profit *Fundació Tot* Raval (2018) is an organization that itself brings together 68 community groups from the neighborhood, the majority of which are explicitly immigrant-serving.

Fostering immigrant integration

El Raval contains a number of bridging and bonding spaces interspersed throughout the neighborhood. It also includes one of the highest shares of foreign-born residents in the country; some attribute the neighborhood's resistance to the wholesale gentrification facing the rest of the city to the presence of immigrants in such large numbers and the stigma that accompanies it (D. Saurí, personal communication, 4 June 2018).

One of the key goals of Bohigas' *esponjamiento* was to open up the neighborhood physically and attract outsiders. Flagship public-space projects were relatively successful in doing so, even as they have been met with deep consternation due to the destruction of housing and devastating displacement of long-time residents, the majority of whom were poor or on fixed incomes. In his book *Matar al Chino*, Fernández (2014) argues that social ties were broken resulting in the death of *el Raval* as we know it. But rents have remained low, and immigration into the neighborhood skyrocketed during this time. I do not contend that these new public spaces *caused* this immigrant influx, but instead argue that they created key opportunities for co-presence and congregation across the diverse groups that have come to

Figure 30.5 Carrer de Joaquín Costa
Image credit: Sensaos

populate the neighborhood today. What is clear is that *el Raval* contains a rich mix of old-timers and newcomers, and the physical spaces of the neighborhood seem to support that mix, providing opportunities for bridging due to the propinquity that I believe is a necessary, albeit insufficient, precondition to building solidarity across difference. One resident interviewed by Subirats and Rius (2006) claims: "Probably you have a lot more to do with your neighbor who's Moroccan than with somebody from two streets over who is from Sant Gervasi [another Barcelona neighborhood] … " (43).

Nevertheless, just because diverse groups occupy the same space does not mean their interactions build trust, decrease stereotypes, or create lasting bonds that transcend contentious histories. In fact, many immigrants in *el Raval* lead parallel lives to native-born residents, rarely interacting but sometimes rubbing shoulders in the streets and sidewalks or in the workplace. A recent study of Pakistani women in *el Raval* partly attributes their high levels of life satisfaction to the dense concentration of immigrant-serving shops in the neighborhood (Valenzuela-Garcia, Parella, and Güell 2017). What is clear is that opportunities for bridging and bonding abound, and residents slip in and out of these spaces day and night. Yet just because immigrants live in *el Raval* does not mean their entire lives are experienced in the neighborhood itself; some work outside the neighborhood or regularly visit friends or family who live elsewhere.

Three key lessons emerge from this study of *el Raval*. First, scholars and practitioners alike must move beyond the either/or focus of assimilation vs. isolation toward a both/and approach that recognizes the importance of both iconic bridging spaces and everyday bonding spaces. Instead of focusing on individual spaces, planners and designers seeking to support immigrant integration should turn their focus to how *series* of public spaces can together shape the experiences and perceptions of users. Sociologists have long recognized the importance of ethnic enclaves and co-ethnic engagement for the recently-arrived immigrant, and this case study bears this out as immigrant groups in *el Raval* seem content to maintain a space of their own within their rapidly-changing neighborhood, sometimes long after their initial arrival.

Second, if we believe that place matters, then planners and designers must recognize the key role we play in the immigration debate and conduct detailed longitudinal research to understand how the built environment shapes actual perceptions of integration and belonging. Recent work by Eric Klinenberg (2018) has helped solidify the notion that material place and embodied social networks can foster sociocultural integration in an increasingly fragmented world. Whereas social interaction online is mediated and structured by algorithms and pre-established preferences, the material city is where we are forced to encounter and negotiate difference. Indeed, John Parkinson (2012) argues, "democracy [still] depends to a surprising extent on the availability of physical, public space, even in our allegedly digital world" (2).

Third, we must support policies that allow for the spatial arrangements known to foster immigrant integration. This is particularly important in the U.S., where immigration is becoming an increasingly suburban phenomenon. Hundreds of suburban communities – most of which were originally designed for middle-class, White, nuclear families – have become home to huge numbers of immigrants, who are in turn reshaping the physical and social fabric of their new communities. Local governments and service providers are often unprepared for the influx of new immigrant populations and the backlash that often ensues from long-time residents who feel their way of life is being threatened (Lung-Amam 2017). Proactive planners and designers should anticipate these transformations and take steps to address the concerns of long-time residents while creating positive contexts of reception for new arrivals.

This chapter is intended as an entry point into the relationship between immigration and public space in a thriving immigrant arrival neighborhood. It is now up to planning and design scholars and professionals to start playing a more powerful role at this critical juncture.

References

Ajuntament de Barcelona Department d'Estadística (2018) Visualitzador de dades estadístiques. *City of Barcelona.* Accessed 15 July 2018. http://estadistica.bcn.cat

Amin, A. (2002) Ethnicity and the Multicultural City: Living with Diversity. *Environment and Planning A*, 34: 959–980.

Arbaci, S. and Tapada-Berteli, T. (2012) Social Inequality and Urban Regeneration in Barcelona City Centre: Reconsidering Success. *European Urban and Regional Studies*, 19: 287–311.

Bader, M. (2016) *Diversity in the DC Area: Findings from the 2016 DC Area Survey.* Washington, DC: Metropolitan Policy Center and the Center for Latin American and Latino Studies, American University.

Bissoondath, N. (1994) *Selling Illusions: The Cult of Multiculturalism in Canada.* Toronto: Garamond Press.

Bittner, J. (2018) "How the Far Right Conquered Sweden." *New York Times*, 6 September. Accessed 6 September 2018. www.nytimes.com/2018/09/06/opinion/how-the-far-right-conquered-sweden.html

Bloemraad, I. (2006) *Becoming a Citizen: Incorporating Immigrants and Refugees in the United States and Canada.* Berkeley: University of California Press.

Bloemraad, I., Korteweg, A., and Yurdakul, G. (2008) Citizenship and Immigration: Multiculturalism, Assimilation, and Challenges to the Nation-State. *Annual Review of Sociology*, 34: 153–179.

Chetty, R. et al. (2014) Where is the Land of Opportunity? The Geography of Intergenerational Mobility in the United States. *The Quarterly Journal of Economics*, 129: 1553–1623.

Chetty, R., Hendren, N., and Katz, L. (2016) The Effects of Exposure to Better Neighborhoods on Children: New Evidence from the Moving to Opportunity Experiment. *American Economic Review*, 106: 855–902.

Coates, T. (2014). "The Case for Reparations." *The Atlantic* (June). www.theatlantic.com/magazine/arch ive/2014/06/the-case-for-reparations/361631/

Degen, M. (2003) Fighting for the Global Catwalk: Formalizing Public Life in Castlefield (Manchester) snd Diluting Public Life in El Raval (Barcelona). *International Journal of Urban and Regional Research*, 27: 867–880.

Degen, M. (2017) Urban Regeneration and "Resistance of Place:" Foregrounding Time and Experience. *Space and Culture*, 20: 141–155.

Fernández, M. (2014) *Matar al Chino: Entre la revolución urbanística y el asedio urbano en el barrio del Raval de Barcelona*. Barcelona: Virus Editorial.

Gelbke, U. (2014) Zero Points: Urban Space and the Political Subject. In: *Architecture Against the Post-Political: Essays in Reclaiming the Critical Project*, Lahiji, N. (ed). New York: Routledge, pp. 167–179.

Güell, B. (2016) The Backstage of Pakistani Businesses in Barcelona: Unravelling Strategies from the Ground. *South Asian Diaspora*, 8: 15–30.

Horta, G. (2010) *Rambla del Raval de Barcelona*. Barcelona: El Viejo Topo.

Kishi, K. (2017) "Assaults against Muslims in US Surpass 2001 Level." *Pew Research Center*. Accessed 3 November 2018. www.pewresearch.org/fact-tank/2017/11/15/assaults-against-muslims-in-u-s-sur pass-2001-level/

Klinenberg, E. (2002) *Heat Wave: A Social Autopsy of Disaster in Chicago*. Chicago: University of Chicago Press.

Klinenberg, E. (2018) *Palaces for the People: How Social Infrastructure Can Help Fight Inequality, Polarization and the Decline of Civic Life*. New York: Random House.

Lopez, M., Gonzalez-Barrera, A., and Krogstad, J. (2018) "More Latinos Have Serious Concerns about their Place in America under Trump." *Pew Research Center*. Accessed 3 November 2018. www.pewhis panic.org/2018/10/25/more-latinos-have-serious-concerns-about-their-place-in-america-under-trump/

Lung-Amam, W. (2017) *Trespassers? Asian Americans and the Battle for Suburbia*. Berkeley: University of California Press.

Marshall, T. (2004) *Transforming Barcelona: The Renewal of s European Metropolis*. New York: Routledge.

Massey, D. and Denton, N. (1993) *American Apartheid: Segregation and the Making of the Underclass*. Cambridge: Harvard University Press.

May, R. (2001) *Talking at Trena's: Everyday Conversations at an African American Tavern*. New York: NYU Press.

Mehta, V. (2013) *The Street: A Quintessential Social Public Space*. New York: Routledge.

Nakamura, D. and Witte, G. (2018) "'The Good Times for Illegals are Over:' Trump Finds Allies in Europe's Anti-Immigration Movement." *Washington Post*, 20 June. Accessed 15 August 2018. www. washingtonpost.com/politics/the-good-times-for-illegals-is-over-trump-finds-allies-in-europes-anti-immigration-movement/2018/06/19/1a1da27a-73d5-11e8-805c-4b67019fcfe4_story.html

National Academies of Sciences, Engineering, and Medicine. (2015) *The Integration of Immigrants into American Society*. Panel on the Integration of Immigrants into American Society, Committee on Population, Division of Behavioral and Social Sciences and Education. Washington, DC: The National Academies Press.

Németh, J. (2019) Designing for Difference in Barcelona's El Raval. In: *New Companion to Urban Design*, Banerjee, T. and Loukaitou-Sideris, A. (eds). New York: Routledge, pp. 122–134.

Ortiz, A., Garcia-Ramon, M., and Prats, M. (2004) Women's Use of Public Space and Sense of Place in the Raval (Barcelona). *GeoJournal*, 61: 219–227.

Packer, G. (2015) The Other France. *The New Yorker*, 31 August.

Parkinson, J. (2012) *Democracy and Public Space: The Physical Sites of Democratic Performance*. Oxford: Oxford University Press.

Pastor, M. et al. (2012) "California Immigrant Integration Scorecard." *Center for the Study of Immigrant Integration.* Accessed 10 December 2018 https://dornsife.usc.edu/assets/sites/731/docs/California_Immigrant_Integration_Scorecard_web.pdf

Popay, J., Escorel, S., and Hernandez, M. (2008) *Understanding and Tackling Social Exclusion: Final Report to the WHO Commission on Social Determinants of Health.* Lancaster, UK: Social Exclusion Knowledge Network.

Portes, A. and Manning, R. (1986) The Immigrant Enclave: Theory and Empirical Examples. In: *Ethnicity: Structure and Process*, Nagel, J. and Olzak, S. (eds). New York: Academic Press, pp. 47–68.

Portes, A. and Rumbaut, R. (2014) *Immigrant America: A Portrait.* Berkeley, CA: University of California Press.

Putnam, R. (2000) *Bowling Alone: The Collapse and Revival of American Community.* New York: Simon and Schuster.

Putnam, R. (2007) E pluribus Unum: Diversity and Community in the Twenty-first Century the 2006 Johan Skytte Prize Lecture. *Scandinavian Political Studies*, 30: 137–174.

Fundació Tot Raval. (2018) "Home Page." Accessed 1 August 2018. http://totraval.org

Rumbach, A., Makarewicz, C., and Németh, J. (2016) The Importance of Place in Early Disaster Recovery: A Case Study of the 2013 Colorado Floods. *Journal of Environmental Planning and Management*, 59: 2045–2063.

Saunders, D. (2010) *Arrival City: How the Largest Migration in History is Reshaping Our World.* New York: Pantheon Books.

Staeheli, L., Mitchell, D., and Nagel, C. (2009) Making Publics: Immigrants, Regimes of Publicity and Entry to 'The Public.' *Environment and Planning D: Society and Space*, 27: 633–648.

Subirats, J. and Rius, J. (2006) *From the Xino to the Raval: Culture and Social Transformation in Central Barcelona.* Barcelona: Centre de Cultura Contemporània de Barcelona.

Talen, E. (2006) *Design for Diversity: Exploring Socially Mixed Neighborhoods.* Oxford: Architectural Press.

United Nations Human Settlement Programme. (2017) "Data Explorer." Accessed 12 July 2018. http://urbandata.unhabitat.org/explore-data/

Valentine, G. (2008) Living with Difference: Reflections on Geographies of Encounter. *Progress in Human Geography*, 32: 323–337.

Valenzuela-Garcia, H., Parella, S., and Güell, B. (2017) Revisiting the 'Ethnic Enclave Economy:' Resilient Adaptation of Small Businesses in Times of Crisis in Spain. *International Journal of Anthropology and Ethnology*, 1: 5–19.

Wessendorf, S. (2019) Migrant Belonging, Social Location and the Neighbourhood: Recent Migrants in East London and Birmingham. *Urban Studies*, 56: 131–146.

Young, I. (1990) *Justice and the Politics of Difference.* Princeton: Princeton University Press.

Young, I. (2000) *Inclusion and Democracy.* Oxford: Oxford University Press.

31

PUBLIC SPACE CHALLENGES AND POSSIBILITIES IN LATIN AMERICA

The city's socio-political dimensions through the lens of everyday life

Andrei M. Z. Crestani and Clara Irazábal

Introduction

This chapter analyzes three urban socio-spatial and political dynamics in Latin America helping us to understand the production and reproduction of contemporary public space in the region. The study is approached with three dispositions in mind. First, public space is considered beyond its physical attributes. The physical dimension of public space is important, not solely as a result or tangible manifestation of design and legal determinations, but especially because it supports and is an active element embodying its socio-cultural and political dimensions (Harvey 2012; Crestani 2017; Crestani and Brandão 2018). Public space is a socio-spatial and political realm that favors encounters, shared presence, expressions and interactions of differences, possibilities of appropriations and contestations, and manifestations of collectively built systems of evolving symbols and meanings (Arendt 1958; Sennett 1977; Lefebvre 1991; Delgado 1999; Deutsche 2007).

Second, public space is examined according to contingent conditions of time and space, history and institutions (Bhabha 1994; Adichie 2009). Public space acts as a place of mediation between civil society and the state, remaining an "irreplaceable orbit of the democratic constitution of collective opinion and will" (Avritzer and Costa 2004, 708). Evidence of this is the emergence of growing demonstrations across Latin America, for example in Brazil (2013, 2016, 2018), Bolivia (2013), Venezuela (2002, 2014, 2019), Buenos Aires (2016), Mexico (2017), and Nicaragua (2018). These demonstrations demand collective rights, occupying the public space as both an arena for political resistance and a place of hope for the emergence of more just socio-economic and spatial orders (Irazábal 2008).

Third, the study focuses on everyday micro-relations in public space. While mass protests are extraordinary events that reveal large-scale struggles to defend and/or strengthen democracy, the ordinary moments of everyday life also contain practices that nurture, publicize, and animate the social and political dimensions that characterize urban life and sustain the city as the place of the possible (Lefebvre 1991; Delgado 2007; Irazábal 2008; Monter,

Huffschmid, and García 2010; Netto 2012; Vommaro 2014). Yet, the scale of everyday micro-relations still occupies a timid place within the scholarship on public space.

Although these everyday dynamics are difficult to capture (because of their scale, nature, and specificity), the study offers qualitative and descriptive-exploratory reflections focused on everyday occurrences in public space in three Latin American examples, interpreting how they show interconnections between voices of different actors, such as civil society, the state, and the market surrounding the production of and demands for the city. The study also reflects on the extent to which different voices and modes of appropriating and reclaiming the city can help us understand challenges, possibilities, and necessary changes in the disciplinary approaches to Latin American public space's research and practice.

The socio-political dimensions of public space through the lens of everyday life

Latin America inherited from the European context the structure of political institutions and their organizational principles (Andar and Abraham 2009); yet not how people historically organized as citizens in public space. While in the European case the movements which gave rise to the modern state, public life, and the distinction between the public and private spheres derive from a specific organization of civil society (albeit a selective one), in Latin America public space was defined primarily as a function of colonial powers, as a physical expression of legal and political conditions of military and religious domination (Andar and Abraham 2009; Cobos 2014).

This section offers three representations of public space dynamics in Latin America that are an alternative to the conventional forms of standardization and control that weigh upon Latin American cities. These vignettes provide tangible evidence of how "ordinary" experiences can result in positive and lasting impacts on the production of public spaces (Irazábal 2008); yet, they do not exhaust the modalities of public space's everyday life in the region.

Cultural (re)invention of "unused" space: the Santa Tereza overpass

The Santa Tereza overpass in Belo Horizonte, Brazil has emerged as a living territory for everyday resistance movements. Their main tactic in this case has been the use of bodies as a performative form of collective occupations to claim for the right to public space. The space under the overpass has been the target of interventions and regulations by the state, which has attempted to determine how it can be used.

The space was abandoned for many years, yet from 2007 it became a site dedicated to increasingly visible collective activities questioning the representativeness of the state with regard to matters of public interest. The space established itself as a socio-political and cultural arena, hosting a variety of activities that unite a diversity of participants, such as hip-hop MCs, street dwellers, vendors, college students, collectives, and artists.

In 2014, the space was blocked off with construction fencing for public renovations as part of urban projects that (according to the users of this space) were meant to disrupt the social dynamics there, which ran counter to the increasing top-down "sanitization" of the city. The demonstrators claim that the city and state governments' proposed interventions that did not meet their cultural demands:

(a) Replacing the Miguilim (a space that serves homeless children and young people) with a Reference Center for Youth that considers neither the population that previously lived in this space nor the suggestions made by the Greater BH Youth Forum;

(b) Physical renovations of the overpass, to be conducted through an accord signed with the state government (not presented in detail);

(c) Execution of an "Overpass Upgrade Project," which includes management activities by several municipal departments, and according to the population, is disconnected from the social and cultural richness and diversity witnessed in the space;

(d) Specific legislation addressing the use of spaces under overpasses.In an episode in 2014, the population that used the space removed the construction fence that closed off the space under the overpass and reoccupied it, sharing and discussing the consequences of the interventions that had been planned in a top-down manner for the space, along with the need to struggle for the right to the space and to the city.

According to newspaper reports from the people who frequented this space, work began without notice, and the construction fence that appeared around the space surprised many groups who had used the space. The people stated that their opinions about the project had gone unheard, which ran counter to the appropriations and contents hosted in the space. The state strategy aimed to delegitimize this area as a public space for spontaneous socio-cultural significations and appropriations free from homogenizing standards.

During protests against the project, the space under the overpass became a place for every-day collective activities ranging from political meetings, debates, concerts, and swaps, to hip-hop gatherings and Samba Nights, events celebrating Afro-Brazilian religious practices like *candomblé* and *umbanda*, with various urban tribes congregating through samba and extending the culture of *terreiros*, sites where these religions are practiced, into the urban space.

The space under the Santa Tereza overpass is more than just a location for protests: it became a territory of the unpredictable, inseparable from the practices and bodies that mobilized meanings there, with no centralized organization but constant updating as a strategic space for both resistance and project making.

These autonomous manifestations create "pathways which, rather than coming together to standardize the public spaces in the area, foster tensions or indicate possibilities toward experiences of living (and production) which are truly open and democratic" (Berquó 2015, 113). The different ways in which the population resisted the imposed transformation of the space enacted their own reveal tactics that question the production of the city on a more structural scale. They also convey the evolving need for (re)occupations of everyday routines in space with acts related to what it means to participate in the city. These tactics are both individual and collective experiments that create relationships with peaks of intensity in re-signifying the overpass as a platform for a multifaceted development of the city and the individuals who inhabit it. From this perspective, the daily events in this space were able to rescale and transform their symbolic and subjective aspects, recovering the role of material space as an active element in creating the cultural and socio-political fabric of the city (Berquó 2015).

Public space for defiant denunciation and resistance: Pixo

I think that just seeing *pixo*, wherever it is, awakens an internal questioning within any person about the city, space, limits, freedom, or even about the violence of the city that overlies our everyday experiences, our day-to-day lives. Sometimes you never really noticed a house, or a wall that went up. So you go there and you paint it, it lights up the place [...]. The message is: "The city is ours, the people's!"

(Jacobini 2015, 29)

Pixo is a form of visual expression on city surfaces that uses symbols drawn in a specific format, done individually or collectively. Unlike the street graffiti that uses artistic references drawn from the graffiti of 1970s New York, *pixo* emerged in Latin America in the 1960s not as artistic language but as subversive messages from leftist groups as a way of fighting dictatorships, (Tavares 2010). It evolved into a tactic for demarcating territory and protesting and was considered as illegal vandalism subject to punishment. Over the last decades it has increasingly appeared in everyday life in Latin American cities, provoking changes in the public space that go beyond its character as visual expression, denouncing inequalities related to the rights and freedoms of movement and ownership in the city.

Although *pixo* appears in other parts of the world beyond, the way this kind of spontaneous speech takes shape in Latin American public space is unique as the expression of a specific social class that is excluded territorially (generally on the outskirts, without access to the city) as well as culturally (a criminalized cultural practice alongside many others in Latin American history: samba, *maxixe, umbanda*, capoeira …), as a means of communicating conditions of inequality.

Pixo is a counterpoint to a city that is more and more reclusive, enclosed in by walls, barbed wire, iron gates, and cameras that reinforce the idea of public space as a place of fear. Paradoxically, places associated with fear are the platforms where subjects on the sidelines can leave their mark. Here, the marginalized express themselves in at least three modalities: in the geographical position they occupy in the city, in the claim for public policies that guarantee more equitable access to public assets and services, and in the position of subculture in relation to the institutionally recognized forms of art and urban performance (Caldeira 2012).

Although these marks get public visibility, *pixo* does not attempt to send a clear message to its readers. *Pixo* is generally a set of symbols in codes discernible to its practitioners that consequently excludes others. In this way, *pixo* externalizes the act of protest, not necessarily by the content of the written message, but rather by opening the possibility for exclusion to exist, refusing integration (Mano 2009; Caldeira 2012). Its practitioners do not organize. There is no target, theme, or a priori physical context, and there are also no rules, agreements, or plans about how to do *pixo* and/or flee from the police if performers are caught. *Pixo*'s strength comes from the fact that on its own, it is a sign that destabilizes the rules of public space and reveals that the city has not yet fulfilled democracy's promise to ensure it is a space accessible to all.

There are four ways in which *pixo* markings can be seen as a unique and inventive form of producing Latin American public space in comparison with other places in the world:

> First, they create new visibility and a new type of presence for the peripheral that disrupts a certain order of things that tend to characterize the public space. Second, this disruption is contradictory because it connects with narratives of rights and pleasure while simultaneously expressing itself as danger or tension, and often materializes in illicit or even violent forms. Third, these practices unequivocally help reproduce and deepen gender hierarchies. Fourthly, in short, these are paradoxical interventions. They expand and repeatedly fracture the public space, demanding rights and challenging them, affirming pleasure and finding death, and denouncing injustices but stubbornly rejecting assimilation.
>
> *(Caldeira 2012, 413)*

The fourth aspect may be *pixo*'s greatest transgression, not the fact that it is deviant with regard to the forms of policing and control of urban space, but rather the fact that it literally registers "the presence of those who should be invisible" in the city (Caldeira 2012, 400). Its practitioners prefer to be despised by the general population and city management than to simply be ignored. Consequently, *pixo* does not seek inclusion while denouncing the exclusionary nature of the city. This is because its foundation is transgression: the goal is not to be assimilated as part of the city or accepted as a form of "urban art." Being illegal is what protects the inventive power of *pixo* and reveals it as a form of political agency in public space.

Between mobility and permanence: informal trade in Mexico City

Since the 1990s, a wave of enterprise-focused urban governance in Latin America has led to various forms of social and spatial exclusion and space has been transmuted into physical and symbolic merchandise. This commodification is reified by urban marketing, the privatization of public space, the formation of ghettoized urban enclaves, securitization of the city, economic and political exclusion, and segregation (Caldeira 2000; Leite 2010; Németh and Schmidt 2011; Sobarzo 2011; Crestani 2015). Less frequently, the literature explores socio-spatial forms that emerge as a way to cope with neoliberal urban governance (Letelier and Irazábal 2017; Irazábal 2018).

Here, the case of *Programa Rescate* is briefly analyzed, which was implemented to revitalize the historic center of Mexico City. Street vendors have generated a series of collective tactics to combat neoliberal policies that form the structure of many urban renewal programs like this one in Latin America. As seen in other interventions in historic areas of cities, the presence of activities generally associated with informality is interpreted as a threat to the interests of redevelopment in these areas, and the removal of these activities and populations is a directive commonly found within the scope of these projects (Angotti and Irazábal 2017). These entrepreneurial strategies favor benefits for private agents such as landowners and real estate entrepreneurs, while lower-class populations in the project areas generally have a hard time fulfilling their needs (Crestani 2015).

The economic crisis between 1980 and 1990 triggered a considerable increase in the activity of street vendors in Mexico City, who made up 40.5% of the population that participated in the informal market in Mexico during this period (Jusidman 1992). Today, Mexico City has a total of 2 million informal vendors, only 5% of whom have documents authorizing this activity (based on data from Siscovip–Sistema de Comercio en la Vía Pública). *Programa Rescate* attempted to reduce the presence of these traders "so as not to undermine the capitalization of the historic sector in terms of its tourist and residential potential," since itinerant vendors are generally associated with a negative image of "insecurity, poverty, and deterioration" (Crossa 2009).

The program, however, did not escape spontaneously organized acts of resistance and subversion by street vendors, which proved powerful in undermining the state's regulatory frameworks. Crossa (2009) examined how these processes occurred, identifying mechanisms that vendors created from their everyday associations. In addition to directly confronting repressive relocation measures and policies prohibiting informal activities, vendors established cooperative ties between different groups who worked in itinerant sales in the historic center.

One of these agreements was the loosening of "territories," which had traditionally been negotiated between different groups of vendors. The different groups who sold goods on the streets established agreements allowing members of one group to also sell in a second group's area while the first group's territory was subjected to removal efforts. In that way, public space was consciously managed as an active arena to facilitate the continuity of productive activities that supported this population, which were initially hindered by institutional forces. Another tactic the vendors adopted was mobility as a means of avoidance. In this case, after being removed from their original locations, vendors found a way to sustain their activities through a modality known as *toreo* (bullfighting):

> *torear* means selling goods while remaining mobile; *toreros* are the nomadic suppliers of Mexico City. Some walk the streets carrying their products, or with them attached to their bodies. Others place their products on a blanket or piece of plastic sheeting on the ground. If the police appear in the area, the *toreros* pick up the four corners of the blanket or sheet, grab their merchandise, and run to a safe area.
>
> *(Crossa 2009, 56)*

In addition to the mobility provided by *toreo*, other informal traders take turns monitoring the streets for police presence, alerting each other with walkie-talkies so that vendors can gather their goods and move to other streets when necessary. This strategy involves organization requiring geographical coordination, communicative skills, and investment in equipment like walkie-talkies (Crossa 2009). The workspace is the product of collective efforts and reiterations of its use, despite and around the offensive activities of government management. The informal vendors mobilize a form of experiencing the city that mixes suffering, fears, uncertainties, conflicts, solidarity, hopes, and triumphs, finding their own mechanisms of struggle to exercise and recover concrete socio-political space in the streets and transform reality (Gayosso 2019).

The informal vendors' strategies open up a politics of the everyday through their collective and conscious associations and collaborations to reaffirm their right to the city, defying the regulations and practices imposed by renovation programs and their attempts to displace these people. The nature of such resistance transcends the issue of territory and becomes a struggle for the right to be recognized as citizens and as part of the physical and symbolic landscape of the historic center's identity, an identity consisting of "counter-spaces" (Lefebvre 1991; Leite 2002; Crestani 2017) that continuously challenges the top-down disciplining and regulating of that reality.

Final considerations

The vignettes above present autonomous socio-spatial dynamics that question existing conceptions, uses, and meanings of public space and offer alternatives. These everyday-life social (re)productions of space expose and value differences while making visible the broad regime of forces that operate upon Latin American public space.

Although similar practices contesting public space exist in other locations around the world, the vignettes of daily life in Latin American cities presented here suggest distinctive ways of reconquering the city, and despite materializing in multiple formats, they challenge similar mechanisms for intervening in and controlling urban space that have distressingly spread in cities across the region. This group of practices reveals a search for alternatives to structural changes in the conditions of ownership and production of Latin American public space, increasingly based on tactics of government and market control. Examples of these tactics include the expulsion of street dwellers in Buenos Aires, with the police hosing down these people's spaces and forcing them to move (Rosa 2018); the São Paulo city government's decision to paint over urban artwork; the displacement of street vendors in Querétaro, Mexico and Cuzco, Peru; a recently proposed bill that would ban street art in Curitiba, Brazil; constant disputes between *favela* dwellers and the police against removals and destruction of these communities in São Paulo and Rio de Janeiro; restructuring of the social fabric of historic centers in San Victorino, Bogotá, Colombia; Pelourinho, Salvador, Brazil; and Buenos Aires, Argentina; among others.

Obviously, the three episodes presented in this essay do not encompass all the possible formats of inventive schemes that are currently taking shape in Latin American public space, nor all the possible ways of reclaiming the city. The highlights of each are the intensity of the relationships they produce and their strength in offering alternatives to the production of the city when they suspend or at least destabilize material, symbolic, and legislative dimensions, indicating other possibilities for public space.

From the point of view of understanding how we live and produce the city, the study of everyday experiences in public space is fundamentally important, since this is where concrete agencies mobilizing the socio-political dimension in the city can be found. This implies recognizing public space in motion and manifested in everyday practices that present conflict and dissent as productive forces, not necessarily as undesirable manifestations that require some fixing.

Setting aside the differences between empirical episodes, one point of convergence is that the socio-political dimension of public space is recognized when it is appropriated as a platform for discourses and expressions that can range from discontent to feelings of appreciation and collective desires to protect the commons. Paradoxically, this dimension seems to be driven by situations that express what Sharon Zukin (1996) calls "landscapes of power," which are marked by hegemonic forces that seek to generate exclusionary spaces, social homogenization, and the inhibition of significant sociabilities. What each case demonstrates is that these forces are not irrefutable. They can be destabilized and even suspended by ephemeral or longer-lasting creative interventions that deviate from the imposed expectations. In doing so, these appropriations give life to the political meaning of public spaces precisely because they challenge a status quo that favors the interests of a few, evidencing the limitations of the institutions responsible for protecting the city as a "common good" and practicing democracy (Angueloski, Connolly, and Irazábal 2018).

The practices explored in this study do not resolve the inequalities, tensions, intolerance, and exclusion they bring to light. Instead, their function is to question stagnant democracy and raise awareness of the need to create new political languages to live in and to consider the Latin American city (Irazábal 2009; Caldeira 2012). This calls the very concept of public space into question. When groups in Belo Horizonte convert the obsolete space under an overpass into a celebration of multiple cultures, when *pixo* painters bring the exclusionary and unequal nature of our cities into the foreground, when informal street vendors find ways to support themselves in the face of municipal governments that try to neutralize them, these players all comprise a framework of simultaneities in Latin America that question to what degree public space guarantees the right to the city.

Among the challenges posed by the processes of producing the contemporary city associated with (but not limited to) globalization, this study sought for approaches to reality from distinct viewpoints, attentive to emerging opportunities. The study shows that by attentively examining everyday life, the multiple geographies of power that compete for the materiality and immateriality of the city can be recognized. The study invites us to de-territorialize our preconceived notions of public space in order to broaden our capacity to interpret historically, culturally, and geographically situated realities. This task is especially important at a time increasingly marked by socio-spatial fragmentation, disciplinarization, political polarization, and other challenging manifestations widespread in the production and reproduction of Latin American public space.

References

Adichie, C.N. (2009) "The Danger of a Single Story" *TEDGlobal*. Streamed video. www.ted.com/talks/chimamanda_adichie_the_danger_of_a_single_story?language=pt#t-19712.

Andar, E. and Abraham, M. (2009) La formación del espacio público en América Latina. anuario GRHIAL. Universidad de Los Andes. Mérida. Enero-Diciembre, 3: 17–38.

Angotti, T. and Irazábal, C. (2017) Introduction: Planning Latin American Cities: Dependencies and 'Best Practices'. *Latin American Perspectives Special Issue 2: Planning Latin American Cities*, 44(2): 4–17.

Angueloski, I., Connolly, J., and Irazábal, C. (2018) Grabbed Landscapes of Pleasure and Privilege: Socio-Spatial Inequities and Dispossession in Infrastructure Planning in Medellín. *International Journal of Urban and Regional Research*, 43(1): 133–156.

Arendt, H. (1958) *The Human Condition*. Chicago: University of Chicago Press.

Avritzer, L. and Costa, S. (2004) Teoria Crítica, Democracia e Esfera Pública: Concepções e Usos na América Latina. *DADOS-Revista de Ciencias Sociais, Rio de Janeiro*, 47(4): 703–728.

Berquó, P.B. (2015) "A ocupação" e a produção de espaços biopotentes em belo horizonte: entre rastros e emergências. Dissertação de Mestrado apresentada ao Núcleo de Pós-Graduação em Arquitetura e Urbanismo da Universidade Federal de Minas Gerais.

Bhabha, H.K. (1994) *The Location of Culture*. London: Routledge.

Brighenti, A.M. (2010) *The Publicness of Public Space: On the Public Domain*. Quaderni del Dipartimento di Sociologia e Ricerca Sociale; 49. Trento, Italy: Università di Trento.

Caldeira, T.P.R. (2000) *City of Walls: Crime, Segregation, and Citizenship in São Paulo*. Berkeley: University of California Press.

Caldeira, T.P.R. (2012) Imprinting and Moving Around: New Visibilities and Configurations of Public Space in São Paulo. *Public Culture*, 24(2): 384–419.

Castells, M. (1989) *The Informational City*. Oxford: Blackwell.

Cobos, E.P. (2014) La Ciudad Capitalista En El Patrón Neoliberal de Acumulación en América Latina. *Cadernos Metrópole, São Paulo*, 16(31): 37–60.

Crestani, A.M.Z. (2015) As Faces (In)visíveis da Regeneração Urbana: Rua Riachuelo e a Produção de um Cenário Gentrificado. *Cadernos Metrópole*, 17: 179–200.

Crestani, A.M.Z. (2017) "In-Between Zones: Other Possibilities of Investigation of the Contemporary Public Space." PhD Diss., University of São Paulo, São Carlos.

Crestani, A.M.Z. and Brandão, B. (2018) The Public Space (In)visible to the Eyes of Jane Jacobs. In: *Jacobs Is Still Here: Jane Jacobs 100, Her Legacy and Relevance in the 21st Century*, Rocco, R. (ed). TU Delft, pp. 48–54. https://books.bk.tudelft.nl/index.php/press/catalog/view/isbn.9789461869005/742/586-3.

Crossa, V. (2009) Resisting the Entrepreneurial City: Street Vendors' Struggle in Mexico City's Historic Center. *International Journal of Urban and Regional Research*, 33(1): 43–63.

Delgado, M. (1999) *El Animal Público*. Hacia Una Antropología de Los Espacios Urbanos. Barcelona: Anagrama, 229.

Delgado, M. (2007) *Sociedades Movedizas*: Pasos Hacia Una Antropología de Las Calles. Barcelona: Anagrama.

Deutsche, R. (2007) *Evictions, Art and Spatial Politics*. Cambridge: MIT Press.

Deutsche, R. (2008) *Agorafobia*. Cambridge: MIT Press.

Gayosso, J.L. (2019) The Collective Action of Informals. *Revista Mexicana de Estudios de los Movimientos Sociales*, 3(1): 17–32.

Harvey, D. (2012) *Rebel Cities: From the Right to the City to the Urban Revolution*. London: Verso.

Irazábal, C. (2008) *Ordinary Places, Extraordinary Events: Citizenship, Democracy and Public Space in Latin America*. Irazábal, C. (ed). New York: Routledge.

Irazábal, C. (2009) Revisiting Urban Planning in Latin America and the Caribbean. *Global Report on Human Settlements 2009 Planning Sustainable Cities: Policy Directions*. United Nations Human Settlement Programme. Available from www.unhabitat.org/wp-content/uploads/2010/07/GRHS2009RegionalLatinAmericaandtheCaribbean.pdf.

Irazábal, C. (2018) Counter Land Grabbing by the Precariat: Housing Movements and Restorative Justice in Brazil. *Urban Science*, 2(2): 1–18.

Jacobini, M.G. (2015) Na Cidade Muro Implora: O Que Dizem As Pichações Em Meio À Paisagem Urbana. www.paineira.usp.br/celacc/sites/default/files/media/tcc/tcc_na_cidade_muro_implora_marina_jacobini.pdf.

Jusidman, C. (1992) The Informal Sector in México. *Secretaria dei Trabajo y Previsión Social*/United States Department of Labor.

Lefebvre, H. (1991) *The Production of Space*. Oxford: Blackwell.

Leite, C. (2002) Counter Uses of Pulbic Space: Notes on the Social Construction of Places in Manguetown. *Brazilian Journal of Sciences*, 17(49): 116–172.

Leite, R.P. (2010) A Exaustão Das Cidades Antienobrecimento E Intervenções Urbanas Em Cidades Brasileiras E Portuguesa. *RBCS*, 25: 72.

Letelier, F. and Irazábal, C. (2017) Contesting TINA: Community Planning Alternatives for Disaster Reconstruction in Chile. *Journal of Planning Education and Research*, 38(1): 67–85.

Mano, M.K. (2009) Graffiti, the Mark of Social Inequality, Interview with Joao Wainer and Roberto T. Oliveria. *Le Monde Diplomatique Brasil*, 3 December. www.diplomatique.org.br/pichacao-a-marca-da-desigualdade-social/.

Monter, I.A., Huffschmid, A., and García, A.C. (2010) Ciudades Líquidas: El Hacer Ciudad Y El Poder En El Desbordamiento Metropolitano. In: *Metrópolis Desbordadas: Poder, Culturas Y Memoria En El Espacio Urbano*, García, A., Huffschmid, A., Monter, I., and Rinke, S. (eds). México: UACM, pp. 11–72.

Németh, J. and Schmidt, S. (2011) The Privatization of Public Space: Modeling and Measuring Publicness. *Environment and Planning B: Planning and Design*, 38(1): 5–23.

Netto, V.M. (2012) A (Re)Conquista Da Cidade: Polis E Esfera Pública. Porto Alegre. *Cadernos Proarq*, 19(12): 266–289.

Rosa, P.C. (2018) Exclusions of the Public Space of the Inhabitants of the Street in the City of Buenos Aires. *Territories*, 39: 157–173.

Sennett, R. (1977) *The Fall of Public Man*. New York: Knopf.

Sobarzo, O. (2011) A Produção do Espaço Público: Da Dominação À Apropriação. *Geousp: Espaço E Tempo*, 10: 93–111.

Tavares, A. (2010) Urban Fictions: Strategies for City Occupation. *ARS (Sao Paulo)*, 8(16): 21–30.

Vommaro, P. (2014) The Dispute for Public Space in Latin America: Youths in Protests and Common Building. *New Society*, 251(5): 55–69.

Zukin, S. (1996) *The Cultures of Cities*. Hoboken: Wiley-Blackwell.

32

TYPOLOGIES OF THE TEMPORARY

Constructing public space

Mona El Khafif

Introduction

Analyzing theories and projects dedicated to the Temporary City, this essay will explore case studies in Vienna, Berlin, New York, and San Francisco that use four different strategies to initiate temporary urban spaces. Positioned between top down and bottom up, these projects encourage collaboration between citizens and public agencies, utilizing the temporary to expand, adjust, and test public spaces or to use available land resources for the public while bridging between different development phases. The theoretical discourse concentrates on the understanding that public spaces, despite their physical materiality, are also defined by their ephemeral social appropriation. Process-oriented design and temporary spaces, therefore, are not antonyms of the permanent city, but important tools in constructing public spaces.

The urban utopia that serves as the most historically referenced model for the contemporary Temporary City is most likely the Situationist City. Archigram's work from the Living City exhibition of 1963, for example, focused on the definition of space through time-based interaction, rather than demarcation and more permanent physical boundaries. The city was declared as the *"sum of its atmospheres"* and, emerging from this attitude in the early 1960s, architectural, urban, and artistic interventions conceptualized temporary architecture as a means of moving away from objectification—generating an understanding of space-environment prioritizing action above form (Shepard 2011).

While the approach of the 1960s was clearly a counter position to the technocratic top-down agenda of the modernist era contemporary interventions—while very often informal and identified as DYI urbanism—tend to foster a collaborative ecology (Lydon and Garcia 2015). Planning departments started to understand the value of informal appropriations, and designers, artists, citizens, and—more recently—city administrations are exploring temporary strategies to test a variety of social-spatial set-ups (El Khafif and Del Signore 2014).

A series of publications and theories have been dedicated to its strategies and while earlier attempts tried to argue for trial and error methods, opposed to long-term planning approaches, more recent works were dedicated to the establishment of an inventory of temporary urban projects to further identify and understand different applications (Haydn and Temel 2006). *Urban Pioneers*, for example, investigated the potential of urban development through temporary uses that emerged in Berlin at the turn of the century (Senate of Berlin 2007). Looking at temporary projects partly fostered through vacant land resources after the re-unification, the publication is an in-depth catalogue of more than 40 projects, analyzing legal frameworks like landownership and applied zoning codes, in combination with time-based interim uses. The authors argue that interim use can be understood as process-oriented planning, filling the gap between two different stages of land use. As such, it can synchronize stages of the formal planning process with phases of informal use. Following this first publication, *Urban Catalysts* unveils the logic of temporary uses in a series of Euro-pean-wide examples, featuring informal processes as a part of urban planning and design (Oswalt, Overmeyer, and Misselwitz 2014). Here, projects are categorized according to typologies of temporary uses and patterns of the unplanned. Identifying a wide range of project-based attributes, the typologies vary from free-flow, co-existence, parasite, and consolidation, among others, to establish site-specific behaviors.

The concept of The Temporary City, as outlined by Peter Bishop and Lesley Williams (2012), unfolds out of an understanding that the city—as our social, economic and ecological environment—is rooted in a four-dimensional, highly dynamic scenography: a space that changes on a daily or even hourly basis while also consisting of more permanent elements. Bishop and Williams argue that The Temporary City today has its own legitimation. In a time of increasing pressure on scarce resources, temporary uses are increasingly legitimate and power-ful to feed initiatives for incremental change, while long-term investments can occur over time. The authors present 68 case studies categorized through the program, but also the project's agency, arguing that long-term solutions can be implemented when resources are available, while being guided by a temporary version. This position aligns with urbanist Jan Gehl (2012), who proved through a series of realized projects that temporary public spaces, when integrated as a part of a wider street-design process, can act as public consultation at actual scale and in real time, offering a model for citizen engagement while simultaneously reshaping the public realm on a temporary basis until the "permanent" design will be implemented.

To paraphrase Bishop and William's book, this contemporary model encourages a practice within the design profession that understands the temporary city as an ongoing component of the permanent city, allowing for the integration of multiple actors and the negotiation of top-down and bottom-up processes (El Khafif and Del Signore 2014).

While the temporary is often celebrated as the vibrant informal and unplanned urban space, we need to recognize, however, that these temporary projects are a tool for process-oriented urban planning and design. Hence, the mechanisms of temporality and the produc-tion of social space within public spaces need further investigation.

City and temporality

The most compelling theoretical framework for The Temporary City is *Cities in Time: Tem-porary Urbanism and the Future of the City* (Madanipour 2017). Madanipour analyzes temporary urbanism as a pattern of events at the intersection of three forms of temporality: instrumental, existential, and experimental, starting with the thesis that the city is itself a temporary phe-nomenon. Instrumental temporality is defined by a utilitarian approach to time and existential temporality reflects on its intuitive understanding. Interesting in the context of this essay is Madanipour's experimental temporality, which identifies events as moments of "questioning, experimenting and innovating" urban spaces (Madanipour 2017, 5).

Experimental temporality provokes—in a formal and informal manner—the status quo, through displacement and the break-up of structures, fundamentally changing the perspective on a given physical or social urban condition. It creates a set of new possibilities and is therefore significant aspect in transformation processes. Madanipour (2017) further argues that displacement does not necessarily offer solutions but questions the conventional perspective. Hence, the creative potential of temporary urbanism may be explored through its capacity for experimenting with ideas and practices, fostering innovation.

As the city, and social interaction in public spaces, is permanently changing, we must ask how temporary urbanism distinguishes itself from its likewise temporary context. Madanipour offers two important perspectives on this, which are multiple types and different characters of temporalities. For him,

> temporary urbanism is the range of short-term actions and events which take place in time, but their timing may not be in line with the predictable patterns […]. They may appear to run counter to the highly regulated structures […] and the temporary routines of the city.
>
> *(Madanipour 2017, 143)*

Madanipour further appropriates terminologies of the ancient Greece that differentiate between regulated time (*Chronos*) and the unregulated time of the occasional event (*Kairos*), establishing different characters of ephemeral conditions. In this context, temporary urbanism may be interpreted as emphasizing Kairos—formed of recognizable moments and fractures—in contrast to the continuity of Chronos.

Following Madanipour's argument, we can comprehend that cities consist of multiple temporalities, and that the material public space can be understood as the physical container of immaterial events. Public space is not only characterized by its physical boundaries and morphology but also defined by ephemeral appropriation, changing narratives, and identities.

Constructing public space and public agendas

With this understanding in mind, it remains open if and how public space—with its temporary character, mechanisms of events, and appropriation—can be composed. Four case studies that utilize different typologies of the temporary to encourage spatial appropriation and the construction of public space are presented below (see Figure 32.1).

Re-programming public space: museums quartier vienna

Spatial typology

The Museums Quartier in Vienna (MQ) is an example of an existing physical space with fixed boundaries whose narrative was re-shaped through the implementation of new experiences, in the form of predictable and unpredictable urban events. These events stimulated social appropriation and with this a new identity of the public space.

Project background

The MQ, as it exists today, is a piece of the city: a material collage of 250 years of history, starting with the fabric of Fisher von Erlach's Baroque architecture and its subsequent architectural implementations, redeveloped as the new Museums Quartier in 2001. Much as the built space articulates a landscape of historical fragments; the programming of the surrounding

Figure 32.1 Site map of precedent projects
Image credit: Mona El Khafif

buildings reflects on diverse cultures, hosting more than 60 institutions ranging from subculture, everyday culture, high culture, fashion, and design to kids programs. However, the urban strategy developed by Ortner and Ortner, meant to initiate a cultural urbanism in the district, did not unfold directly after the opening in 2001. Specifically, the outdoor areas described at this time as "Vienna Vacui," suffered from a lack of public occupation.

Temporary strategy

To generate a new identity of the site, the MQ directorship initiated a place branding strategy that transformed the MQ and its outdoor areas into a vibrant urban space with rising visitor numbers. Known today as the "Viennese Public Living Room," it is an all ages destination for tourists and locals.

To support the generation of this new identity, a branding campaign—featuring the outdoor spaces of the MQ as a container for cultural events and experiences—rolled out over a five-year period. Starting with a neutral label, the place was loaded with new narratives, organized as events and urban programming to transform the spatial experiences. Consequently, a novel mental map of the public space was generated. The initiated seasons, MQ Summer (outdoor lectures, boule, performances, and restaurants) and MQ Winter (Christmas market and ice sports), allowed the public to participate in the event cycle. A network of partners and contributors was generated, comprised of local institutions, external partners, and neighboring communities.

The analysis of the event cycle throughout the year 2005 shows diverse activities, from leisure to cultural programs, activating the space in different scales, rhythms, and sites. The programs and events were endogenous, organized by multiple groups and occasionally simultaneous, fostering a pattern of the unplanned. While programming increased over time, the space still allowed for un-programmed appropriations (El Khafif 2010).

Using Madanipour's concept of different temporalities, the events were perceived as stimulations and initiated spontaneous activities of second order, resulting in a vibrant public space, inhabited by a diverse group of people. A key element of success was the design of flexible and temporary outdoor architecture, easily rearranged to support different scenarios. The ENZIES —designed by PPAG Architects—are an emotional touchstone for the public. The color changes with the seasons, voted by users online. It adapts its physical form to accommodate different activities, giving users a physical connection to the public space and a sense of agency.

The time-based notation of MQ urbanism further shows the synthesis of four layers of space (built, programmed, organized, communicated), and their time-related choreography. The process of place making through temporary programming navigates between top-down (place branding) and bottom-up (user generation), allowing the public to be an audience and an active partner in the production of the social and physical space (El Khafif 2009).

For success as a place-making strategy, it was critical to confirm the new identity over time. The programming of events and appropriation of public spaces repeated and intensified each year between 2000 and 2005 (see Figure 32.2). This process of place branding and place

Figure 32.2 Programming strategy Museums Quarter Vienna
Image credit: Mona El Khafif 2009

management depended on monitoring the public space to ensure that the new narratives and events could stimulate the construction of the public space without dominating its social appropriation.

Converting streets: the New York City Broadway and the Public Plaza program

Spatial typology

The New York City Broadway project, along with San Francisco's "Pavement to Park" initiative and Toronto's recent King Street Pilot project, re-thinks infrastructural street spaces dedicated to automobile traffic to gain public space for pedestrians and bicyclists. In all three cases, temporary strategies in form of surface treatments, mobile furniture, and landscape elements test the modification of the public realm before a more permanent transformation occurs.

Project background

In 2008, Mayor Michael Bloomberg and Department of Transportation (DOT) collaborated to address the worsening traffic situation along Broadway and Times Square specifically. Working with local advocates and businesses, the idea was to improve New York's most iconic public space.

Temporary strategy

In collaboration with Danish urbanist Jan Gehl, known for his people centered approach, a survey was conducted showing that 90% of the plaza's users are pedestrian, while nearly 90% of the physical space was dedicated to car traffic—confirming that Times Square should indeed be a square and not a traffic intersection (Gehl, Risom, and Daily 2015).

Gehl's team first explored weekend street closures called "Sunday Streets," inviting residents, visitors, and administrators to re-imagine Broadway and Times Square as a public space. Other pilot projects were initiated, using inexpensive materials to mark new bike lanes and pedestrian zones (see Figure 32.3). As stakeholder support grew, more expensive investments were made, and later pilot projects like Herald Square used street furniture and planters to claim streetscapes as new public spaces (Gehl and DOT NY 2014).

The temporary redefinition of the physical space was evaluated, showing mixed results in terms of improving traffic conditions, but enormous improvements regarding collisions and pedestrian volumes. Over a six-month observation period, there was a 39% reduction in injuries to pedestrians, motorists, and cyclists, and Times Square converted into a vibrant public space (Goldwyn 2014). In 2016 the site was redesigned by Snohetta and transformed from a temporary pilot project to a new public space with more permanent material conditions. According to Gehl, Risom, and Daily (2015), the conversion of Times Square from a traffic-dominated place to a public space served as a flagship for the NYC Plaza program, creating more than 50 new plazas in all five boroughs. Along Broadway alone, five new adjacent sites transformed to public spaces, equaling in area to Bryant Park.

In Gehl's approach to test the city through temporary strategies, public discussion becomes a key element of city making. Today the changes have been largely accepted. However, while Times Square is occupied by nearly half a million people daily, and the plaza program has made New York a leader in the international movement of people-friendly city making, other

Figure 32.3 Before (a) and after (b): Herald Square

Image credit: https://www.flickr.com/photos/nycstreets/9135791611

occurrences showed that the construction of public space is not a simple accomplishment, as its physical transformation extends into its social dimension (Low 2005). In 2015, public controversy emerged when Times Square also became home to topless body-painted women. Accused of polluting the public space with aggressive panhandling, city officials called the pilot project a failure due to inappropriate behavior in the newly gained space (Grynbaum and Flegenheimer 2015). However, Gehl clearly articulated that a traffic-jammed Times Square is not the preferred alternative. Instead it needs to be understood that reclaiming space requires stewardship and cultivation over time (Gehl, Risom, and Daily 2015).

Transitioning operational space: Waiting Lands and Berlin Tempelhof

Spatial typology

The spaces identified in this typology can be described as former industrial landscapes or "Waiting Lands," as coined by Kees Christiaanse and KCAP (2005). These sites are for various reasons not ready for immediate development and form potential reservoirs for future urban expansions. The aggregated surface of these land resources according to Christiaanse can be compared to the footprint of suburban expansions, featuring a massive potential for internal growth. While "waiting" for their future transformation, these lands could begin the transition early, through the initiation of programming or serving as a link between two different stages of urban development. In the latter, these former private industrial or infrastructural lands are temporarily opened to serve as borrowed public spaces.

Project background

One example of a temporary public landscape is the Airfield Berlin Tempelhof, which due to its size, inner-city location, and history, is one of Berlin's most prominent temporary projects. Meant to serve as an expansion zone for future urban development, the city identified its potential to perform on an interim basis as an inner-city territory temporarily opened to the public until further development took place.

Respecting the socio-economic circumstances that would require development in the future—specifically offering housing for a growing population—an incremental development process was sought to open the site to the public after more than 100 years. Challenged by the interest of diverse stakeholders, a framework was needed for temporary public usage of the site without hindering future development. In 2009 the land was transferred to the city of Berlin and the Senate is responsible for the coordination of the temporary public realm entitled "Tempelhofer Freiheit." After gathering feedback through community engagement workshops, a landscape planning competition generated a master plan suggesting mixed-use development at the edge in development zones and a reuse of the old airport entrance halls for creative industries. The airfield itself was meant to stay open as an inner-city green space (Metropolis 2018).

Temporary strategy

The process-oriented dynamic master plan that would reanimate the open spaces and facilitate the temporary construction of public space on the airfield was designed by Raumlabor, Klaus Overmeyer, and Michael Braum & Partners (Raumlabor 2009). The transformation was based on a ten-step process (see Figure 32.4) to facilitate future interim uses for the exhaustive

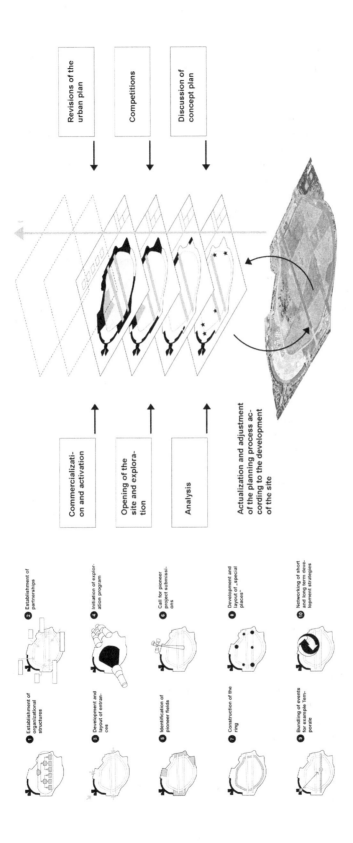

Figure 32.4 Process oriented planning Berlin Airfield Tempelhof

Image credit: Raumlabor Berlin, Christof Mayer, Markus Bader in cooperation with Studio UC Klaus Overmeyer, Michael Braum & Partner, English translation Mona El Khafif

landscape, existing building structures, and surfaces. The plan was based on an operational strategy that created sequences of public occupation and initiated a new social ecosystem for the usage of the airfield. The process started with the establishment of organizational structures initiating the interim uses and partnerships needed to establish collaborations with the neighboring areas. Opening the site to the public through the identification of gates and the initiation of discovery days was critical to the plan. The temporary strategy worked with the future development zones and identified these areas as "pioneer fields." Following a call in 2010, anyone was able to submit project ideas for the pioneer fields. While economical surplus was undesired, pioneers could suggest diverse activities as long as this new program would benefit the general public. Selected projects had to fulfill further criteria for sustainability and address one or more of the topics chosen for the development: "knowledge and learning," "clean future technologies," "sports, health, and wellness," "integration of neighborhoods," "inter-religious dialogue," and "stage of innovations." Pioneer rights were granted for one year, with the possibility of extension after a review process, but ideally successful pioneer programs would be integrated into the future development. Out of more than 200 project submissions, 22 urban pioneers were chosen to contribute towards upgrading and enriching the open space. Among those urban gardening projects, Shiatsu and Qigong, new sport activities like kite surfing (Heilmeyer 2011), and art projects like Arche Metropolis (Knueppers 2011), who organized citizen engagements on site.

While the selection process of the pioneer programs continued, the design team developed a graphic wayfinding concept and infrastructural ring that would connect the pioneer fields with the existing programs, gates, and airfields that were used for leisure activities. Along with this infrastructure, the plan promoted the initiation of hot spots and anchor places. The moderation of events, program cycles, and pioneer activity until 2020 and beyond, as well as the synchronized process of long-term and short-term development, was identified as a key component for the process-oriented dynamic master plan.

Conceptually, the temporary strategy deployed at the Tempelhofer Freiheit has similarities with Madanipour's "*experimental temporality*," questioning the status quo through displacement and the break-up of structures. In contrast to earlier projects discussed here, the ultimate goal was not to permanently transform the area into public spaces, but to bridge the time until development would take place. The seeding of temporary pioneer programs —while conceptually very clear—however had to face a series of challenges. According to the designers, the density of pioneers on the open field was not enough to establish a vibrant public landscape, and the time-based contracts of one or three years were too short to allow for experiments and the longer-term investments that landscape projects required (Raumlabor 2009). Further, anchor programs run by network partners, like universities or schools, would have helped to intensify and stimulate public appropriation. In addition, a mistrust in the future development plans of the city birthed "100% Tempelhofer Feld," a movement that opposed the master plan development. On May 2014 a referendum was held and resulted in a civilian decision to leave the field open to the public. No development at the edge or any sale can occur, and public accessibility during opening hours has to be granted (Juergens 2014).

While we might argue that this is a "success" for the project—as public and political engagement was truly unleashed—it is important to see that the success of transitional strategies lies in crafting a bridge between two development stages. These temporary projects offer tremendous resources and request collaboration between different stakeholders to balance interest. The Tempelhof case—providing conceptually a well thought through framework that could be implemented on other sites—might close the door to following projects.

The fact that the city opened a refugee camp on the airfield in 2015 could be seen as another counterproductive step in a public collaboration process that needs trust and commitment.

Testing micro public spaces: San Francisco urban prototyping event

Spatial typology

Based on the success of Rebar's Park(ing) installation from 2005, that evolved into the global annual Park(ing) Day event and a new micro public space typology named Parklet the Urban Prototyping Events from 2012 to 2016 use temporality for quick-fires to test the design of small urban implementations in the public realm (Schneider 2017).

Project background

In 2012, Peter Hirshberg and Jake Levitas of the Gray Area Foundation began to refine a model for civic engagement around smart technology, launching the Urban Prototyping Event in San Francisco. Urban prototyping is a new global movement exploring the potential of small-scale art and design projects in the public realm, and their impact on urban transformation, civic participation, and sensing—enhancing livability. The motivation was to strengthen the collaboration between local agencies, actors, and designers to improve the top down process of urban regeneration. The process began with an open call for projects combining digital and physical elements of the city that could be replicated in other places, resulting in over 100 submitted proposals. Teams received seed funding, technical assistance, and support from the city to deploy their prototypes in San Francisco's mid-market neighborhood. The event established a new model for administrators, designers, and citizens to work together and use technology to address pressing needs (Townsend 2013).

Temporary strategy

While the event honored the spirit of tactical urbanism, it also addressed the need for long-term conversations and change. To direct this need, the initiators developed an approach consisting of three steps: prototyping, replication, and adoption (Urban Prototyping 2012). The prototypes were meant to be mock-ups of a pilot scheme, developed, and tested in partnership with local agencies. After collecting feedback during the event, the final product underwent further development for potential global replication and mutation. The event took place on a Saturday afternoon with high pedestrian traffic, with four panel discussions as a stage for exchange and feedback. While the 24 selected projects were all very different, they can be categorized into strategies for urban sensing, urban projections, urban play, urban ecology, and urban appropriation (see Figure 32.5).

After a success in 2012, the festival continued in 2015 and 2016 as a collaborative event between the Planning Department and Yerba Buena Center for the Arts, rebranded as the Market Street Prototyping Festival. A physical/digital interface was recommended but not required. Instead, the city concentrated on shortening the distance between citizens and the decision makers, while unlocking the potential of civic assets and envisioning the unimaginable.

In this new partnership, the festival was tied to the Better Market Street project, a five-year effort to redesign and identify Market Street as a vibrant public destination. The project

Figure 32.5 Urban Prototyping Event 2012

Image credit: Grey Area Foundation Jake Levitas, 2012

introduced a new typology called "Street Life Zone," developed by Gehl Architects, to enhance Market Street's identity through the small-scale design of public spaces and programming along the entire length of the street (see Figure 32.6). The festival was meant to prototype the Street Life Zone and transform the public's relationship with design through community engagement, investment in civic innovators, and fostering connections through the design of urban prototypes. These goals were based on the idea that civic innovators and residents can help to create a culture of engagement, and that thoughtfully designed and programmed places strengthen a diverse public realm as the basis for a better city. The set-

Figure 32.6 Urban Prototyping Event 2015

Image credit: Gehl Studio 2015

up of the event allowed for open calls of project submissions that built new partnerships and invited established actors to test ideas (Cheng, Ho, and Sarmiento 2016).

The Urban Prototyping Festival is a new take on urban design, in which temporary implementations of small-scale public spaces offer within a contained time frame a multitude of design strategies. In contrast to initiatives like the New York Plaza Program, the immediacy of the event allows for an intense dialogue between residents, designers, and administrators. Under the guidance of Gehl Studio, the organizers evaluated the performance of the prototypes—a critical element that stands in contrast to the trial and error tactics of DIY urbanism. Besides visitor counts and interviews, the team identified seven evaluation criteria considering questions, if the design promotes a new idea that connects the implementation to larger urban strategies, or if the prototype could support community capacity and promote a diverse social life (Gehl Studio 2015). As another remarkable novelty, the Urban Prototyping initiative in 2012, 2015, and 2016 required that every prototype be posted in the UP library on Instructables. These projects are now open source, supporting adaptations and replications as a critical component of the temporary strategy.

Design of the temporary

As Kees Christiaanse (2005) argues, stimulating development within the context of time-based transformations requires the designer to shift their thinking from physical form to complex interactive and responsive processing, transforming the role of the designer to that of a curator, negotiator, and collaborator.

The presented case studies vary in their scale, physical morphology, and temporary strategy. The MQ in Vienna constructed a new identity of an existing plaza over a long period of time through programming and place branding; The Broadway project converted street infrastructures by redrawing physical boundaries and allowing for the appropriation of the newly gained space. The Airfield Tempelhof sought to temporarily extend the public realm through curated pioneer programs and new infrastructures. The Prototyping Event—itself temporary—stands on the other side of the spectrum, offering a whole series of small-scale prototypes to enrich future public spaces. Despite their differences, however, they all have their similarities as they utilize time for the transformation process, and the typologies emerge from embedded design strategies composed of four layers of space (built, appropriated, organized, and imagined) to construct a new public realm (see Figure 32.7).

These case studies are all defined by negotiations between top-down and bottom-up processes to establish new ecologies: public space, with its physical and immaterial components, is initiated and edited in its daily production, and defined by its spatial boundaries, social appropriation, temporary events, cultural frameworks, and the resulting emerging identity. The design of the temporary inherently interacts with the ephemeral character of public spaces. While monitoring spatial construction is key to the success of these projects, it is also the most delicate aspect, as public space must unfold as an authentic product of its diverse, undirected, and multifold urban context.

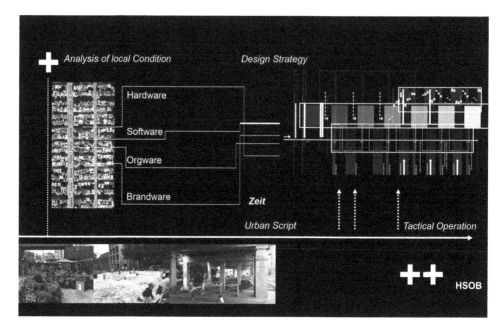

Figure 32.7 The Urban Script

Image credit: Mona El Khafif 2009

References

Bishop, P. and Williams, L. (2012) *The Temporary City*. London: Routledge.

Cheng, K., Ho, W., and Sarmiento, R. (eds). (2016) "Market Street Prototyping Festival." Accessed October 2018. marketstreetprototyping.org

Christiaanse, K. (2005) Waiting Land. In: *Situation KCAP*, P. Ursprung, M. Michaeli, W. Sewing, W. Van Stiphout, F. Von Borries, and F. Marshall (Eds.). Amsterdam: NAI Publisher, p. 152.

El Khafif, M. (2009) *Inszenierter Urbanismus. Stadtraum Fuer Kunst, Kultur, Und Konsum*. Saarbruecken: VDM Publisher.

El Khafif, M. (2010) Staged Urbanism. Urban Regeneration between Place Branding and User Generation. *2A Arts and Architecture, Special Issue: Women's Voices in Architecture and Design: Geographic and Professional Regions*.

El Khafif, M. and Del Signore, M. (2014) "Allographic Urbanism. From Guerilla to Open Source", Miami: ACSA Conference Proceedings.

Gehl, J. (2012) "Urban Prototyping – Exploring Temporary and Permanent", October 2012. Accessed Februray 2020. https://gehlpeople.com/blog/urban-prototyping-exploring-temporary-permanent/

Gehl, J., Risom, J., and Daily, J. (2015) "The Naked Truth." *The New York Times*, 31 August. Accessed October 2018. www.nytimes.com/2015/08/31/opinion/times-square-the-naked-truth.html

Gehl Architects and DOT New York (2014) World Class Streets: Remaking New York City's Public Realm. Published on January 7, 2014 and accessed in 2019 https://issuu.com/gehlarchitects/docs/issuu_561_new_york_world_class_stre

Gehl Studio. (2015) *Makers on Market. Lessons from San Francisco's Market Street Prototyping Festival, Volume 1*. http://marketstreetprototyping.org/wp-content/uploads/2016/04/MSPF-Report_Vol1_Makers-on-Market.pdf

Goldwyn, E. (2014) "How "People-centered" Design Made Times Square the Place to Be on New Year's Eve." *NEXT CITY*. Accessed December 30, 2014. https://nextcity.org/daily/entry/urban-design-times-square-concrete-bowtie-density-gehl-new-years-eve

Grynbaum, M. and Flegenheimer, M. (2015) "Mayor De Blasio Raises Prospect of Removing Times Square Pedestrian Plazas." *The New York Times*, 20 August.

Haydn, F. and Temel, R. (eds). (2006) *Temporary Urban Spaces: Concepts for the Use of Cities.* Basel: Birkhaeuser.

Heilmeyer, F. (2011) "Pioneernutzung Als Modell? Zwischennutzungen" *Bauwelt 36.201.* Accessed May 2019.www.bauwelt.de/themen/pioniernutzungen-als-modell-tempelhofer-feld-entwicklung-masterplan-2118018.html

Juergens, I. (2014) Wie Es Jetzt Auf Dem Tempelhofer Feld Weitergeht. *Berliner Morgenpost*, 27(May). www.morgenpost.de/berlin/article128441108/wie-es-jetzt-auf-dem-tempelhofer-feld-weitergeht. html

Knueppers, B. (2011) "Arche an Der Stadtbahn. Auf Der Experimentierwiese Tempelhofer Feld in Berlin Wird Kunst Anfassbar." *Oya.* September. https://oya-online.de/article/read/440-arche_an_ der_startbahn.html

Low, S. (2005) *The Erosion of the Public Space. Paranoia, Surveillance and Privatization in New York City.* London: Routledge.

Lydon, M. and Garcia, A. (2015) *Tactical Urbanism: Short-Term Action for Long-Term Change.* Washington, DC: Island Press.

Madanipour, A.R. (2017) *Cities in Time: Temporary Urbanism and the Future of the City.* New York: Bloomsbury, Kindle eBook, purchased 2018 from Amazon.com.

Metropolis (2018) "Activation of an Urban Space through Citizen Participation, Berlin, Germany." *Urban Sustainability Exchange.* Accessed October 2018. https://use.metropolis.org/case-studies/ger many-berlin-tempelhofer-freiheit-urban-open-space

Oswalt, P., Overmeyer, K., and Misselwitz, P. (2014) *Urban Catalyst: Mit Zwischennutzung Stadt entwickeln* second edition. Berlin: DOM Publishers.

Raumlabor, with Studio UC. (2009) "Airfield Tempelhof a Dynamic Master Plan." Accessed October 2018. http://raumlabor.net/aktivierende-stadtentwicklungflughafen-tempelhof/

Senate of Berlin. (2007) *Urban Pioneers: Stadtentwicklung durch Zwischennutzung.* Berlin: Jovis Verlag

Schneider, B. (2017) How Park(ing) Day Went Global. *CITY LAB*, 15 September. Accessed May 2017. www.citylab.com/life/2017/09/from-parking-to-parklet/539952/

Shepard, M. (2011) *Toward the Sentient City.* Cambridge: MIT Press.

Townsend, A. (2013) *Smart Cities: Big Data, Civic Hackers, and the Quest for a New Utopia.* New York: Norton.

Urban Prototyping (2012) "A Global Movement Exploring How Participatory Design Can Improve Cities", Urban Prototyping Research Lab, accessed November 2018. https://grayarea.org/initiative/ urban-prototyping/

33

BRINGING PUBLIC SPACES TO LIFE

The animation of public space

Troy D. Glover

Introduction

Public spaces ought to be understood, planned, designed, and managed as places. More than geographic locations and material form, public spaces are places invested with meaning and value (Gieryn 2000). In pursuit of greater meaning and value, then, public spaces require active efforts to make them into places. Placemaking aims to enhance sense of place, the socio-cultural meanings and attachments held by individuals or groups for a spatial setting, deliberately (Gieryn 2000; Glover 2015; Silberberg et al. 2013; Stokowski 2002). It does so through the process of *making*, a term that has multiple meanings in such a context: making can denote the construction or alteration of material form, imply the use of democratic and civic processes, and/or indicate a DIY or hacker orientation to create something new or tinker with something existing (i.e. maker culture) (Silberberg et al. 2013). Whatever the approach, "making" points to continuous placemaking over the life's course of a public space. No public space is an end product, in other words. A public space always represents an unfinished project, a work in progress.

Placemaking becomes a transformative endeavour when it refreshes public spaces and revitalizes them, usually in creative ways. Sure, vast sums of capital can be sunk into a public space to make it spectacular, but public investments can sometimes be viewed as a luxury, especially in an age of austerity (Lashua 2013). As a result, low cost, temporary strategies that "animate" public space have emerged as one of the key practices of 21st-century placemaking (Reid 2012). The *animation of public space* refers to "the deliberate, usually temporary, employment of festivals, events, programmed activities, or pop-up leisure to transform, enliven, and/or alter public spaces and stage urban life" (Glover 2015, 96). Irrespective of the global proliferation of animation practices, however, no formal categories of animation tactics have been articulated either academically or professionally, despite their potential to inform (and transform) the programming and design of public spaces.

This chapter aims to fill this gap by identifying and describing eight animation strategies: (1) naturalizing, (2) activating, (3) culinizing, (4) spectaclizing, (5) festivalizing and eventifying, (6) aestheticizing, (7) convivializing, and (8) gamifying and whimsicalizing. Each strategy will be explained and illustrated with current applications from around the globe. All told, these categories will be relevant to activists, planners, and policymakers who intentionally aim to inject public spaces with new functions and meanings and facilitate the ongoing alteration of places to better meet the needs of members of their communities. Before explaining the various strategies, I begin by underscoring the importance of programming in the built environment.

Beyond construction: the importance of programming public space

While a public space may be regarded as "a sculpture worth displaying," it more importantly embodies "a slab of marble that will be carved anew by the interaction of whatever assortment of people pass through it each day" (Willis 2018, 21). In other words, people imbue a public space with meaning through their everyday interactions within it. So-called "design it and leave it" approaches to public space do nothing to actively address these interactions. While front-end community engagement with its exclusive focus on the initial creation of a public space ought to necessarily happen, continuous placemaking through programming warrants just as much, yet receives not nearly enough, attention (Montgomery 1995; Silberberg et al. 2013).

Interactions in public space matter because they have the potential to deepen social ties and affect sense of community (Demerath and Levinger 2003; Friedmann 2010; Putnam 2000). Admittedly, we need to know more empirically about the level of intensity of inter-action that makes a significant difference in this regard (i.e. the impacts of friendly greetings vs. engaged activity), but the most significant interaction opportunities clearly result from the presence of others (Jacobs 1961). As Montgomery (1995) noted, "it is the public realm and associated semi-public spaces which provide the terrain for social interaction and therefore transactions" (107). Bringing people into physical contact within public space, therefore, matters, especially in an age when reported feelings of social isolation are on the rise (Glover 2018).

Public spaces offer urban inhabitants a familiar spatial setting in which the rhythms of city life unfold and in which socio-cultural transactions and public sociability take place. As Paul Knox (2005) explained:

> the density of routine encounters and shared experiences that underpin the inter-subjectivity that is the basis both for a sense of place and for a structure of feeling within a community. The same is true of elements of weekly rhythms, such as street markets and farmers' markets; and of seasonal rhythms, such as food festivals, craft shows and arts festivals.
>
> *(8)*

Whether associated with mundane routines and practices or special events that interrupt and challenge those routines, public spaces represent important sites of everyday social engagement (Borer 2013). And the capacity of these spaces to promote sociability can easily be enhanced through animation practices. As Silberberg et al. (2013) made clear, "There is endless opportunity to improve existing places through programming" (10).

All told, animating public space represents an active effort to draw people to public spaces: to socialize, to linger, to be together. To animate means to bring something to life. Taken within the context of public spaces, then, animation efforts ultimately endeavor to boost urban vitality (Montgomery 1995). Of course, the range of strategies used to add vital-ity to a public space is expansive. Nevertheless, this chapter attempts to profile the variety of efforts evident in current practice. I move next to the categorization of the various tactics.

Eight strategies to animate public space

What follows is a list of eight animation practices used across the globe to add greater vitality to public spaces. These categories are not mutually exclusive, however. Many efforts to animate public space, as you will see, can be combined or may span more than one strategy. Nevertheless, the categories offer an effort to articulate the breadth of intentional and flexible, often temporary, low-cost, scalable, and (typically) bottom-up practices available to individuals, groups, and organizations to add vitality to a public space. Because animation strategies are action oriented, I describe them accordingly and deliberately to provide direction for what can be *done* to a public space. In so doing, I take some liberties in fashioning words (i.e. verbifying) that represent the efforts involved. My hope is that this new(ish) lexicon will be useful to those involved (in whatever capacity) to make public spaces into better places.

Naturalizing

Naturalizing refers to enhancing public spaces with natural features, including plants, flowers, gardens, water, rocks, and/or grasses, to "green" a landscape and make spaces more appealing as places for visitors to spend time. Extensive evidence of the health benefits of greening documents the psychological and subjective impacts of natural environments, ties the presence of natural features to healthy functioning, and positions exposure to nature as a precursor to wellbeing and links it with objective measures of health (Kuo 2013). Naturalizing a public space, particularly in urban environments, takes aim at so-called nature deficit disorder in contemporary society by drawing people to public spaces who desire a connection to nature (Louv 2005).

Planting trees, natural grasses, and flowers, of course, add vitality to a listless public space by making the space more appealing to visitors. Even unsanctioned attempts to do so, such as the practice of guerrilla gardening—the unlawful act of targeting spaces of neglect to transform the environment without the landowner's consent—beautify and increase biodiversity in public spaces that suffer from neglect (Adams, Hardman, and Larkham 2015). However, examples of naturalizing public space can easily transcend the simple addition of natural features. As Kuo (2013) noted, "providing immediate access to green spaces will have limited health benefits if there is nothing in those spaces and therefore nothing to hold people in them. Programming and activity infrastructure are key here" (181). In the context of naturalizing, the programming itself can involve greening activities. Community gardens, for example, represent efforts by communities to not only make a landscape more useful and productive, but also communal and social (see Glover 2003). Similarly, events such as Park(ing) Day, a popular global event during which parking spaces are transformed temporarily into "parks," often with green grass (i.e. sod rolled out), beach chairs, and umbrellas, draw attention to the privileging of cars in society and encourage the public to rethink the purpose of public space (see Coombs 2012). The imaginative nature of events like Park(ing) Day draw passersby and encourage them to assemble. Whatever the example, naturalizing public space connects users to natural elements to enhance their lived experience of that space.

Activating

In the context of public space, the term *activating* is often conflated with *animating*. Like Malgorzata Kostrzewska (2017), I prefer to narrow its meaning by referring to it more specifically as an intentional strategy to use physical activity to draw active participants and/or spectators to a public space. That sport takes place in public spaces is no surprise, given its common use of parks, playgrounds, fields, and community centers, spaces often built and maintained by local authorities. In the context of animating public space, though, physical activity and sport is used more deliberately within nontraditional sport spaces to enliven streets, sidewalks, public squares, and alleyways (among others). Much like most other animation strategies, activating aims to attract people to public spaces, albeit by using physical activity as a draw.

Visual displays of physical activity, such as sport demonstrations, pop-up labs (i.e. sport tutorials), or the availability of athletic equipment (e.g. portable basketball nets, fitness stations) or infrastructure (e.g. goal posts) attract crowds and participants. Streets can be converted into spaces for formal running events (e.g. 5K Run) or everyday physical activity (e.g. jogging, walking), and using the street as a playing surface transforms an otherwise uninviting space into a game of street hockey or soccer. Several initiatives, such as the "Play Street Movement" (USA), Playborhood (USA/Canada), and Playing Out (UK), seize on this potential by encouraging neighborhoods to close street blocks to cars and open them up to play. And unsanctioned use of spaces for physical activity by practitioners of parkour— the activity of negotiating obstacles within an urban environment, by running, jumping, and climbing—and skateboarders can provide an attention-grabbing visual display to onlookers or perhaps offend users of public space if participants' actions are interpreted as rude or invasive. All told, activating public space animates people's experience through its deliberate use of physical activity or sport.

Culinizing

Culinizing refers to incorporating food, particularly its preparation, distribution, and consumption, into a public space. Used in this sense, food becomes an "attraction asset" that contributes to the sensory experiences of a public space, making it into a "foodscape" of sorts (see Richards 2015). When visiting a public space, users not only see and hear, but also smell, taste, and experience, so food can be used deliberately to produce, maintain, and/or change the "landscape of senses" available to visitors (Quan and Wang 2004). Culinizing a public space contributes to an atmosphere that attracts people and potentially protects or amplifies a sense of local identity (e.g. through the incorporation of food claimed to be authentic expressions of the history or heritage of the place) (Berg and Sevón 2014). Whatever the aim, culinizing public space ultimately brings people together as a community. More than a simple service encounter, in other words, food serves as a powerful cultural medium that transcends its value and meaning as a tangible object (Lashley, Morrison, and Randall 2005) to nurture collective spaces of encounter (Marovelli 2018). Material and spatial aspects of public space form features used to facilitate a social atmosphere, and types of food made available in public space, as well as the way food is prepared, have implications for the quality of interactions among users of public space.

Examples of culinizing public space include the presence of food carts, such as hotdog or lemonade stands, that sell foodstuffs on sidewalks, public squares, or parks. Food trucks, an increasingly popular strategy to bring food to city centers for a limited period of time during each workday or during an event, represent a growing service option in most North American cities. Hawker centers, open-air complexes found in many Asian countries, offer food stalls for the sale of local cuisine in public spaces located in close proximity to transportation hubs. In more of a bottom-up example of culinizing public space, community members host cookouts, barbeques, and picnics in public spaces to build connections among neighbors/community members. And farmers' markets offer a festive environment in which to meet friends, taste new foods, and enjoy a meal while shopping for produce. These examples represent some of the many ways food animates public space.

Spectaclizing

Spectaclizing refers to the use of featured entertainment to attract people to public space. As a form of amusement, live music, dance, concerts, movies, parades, fireworks, and theatre, whether stage-managed (i.e. formal) or improvised (i.e. casual), capture the attention and interest of an audience and/or provide a backdrop to support sociability in public space. Simpson (2011) wrote that such performances "maintain within [them] the possibility of liminality, of producing a period people are 'betwixt and between' and so open to change" (415–416). That is, they capture the attention of those in public space, even if for only a brief moment, thereby affecting the space in terms of "density, accretion, durations, dispersal, and flow" (Harrison-Pepper 1990, 131).

Examples of spectaclizing public space include pop-up entertainment and street performances, both musical and nonmusical, sometimes in pursuit of donations from passersby. Buskers, for instance, perform so-called "circle shows" that draw lucrative audiences or "walk-by acts" that attempt to appeal to pedestrians as they walk past. Pop-up cinema uses the walls of abandoned buildings on which to project films (Lashua 2015). Whatever the example, Vivian Doumpa and Brian Doucet (2013) suggested that street performances can "create a sense of pleasure and eventually comfort due to the symbolisms and emotions that it carries" (3). The presence of entertainment, in other words, has the potential to produce emotions in users of public space and therefore influences the perception of attractiveness and pleasure in the space in which the entertainment is observed. "Counter-spectacle," the appropriation and subversion of performances, re-appropriate scripted narratives of public space to suit the purposes of the user, challenge the way people think about public space by promoting spontaneous encounters in routinized and alienating urban environments (Tanenbaum 1995).

Festivalizing and eventifying

Festivalizing and eventifying public space refer to the use of organized festivals and events to encourage public gatherings and bring a festive atmosphere to a public space. In Doreen Jakob's (2013) words, events represent "the deliberate organization of a heightened emotional and aesthetic experience at a designated time and space." To be sure, festivals and events transform the way we use and imagine public spaces. Andrew Smith (2017) noted, for example, by closing off a street to traffic, an event alters the rhythm of the street by encouraging pedestrians to slow down and interact with each other. In this way, events produce public space through the social interactions they encourage. Put another way, festivalizing and eventifying public space aims to loosen the space in which it is practiced, yet it can often tighten restrictions on the space by introducing new controls, while potentially contributing to the processes of commercialization and privatization (Smith 2017). The production and promotion of festivals represents a form of "experience planning," a prominent planning tool used in cities to advance local urban and economic development, consumer experiences, and city images (Jakob 2013). Nevertheless, bottom-up festivals introduce organic attempts by urban inhabitants to drive social engagement within their community groups or local settings.

Ciclovia, an event that temporarily closes major downtown streets (as opposed to neigh-borhood streets) to cars to provide a safe space for walking, bicycling, and social activities, embodies the idea of festivalizing and eventifying through its efforts to advance livability in the city. Known in many North American cities as an Open Streets Initiative, these events can build broader political support for undertaking more permanent improvements to street-scapes, making them more pedestrian-friendly (Lydon and Garcia 2015). These events resemble a larger scale Block Party, which can transform neighborhoods into social experi-ences. Along these lines, the Festival of Neighborhoods in Kitchener (Ontario, Canada) encourages neighborhoods to host events, broadly defined, that bring neighbors into contact with each other to strengthen social ties (Yuen and Glover 2005). Porchfest, a grassroots event that began in Ithaca, New York, transforms neighborhood landscapes into event spaces that bring people together to reacquaint themselves with what it means to be a good neighbor. These sorts of temporary events animate spaces to transform the everyday into something special.

Aestheticizing

The use of the arts in all of their varieties and forms can beautify and/or humanize mundane landscapes and encourage greater activity in public space. Temporary art installations, painted surfaces (e.g. graffiti, painted walkways), and murals aestheticize drab places, making public spaces more attractive or more interesting than otherwise. These works enhance public spaces aesthetically, thereby increasingly establishing them as cultural strategies of urban (re)develop-ment and (re)generation that have gained economic, social, and cultural status over time. Whether interactive or curated, they facilitate pausability in public spaces by temporarily inter-rupting the flow of users in the space, yet allow users to resume their initial activity (Demerath and Levinger 2003). In some cases, these art forms represent community efforts and vehicles for participatory and co-operative activity that instill civic pride, foster social interaction, promote a sense of community, contribute to local identity, and address social exclusion (Schuermans, Loopmans, and Vandenabeele 2012). In others, they act as a medium for the communication of symbolic meanings, largely driven by the work of artists, entrepreneurs, and cultural industries (Hall and Smith 2005; see Markusen and Gadwa 2010). More than simply "prettifying" a space, though, aestheticizing public space can be "a process in which dissensus emerges as a consequence of the capacity of aesthetics to create new senses" (Pan 2015, 10). What Pinder (2008; see also 2005) termed "arts of urban exploration"—artistic practices that disrupt everyday urban life, question and explore social problems and conflicts without necessarily prescribing solutions, and resist the processes through which urban spaces are currently produced to encour-age more democratic alternatives—treat art as a form of appropriation of public space through which artists articulate and communicate interests and identities (731).

Examples of aestheticizing are plentiful: painting intersections creates vibrant, colorful spaces that calm traffic and attract people; yarn bombing (or a yarnstorm) provides non-permanent feminized graffiti installations made of yarn (e.g. tree or hydrant coverings) aimed at reclaiming and personalizing lifeless public spaces; adding a strong punch of color to back alleys and lane-ways, including the street, walls, and dumpsters, transforms unsafe spaces into a brighter, safer space that encourages social uses; providing sanctioned opportunities to interact with installations or add content to public spaces can serve to engage users; chalking the sidewalk or boards set up for such purposes, or permitting people to "vandalize" a wall or piece of public art, offer users of public space opportunities to interact with the built environment and feel more engaged. In these ways, practitioners of animating public space aestheticize the built environment.

Convivializing

Providing features that enable individuals to gather, sit, and socialize enliven otherwise banal public spaces. Amanda Wise and Selvaraj Velayutham (2014) described conviviality as an atmosphere and an affect in which social dimensions coalesce with material, sensory and spatial ones. Thus, convivializing public space aims explicitly to make spaces more sociable, for "convivial relations rest as much on material environs as they do on interpersonal and social relations. The physical organization of social space, and the ways humans make use of this space, are fundamental to the logic of connection or discrimination" (Wise and Noble 2016, 427). Simply put, "What attracts people most [to a public space], it would appear, is other people" (Whyte 1980, 105). Indeed, visual assurances of sociability encourage people to socialize themselves. Correspondingly, conviviality marks an important attribute of the quality of a public space. Convivial public spaces serve as meeting places, encourage their use as a means of socializing, and create welcoming environments.

Simple convivializing tactics such as strategies or policies that support the provision of permanent seating and the creation of spaces for pause encourage people to loiter. Street furniture of any kind welcomes people to sit and take in the social scene. "Chair-bombing," the addition of temporary seating to public spaces that need them, or adding tables with umbrellas, represent more flexible efforts to make a space more inviting and welcoming for sociability. Allowing businesses to establish temporary seasonal patios by using on-street parking spaces has emerged recently as a popular strategy by many municipalities to encourage greater convivial interactions in neighborhood retail streets or commercial areas. Referred to as parklets, street seats, or curbside seating, these public seating platforms usually involve a coordinated effort by cities, local businesses, residents, and neighborhood associations. Considered "streateries" because they tend to be open to the public during business hours only, some communities opt alternatively for "pavement parks," social spaces created in awkward three-way intersections where the corner of two angled streets connect with a third. As well as serving as strategy to improve safety, barriers and planters can be added to either side of the intersection, the surface of the space, usually pavement, can be painted, and tables and chairs can be installed to provide a public open space to neighborhoods lacking in convivial areas (see Cohen 2015). Sometimes the simple addition of seating can transform a space for greater conviviality.

Gamifying and whimsicalizing

Gamifying refers to the application of game-design elements and game principles to public space to motivate participation and engagement. Adding illustrations to pavements that signify game play encourages passersby to engage with public space. A hopscotch board or painted maze leading to a garbage bin, for example, aims to make the act of throwing out refuse feel like a game to users of public space. Similarly, the use of augmented reality associated with apps such as Pokémon GO encourage users to wander through public spaces as they tag, collect, trade, and battle for digital artefacts via their mobile devices that provides a digital overlay of game objects and virtual locations across the actual public space (Hjorth and Richardson 2017). Through this augmented layering of the digital onto place, banal and familiar surroundings become significant game locations. All told, these efforts aim to create the playful city, *a recognition that* one of the fundamental functions of public space is to serve as a setting for informal, noninstrumental social interaction or play (Stevens 2007).

Similarly, adding imaginative and sometimes quirky features to everyday landscapes can capture people's imaginations and encourage use of public space. As Scolere et al. (2016) explained, whimsical features

> act as 'social signifiers' that signal a certain type of atmosphere or ambience apart from the 'serious' landscape—a cue that social connectedness is valued, that there are moments that offer respite from the office or the lab, and that playfulness and creativity are encouraged.
>
> *(210)*

Recently, many North American cities have added pianos to public spaces, swings in bus shelters, hammocks to street posts, and little library boxes to yards accessible to pedestrians on sidewalks. These tactics, like the others mentioned, create a playful environment that encourages users of public space to interact more meaningfully with the landscape around them.

Possibilities and pitfalls

In this chapter, I list a variety of tactics aimed at animating public space under eight categories of strategies. Nevertheless, the number of interventions available for animation are limited only by our imagination. Indeed, more and more creative ideas emerge every day as people around the world seek to transform lifeless, meaningless spaces in their communities into vibrant, meaningful places. Cities can and do use these strategies (i.e. top-down), but most of the creativity stems from the innovations of grassroots efforts (i.e. bottom-up), which can either be sanctioned or unsanctioned. Almost any of these strategies can be driven by a top-down or bottom-up process (e.g. a city vs. resident-built flower garden, city sanctioned food venders vs. neighborhood BBQ, Cultural planning department organized concert vs. community porch party, commissioned art vs. unsanctioned graffiti), so assigning them to a particular institution, group or individual would be overly simplistic and unnecessarily prescriptive. Approval by appropriate authorities to engage in such activity (i.e. sanctioned vs. unsanctioned), however, will often determine the acceptability of the strategy (by authorities). Skateboarders who apply wax on rough surfaces to smooth ledges and enable speed, though "activating" a public space, are routinely vilified by authorities for defacing public property, while children playing street hockey on the road engage in seemingly acceptable behavior according to local by-laws). To be sure, unsanctioned animation strategies push authorities to think beyond their regulated and bureaucratic world to allow bottom-up development, often in a way that leads to new innovations (Lydon and Garcia 2015).

Irrespective of who drives the effort and whether it is sanctioned or unsanctioned, animating public space ultimately aims to "subvert, loosen, or transform presupposed rules of social conduct" (Pinder 2005, 400). That is, they introduce users of public space to new spatial patterns by shifting practices and encouraging different ways of relating to others (Simpson 2011). For example, adding flowers to a boulevard beautifies an unseemly space, augmenting a bike trail with a whimsical "fun box" (a combination of jumps or wedges) offers cyclists a material feature to add to their ride experience, and painting a crosswalk in rainbow colors signifies a welcome space to the LGBTQ+ community. Whatever the animation strategy, these tactics provide valuable cues as to the character of public spaces and of potential social activities within them. Consequently, animation tactics that provide a focal point or targeted group activity appear to help to promote social cohesion between different groups and overcome constraints that may prevent some people (in marginalized groups) from taking part (Bagnall et al. 2018). Animation and programming, in other words, help public spaces be seen. They recognize, as argued above, that just building a public space is not enough.

Even so, animating public space is no panacea. In many cases, it can have potentially negative effects in terms of some community members feeling excluded, particularly in relation to strategies that target or celebrate certain groups at the expense of others (Bagnall et al. 2018). Defensive animation practices may include modifying existing infrastructure to make it impossible to use in certain ways (e.g. a planter box that blocks entry), adding elements to a public space to discourage possible uses (e.g. bolting metal pieces to curbs to prevent skateboarders from grinding), removing or displacing objects from public space so that certain functions disappear or relocate to more desirable areas (e.g. building youth drop-in centers away from the downtown core), using off-putting nudges (e.g. colors or sounds) to encourage unwanted individuals to gather elsewhere, or closing off areas to certain uses, prohibiting certain activities, or restricting access (de Fine Licht 2017). These defensive measures give a clear indication of who belongs and who does not belong in a public space.

As critical scholars of space warn, all spaces, including spaces of (unintentional or intentional) resistance, must be problematized. The qualities that define an animated public space can be at once liberating (for animation practitioners and like-minded supporters) and exclusionary (for others) inasmuch as they support one social group's entitlement, while physically and/or symbolically evicting its "others" (Glover 2017). In contrast to what some commentators see as a means to advance the right to the city, the animation of public space can be oppressive, sometimes embodying sites of discriminatory practices wherein marginalization is (re)produced and (re)enforced.

To be sure, concerns about social justice pervade in the growing literature that exists on the animation of public space. Heim Lafrombois (2017) highlighted the racialized, classed, gendered, and sexualized biases found within the conceptualization of the topic. Moreover, Mould (2014) argued "urban neoliberal development discourse" co-opts the emancipatory nature of animation practices by subordinating them to Florida's (2002) Creative City language and practice to re-establish economic and political hegemony. Similarly, scholars have expressed concern that animation practices serve as catalysts in the process of gentrification and the displacement of certain residents, often minority or low income, who can no longer afford to live in the neighborhoods in which animation has taken hold (see Tardiveau and Mallo 2014). Given the potential for negative outcomes, Douglas (2014) underscores the pressing need to measure the impacts of the animation of public space.

As Zukin (1998) reminds us, "Whoever controls public space sets the 'program' for representing society," an observation consistent with Lefebvre's (1991) belief that space, whether public or private, encourages and discourages certain forms of interaction and gives form to social structures and ideologies (1). Those who study, practice, or support the animation of public space, therefore, must be sensitive to these potentialities and be sure to address them to ensure greater outcomes of inclusivity, and belonging. While every community may possess spaces that lack character, there is no reason animation tactics cannot be employed with character.

References

Adams, D., Hardman, M., and Larkham, P. (2015) Exploring Guerrilla Gardening: Gauging Public Views on the Grassroots Activity. *Local Environment*, 20(10): 1231–1246.

Bagnall, A., South, J., Di Martino, S., Southby, K., Pilkington, G., Mitchell, B., Pennington, A., and Corcoran, R. (2018) *A Systematic Review of Interventions to Boost Social Relations through Improvements in Community Infrastructure*. Leeds: What Works Centre Network.

Berg, P.O. and Sevón, G. (2014) Food-Branding Places–A Sensory Perspective. *Place Branding and Public Diplomacy*, 10(4): 289–304.

Borer, M.I. (2013) Being in the City: The Sociology of Urban Experiences. *Sociology Compass*, 7(11): 965–983.

Cohen, J. (2015) "Stop Building Mediocre Parklets, Start Building Pavement Parks." *Next City*. Accessed May 18, 2019. https://nextcity.org/daily/entry/parklets-stop-build-pavement-parks

Coombs, G. (2012) Park(ing) Day. *Contexts*, 11(3): 64–65.

de Fine Licht, K.P. (2017) Hostile Urban Architecture: A Critical Discussion of the Seemingly Offensive Art of Keeping People Away. *Etikk I Praksis-Nordic Journal of Applied Ethics*, (2): 27–44.

Demerath, L. and Levinger, D. (2003) The Social Qualities of Being on Foot: A Theoretical Analysis of Pedestrian Activity, Community, and Culture. *City & Community*, 2(3): 217–237.

Douglas, G.C. (2014) Do-It-Yourself Urban Design: The Social Practice of Informal "Improvement" through Unauthorized Alteration. *City & Community*, 13(1): 5–25.

Doumpa, V. and Doucet, B. (2013) "Music Performance in Public Space: Changing Perception, Changing Urban Experience." Echopolis-Days of Sound-International Conference, Athens, September.

Florida, R.L. (2002) *The Rise of the Creative Class and How It's Transforming Work, Leisure, Community and Everyday Life*. New York: Basic Books.

Friedmann, J. (2010) Place and Place-Making in Cities: A Global Perspective. *Planning Theory & Practice*, 11(2): 149–165.

Gieryn, T.F. (2000) A Space for Place in Sociology. *Annual Review of Sociology*, 26(1): 463–496.

Glover, T.D. (2003) The Story of the Queen Anne Memorial Garden: Resisting a Dominant Cultural Narrative. *Journal of Leisure Research*, 35(2): 190–212.

Glover, T.D. (2015) Animating Public Space. In: *Landscapes of Leisure: Space, Place, and Identities*, S. Gammon and S. Elkington (eds.). London: Palgrave Macmillan, pp. 96–109.

Glover, T.D. (2017) Leisure, Social Space, and Belonging. In: *The Palgrave Handbook of Leisure Theory*, K. Spracklen, B. Lashua, E. Sharpe, and S. Swain (eds.). London: Palgrave Macmillan, pp. 873–890.

Glover, T.D. (2018) All the Lonely People: Social Isolation and the Promise and Pitfalls of Leisure. *Leisure Sciences*, 40(1–2): 25–35.

Hall, T. and Smith, C. (2005) Public Art in the City: Meanings, Values, Attitudes and Roles. In: *Interventions. Advances in Art and Urban Futures*, M. Miles and T. Hall (eds.). Portland, OR: Intellect Books, pp. 175–179.

Harrison-Pepper, S. (1990) *Drawing a Circle in the Square: Street Performing in New York's Washington Square Park*. Jackson: University Press of Mississippi.

Heim Lafrombois, M. (2017) Blind Spots and Pop-Up Spots: A Feminist Exploration into the Discourses of Do-It-Yourself (DIY) Urbanism. *Urban Studies*, 54(2): 421–436.

Hjorth, L. and Richardson, I. (2017) Pokémon GO: Mobile Media Play, Place-Making, and the Digital Wayfarer. *Mobile Media & Communication*, 5(1): 3–14.

Jacobs, J. (1961) *The Death and Life of Great American Cities*. New York: Random House.

Jakob, D. (2013) The Eventification of Place: Urban Development and Experience Consumption in Berlin and New York City. *European Urban and Regional Studies*, 20(4): 447–459.

Knox, P.L. (2005) Creating Ordinary Places: Slow Cities in a Fast World. *Journal of Urban Design*, 10(1): 1–11.

Kostrzewska, M. (2017) October. Activating Public Space: How to Promote Physical Activity in Urban Environment. In: *IOP Conference Series: Materials Science and Engineering*, 245(5).

Kuo, F.E. (2013) Nature-Deficit Disorder: Evidence, Dosage, and Treatment. *Journal of Policy Research in Tourism, Leisure and Events*, 5(2): 172–186.

Lashley, C., Morrison, A., and Randall, S. (2005) More than a Service Encounter? Insights into the Emotions of Hospitality through Special Meal Occasions. *Journal of Hospitality and Tourism Management*, 12(1): 425.

Lashua, B.D. (2013) Pop-Up Cinema and Place-Shaping: Urban Cultural Heritage at Marshall's Mill. *Journal of Policy Research in Tourism, Leisure and Events*, 5(2): 123–138.

Lashua, B.D. (2015) Zombie Places? Pop up Leisure and Re-Animated Urban Landscapes. In: *Landscapes of Leisure: Space, Place, and Identities*, S. Gammon and S. Elkington (eds.). London: Palgrave Macmillan, pp. 55–70.

Lefebvre, H. (1991) *The Production of Space*. Malden, MA: Blackwell.

Louv, R. (2005) *Last Child in the Woods*. Chapel Hill: Algonquin Books.

Lydon, M. and Garcia, A. (2015) *Tactical Urbanism: Short-term Action for Long-term Change*. Washington, DC: Island Press.

Markusen, A. and Gadwa, A. (2010) *Creative Placemaking*. Washington, DC: National Endowment for The Arts.

Marovelli, B. (2018) Cooking and Eating Together in London: Food Sharing Initiatives as Collective Spaces of Encounter. *Geoforum*, 99(2): 190–201.

Montgomery, J. (1995) Editorial: Urban Vitality and the Culture of Cities. *Planning Practice & Research*, 10(2): 101–110.

Mould, O. (2014) Tactical Urbanism: The New Vernacular of the Creative City. *Geography Compass*, 8(8): 529–539.

Pan, L. (2015) *Aestheticizing Public Space: Street Visual Politics in East Asian Cities*. Bristol: Intellect Books.

Pinder, D. (2005) Art of Urban Exploration. *Cultural Geographies*, 12(4): 383–411.

Pinder, D. (2008) Urban Interventions: Art, Politics and Pedagogy. *International Journal of Urban and Regional Research*, 32(3): 730–736.

Putnam, R.D. (2000) *Bowling Alone: The Collapse and Revival of American Community*. New York, NY: Simon & Schuster.

Quan, S. and Wang, N. (2004) Towards a Structural Model of the Tourist Experience: An Illustration from Food Experiences in Tourism. *Tourism Management*, 25(3): 297–305.

Reid, D. (2012) Urbanism for a New Century. *Spacing*, 23 (Winter).

Richards, G. (2015) Evolving Gastronomic Experiences: From Food to Foodies to Foodscapes. *Journal of Gastronomy and Tourism*, 1(1): 5–17.

Schuermans, N., Loopmans, M.P., and Vandenabeele, J. (2012) Public Space, Public Art and Public Pedagogy. *Social & Cultural Geography*, 13(7): 675–682.

Scolere, L.M., Baumer, E.P., Reynolds, L., and Gay, G. (2016) Building Mood, Building Community: Usage Patterns of an Interactive Art Installation. In: *Proceedings of the 19th International Conference on Supporting Group Work*: 201–212.

Silberberg, S., Lorah, K., Disbrow, R., and Muessig, A. (2013) *Places in the Making: How Placemaking Builds Places and Communities*. Cambridge: MIT Press.

Simpson, P. (2011) Street Performance and the City: Public Space, Sociality, and Intervening in the Everyday. *Space and Culture*, 14(4): 415–430.

Smith, A. (2017) Animation or Denigration? Using Urban Public Spaces as Event Venues. *Event Management*, 21(5): 609–619.

Stevens, Q. (2007) *The Ludic City: Exploring the Potential of Public Spaces*. New York: Routledge.

Stokowski, P.A. (2002) Languages of Place and Discourses of Power: Constructing New Senses of Place. *Journal of Leisure Research*, 34(4): 368–382.

Tanenbaum, S.J. (1995) *Underground Harmonies: Music and Politics in the Subways of New York*. Ithaca: Cornell University Press.

Tardiveau, A. and Mallo, D. (2014) Unpacking and Challenging Habitus: An Approach to Temporary Urbanism as a Socially Engaged Practice. *Journal of Urban Design*, 19(4): 456–472.

Whyte, W.H. (1980) *The Social Life of Small Urban Spaces*. Washington, DC: Conservation Foundation.

Willis, B. (2018) "How Public Space Can Build Community and Rescue Democracy." *Common Edge*, 6 March. Accessed 12 October 2018. http://commonedge.org/how-public-space-can-build-community-and-rescue-democracy/

Wise, A. and Noble, G. (2016) Convivialities: An Orientation. *Journal of Intercultural Studies*, 37(5): 423–431.

Wise, A. and Velayutham, S. (2014) Conviviality in Everyday Multiculturalism: Some Brief Comparisons between Singapore and Sydney. *European Journal of Cultural Studies*, 17(4): 406–430.

Yuen, F. and Glover, T.D. (2005) Enabling Social Capital Development: An Examination of the Festival of Neighbourhoods in Kitchener, Ontario. *Journal of Park and Recreation Administration*, 23(4): 20–38.

Zukin, S. (1998) Politics and Aesthetics of Public Space: The "American" Model. In: *Real City, Ideal City: Signification and Function in the Modern Urban Space*, Barcelona: Center of Contemporary Culture of Barcelona, www.cccb.org/rcs_gene/politics_aesthetics.pdf

PART 5

Futures

In the twenty-first-century city, the boundaries of public space have been blurred in more than one way. On one hand, public life now commonly exists in public as well as quasi-public and even privatized space. On the other, more and more publicly owned, yet traditional space within the city is being appropriated for public use. Cities are confronted with several major trends that pose challenges and bring opportunities. The global economy and persistent competitiveness dominate over local issues of equality, diversity, health, education, and economy; the commoditization of public goods is almost ubiquitous; decades of bad ecological practices have made our settlements vulnerable; changing demographics bring new challenges and meanings to space; and technology is revolutionizing communication, information flows, and the right to privacy. These phenomena and changes call for a new interpretation of public space as a place of diversity and they should equally inspire us to critically defend the role of public space as a place for assembly. As a barometer of a city, these tendencies are visible in public space. The chapters in this section consider the futures of public space in the context of these trends and major shifts.

One of the primary expectations of good public space is the assumption of safety. But this quest for safety, resulting in oversecuritization of public space, has also led to diminishing publicness. Nevertheless, as compared to the seemingly safe private, public space, with its openness, exposure, and unpredictability, comes with unknown possibilities but also risks. Mark Kingwell's chapter presents the nature of risk inherent in public space, including how public space, perversely, becomes an enabler of violence. Reflecting on the recent acts of terrorism in public spaces, the author questions how the threat of violence affects our sense of time as we move between the spaces of seemingly safe private to the insecure public. Kingwell concludes by asking, "Is there any valid response to this brutal inversion of the public ideal as non-rival and non-excludable?"

Exploring the publicness of public space is a common theme among scholars to critically appraise public space. Margaret Kohn's chapter presents the tension between the public trust doctrine that is interested in protecting of the natural environment such as parks, and public access to parks. The author asks readers to look at urban parks as a way of compensating for the loss of nature in the city. Using solidarism, a theory of compensatory justice, Kohn shows how a solidarist approach can accommodate the goals of public trust doctrine by limiting its potentially anti-democratic implications, while at the same time providing public access to parks.

The ubiquity of Public Private Partnership (PPP) regimes in the creation and management of public space is one of the most visible signs of the neoliberal capitalism in the city. The dynamic relationships among government officials, private actors, and parks advocates in the governance of parks as public spaces is critically assessed by Susanna Schaller and Elizabeth Nisbet. The authors critique the broader placemaking ethos in public space by examining the dialogues around the use of public parks for large events. Through a narrative analysis of public hearings and interview scripts, their chapter examines how government officials, private partners, and parks advocates frame New York City's eventisation of urban parks and how these narratives express support for or a critique of the larger PPP regime.

The complex and intricate structures of public space that emerge from a response to rules, laws, customs, tacit agreements, and adaptations over time are discussed by Michael Mehaffy and Peter Elmlund. The authors help us understand the wide range of differentiations in the formal and informal territories of public space. This knowledge can aid planners and designers with producing public spaces that result in fewer conflicts by carefully considering and mediating between numerous and diverse individual, group and public needs in public space.

In their chapter, Avigail Vantu and Kristen Day, discuss how big data can support and transform public space research and design by addressing several challenges including limited objectivity, high cost in labor and time, restricted scope to local orientation and small scale, and lack of temporal fluidity and separation of behavior from precise geographical location. The authors recognize ethical issues regarding big data research on public spaces, including questions of privacy and fairness. They conclude by suggesting that researchers and practitioners should encourage multidisciplinary research, practice and spatial analysis into urban design and planning curricula in order to advance the use of big data in public space research and design.

Tim Jachna's chapter presents a historical story of the uses of digital technologies in public spatial practice in Chinese cities in the twenty-first century. Through numerous examples, the author illustrates the relationships of these evolving practices with the urban social, political, and economic developments in contemporary China. The chapter illustrates how the co-existence and co-evolution of spatial and digital practices and networks introduce new modes of control of public life, how people circumvent these controls, and ultimately how these practices test and bend cultural and societal norms in public space.

Borrowing from the literature in forecasting, and seeking inspiration from art and literature, Mark Childs' chapter looks at the futures of public space. Based on concepts such as *framing, drivers of change, causal layered analysis, scenario development*, and *signals of change*, the author proposes informed visions of public space.

The final chapter by Tridib Banerjee explores the contemporary challenges to the concept of public space, particularly when it is conceived as integral to the notion of the urban commons. Drawing from comparative examples from cities around the world, the author presents these as challenges of enclosure, encroachment, and the exclusion of the common (the public) in urban transformations in a global economy. Banerjee concludes by suggesting possible imperatives for urban design and urges urban designers and planners to advocate for the urban commons as the principal "protagonist" in urban design.

34

PUBLIC SPACE AND THE TERRORISM OF TIME[1]

Mark Kingwell

Revision

Some familiar studies of architecture and its phenomenology – Harries (1982), following Bachelard, is prominent – examine the "terror of time" embodied in built forms. In addition to setting off places from space at large, architecture likewise attempts to negotiate our temporal existence through, among other things, the creation of durability. In fact, though, such attempts to control time inevitably succumb to pressures of desolation even as they, sometimes, offer transcendental *kairotic* release from the linear, chronological time of modernity. Here Debord, Mumford, and others provide the requisite intellectual heft. Time is not just an arrow in the metaphorical sense of being unidirectional; it is also an arrow designed to pierce the hearts and minds of all mortal beings who can know its trajectory – though never its precise striking point.

In what follows I propose to do two things: first, to extend this argument by situating the idea of public space in this same "terror of time" logic of place, beginning with the most basic of private buildings, the hut; and second, to raise the specter of actual terrorism – parasitic violence that is sometimes perversely enabled by public space. We might think here of the brutal infliction of murder using trucks or cars in public parks or promenades. How does the potential for real violence affect our sense of time as we emerge from the fragile safety of the private building in order to enter, occupy, and enjoy public spaces? These are spaces that, unlike some buildings, can never fully exclude themselves from outside penetration; their potential for terror is perpetual.

When we speak *of* time we also, necessarily, speak *in* time, casting bottled messages into uncertain futures. The past overcomes every present, it is said; but the future equally overcomes every past.

Since I wrote the two paragraphs above, some things had happened, and others had not yet. A terrorist had, on July 14, 2016, used a rented truck to wreak havoc in the French city of Nice. The 19-tonne cargo van was used by Mohamed Lahouaiej-Bouhlel to kill 86 people and injure 458 others who were celebrating Bastille Day on the Promenade des Anglais in the seaside city. This was not the first time, or the last, in which a man on a self-imagined mission weaponized a utilitarian vehicle. Not quite two years later, a deranged young man called Alek Minassian used a rented panel van to fell ten

429

people and injure dozens more in an apparently unmotivated attack in the suburb of my city, Toronto, known as North York. His motives were terroristic in the broad sense – that is, he wanted to inspire fear – but not necessarily political. The nearest we can say is that he may have been motivated by sexual resentment and the sense of murderous anger fomented by the so-called "incel" (involuntary celibate) community of male online saddies who can't get a date.

There have been dozens of attacks of various kinds carried out over recent years with rented trucks or vans, and police in every jurisdiction acknowledge that they are almost impossible to stop. The Nice attack, with its large vehicle and appalling body count, is still at the high threshold, but police services around the world report that this style of action is both frequent and deadly. One can easily see why: renting a truck is not procedurally difficult, and requires few legal documents, let alone a security check; driving into populated areas, often recreational public spaces, is not controlled; and once the attack begins, especially if the perpetrator has no regard for his own life, the damage in human life is likely to be high. One can kill more people more efficiently with an automatic rifle, to be sure, as in Las Vegas or a dozen other places; but for sheer convenience and ease of execution, a person bent on killing can't really beat a rental van.

Is this any different from other forms of depredation in our public spaces? I think it is, because the very innocuousness of the weapon is the new distinguishing feature. Pedestrian and cyclists are used to calling out motor vehicles as mechanisms of destruction both direct and indirect; the truck or van as murder weapon is just the logical extension of the argument. A pedestrian literally *has no chance* when a murderous, and not merely careless or stupid, driver decides to target them. One might recall here, in morbid register, the cheesy Roger Corman dystopian science-fiction film *Death Race 2000* (1975), in which a cross-country road rally is enlivened by side-bets on how many people a given driver manages to run over, with extra points for brutality and targets (old people, picnicking families with young children, etc.).

The serious point from this combination of real-life and speculative sources is the nature of public space, especially when that space must for the foreseeable future accept an admixture of humans and vehicles, freedom, and enforcement. Between real-life terrorism and fictional visions of spectacular evil entertainment lies the question of whether we can enjoy public spaces in the ways our theoretical forebears imagined. For myself, I have asked before in talks and papers *whether public space is a public good*. This is no idle inquiry. A public good, on basic economic understanding, is a good that is non-rival and non-excludable. That means that its routine enjoyment by me should not prevent the same enjoyment by you, or by anyone else so inclined; also that there should be no barriers in principle on who may even attempt to enjoy the good. Fresh air is a genuine public good, or should be, bordered by the usual caveats concerning pollution and pathogens. So are regulated spaces such as municipal parks and cemeteries. Even here there are of course limits: in the park near my house, families routinely use open-air grills to make a communal dinner, even though this may be close to flouting city ordinances; and dog-owners walking their pets between the headstones don't always stoop and scoop, a clear violation. Nevertheless, each of them believes they are enjoying public spaces in a non-rival, non-excludable manner.

But what about the more controversial spaces of our built environment, the squares, and plazas of a dense downtown? These, after all, are frames both for architecture and for the destruction of human tissue that passes for political expression by the evil ones among us, whatever their specific motives. To reveal the special challenges of public space and time we must go back to basics.

Building

In his now-classic essay "Building and the Terror of Time," Karsten Harries (1982) offers a subtle account of the relationship between the enclosures and edifices of human construction and the underlying anxieties that both prompt such construction and render it forever unstable. The most obvious and atavistic motive for building is the creation of material safety in a world of hostile outside forces. The origin of building is *The Fall* itself.

"Building has been understood to be a domestication of space," Harries (1982) argues. "Every house may be considered an attempted recovery of some paradise" (59). From the start, then, the project of building is intimately tied to our needs and fears. The most basic of these may be counted as the fear of death. As in Hobbes' transition from a state of nature where life is "solitary, poor, nasty, brutish, and short" to a stable sovereign condition, there can be no possibility of commerce, art, literature, or anything else we associate with human civilization *unless and until* there is relative freedom from death's dominion. Building is essential here: we *buy time* with our structures, so that we may exploit the possibilities of a mortal span whose precise length is unknown to us. And yet, all such temporal purchases are ultimately revealed as high-interest loans from the toughest lender there is, time itself. "Thus, if we can speak of architecture as a defense against the terror of space," Harries (1982) suggests, "we must also recognize that from the very beginning it has provided defense against the terror of time" (59).

And this is true for any "building," from the most primitive huts and to the complex mega-projects of supermodernity. As Harries (1982) notes, citing an uncompleted Kafka story, "Der Bau," all creatures attempt to create a shelter of some kind, to find safety in structure. In German, *Bau* can mean both building and burrow, and we are reminded, as in Heidegger, how primordial is the task of building. "Unable to possess the world, [the creature] tries to withdraw into its artificial environment," Harries notes. "It intends to replace nature with artful construction. But the threatening outside cannot be eliminated" (60). To see this point more clearly, let us consider a very basic structure and its relation to time: the hut.

What, after all, is a *hut*? The etymological trail suggests it is not just the idea of shelter, but the intimation of hiding – a hut conceals and covers, and its somewhat tortuous entry into the great assimilative maw of English usage was via military pragmatics. A hut, in French or German, might mean a cottage or some similar abode of retreat and respite. For the earliest English users of the imported monosyllable, it represented shelter and camouflage. Deep in the Old English linguistic roots, *hut* and *hide* are linguistic cousins.

Martin Heidegger is perhaps the most famous hut-dweller in the Western tradition of philosophy. Like Ludwig Wittgenstein, the Austrian genius with decidedly different philosophical proclivities, Heidegger found solace, and the opportunity to think deeply, in a rough woodside structure. Wittgenstein's favorite location was Norway; for Heidegger, the philosopher of *Blut und Boden*, it was inevitably the *Schwarzwald* – the Black Forest. Where once had disported the characters from the Brothers Grimm, Heidegger now made his own off-grid retreat. In the 1930s and after, this densely treed and mountainous region provided the philosopher with a magical home.

The hut was situated on a hillside near Todtnauberg. The basic structure is a small four-square log house, its most arresting feature a long roof that extends back and joins the building organically to the sloping hill. The hut is rooted in place, a material illustration of the later Heidegger's notion of building (*Bauen*), explored most completely in the essay *Building Dwelling Thinking* (Heidegger 1971a). According to the dense poetic idiom of Heidegger's work, much-quoted by architectural theorists, conventional notions of building as mere structural achievement fail to heed the linguistic "call" embedded in the verb-form of *bauen*. Most importantly, the notion of *building* also embraces *dwelling* – being at home, in place – and *cherishing* or *preserving* the world.

To build, therefore, is to engage the prospect of dwelling, which in turn is a preserving or holding. Here, and only here, is thinking made possible: the true thinking of our essence, which is the fact of *being here*. Thus, a hut like Heidegger's own gathers the "fourfold" by sitting on the earth, by being situated under the sky. The mortals who dwell there, Heidegger and his wife Elfride, sporting their traditional costumes and drawing water or chopping wood, commune with divine presences that are more immanent than transcendent. "The mere object is not the work of art," Heidegger wrote in another influential essay (Heidegger 1971b). We might echo this by saying the mere structure is not the building. A building must stand up, and it must also last – commodity, firmness, and delight as the Vitruvian adage has it, with firmness firmly bracketed.

In short, buildings are *meant for us* in intricate and looping ways. We build them but in essence they build us, forcing their way into the world through the minds, imaginations, and skills of human builders. They, like all of us who see and experience the building, do not wish simply to exist, they desire to *live*. Holding the hostile external world at bay is necessary yet insufficient. In addition, it cannot really achieve its own goal: the natural world comes inside with us, *as* us. The artifice of artful construction, as Harries describes Kafka's safety-seeking creature, is artificial in another, hidden yet obvious sense: security is an illusion.

And so, we don't simply build to shelter, but also to speculate and remember. Monuments are testaments, often compromised, to historical time. Follies are *memento mori*, the large-scale equivalents of the skull upon the desk in a Dutch still life. Ruination is as essential to post-Edenic condition as the quickness of living and breathing.

Leisure times, leisure places

It was Aristotle, in the *Nicomachean Ethics*, who first argued for the central role of leisure in human life, and for the idea of public spaces in which to pursue its special gifts. The Greek word he employed, *skhōle*, has no exact English equivalent. It is retained as the root of *school*, but you would need time and some fancy intellectual footwork to convince school-children that they are at leisure. "We work to have leisure," Aristotle says. It is the condition "on which happiness depends." Leisure, he elaborates, is "a state of being free from the necessity to labor." In his monumental *Politics*, the philosopher calls it "the first principle of all good action."

Clearly leisure means something quite different, on this account, from the typical understanding we accord to it. For us today, leisure might connote rest from work, yes, but almost always with an associated resonance of rest and recreation. Time off from work is still effectively colonized by work, since it is leisure in the service of restoring the worker's mental and physical fitness for the tasks of employment. Aristotle's leisure resembles cognate notions such as the Sabbath: this is time out of time, set off from workday principles and expectations. Here we are meant to contemplate the highest parts of ourselves, the divine presence within – or at least to set aside the cares and preoccupations of quotidian life in favor of a more transcendent form of activity.

How does all this relate to the simple hut and its extensions into more complex buildings and spaces? Well, consider again Aristotle's claim that genuine leisure is the basis of happiness because it calls out the best in our natures. Not-working, or genuine leisure, is an activity without profit or reward that nevertheless remains in some sense productive and public.

That is of course the optimistic vision of the time out of time which philosophers imagine as true leisure. I won't speak further here of the many depredations of work and culture that colonize and hollow out this vision (see Kingwell 2008; Kingwell and Glenn 2008). I am, rather, interested in what the seeking of leisure time in specific places – public spaces – means for us today.

For let us be clear: if building itself, from the hut on up, was a way to try and conquer the existential terror of time, it is clear that the creation of enclosures also opened up new rifts and inversions in our seeking after diversion from death. We often speak of breaches into the domestic enclosure as "home invasion," setting out in linguistic form the sense of warfare – the return to natural conditions of might – that characterizes such violations of the private. But what of those moments when we choose to enter the public space, to spend the time that we have bought against our existential going in enjoyment of what we share, not just what we shelter?

It is not just the space that is different when we venture into the public realm. Our sense of time, so often tied to work and the demands of capital, is freed to new possibilities. We walk more slowly, we tarry and laze, we are off the clock. The afternoon stretches before us as a clear horizon, not an agonized duration of measured existence. We may even, if we are lucky, briefly slip the bonds of mortal reckoning: not in reality, but in an oneiric state of transcendence that is surely what Aristotle meant when he said that we touch the divine in genuine leisure. In a French idiomatic slippage, we move from a sense of confronting *la future*, that wall-like inevitable future of capital and technology, to dwelling in *l'avenir* – what will come.

In order to understand our shifts of time-sense in leisure, or anyway this idealized notion of leisure, let us consider how time works when it comes to the realms of human existence.

In his brisk history of secular political consciousness and the public sphere, *Modern Social Imaginaries*, Charles Taylor (2004) remarks in passing how different concepts of time, or time-consciousness, are necessary for the emergent modern political order. "The 18th century public sphere thus represents an instance of a new kind: a metatopical common space and common agency without an action-transcendent constitution, an agency grounded purely in its own common actions" (96). This "metatopical common space" was, crucially, continuous, and evenly distributed across its participants; that is why it could become the basis for what we now recognize as democratic civil society.

The distinction Taylor suggests here has many forebears. One clear way of capturing its impact hinges on the fact that Greek has two words that both translate as "time:" *Chronos* and *Kairos*. Chronological time is the time of measurement and portioning, the time that passes. In its modern manifestation, the history of chronological time is nicely traced by both Taylor and other, more radical thinkers such as Guy Debord, in *Society of the Spectacle* (1967). This conception of time is crucial to the emergence of a shared public sphere—but also to the emergence of a work-world in which time can be subject to transaction. This is especially true in the special set of social relations that Debord calls "the spectacle," in which everything and everyone is a commodity. This is secular time, in the sense that it is "of the age:" the space of everydayness, work, and exchange.

We can summarize the qualities of secular, chronos time this way: it is (i) everyday, (ii) profane, (iii) homogeneous, (iv) linear, (v) horizontal, and (vi) egalitarian. We constantly encounter this time, measuring it and meeting its demands by being on time, matching our movements and achievements to its punctums, saving time and spending time, each of us equally available to time, and having it available to us. Debord closely associates this time with the emergence of labor mechanisms and the bourgeois conception of society, taking time away from the more natural cyclical rhythms of seasonal agriculture and, before it, hunting-gathering to create a time-world in which production is potentially constant. Workers may now punch into the line *twenty-four-seven*, as we would now say, making the relation to the time-clock explicit. Consistent with orthodox Marxist critique, Debord argues that this process is inseparable from the emergence of class, and so class conflict

We are now in a position to characterize the useful diacritical opposite of chronos time, namely *Kairos* or transcendental time. It is (i) mysterious, (ii) divine, (iii) eternal, (iv) infinite, (v) vertical, and (vi) hierarchical. In many cases, of course, precisely this kind of time—the time of divine intervention or communion with the eternal realm—is familiar as part of an anti-democratic social order, in which privileged access, or anyway claims thereto, keeps a steeply hierarchical class division firmly in place, ostensibly as part of a Great Chain of Being or Divine Universal Scheme. The forerunner here might be the Platonic Theory of the Forms, with realms of knowledge and reality arranged in rigid order. The upward ascent of the self-freed slave of Plato's Cave, struggling through blindness and pain toward the sun's light, is an ascent to eternity as well as reality—for they are the same.

But these towering religio-philosophical edifices have their less grandiose analogues in our own world. I mean the sense of "time beyond time" that still marks genuine leisure, play, and idleness in public spaces, the *skholé* of Aristotle even now to be found in our aim-less games and blissful moments of "flow"; or the true holiday, where the usual tyranny of work and use-value is suspended in the name of carnival or sabbath. The common desire for what the Germans call *Freizeit*—time free of obligation—is united with the transcend-ence of time available to almost any North American urban dweller in a baseball game, say, where time is told only in outs and innings, in a pastime that is played in what is usually called a park. (The cognate game of cricket arguably offers even more in the way of time out of time!)

These and other *ludic episodes* to be found within everyday existence are portals to the gift of public space, for they remind us of the resistance to transactional reduction that grounds the most valuable features of our common life. But they are also, as we have seen, high-risk propositions in an age of terror.

Three problems

And precisely here danger lurks. This is not the danger of an external natural world, whose savage indifference prompted our building in the first place; it is instead the danger within – human-caused violence, worse than indifferent even when its motives are unclear and unleashed on an innocent public. We shoulder these risks every time we leave the safety, however illusory, of the private space. In economic terms, we balance off the positive exter-nalities of public space (stimulation, cultural opportunity, perhaps new relationships) against the negative externalities created by the very same material forms (crime, harassment, rude-ness, violence). Every traversal of a threshold, performed however unconsciously, strikes a new bargain with time and terror.

Thresholds serve as liminal alteration sites, neither inside nor outside but establishing the demarcation thereof. Traditionally, the "thresh" was fresh hay or grass, placed in the door-way to offer a cleansing opportunity – we wipe our shoes or boots before entering the interior from the dirty outside. The metaphorical extension is then obvious: over a threshold we move from the profane to the sacred, from the public to the private, from the world of commerce and exchange to the protected space of domestic activity. Typical interiors are temples, hearths, or bedrooms, offering warmth and sustenance both material and spiritual.

Importantly, the distinction between public and private space is itself a public achievement: an agreement, either in principle or by long practice, of what will be excluded from community view and even, in some cases, from the reach of the law. If the threshold were not so publicly agreed, the fragile balance of these realms of human existence would not hold. I may lock my doors at night to enclose the private realm, but if there is no punishment for those who breach the lock, there is in effect no lock. There must be at least some measure of consensus on where the public leaves off and the private begins – and vice versa – so that each may thrive. The very idea of public space is predicated on the Aristotelian-Arendtian notion that *action* takes place in public, however much it relies on private resources and preparation. We may repair to the home to recuperate, regroup, rethink; but a sustained retreat from public life is, many of us believe, antithetical to community and even justice.

In the current context, then, two problems immediately arise which are distinct from a more familiar third, what I will call The Standard Problem. The Standard Problem of public space has been understood to be one of limits on access to such space. That is, given what we know about controlled resources and the depredations of capital markets on all aspects of human life, how can we *open up* public spaces to make them more inclusive, diverse, and universally available. The logic of The Standard Problem is, as mentioned, that public spaces should ideally be public goods: non-rival (nobody's enjoyment of them impairs anyone else's) and non-excludable (nobody can be denied access). One way of summarizing The Standard Problem is to reference Lefebvre's (1967) celebrated "right to the city." We desire cities that are open in the sense (at least) that their public spaces should be available to everyone who wishes to enjoy them, with no countervailing costs to others who likewise wish to do so.

The two new problems subvert the basic logic of The Standard Problem, and they are thus the two related prongs of the new terrorism of time in public spaces. The first is perhaps the more obvious: what we may call *defections* or *buy-outs* from public space.[2] By these terms I mean systematic retreats from public spaces, which have become a common luxury good in the age of postmodern capitalism. Those with the relevant resources can choose to live their lives almost entirely in private, disdaining both the noise and bustle of public spaces and the injunction that action – political life as such – can only be enacted publicly. Such people are really no longer citizens of a democratic polity; they exist on a transnational plane where taxes, say, when they cannot be avoided altogether, are regarded as tolerable fines on the margins of their freedom.

Defections and buy-outs are distinction modalities of this public-space problem. The typical defection involves a retreat behind barriers both real and notional: gated communities and elaborate home security, but also velvet ropes, first-class lounges, and the strange non-existence of owning multiple homes (Manhattan, Hamptons, Miami, Aspen is the current "Four-Pack" dream). When someone may be anywhere, he is essentially nowhere. Small-time defections are also possible, however, as when we spend more and more time online, watching Netflix, playing video games, or dallying with social media. Every retreat from public space in favor of the cosseted private realm is a small blow against democratic politics – once more, at least on this idealized model.

Buy-outs are more explicit, running a gamut from the silly (paying someone to wait in line for a seat on a roller coaster or at a theatre premiere) to the comprehensive (using accumulated capital to influence politics well beyond legal one-vote personhood).[3] A totalized combination of defection and buy-out might be exemplified by a Silicon Valley figure such as Peter Thiel, who has used accumulated wealth to opt out of democratic politics altogether, viewing them as antithetical to his goals of "innovation" and "disruption." Even those multi-billionaires who donate portions of their wealth to high-profile charities are playing a private game that represents a more nuanced buy-out of publicness. When I put my vast wealth in the service of specific causes, I act to hollow out of the general principle of the public as such.

The second problem is harder to grasp, since its contours are necessarily hazy. But with it, we return to our starting point: the terrorism of time in public spaces. Because the second problem is the matter of risk in public. There are, after all, good reasons for defections on a limited-rationality calculation. Bad things can happen when I expose myself to the uncertainties and negative externalities of public space. These risks once more may run from the trivial (being confronted by people I find strange on the sidewalk or in the subway) to the mortally serious (being struck down by a madman while taking my easy lunchtime stroll). The temporal element here should be obvious: the longer I spend in public, the longer I expose myself to the myriad risks that lurk there.

A tension is thus set in motion between separate demands of time. On the one hand, leisure suggests that we should enjoy public spaces in a manner that lets go of chronological time, where time is present only in the passing scene and my enjoyment of it. Our literally slower pace is a mark of this different relation to temporality. On the other hand, awareness of risk arouses a fight-or-flight reaction in our brains such that we may feel a desire to scuttle, combat-style, from place to place.

Open spaces, intended to be appreciated for their lack of building – parks, plazas, squares – are transformed into danger zones with inadequate cover, imagined snipers on every surrounding rooftop, and open roadways that invite rampaging vehicles. The video game that we left inside in order to explore the outside has somehow reached out and colonized the space, pulling us inside the game, unwilling first-person shooters who don't know how to level up.

This is a dark fantasy, to be sure, but it is sufficiently real, given recent events, to make us rethink our devotion to the right to the city. We might now speak, instead, of the *precarity of the city*. The urban death-maze imagery is inseparable from all the positive externalities that bring us into public spaces in the first place. It is not possible to render the site of leisure entirely immune to the threat of violence, for any measures that sought to do so would, at some point, tip the project into self-defeat. The public space would be destroyed, converted into yet another bunker or zip-wire enclosure, when its permeability is precisely the virtue that makes for genuine publicness.

Conclusion

And so, what to do? Defections and buy-outs will not resolve the issue of terror, only exacerbate it, since we know that the safety they vouchsafe is itself precarious even as they hollow out existing public spaces. And all the while, some version of The Standard Problem still holds: we should always wish for, and work for, greater access to public spaces for everyone. But just as we cannot solve the problem of terrorism generally with conventional tactics, so we cannot hope to solve it in respect of public spaces with standard wishes.

In other words, the proper response to terrorism is the simple, if dangerous, one we have heard all too often: we must carry on as before, or the terrorists win. Time will carry all of us away eventually. There is no avoiding this. Nor is there any avoiding risk in life, for neither buildings nor space between buildings offer anything except temporary shelter, and solace. This is as it must be. So be an everyday public hero: cross the threshold, accept the risk, go outside and play.

Notes

1 Title influenced by Martin (2018).
2 I thank Diana Boros for emphasizing the notion of buy-outs in thinking about wealth and public space.
3 The *Citizens United v. Federal Election Commission* decision of the United States Supreme Court (558 U.S. 310 [2010]) is an example of how an otherwise democratic nation can consume itself from within, a version of Derrida's "auto-immunity." The decision grants money the status of political speech, therefore under protection of the First Amendment to the U.S. Constitution. This makes it possible for the wealthy to exercise massive force multipliers on their political views: money now literally talks.

References

Debord, G. (1967) *Society of the Spectacle*. Kalamazoo: Black & Red.
Harries, K. (1982) Building and the Terror of Time. *Perspecta: The Yale Architecture Journal*, 19: 58–69.
Heidegger, M. (1971a) Building Dwelling Thinking. In: *Poetry, Language, Thought*, Hofstadter, A. (trans). New York: Harper & Row.
Heidegger, M. (1971b) The Origin of the Work of Art. In: *Poetry, Language, Thought*, Hofstadter, A. (trans). New York: Harper & Row.
Kingwell, M. (2008) Idling Toward Heaven: The Last Defence You Will Ever Need. *Queens Quarterly*, 115(4): 569–585.
Kingwell, M. and Glenn, J. (2008) *The Idler's Glossary*. Emeryville: Biblioasis.
Lefebvre, H. (1967) *The Right to the City*. London: Wiley-Blackwell.
Martin, A. (2018). "The Incel Rebellion—How Involuntary Celibates are Dangerous in Their Desires." *The Independent*, 4 May. www.independent.co.uk/news/long_reads/incel-what-is-involuntary-celi bates-elliot-rodger-alek-minassian-canada-terrorism-a8335816.html
Taylor, C. (2004) *Modern Social Imaginaries*. Durham: Duke University Press.

35

RENEWING THE PUBLIC TRUST DOCTRINE

A solidarist account of public space

Margaret Kohn

Introduction

In 2018 President Donald Trump issued an Executive Order that scaled back the size of national monument land in Utah by 2 million acres (Dawsey and Eilperin 2017). Large parts of Bears Ears and Grand Staircase-Escalante national monuments were opened to mining, and other protected areas are being considered for privatization. Much of the press coverage of the policy change focused on the debate over resource exploitation and the constitutionality of Trump's use of an Executive Order (see Korte 2017; Regan 2017). These are theoretically significant questions with wide-ranging practical implications, but we should also ask two more basic questions: why should space be public and what does it mean to describe space as public? Local debates over real estate development or commercialization in urban parks raise similar issues. For example, the Brooklyn Bridge Park is a flourishing public space on 85 acres of prime waterfront real estate in New York City, yet, in order to ensure a revenue stream for park maintenance and programming, the project dedicated large parcels of park land for high-priced, high-rise condominium towers (see Gregor 2013; Brooklyn Bridge Park 2018).

Parks are among the most important public goods, yet the dominant theories of public goods fail to explain why parks should be public. After examining the three dominant theories of public goods, I turn to the history of the public trust doctrine, which has been advanced as a legal tool for protecting the natural environment and parkland. Proponents see the doctrine as a way to protect public access to land, but critics point to the anti-democratic character of a mechanism that binds future generations and limits the ability of citizens to balance competing priorities. I introduce the solidarist approach to public space and show how this helps realize the normative aspirations of the public trust doctrine while limiting its potentially anti-democratic implications.

Public space and public goods

The scholarly literature includes three different answers to the question why the state should provide public goods: market failures, basic needs, and democracy. According to the influential market failures approach, goods should be provided by the state when the provision of benefits is non-excludable and the enjoyment of the goods is non-rivalrous (e.g. the enjoyment by one person does not prevent someone else from accessing the same good) (Samuelson 1954). Clean air is described as "non-excludable" because it is impossible to provide it to some people without others benefitting at the same time. According to the market failures approach, markets are not able to produce such goods efficiently because of the free rider problem. If the state does not compel everyone to contribute, then such goods will be under-produced, even though they are highly valued.

The range of public goods that falls into this category is small, but it expands if we include cases in which private provision involves significant transaction costs. When the additional cost of gatekeeping makes the final price prohibitively high, the market will not be able to provide the good (Touffut 2006). Until recently, when technological innovations significantly reduced the cost of gatekeeping, roads were classic public goods because they were valuable things that could not be efficiently produced by private actors. Where do parks fit in this schema? It is tempting to include parks with roads in this broader category of market failures, but this would not be accurate. Gated parks can be accessed by distributing keys, and the flourishing commercial market in recreational spaces such as amusement parks, country clubs, and gyms demonstrates that there is no systematic market failure.

The market can provide "club goods" (e.g. goods that are consumed collectively), but it only provides them to those people who can afford to pay for them. This leads to the second approach, which treats public goods as the solution to a very different kind of market failure: the market's inability to provide for the needs of the poor. This approach to public goods is a normative theory that holds that state provision is justified when it is necessary to supply primary goods (Klosko 1987; Shue 1996). If individuals cannot secure their own basic needs, then the responsibility to do so devolves other people and the state. Education and healthcare are treated as public goods because they are essential elements of a decent life (Weinstock 2011). The normative rationale for these kinds of public goods is the decommodification of basic necessities.

The democratic approach to public space expands the definition of public goods to include all of those things that citizens choose to pay for collectively. It is also closest to the conventional use of the term "public good" to describe anything provided by the government This purely descriptive approach, however, does not help us answer the crucial questions: why *should* parks be public and how public should they be? The democratic approach does not provide substantive grounds to object to city councils when they decide to charge user fees to access parks. Nor does it provide substantive grounds to challenge a legislative body's decision to allow commercial exploitation of federal lands.

Given the limitations of the three most influential theories of public goods, what resources can help us think through the significance and justification of public space? To answer this question, I turn to the common law doctrine of public trust, which has been taken up by environmental lawyers and activists in the United States as a way to challenge the key elements of modern property rights: the doctrines of exclusive access, control, and alienability of land. This doctrine contains an answer to the question why certain spaces such as rivers and shorelines are naturally communal, and this answer helps us formulate a more general theory of the publicness of physical space.

The public trust doctrine includes the following features: the property must be used for a public purpose and made available to the general public; the property may not be sold; and the property must be maintained for particular types of uses (Sax 1970). These provisions protect public space, but they also place important limitations on the ability of democratic institutions to respond to the preferences of citizens. Since the doctrine seems to depart from democratic principles, it requires justification. The justification of the public trust doctrine is rooted in its origins in the Justinian code and British common law (Sun 2011). Even though the history impact of the public trust doctrine is controversial, its intellectual history reveals a set of principles that can be justified and defended today.

The paradox of public trust

The public trust doctrine has a long history in the United States, and proponents of the doctrine trace its origins back to the Roman Empire (Stevens 1980; Sun 2011). The textual evidence for this origin story is the *Institutes of Justinian*, which states that "by the law of nature ... air, running water, the sea, and consequently the shores of the sea" (Collett Sanders 1922, 2.1.1) are common to mankind. This passage introduces two key features of the public trust doctrine: a concept of common property and a natural rights claim that supersedes civil law (e.g. the law of the city). Furthermore, even though Roman jurists did not have a concept equivalent to "public trust," they did introduce the term *jus publicum* or public right.

In the introduction to his path-breaking article, Joseph Sax (1970) wrote, "The source of modern public trust law is found in a concept that received much attention in Roman and English law – the nature of property rights in rivers, the sea, and the seashore" (474). Carol Rose (2003) has shown that Roman law incorporated legal pluralism and distinguished between a number of forms of non-exclusive property: *res nullius* (things belonging to no one), *res communes* (things open to all by their nature), *res publicae* (public things made accessible by law), *res universitates* (shared property of a corporate body). She concedes that the distinction between these categories was not always rigid. In Roman sources, the oceans could be described as empty or common property. The land of conquered people was described as *res nullius* but also as public property that could be leased to private parties in order to generate revenue for the state. According to Rose, Roman law makes it possible to understand the significance of non-exclusive property and to differentiate between forms of collective property that are lumped together under "the commons." The arguments for treating certain spaces as a public trust are part of an alternative history of property rights that can be uncovered in the past and linked to the present.

A group of revisionist scholars, however, have challenged the claim that Roman and early Common Law discussions of common property should be viewed as the foundation of the modern legal doctrine of public trust. Patrick Deveney (1976) grants that the idea of *jus publicum* emerges late in the imperial period, but he argues that it was not used to describe a source of political legitimacy that is independent of the ruler. In the late Roman empire, private law dealt with disputes between private parties, and public law concerned the actions of the state, but individuals did not have standing to challenge the decisions of state authorities. This means that Roman public law did not fit within our modern understanding of the rule of law. The difference is captured succinctly in the famous dictum, "That which seems good to the Emperor has also the force of law" (Collett Sanders 1922, 1.2.6).[1] Deveney suggests that *jus publicum* was actually employed to strengthen the power of the ruler against private parties. If land belonged to the Roman people, then the ruler could allocate or exploit it as he saw fit, and there would be no private law remedy for any private person who objected.

The revisionist challenge to origin story of the public trust doctrine has three main features. The first one is the claim that the Roman concept of public right was a Stoic moral principle that was taken up by a small number of philosophically inclined jurists rather than a doctrine with the force of law. According to the revisionists, public right should be described as at most an aspirational rather than a justiciable right. The second claim challenges the grounding of public trust doctrine in early modern British common law. For example, James Huffman (2007) has argued there was no suggestion that title to certain lands was inalienable and held by the Crown for common use. Finally, the revisionists examine the legal history of the right to shorelines and waterways, which is the source of more general claims about the public trust doctrine and its implication for environmental protection. They conclude that the right to access waterways and shore-lines was subject to considerable dispute; moreover, claims about open access were often employed strategically by powerful actors challenging existing private uses in order to increase commercial exploitation of these properties. At other times, "public trust" was used by Parliament to prevent the king from selling or enclosing "crown land" as if it were personal alienable property.

From this scholarly debate we learn that the public trust doctrine was not a firm legal principle but rather a contested claim. Proponents of a robust modern public trust doctrine have focused on legal cases in which key elements of the doctrine (e.g. the naturally common character of resources) was used to prevent further privatization and secure public access or benefit. This is a strategic use of history for normative purposes and is similar to the way that the expansion of individual rights was justified by a selective history that stitched partial victories into a story of necessary and inevitable progress. The right to common property held by the people was an aspirational right that was only sporadically recognized.

The crucial question is whether it is a right that should be recognized today. Should we endorse the notion of a public trust, which would effectively limit democratic control over public property and create a new kind of collective right? The historical analysis draws our attention to some normative concerns, by focusing on the relationship between state sovereignty and collective rights. The debate in the British common law focused on the question of who should be the trustee with fiduciary responsibility for managing the commons: the Crown or the representatives of the people? In the absence of monarchical authority, this question loses its direct relevance, but it still draws our attention to an important, related issue. Who is the trustee of a public trust in a democratic state? The justification of trusteeship becomes more complicated when ownership and legal authority are vested in a state whose legitimacy rests on popular sovereignty. For the concept of a public trust to make sense in the present day, we need the following: a rights-based argument for public space; a rationale for distinguishing between the trustor, the trustee, and the beneficiary; and an institution that can enforce the contract.

Back to the future

The earliest argument for a right to common land comes from Roman sources and was based on two sources: natural law and the rights of nations, which was a term used to describe principles known through reason and/or recognized everywhere. The *Institutes of Justinian* contains the following passage:

All rivers and ports are public; hence the right of fishing in a port, or in rivers, is common to all men. The seashore extends as far as the greatest winter [tide] runs up. The use of the banks of a river is public, and governed by the law of nations, just as is that of the river itself. All persons therefore are as much at liberty to bring their vessels to the bank, to fasten ropes to the trees growing there, and to place any part of their cargo there, as to navigate the river itself.

(Collett Sanders 1922, 2.1.1–5)

In this passage there are two interesting arguments. First, there is an explicit claim that land below the winter high-tide mark has a different legal status from other property. Why? The emphasis on shorelines, waterways, and ports makes sense in the context of the limited land-based infrastructure in pre-modern times and the significance of the sea in a Mediterranean empire. It was difficult to construct roads in a mountainous country like Italy. Ships that stayed close to the shore played an essential role in communications and trade. The safety and viability of sea travel depends on unrestricted access to the shore, because without fresh water humans cannot survive at sea; therefore access to the shore is literally a matter of the right to life. In the modern era, the sea is less significant, but the ability to move from place to place remains a fundamental need. Today, we think of roads as the paradigmatic public spaces because roads and public transit play an essential role in facilitating commerce, communication, and social life. Indeed, without access to roads, freedom itself is impossible. You may be the most powerful sovereign in your own castle, but, if you have no ability to travel off your property to interact with others, then you are no better than a prisoner (Ripstein 2009).

A second claim advanced in the quotation is the right to fish. Given the significance of fish and seafood as a source of protein, this right is also dimension of the right to subsistence. Yet, it is hard to imagine a legal treatise that would recognize the right to farm, since this would directly challenge private property and the right of property owners to exclude non-owners. By designating the rivers and harbors as public, the *Institutes* seem to endorse a right to subsistence while limiting it to a circumscribed public realm. In a similar vein, in his essay "Homelessness and Freedom," Jeremy Waldron (1991) argued that the right to existence is secure only as long as there are common spaces that everyone is entitled to access because all people need places where they can carry out functions essential to life.

These arguments force us to revisit the basic needs approach to public goods introduced above. At least for some people, access to public space is a basic need and public roads are essential to the freedom of all. Yet the Roman account is still quite limited because it does not justify the protection of common lands, the creation and maintenance of parks, or the redistribution of resources that have been privatized. In fact, some commentators have argued that the designation of land as common was used to signal its availability for individual appropriation (Deveney 1976). The public character of the shoreline is important for debates about easements on waterfront properties, but it does not provide clear guidance about whether to allow mining on federal lands or condos in urban parks.

What we need is a theory of common property that is not based on the emptiness of land but rather its fullness. By that I mean that a theory of public goods should recognize that in modern, industrial societies, value is produced collectively and therefore even the value of undeveloped land has a social dimension. The modern right to public space rests on its social properties as well as its natural properties. From the Roman and early modern legal literature, we learn that public space is necessary to secure the movement of people and the basic needs of those who do not have access to private space. The theory of solidarism expands this foundation to include a right to nature and physical space that can be shared by the community.

Solidarity and public space

Solidarism is a political theory that emerged in late 19th-century France, and it provides a systematic account of ideas that were also associated with social liberalism, and social democracy, and progressivism in the early 20th century (Freeden 1986). Neo-solidarism has three related features: a descriptive social theory, a normative theory, and a political theory (Bouglé 2010; Hayward 1961). The social theory is an account of interdependence that serves as the foundation for a normative argument in favor of fair allocation of the value produced through the division of labor. The political argument emphasizes that a share of commonwealth is a collective and not an individual right and justifies democratic institutions as the only way to exercise this right. The social theory builds on Durkheim's (1960) notion of organic solidarity. Durkheim argued that modern societies differ from pre-modern societies in two ways. They have higher levels of interdependence due to the division of labor and lower levels of mutual identification and shared values, because of the differentiation of roles. The solidarists aimed to foster shared values by increasing recognition of interdependence and its attendant obligations.

Solidarists such as Leon Bourgois, who briefly served as Prime Minister of France, emphasized the empirical claim that the division of labor, urbanization, and industrial production generate a social product. The value of a piece of urban land is a particularly striking illustration of this claim because it is widely recognized that the value of land reflects a range of social factors: proximity to markets, infrastructure, transit, schools, and population growth. The land may be owned privately but the value is produced socially. The solidarists argued that the social, cultural, political, and technological infrastructure was also a kind of inherited common property, and therefore the products of modern society were also composed of both social and individual shares. From this perspective, the social share of wealth is analogous to the sea: it is something that belongs to everyone.

While urbanization generates an enormous amount of commonwealth, it also makes some things that were once free, such as access to nature, or relatively inexpensive like shelter, into commodities and commodification generates exclusion. To put it more forcefully, attached to the aggregate prosperity of modernity are public bads such as pollution, traffic, unsanitary housing, and alienation from nature. Solidarism is a theory of compensatory justice and, while it might at first seem like an odd way to think about it, public parks are a way of compensating for the loss of nature, and public space is way of compensating for the high cost of land.

For people who live in the heart of densely populated and noisy neighborhoods, parks provide quiet, fresh air, sunlight, and greenery. This was an important theme in the writings of early park promoters such as Frederick Law Olmsted who emphasized the benefits of healthy air, sunlight, and a break from the sensory overload of urban life (Kohn 2011). Today there is a burgeoning empirical scholarly literature on the psychological and health benefits that come from nature (Berman, Jonides, and Kaplan 2008; Bratman et al. 2015; Hartig et al. 2014; Shanahan et al. 2015; Sullivan, Kuo, and Depooter 2004; Taylor et al. 2006). There are also social benefits attached to the creation of decommodified spaces where people can interact with each other and enjoy the benefits of social cooperation. This builds on Jean-Jacques Rousseau's point that in order to cultivate citizens willing to prioritize the common good, a polity had to provide shared public things. According to Rousseau (1972), the proper enjoyment of public space could form civic citizens.

Solidarism provides the first component of a renewed public trust doctrine, a rights-based argument for public space, but it is a collective right rather than an individual right. The core idea is that people have the right to access nature and public space that provides adequate compensation for what they lost through urbanization and commodification. Social property, however, is legitimately controlled by society, for the benefit of its members, and in practice this means the people exercise their control through the state. This raises the second challenge in applying the public trust doctrine: how to distinguish between the trustor, the trustee, and the beneficiary. In other words, can the doctrine justify and institutionalize accountability?

The public trust doctrine made sense in the context of monarchical government. One could imagine the king as the original owner (trustor) who authorizes Parliament or local government (the trustee) to manage property on behalf of the people (the beneficiary). How would this work in democratic states? It seems clear that the beneficiary is still the people and the trustee must be the administrative branch or executive, which leaves the will of the people (expressed through a constitution or enabling legislation) as the trustor. A claim that the public trust doctrine is violated rests on the argument that current administrators are violating the law or constitution. Indeed, several US courts have prevented administrators from selling parkland and/or constructing structures that are not related to park purposes on park lands, and they have justified these decisions on the ground that no law authorized such actions (see Burt 1970; Newman 2000).[2] Guided by the Public Trust Doctrine, courts have required explicit legislative authorization of administrative actions that seem, on their face, to restrict public access to public space.

This principle is directly relevant to the dispute between President Trump and his environmentalist critics, and to a much broader range of disputes over privatization. Opponents of the decision to privatize federal lands emphasized a procedural as well as a substantive point. The procedural point is that even though the President was granted the authority to designate federal lands as part of a national monument, he was not given the authority to remove this designation. The concept of public trust provides a compelling argument in favor of the view that expanding public protections and limiting or withdrawing them should be treated differently. The core of the public trust doctrine is the idea that the trustee (here the President) is obliged to carry-out the trustor (Congress's wishes) to benefit the interest of the public. The public interest, of course, is a contested concept and in these disputes over public space we see two conflicting understandings, one that emphasizes environmental preservation/sustainability and another that emphasizes the promotion of economic growth and resource development. The public trust doctrine holds that any restriction on public ownership, benefit, and access, including the access of future generations, must be explicitly authorized by the trustor. This is not anti-democratic, because the elected representatives still have the authority to define the public interest.

Solidarism provides guidance for these democratic deliberations by reminding citizens and policy makers that the public trust doctrine is not just a relic of the ancient past but rather a response to historical circumstances that have alienated people from nature and commodified social space. Solidarism links a structural theory of compensatory justice to a politics of civic solidarity that emphasizes the importance of public things. Collective enjoyment of public things is a way to compensate for the loss of a shared natural world. Not only at the edge of the sea, as in Roman times, but in the heart of the city itself, public lands can preserve a common right.

Notes

1 Cited in Deveney (1976).
2 Burt (1970): holding that the public trust doctrine prevents the sale of a park; (Newman 2000): holding construction of water treatment plant in a city park required state legislative approval.

References

Berman, M.G., Jonides, J., and Kaplan, S. (2008) The Cognitive Benefits of Interacting with Nature. *Psychological Science*, 19(12): 1207–1212.

Bouglé, C.C.A. (2010) *Le Solidarisme*. Charleston: Nabu Press.

Bratman, G.N., Daily, G.C., Levy, B.J., et al. (2015) The Benefits of Nature Experience: Improved Affect and Cognition. *Landscape and Urban Planning*, 138: 41–50.

Brooklyn Bridge Park (2018) "About Us" (Overview). Accessed 13 July 2018. www.brooklynbridge park.org/pages/aboutbbp

Burt, M.F. (1970) *Paepcke V. Public Building Com., 263 N.E.2d 11 (Ill. 1970)*.

Collett Sanders, T. (1922) *The Institutes of Justinian*. London: John W Parker and Son.

Dawsey, J. and Eilperin, J. (2017) "Trump Shrinks Two Huge National Monuments in Utah, Drawing Praise and Protests." *The Washington Post*, 4 December. Accessed 13 July 2018. www.washingtonpost. com/politics/trump-scales-back-two-huge-national-monuments-in-utah-drawing-praise-and-pro tests/2017/12/04/758c85c6-d908-11e7-b1a8-62589434a581_story.html

Deveney, P. (1976) Title, Jus Publicum, and the Public Trust: An Historical Analysis. *Sea Grant Law Journal*, 1: 13–82.

Durkheim, E. (1960) *The Division of Labor in Society*. New York: The Free Press.

Freeden, M. (1986) *The New Liberalism: An Ideology of Social Reform*. Oxford: Oxford University Press.

Gregor, A. (2013) "Condos that Fund a Brooklyn Park." *The New York Times*, 22 November. Accessed 13 July 2018. www.nytimes.com/2013/11/24/realestate/condos-that-fund-a-brooklyn-park.html

Hartig, T., Mitchell, R., de Vries, S., et al. (2014) Nature and Health. *Annual Review of Public Health*, 35: 207–228.

Hayward, J.E.S. (1961) The Official Social Philosophy of the French Third Republic: Léon Bourgeois and Solidarism. *International Review of Social History*, 6(1): 19–48.

Huffman, J. (2007) Speaking of Inconvenient Truths—A History of the Public Trust Doctrine. *Duke Environmental Law & Policy Forum*, 18(1): 1–103.

Klosko, G. (1987) Presumptive Benefit, Fairness, and Political Obligation. *Philosophy & Public Affairs*, 16(3): 241–259.

Kohn, M. (2011) Public Space in the Progressive Era. In: *Justice and the American Metropolis*, C.R. Hayward and T. Swanstrom (Eds.). Minneapolis: University of Minnesota Press, pp. 81–104.

Korte, G. (2017) "Trump Tries Little-Known Legal Tactic to Protect Controversial Executive Orders from the Courts." *USA Today*, 5 December. Accessed 13 July 2018. www.usatoday.com/story/ news/politics/2017/12/05/trump-tries-little-known-legal-tactic-protect-controversial-executive- orders-courts/101334288/

Newman, J.O. (2000) *232 F. 3d 324 - Friends of Van Cortlandt Park V. City of New York*. 00-6183(L), 6197(CON), 6198CON.

Regan, S. (2017) "Trump's Monument Fight." *National Review*, 5 December. Accessed 13 July 2018. www.nationalreview.com/2017/12/monuments-executive-order-trump-administration-utah-bears- eagle-grand-staircase-escalante-act-antiquities-act-1906/

Ripstein, A. (2009) *Force and Freedom: Kant's Legal and Political Philosophy*. Cambridge: Harvard University Press.

Rose, C.M. (2003) Romans, Roads, and Romantic Creators: Traditions of Public Property in the Information Age. *Law and Contemporary Problems*, 66(1/2): 89–110.

Rousseau, J.-J. (1972) *The Government of Poland*. Indianapolis: Hackett Publishing.

Samuelson, A. (1954) The Pure Theory of Public Expenditure. *The Review of Economics and Statistics*, 36(4): 387–389.

Sax, J.L. (1970) The Public Trust Doctrine in Natural Resource Law: Effective Judicial Intervention. *Michigan Law Review*, 68(3): 471–566.

Shanahan, D.F., Fuller, R.A., Bush, R., et al. (2015) The Health Benefits of Urban Nature: How Much Do We Need? *BioScience*, 65(5): 476–485.

Shue, H. (1996) *Basic Rights: Subsistence, Affluence, and U.S. Foreign Policy*, 2nd Ed. Princeton: Princeton University Press.

Stevens, J.S. (1980) The Public Trust: A Sovereign's Ancient Prerogative Becomes the People's Environmental Right. *U.C. Davis Law Review*, 14: 195–232.

Sullivan, W.C., Kuo, F.E., and Depooter, S.F. (2004) The Fruit of Urban Nature: Vital Neighborhood Spaces. *Environment and Behavior*, 36(5): 678–700.

Sun, H. (2011) Toward a New Social-Political Theory of the Public Trust Doctrine. *Vermont Law Review*, 35: 563–622.

Taylor, A.F., Kuo, F.E., Spencer, C., and Blades, M. (2006) Is Contact with Nature Important for Healthy Child Development? State of the Evidence. In: *Children and Their Environments: Learning, Using and Designing Spaces*, C. Spence and M. Blades (Eds.). Cambridge: Cambridge University Press, pp. 124–140.

Touffut, J.-P. (2006) *Advancing Public Goods*. Northampton: Edward Elgar Publishing.

Waldron, J. (1991) *The Right to Private Property*. Oxford: Clarendon Press.

Weinstock, D.M. (2011) How Should Political Philosophers Think of Health? *The Journal of Medicine and Philosophy*, 36(4): 424–435.

36

EVENTS ON URBAN PARKLAND

Scrutinizing public–private partnerships in parks governance regimes

Susanna F. Schaller and Elizabeth Nisbet

Introduction

The governance of the public realm, including public parks, has increasingly been delegated to Public Private Partnership (PPP) regimes. This means that private actors and the values they articulate influence how the urban public realm is designed, managed, activated, and programmed. Since the early 1980s, conservancies and business improvement districts (BID) have taken on significant management and financing roles in parks. New York City is often cited as the epicenter of PPP governance strategies to "reclaim" the public realm in central cities from crime and grime in order to stem economic decline. Since then, several forms of PPPs have grown more prominent and controversial even as the models are being emulated in cities across the country and indeed internationally. Private policy actors such as the International Downtown Association, which promotes these PPP arrangements and place-making as part of its public policy work, have framed the activation and programming of the public realm as an integral strategy to produce inclusive cities by fostering new spaces for sociability, cultural activity, and consumption.

In this chapter, we critically assess one aspect of this placemaking ethos: the use of public parks for large events and their transformation into "eventscapes" (Furman 2007; Hou 2010). Activists have scrutinized and critiqued the management and programming of public parks by PPP regimes in New York City. By 2013, outsized donations from very wealthy individuals or significant fundraising for specific parks, namely Central Park and the newly created High Line, had moved the PPP governance of parks into the public spotlight, and between 2013 and 2016 public officials held hearings to probe how private involvement in parks management related to decision-making processes and the allocation of resources in the park system (Foderaro 2011; 2012). They also probed funding structures of various PPP arrangements, including those that enabled the use of public parks by private actors for revenue production, and explored how eventization was impacting equitable access. The hearings addressed oversight of Public Private Partnerships (New York City Council, Committee on Parks and Recreation 2013a), the management of events (New York City

447

Council, Committee on Parks and Recreation 2013b), and parks equity (New York City Council, Committee on Parks and Recreation 2014; 2016). Through an analysis of these hearings, we examine the following questions:

1) How did government, private partners, and parks advocates frame New York City's Public Private Partnership parks regime and the eventization of urban parklands?
2) How did their narratives indicate support for and/or critiques of these PPP arrangements?

Background: Public Private Partnerships and parks governance

In New York City, the Central Park Conservancy (CPC) and Bryant Park Restoration Corporation (BPRC) in Midtown Manhattan, both formed in 1980, became the representative PPP models to restore urban parks on the heels of the 1970s fiscal crisis; the aim was to stimulate the revival of the urban tax base (Krinsky and Simonet 2017; Murray 2010). The fiscal crisis, of which New York became the iconic example in 1975, ushered in a new era in urban governance (Beauregard 1998). Federal retrenchment and municipal austerity renewed a focus on PPPs to catalyze and harness urban growth (Beauregard 1998). This paved the way for both "corporate-led" and entrepreneurially-oriented governing coalitions, which meant generating new kinds of coordination to leverage private power, money, and individual and grassroots initiative in the process of restructuring and managing post-industrial landscapes (Lauria 1999, 137; see also Beauregard 1998; Harvey 1989; Martin 2004). As early as the 1950s, business elites were gathering across the country to find urban management strategies to economically reposition the urban core as central cities and downtowns continued to bleed population and businesses in large measure due to the federal policies underwriting "white flight" and suburbanization (Dreier et al. 2004; Isenberg 2004; Schaller 2019). The imperative to rely on private sector actors as part of this reconstituted growth coalition emerged in full force in the 1970s as economic crisis and declining federal funding forced municipal governments into an entrepreneurial mode (Harvey 1989; Molotch 1993). Urban policy innovators, including business people, planners, real estate developers, and non-profit and philanthropic organizations, focused on "salvaging" the city from "crime and grime" by recuperating the city's public realm and public spaces for a middle class visitor, consumer and even potential resident used to a suburban aesthetic of controlled and programmed leisure and consumption spaces (Gillette 2012; Sorkin 1992; Zukin 2011). Historical precedent like the investments in Central Park, which had from its inception in the 19th century served as a key public amenity both to enhance the "quality of life" for urban dwellers and as an economic engine raising property values along its edges, pointed the way for a post-industrial and tourism-oriented repositioning strategy (Judd 1999; Loughran 2014). Centrally located parks figured prominently in this approach (Krinsky and Simonet 2017).

The CPC and BPRC represent two distinct models: the CPC, a conservancy, relies largely on private philanthropy, a predictable public budget obligating the city to contribute a minimum of 25% of operating costs, and to a lesser extent on revenue producing activities (New York City Council, Committee on Parks and Recreation 2013a, 66); the BPRC is a business improvement district and is funded largely by tax assessments on surrounding properties as well as revenue producing events in the park (Krinsky and Simonet 2017; Murray 2010). Since the 1980s myriad organizations have emerged, ranging from large conservancies with multi-million-dollar budgets to small "friends of" organizations that rely on volunteers and in-kind contributions; moreover, while 20 of these PPP relationships are guided by formal contracts, hundreds are under non-contractual agreement. This differentiated landscape of private partners has raised concerns among some city officials and "citizen"[1] advocates because their capacity and resources vary across the city (New York City Council, Committee on Parks and Recreation 2013a). This also means the organizations have differential access to power and political influence.

In 2013, New York City elected officials began to publicly interrogate these PPP arrangements as extra-large private donations to signature parks raised questions of parks equity and a tragic, drug-related death during a large event on Randall's Island precipitated particular scrutiny of events on city parkland (Mascia 2014; Nisbet and Schaller 2019; Yee and Rashbaum 2013). Thus while reviewing PPP arrangements more broadly, the City Council also homed in on how events "affect the urban space [and parks] they occupy" by temporarily reconfiguring the public purpose parks serve as well as interrupting usage patterns and access (Smith 2016, 2).

Eventscapes: capturing urban parkland for private gain?

Among many urban policy advocates, including planners and architects and place-making organizations, programming of urban open or interstitial spaces is viewed as a crucial strategy to increase their "sociability" and "inclusiveness" (Smith 2016). Using temporary landscapes or events to activate streets, urban plazas and parks from this perspective can encourage the "co-creation" of place identities and foster a sense of place attachment that brings people back to engage with the space over time (Furman 2007). Jeffrey Hou (2010) describes how the grassroots efforts by historically marginalized communities to produce culturally significant eventscapes assert a shared "right to the city" and particular urban places: programming and cultural events in the public realm can "loosen up" spaces to soften the potential rigidity inherent in a particular design or place-association (Franck and Stevens 2006). This may make spaces feel more accessible and help break down perceived mental and social barriers (Madanipour 1999; 2003) through the repetition of structured encounters such as cultural festivals and seasonal events (Valentine 2013). In 2014, the New York City Parks Commissioner echoed this sentiment in describing the commitment of the department's Catalyst stewardship program "to developing diverse and multi-faceted programming in parks, fostering open leadership, and creating opportunities and projects, which empower groups" (New York City Council, Committee on Parks and Recreation 2014, 29). This is the positive side of public–private partnerships through which public funding is guided toward supporting locally developed and resonant programming with the aim of drawing different kinds of people and places together (Schaller and Guinand 2019). In addition, private organizations and their champions often see themselves as the innovators in this policy arena. The NYC Parks Commissioner, for example, has highlighted that "The Bryant Park Corporation uses extensive research into the social dynamics of public spaces to develop programs, events and attractions that have been emulated by parks all over the world" (New York City Council, Committee on Parks and Recreation 2013a, 20).

The production of eventscapes has also become highly professionalized and marketized (Smith 2016). In signature parks such as Central Park, Bryant Park, Prospect Park, the High Line, Governor's Island, and Randall's Island, eventscapes have come to be used as "value creation platforms" (Brown et al. 2015; Richards, Marques, and Mein 2015; Suntikul and Jachna 2016) to showcase particular businesses or industries (e.g. Fashion Week in Bryant

Park) and generate revenue not just to support the parks system but to underwrite the PPP governance structure (Smith 2018). Yet, a dilemma emerges when the commercialization of events inhibits the very "publicness" and "looseness" of the staging spaces they occupy in urban parks (Franck and Stevens 2006). In New York City the hearings also highlight concerns that eventization raises about the commercialization of parks and transparency within PPP parks regimes.

New York City: scrutinizing the eventization of parks

In this section we turn to the analysis of the hearing transcripts. Importantly, although the hearings centered on the role of the wealthier organizations that have organized large-scale events and figure prominently in parks governance discussions only the Central Park Conservancy and the Friends of the High Line were directly represented at the PPP oversight hearing; the co-founder and former president of the Prospect Park Alliance also testified. No major conservancy representative testified at the events hearing. Instead, myriad mid-sized and smaller organizations and prominent advocacy organizations as well as the Department of Parks and Recreation (DPR) testified. The advocacy organization New Yorkers for Parks for example testified at the oversight hearing, and the DPR predominated at the events hearing. Additionally, a few well-known activists, whom we categorize as non-aligned, private stakeholders, offered highly critical assessments of the PPP paradigm at both. We highlight three themes that emerged from the narrative analysis, relating to 1) the exercise of "authority" or "control" over parks, which raises issues of accountability in the PPP regime, 2) the use of public lands for events and revenue production, 3) the contraction of access to public parks. At the same time, we show that Council Members also had to engage in a delicate political dance to reassure the City's private partners and their donors.

In order to temper their critique of PPPs, New York City Council Members, parks administrators, and representatives of influential non-profit parks organizations engaged in "persuasive story-telling" to generally reinforce the origin mythology used to legitimize the public–private governance of parks (Lieto 2015; Throgmorton 2003). In her opening statement at the 2013 PPP oversight hearing, Council Member Mark-Viverito, the chair of the Committee on Parks and Recreation, evoked a familiar narrative used to support PPPs and by extension the neoliberalization of urban governance (Brash 2011). She rooted the birth of the City's park's PPP regime in the "fiscal crisis" when "its parks were in a serious state of disrepair," and praising the PPP model, she added, "as the success of CPC [Central Park Conservancy] became apparent, the conservancy model spread to other large parks including Prospect Park, Bronx River Park, Battery, Randall's Island, and more recently the High Line" (New York City Council, Committee on Parks and Recreation 2013a, 7). In 2015 and 2016, Council Member Levine, now serving as chair, also used this kind of "constitutive story telling" to open hearings related to parks equity and reporting on private funds (Throgmorton 2003). He noted that Central Park had "felt" the fiscal crisis "acutely, suffering decades of neglect by the public sector." He emphasized that "a legion of volunteers, activists, and donors jumped in to turn around this treasured green space" and reminded attendees at the hearing of the "success on a truly spectacular scale" that the conservancy had achieved (New York City Council, Committee on Parks and Recreation 2015, 2). These origin stories and the homage to prominent private partners that policymakers, advocate organizations as well as smaller parks organizations seemed compelled to repeat illustrates the political influence powerful private actors exercise in the PPP regime and signaled a firm belief in the PPP paradigm (Lieto 2015). It also aligns with globally salient austerity and crisis narratives shaping public policy since the 1970s (Peck 2012). There is an implicit warning embedded in this narrative, namely that reliance on the public sector had spelled the "neglect" of these public amenities and had led to the parks' terrible "state of disrepair." Thus, the narrative suggests that should donors not be able to allocate monies to the park or project they choose, this could jeopardize the inflow of private funds and thus fundamentally threaten the PPP regime (New York City Council 2014).

At the same time, the hearings surfaced that PPP arrangements are sometimes opaque to the public. Even New York City's elected officials inquired about the nature of the various and apparently not clearly defined PPP arrangements. Council members specifically delved into decision-making authority with regard to the allocation of resources and the use of parklands. In the case of events, the hearings brought to the forefront how the PPP regime may have blurred boundaries of authority even as the eventization of NYC parks had increased. While striking a conciliatory tone as noted, city council members also asked hard questions that placed DPR representatives on the defensive.

In opening the events hearing, Council Member and Chair Mark-Viverito linked decision-making and control to problems of accountability:

> concerns have been raised about the size of some of these festivals, the amount of parkland being used for ticketed events, the amount of revenue being generated for the event and the condition the park is left in after the event.
> *(New York City Council, Committee on Parks and Recreation 2013b, 23; 78)*

In an apparent attempt to inoculate the public agency against critiques of creeping privatization or loss of control, DPR Commissioner White testified during the oversight hearing that the department "does not cede its authority to determine policy or activity on city property" and emphasized the shared governance of private entities, noting that "the Parks Commissioner sits on the board of the Conservancy," for example (New York City Council, Committee on Parks and Recreation 2013a, 13; 17). Two weeks later, a representative of the Parks Department and former president of the Prospect Park Alliance also reassured council members that "these people who give so much of their time and energy who are such great leaders ... don't sit down and make decisions that are city decisions" (22). A self-identified sergeant with the Park Enforcement Division, however, contradicted this interpretation, hinting at the influence if not control that private organizations exercise over the parks they govern, such as "being able to pick what laws are enforced, what types of concerts are heard, what types of sports are played, the designs of the park recreation centers and prohibiting the public access" (134). Parks advocates, particularly those not aligned with larger conservancies, similarly questioned the DPR perspective. One Manhattan advocate and blogger argued,

> It's about the ways these private entities transform the very spaces they are charged with ... if it was just money is raised and volunteers are galvanized, that's one thing, but the issue ends up being control. In some cases, the spaces become overly programmed and overly sanitized among other things.
> *(New York City Council, Committee on Parks and Recreation 2013a, 180–183)*

Another non-aligned activist speaking of parks events argued "these deals also hand over enormous power and decision-making authority to conservancy groups with little transparency and accountability on what is supposed to be public land" (New York City Council, Committee on Parks and Recreation 2013b, 116). The Deputy Commissioner of DPR, however, maintained that "final authority to issue any permit [for an event] always resides with the Parks Department."

The line of inquiry council members also pursued sought to illuminate the true costs and benefits of events to the public, which the reigning governance arrangements seemed to obfuscate. DPR staff and commissioners, a former parks employee, and aligned private partners framed events in exclusively positive terms and justified the current funding structure undergirding event management on parkland. The DPR Assistant Commissioner for Marketing and Revenue noted: "Large events enliven public space, introduce a park to new users and provide cultural and entertainment amenities to a community. And lest we forget … New Yorkers demand them, they love them and they turn out in large numbers to enjoy them" (New York City Council, Committee on Parks and Recreation 2013b, 33). A former assistant commissioner, moreover, highlighted the key importance signature parks play as urban amenities to boost the City's economic position (see New York City Council, Committee on Parks and Recreation 2013a, 186). Applauding their potential for value creation, he noted:

> Event money or concession monies, is flowing into about 18 parks. We have 5,000 properties … What do they have in common, those parks? All of them are at least regional if not citywide parks. They serve a broad diverse constituency of New Yorkers.
>
> *(150)*

This narrative positions eventscapes in terms of their emotional attraction and inclusionary nature and their economic value to the city. Yet, DPR also acknowledged that the reporting mechanisms did not fully analyze the costs and economic benefits they represented for the city and the public (New York City Council, Committee on Parks and Recreation 2013b).

According to DPR, the "principal purpose" of fees the DPR negotiated for events was "not to generate revenues per se, but to offset the impact to the park and the public's use and to cover costs" (New York City Council, Committee on Parks and Recreation 2013b, 28–29). While formal agreements allow some of these large "corporate" parks organizations such as Bryant Park, Randall's Island, and the other BID-like organizations to retain all of the revenue, others like the Central Park Conservancy are reimbursed for costs associated with events (Krinsky and Simonet 2017). Due to the complexity and varying partnership structures as well as lack of standard or easily traceable reporting rules, council members had a hard time ascertaining the specific flow, use and distribution of money within this system. For example, Committee Chair Mark-Viverito emphasized that public amenities were being "taken off line" for private, revenue-producing events and asked DPR representatives if they "kept track of revenue that these events generate versus what they're paying in fees" to understand what the city was "getting back" (New York City Council, Committee on Parks and Recreation 2013b, 34). After a long back and forth in which DPR representatives equivocated about how different revenue streams from events are collected and reported, the DPR Assistant Commissioner for Revenue and Marketing noted: "I might make the related but not directly responsive comment that the way we price these events actually has more to do with impact … it's an impact-based fee analysis. So that is not related actually to what they earn" (New York City Council, Committee on Parks and Recreation 2013b, 34–39). It appeared that the eventization of parks had operated in the shadows without standardized oversight in which the public (including city council members) beyond DPR, which is a mayoral agency, might gain an effective understanding of the true costs, including the expansive policing, or "over-policing" in the words of one council member, of these events by the city's police department, the events might produce (61–62).

The hearings also highlighted the impact on access as events effectively remove areas from public use, if temporarily. Sidestepping a conversation about ticketing revenues, DPR executive staff discussed "fencing" off areas and "ticketing" primarily as a way to control crowds. The Commissioner of Marketing and Revenues, for example, explained, "So we don't actually see ticketing as privatizing. Ticketing to us is controlling the number of people there for safety" (New York City Council, Committee on Parks and Recreation 2013b, 59). This focus on crowd control was used to dispel the idea that fencing and/or ticketing represented forms of restricting access or privatizing park space. Chair Mark-Viverito and Council Member Lander, who had homed in on the revenue question, however, centered their inquiry on access and framed it in terms of equity. While DPR defended the idea of tickets made available through a lottery as inclusive by geographically broadening who might gain access to an event because "anyone can get a ticket and … tickets are available first come first serve" (59). Council Member Lander countered that in his experience from talking to constituents, "New Yorkers don't feel it that way … your park … it's a little like family," he commented, "it's like when you show up they have to take you. The parks feel like a place you don't have to make an advance reservation" (59). In this case, New Yorkers for Parks, which testified in support of the hundreds of smaller events across the parks systems that "provide a broad spectrum of opportunities many New Yorkers would not otherwise have," also underscored that "[e]vents in parks should not prohibit nonparticipants' access to and use of the park" (95). Thus, while events *per se* were positively framed, the hearings questioned how the use of private partners and programming to foster local commitment to and stewardship of specific parks within the parks system can be balanced with the privatism and exclusionary tendencies associated with the growth orientation embedded in the increasing eventization of parks and the current PPP model.

Concluding thoughts

PPP parks regimes often operate in the shadows without much public scrutiny until high profile news stories draw public attention, as in the NYC case where extraordinary donations and recent tragedies resulted in increased public scrutiny (Nisbet and Schaller 2019). Yet, the eventization of urban parks is increasing often with the aim of attracting large crowds of visitors, especially to signature parks. This kind of eventization threatens to undermine the very notion of "publicness" or the right to the kind of public and free access people connect with their urban parks (Smith 2018). In fact, as evident from the hearings, elected public officials relied on what Lieto (2015) has called an "origin narrative" or "mythological narrative" to "refresh their own knowledge and traditions each and every time an idea is put into play" (115). In this case, the needed critiques, which were responsive to activist and broader public focus on the PPP regime by the press, were tempered in constitutive narratives that reaffirmed the very necessity of the PPP regime. It is striking that so little of the conversation about events was directly held with the largest event producers, that is the regional parks conservancies or BIDs with large budgets. The Central Park Conservancy (CPC) representative, for example, made a perfunctory statement at the hearing on partnerships that did not allude to any of the controversies or critiques of these organizations. Friends of the High Line's statement focused primarily on the economic development benefits the city derives from parks, and the Bryant Park Corporation was represented at neither hearing. In articulating his thoughts on the PPP regime and the eventization of parks, Council Member Lander articulated this dilemma most directly:

> there really is risk because if people come to conclude in different parts of this city that … [the] hearing doesn't attend to transparency, doesn't attend to issues of equity, then there will be a reaction against the model that we're talking about here, but if we step up together, the Council, the Parks Department, the administration and find better ways to share information and reflective collective decision-making.
>
> *(New York City Council, Committee on Parks and Recreation 2013b, 40)*

Yet, given the limited presence of large conservancies, the city's parks agency rather than the large private actors served as the target of criticism. As such accountability was shifted back onto the public sector. This dynamic in effect means that the most powerful private entities abjured an obligation to engage in a broader public conversation.

If planning or policy-making and maintenance can be understood as "persuasive storytelling" or "constitutive storytelling," then narratives can be deployed to both unsettle and support efforts to legitimize policy paradigms, such as PPPs (Throgmorton 2003). In NYC, the Council Members' line of questioning and statements pointed to a recurring, low-level power clash between the city council and the mayoral agency (DPR), which is embedded in a higher-level power dynamic within a broader urban growth coalition (Molotch 1993). Focusing on the eventization of urban parks in New York City has allowed us to scrutinize PPP arrangements that govern many of our urban parklands and open spaces. They reconfigure usage patterns, create access restrictions and leverage public lands for private gain, often through revenue producing activities. The hearings illustrate that public officials perturbed by these arrangements raised crucially important concerns, but in the end, they did not seek to fundamentally threaten the PPP paradigm. In fact, they seemed uncomfortable confronting the PPP regime's overall legitimacy. This hesitancy to confront and question the powerful position of wealthy private partners such as the CPC and BPRC in the regime emerged clearly in a hearing focused on the challenges that private fundraising pose for equity within the broader parks system, which postdated the event's hearing (Nisbet and Schaller 2019). Council members highlighted that private monies flowing into select parks exacerbated inequities but backed away from supporting a proposal advanced by State Senator Squadron to redistribute a proportion of the funds raised by the large signature parks organizations across the system (New York City Council, Committee on Parks and Recreation 2014). Council members raised the fear that "at some point you do lose donors, and then we're not helping ourselves. We might be more equitable, but have fewer resources" (87). The solution to this dilemma thus far has been two-pronged. Council members approved a bill to require clearer reporting of private funds (New York City Council, Committee on Parks and Recreation 2015). Additionally, the Community Parks Initiative, an equity initiative advanced by the de Blasio administration, has increased public investment to historically disinvested parks; additionally, large, high-capacity private parks organizations have pledged to voluntarily provide in-kind contributions to smaller underresourced parks (Nisbet and Schaller 2019). These measures, however, do not reshape the balance of power among the public and private partners or remedy the lopsided resource distribution within the parks system.

Note

1 We use the term here not to connote legal citizenship but to describe active inhabitants of the city.

References

Beauregard, R. (1998) Public Private Partnerships as Historical Chameleons: The Case of the United States. In: *Partnerships in Urban Governance: European and American Experience*, J. Pierre (ed.). New York: Macmillan, pp. 52–70.

Brash, J. (2011) *Bloomberg's New York: Class and Governance in the Luxury City*. Athens: University of Georgia Press.

Brown, G., Lee, I.S., King, K. and Shipway, R. (2015) Eventscapes and the Creation of Event Legacies. *Annals of Leisure Research*, 18(4): 510–527.

Dreier, P., Mollenkopf, J.H. and Swanstrom, T. (2004) *Place Matters: Metropolitics for the Twenty-First Century*. Lawrence: University Press of Kansas.

Foderaro, L.W. (2011) "Record $20 Million Gift to Help Finish the High Line Park." *The New York Times*, 26 October. www.nytimes.com/2015/11/14/nyregion/new-york-citys-low-profile-parks-to-get-conservancies-help-and-some-cash.html.

Foderaro, L.W. (2012) "A $100 Million Thank-You for a Lifetime's Central Park Memories." *The New York Times*, 23 October. www.nytimes.com/2015/11/14/nyregion/new-york-citys-low-pro file-parks-to-get-conservancies-help-and-some-cash.html.

Franck, K.A. and Stevens, Q. (2006) *Loose Space: Possibility and Diversity in Urban Life*. New York: Routledge.

Furman, A. (2007) The Street as a Temporary Eventscape. *International Journal of the Humanities*, 5(9): 77–84.

Gillette, H. (2012) *Civitas by Design: Building Better Communities, from the Garden City to the New Urbanism*. Philadelphia: UPenn Press.

Harvey, D. (1989) From Managerialism to Entrepreneurialism: The Transformation in Urban Governance in Late Capitalism. *Geografiska Annaler. Series B, Human Geography*, 71(1): 3–17.

Hou, J. (2010) *Insurgent Public Space: Guerrilla Urbanism and the Remaking of Contemporary Cities*. New York: Routledge/Taylor & Francis Group.

Isenberg, A. (2004) *Downtown America: A History of the Place and the People Who Made It*. Chicago: University of Chicago Press.

Judd, D.R. (ed.). (1999) *The Tourist City*. New Haven: Yale University Press.

Krinsky, J. and Simonet, M. (2017) *Who Cleans the Park? Public Work and Urban Governance in New York City*. Chicago: University of Chicago Press.

Lauria, M. (1999) Reconstructing Urban Regime Theory: Regulation Theory and Institutional Arrangements. In: *The Urban Growth Machine: Critical Perspectives Two Decades Later*, A.E.G. Jonas, D. Wilson and Association of American Geographers. Albany, N.Y: State University of New York Press,SUNY Series in Urban Public Policy, pp. 125–39.

Lieto, L. (2015) Cross-Border Mythologies: The Problem with Traveling Planning Ideas. *Planning Theory*, 14(2): 115–129.

Loughran, K. (2014) Parks for Profit: The High Line, Growth Machines, and the Uneven Development of Urban Public Spaces. *City & Community*, 13(1): 49–68.

Madanipour, A. (1999) Why are the Design and Development of Public Spaces Significant for Cities? *Environment and Planning B: Planning and Design*, 26(6): 879–891.

Madanipour, A. (2003) Social Exclusion and Space. In: *Social Exclusion in European Cities: Processes, Experiences, and Responses*, A. Madanipour, G. Cars and J. Allen (eds.). Oxon: Routledge, pp. 237–245.

Martin, D.G. (2004) Nonprofit Foundations and Grassroots Organizing: Reshaping Urban Governance. *The Professional Geographer*, 56(3): 394–405.

Mascia, K. (2014) "Olivia Rotondo Was Not Just the Girl Who Overdosed at Electric Zoo." *Cosmopolitan*, 13 July. www.cosmopolitan.com/lifestyle/news/a28420/olivia-rotondo-electric-zoo/.

Molotch, H. (1993) The Political Economy of Growth Machines. *Journal of Urban Affairs*, 15(1): 29–53.

Murray, M. (2010) Private Management of Public Spaces: Nonprofit Organizations and Urban Parks. *Harvard Environmental Law Review*, 34: 180–253.

New York City Council, Committee on Parks and Recreation (2013a) *Partnerships Oversight Hearing*. New York: The City Council of New York.

New York City Council, Committee on Parks and Recreation (2013b) *Oversight Hearing on Private Events*. New York: The City Council of New York.

New York City Council, Committee on Parks and Recreation (2014) *Equity Hearing*. New York: The City Council of New York.

New York City Council, Committee on Parks and Recreation (2015) *Hearing on Reporting Bill*. New York: The City Council of New York.

New York City Council, Committee on Parks and Recreation (2016) *Partnerships for Parks Hearing*. New York: The City Council of New York.

Nisbet, E. and Schaller, S. (2019) Philanthropic Partnerships in the Just City: Parks and Schools. *Urban Affairs Review [Online]*.

Peck, J. (2012) Austerity Urbanism: American Cities under Extreme Economy. *City*, 16(6): 626–655.

Richards, G., Marques, L. and Mein, K. (eds.). (2015) *Event Design: Social Perspectives and Practices*. New York: Routledge.

Schaller, S. (2019) *Business Improvement Districts and the Contradictions of Placemaking: A Case Study of BID Urbanism in Washington, D.C.* Athens: University of Georgia Press.

Schaller, S. and Guinand, S. (2019) Pop-Up Landscape Design and the Disruption of the Ordinary. In: *Public Space Design and Social Cohesion*, P. Aelbrecht and S. Quentin (eds.). New York: Routledge, pp. 242–259.

Smith, A. (2016) Eventalisation: Events and the Production of Urban Public Space. In: *Events in the City: Using Public Spaces as Event Venues*, A. Smith (ed.). New York: Routledge.

Smith, A. (2018) Paying for Parks-Ticketed Events and the Commercialisation of Public Space. *Leisure Studies*, 37(5): 533–546.

Sorkin, M. (1992) See You in Disneyland. In: *Variations on a Theme Park: The New American City and the End of Public Space*, M. Sorkin (ed.). New York: Hill and Wang, pp. 205–232.

Suntikul, W. and Jachna, T. (2016) The Co-Creation/Place Attachment Nexus. *Tourism Management*, 52(February): 276–286.

Throgmorton, J.A. (2003) Planning as Persuasive Storytelling in a Global-Scale Web of Relationships. *Planning Theory*, 2(2): 125–151.

Valentine, G. (2013) Living with Difference: Proximity and Encounters in Urban Life. *Geography*, 98: 4–9.

Yee, V. and Rashbaum, W.K. (2013) "Weekend Revelry Cut Short after 2 Die at Electronic Music Festival." *The New York Times*, 1 September. www.nytimes.com/2013/09/02/nyregion/electric-zoo-music-festival-is-canceled-after-2-deaths.html.

Zukin, S. (2011) *Naked City: The Death and Life of Authentic Urban Places*. Oxford: Oxford University Press.

37

THE PRIVATE LIVES OF PUBLIC SPACES

Michael W. Mehaffy and Peter Elmlund

Introduction

The idea that public spaces are of the essence of cities is an ancient one. Yet only recently, new research has given us a deeper picture of just how important public space actually is for city functioning. Perhaps most important, we are learning how public spaces serve as essential city-wide connective networks at many scales, through key structural characteristics. Along with that understanding has come a clearer picture of the intimate connection between the public aspects of urban space, and the more private aspects – not only the fully private built spaces that often line public spaces, but also the more subtle and complex tissue of regions within and adjacent to public spaces, transitioning from the most public realms into the more private ones. It turns out that public space is far from one undifferentiated thing, but has its own complex structure of regions, adjacencies, formal and tacit rules, and dynamic evolutions. We can truly see that public space is a complex mix of public life and private life, or if you like, private *lives* – with important lessons for policy and design.

The research on public space has begun to provide a clearer picture of these benefits and how they work. In the process, the research has provided helpful evidence to influence and guide policymakers and designers – to help to answer the questions: "why should citizens and practitioners care about creating and maintaining better-quality public spaces?" and "how can they actually do that, in a way that is feasible, economically viable, and maximally beneficial?"

Surveying the existing literature

One of the chief problems with the research on public space is that it has existed for the most part in fragmentary form within separate disciplines, lacking the unified picture needed to guide policy and design. Instead, knowledge is fragmented, and each discipline has an incomplete picture. The problem is a little like the proverb of the elephant and the blind men, each of whom encounters a different part of the elephant: one of them finds a tree trunk, one finds a snake, another a leaf and so on – but none of them finds the elephant.

The problem was not always so extreme. The decades of the 1960s and 1970s saw promising collaborations emerging between disciplines, as they began to combine research into human interactions with the built environment and its spaces. During that time, architectural research

457

was strongly connected to behavioral and cognitive psychology as well as sociology (Gutnam 1972; Petrović et al. 2015).

As Robert Gutnam (1972) wrote in his preface to the book *People and Buildings,*

> There is at the present time an enormous interest in relating the behavioral sciences to the design disciplines. Most schools of architecture now require their students to take courses in the behavioral sciences, with particular attention being paid to urban sociology. Sociologists and social psychologists are being added to the faculties of architecture schools, where they offer lectures and seminars and participate as programming specialists and design critics in studio courses.
>
> *(i)*

Unfortunately, this period of fruitful collaboration between the fields of architecture and behavioral sciences was short-lived. In the 1980s, as modernist approaches within town planning and architecture fell into disfavor, so too did the pursuit of an integrated understanding emerge from social, economic, and political disciplines. Harvard's new direction at the time was illustrative of this disciplinary re-segregation. Moshe Safdie, director of the Urban Design Program at Harvard from 1978 to 1984 described the trend in his book *The City After the Automobile* (1998):

> At Harvard, following decades of close association between environmental, political, and architectural issues under Walter Gropius and Josep Lluís Sert, the university decided that study and training in architecture and urban policy did not belong together. City planning, now understood primarily as the making of policy, was incorporated into the Kennedy School of Government. Architecture, landscape architecture, and urban design (now understood as the physical design of urban districts, with limited consideration of political, social, and economic factors), remained in the Design School, "purified" of the mundane and "elevated" to the status of art.
>
> *(11)*

Now almost 40 years later, with rapid urbanization and an increased interest of public space, the circle is closing again, and two hopeful trends have arisen to re-unite the disciplines. One trend is that a new interdisciplinary "science of cities" has emerged, investigating overarching structural properties of cities, and discovering that public space networks are central components. Another is that new interdisciplinary centers have arisen to survey the research and to "connect the dots" between the literature in different fields, drawing overarching conclusions for policy and practice.

One example is the Centre for the Future of Places at KTH Royal Institute of Technology in Stockholm, where the authors are researchers. The Centre's "Public Space Database Project" has assembled a coded database of existing research literature, using it to draw out key concepts and conclusions from the literature, and to write a series of policy and design guidance documents, in partnership with UN-Habitat and other partners. In turn, where gaps are identified in the literature, new research can be commissioned, often drawing from project-based field research aimed at implementing the New Urban Agenda.

Our research at the Centre for the Future of Places has clarified the picture of public space as a complex network of territories and governance rules, both tacit and explicit. In this connection, of note is the work of our colleague Ali Madanipour (2003), who describes the territories of control that form within public spaces, often functioning as quasi-private spaces with tacit rules of control.

A practical example from everyday life will be helpful. Imagine that a family has set up a picnic in a public park, with baskets and chairs and so on. Another person happens along, and accidentally steps into this area as they walk by. The family is not likely to be too bothered by such a temporary infringement. On the other hand, if the passerby deliberately walks into the middle of the family's temporary space and sits down, it will likely be painfully evident to all parties that some tacit rules have been broken about who controls that space – even though, properly speaking, they are all in public space, which they all have the legal right to occupy.

Imagine another similar example of a café seating area in the streetscape, which is typically licensed by a city government to the café owners. Nonetheless this seating is occupying, and to some extent displacing, public space. If, say, a person happens to pass by and spots a friend seated at a table, they might perch next to the friend for just a few moments to say hello. But if they linger for too long, a waiter from the café is likely to ask them insistently if they would like to order – and if not, will soon ask them to leave. Yet this space is in the public right of way of the street – a space the passerby typically has every right to occupy. The waiter is clearly not the owner of the space – indeed, the waiter is not even the owner of the adjacent restaurant!

These and many other examples remind us that public space often has, so to speak, its own private lives, and its own complex set of regions where private life goes on. The picture becomes even more complex when we look at the larger network extending into truly private and transitional spaces, with all their complexities – as we will consider in more detail below.

Teasing out the variable publicness of "public" space

As the examples above suggest, we can see that public space is far from one undifferentiated "public" zone. Rather, each local space is a component of a wider network of public and private spaces, and the structures we create (sometimes only temporarily) to delineate these spaces. As we delineate them, we separate them from one another, while still preserving connections between them – connections that, as we will see, can often be adjusted over time.

This series of spaces, starting with the most public and ending with the most private and enclosed buildings, is really of the essence of cities. When we get this wrong – when we fail to delineate these spaces carefully, with adequate public space – we create enormous problems. As Joan Clos (2016), Secretary-General of the UN's Habitat III conference put it:

> In general, the urban community has become lost in strategic planning, master planning, zoning and landscaping ... All these have their own purposes, of course – but they don't address the principal question, which is the relationship in a city between public space and buildable space. This is the art and science of building cities – and until we recover this basic knowledge, we will continue to make huge mistakes.
>
> *Clos (2016)*

Why have we lost the "basic knowledge" about the relation of public space to buildable and private space, as Joan Clos says? Broadly speaking, we have somehow failed to understand what cities actually *are*. They are, in essence, marvelous structures that *mediate between conflicting freedoms*. We need to explore a bit more what that means.

Public spaces and private lives: freedom and conflict

Cities (a term by which we also include towns) pose an intriguing paradox. On the one hand, they open up new freedoms to us to expand our life choices, to become more productive and more prosperous, and to enjoy the city, its resources, and its cultural treasures. At the same time, cities bring us into potential conflict with others – conflicts over adjacencies, over noises and smells, over competition for space in crowded places, and other disruptions. At its heart, we have potential conflicts over the ability to occupy limited public spaces and carry on our potentially conflicting activities there. These potential conflicts pose limitations on our freedom to enjoy the life of the city. We cannot occupy or move through a given space (because someone has put up a wall, for example) or be free from noise (because we live next to a noisy neighbor, say), or access the resources that cities offer us (because we are too isolated, too poorly connected to the rest of the city) – and so on.

But something marvelous has happened in cities over time. In the best cities, we have gradually evolved structures that have mediated between these conflicting freedoms – that have provided both connectivity and separation, freedom, and organization. One could say that these mediating structures are of the essence of urbanism. Conversely, one could say that urbanism represents an evolving form of mediation between these conflicting freedoms.

A simple example can illustrate how these mediating structures work, and how public space plays a central role.

Imagine that one person lives in one dwelling, and another lives in an adjacent dwelling. One person would like the freedom to party until 3:00 AM, while the other would like the freedom to sleep soundly without being awakened by noise. Each of them seeks a form of freedom but finds their freedoms in conflict (see Figure 37.1a).

One way to resolve this conflict is simply to separate our two homes by such a distance that the sleeping resident is no longer disturbed by the noisy resident (see Figure 37.1b). One lives over here, other lives over there, and with enough separation, *voila* – problem solved. One may party happily into the night, while the other may sleep soundly.

That simple formula, conflict plus segregation equals freedom, is the essential strategy of urban sprawl – segregating not only dwellings and their conflicts from each other, but segregating all the other potential conflicts of work, home, commerce, and other uses.

But there is another powerful way to mediate between conflicting freedoms. Imagine that instead of simply segregating the two homes, the residents create structures that control privacy,

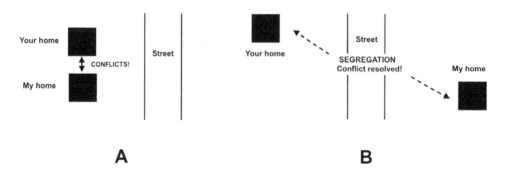

Figure 37.1 Segregation of urban elements can resolve potential conflicts between them, like these two homes (a) and (b)

Image credit: Michael W. Mehaffy and Peter Elmlund

noise and smells, and other areas of potential conflict. The two parties agree to erect, let us say, a wall between the two homes, with sound insulating qualities.

Now they are separated from one another, in a way – but they are not segregated. Indeed, they are still connected in a powerful, mediated way – namely, through the systems of private and public spaces that connect the two homes. One person's bedroom, the most private part of their home, is connected to their living room, and to their front entry, perhaps to a porch or semi-private outdoor zone, where visitors (including their neighbors) may enter, but it is understood that they may not linger, as they are guests on the property. Then that space is in turn connected to the public zones of the street – and over to the adjacent house, where a similarly complex set of zones transitions to *their* most private spaces.

In this way, the citizens are intimately connected to each other through public space, yet they are able to control the conflicts that might restrict their freedom and our privacy (Figure 37.2). They can close a door or window, draw a curtain, lock their doors – or conversely, they can open those things, and welcome visitors in. They can operate a dizzyingly complex network of connections, as can others – connections between spaces that have a membrane-like capacity to let some things through (sounds, sights, people, etc.) and exclude others, at different times and to different degrees.

This ability to control the degree of connectivity – to allow others in, or to exclude others, and thereby to be more or less private – is an essential characteristic of good-quality human and urban spaces. They "afford" to us the ability to control conflicting activities between ourselves and others, and to "mediate between conflicting freedoms." We are able to occupy different spaces, carry on our different activities, and live together within this network of rooms and room-like spaces. This is possible precisely because we can control the connectivity between these spaces, and thereby mediate conflicts.

Understanding the complexity of public space and "place networks"

Nor do these networks of room-like places – which we refer to as "place networks" – stop with built private structures and their literal rooms. Outside of private homes and commercial establishments, one can often observe a dizzyingly complex series of similar room-like spaces, also defined spatially with structural elements or demarcating devices,

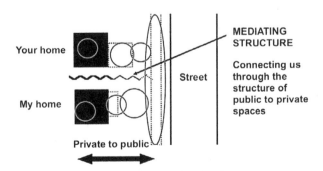

Figure 37.2 Another method to resolve conflicts is to create mediating structures that both separate and connect, usually through adjoining public spaces

Image credit: Michael W. Mehaffy and Peter Elmlund

and often containing clear points of connection to adjacent spaces. These structures may take the form of yards, forecourts, porches, platforms, sidewalk or pavement zones, or other structures. They may even be quite informal – like the family picnic space we discussed earlier.

A more formal example can be seen in the image below, a rather ordinary section of a London "high street" – which, as it happens, was around the corner from where one of the authors once lived (see Figure 37.3a; actually, a montage of two images). It is possible to see many different room-like places where people can be: from the most private spaces, including literal rooms like bedrooms, to the most public spaces along the street, which often have only partially defined room-like spaces – such as the outdoor seating area next to the restaurant. Between these two extremes is a remarkably complex tissue of spaces with mediating structures between them – where one can move, but not if a door is closed; or see, but not if a window blind is drawn, or a hedge obscures the view; or one can hear, but not if there is glass in the window; and so on, with membrane-like connections that can be controlled to suit the situation. The entire network is an intricate web of spaces and connections, and in the second figure (see Figure 37.3b), we only hint at its complexity.

Again, one can see here that we, the users, can modulate the connections between these places and their place networks. We can open doors, close windows, draw blinds, and so on. We can choose our level of public exposure, from a lively place on the street or the square, to the most secluded and quiet place in a bedroom, and many places in between. This capacity for adaptation to varying need (at different times, between different people, etc.) is an important feature.

The example above illustrates another important point. Not only can we make adjustments to our own preferred exposure by closing a blind, opening a window and so on, but by choosing our location within the network, and perhaps moving within it. This kind of adaptive choice can occur over different time spans of our lives as well. Just as one day we might choose to move from a quiet spot in the back of a café over to the sunny window by the street, over a longer period of our lives we might choose to live in a sunny flat above a lively street, and then at a later stage we might prefer to move to a location on a quieter street. The best places afford to us the control over our environments, so that we can "dial in" the degree of exposure we want. In turn, the choices we make will contribute to the shaping of the adjacent public spaces – the degree to which they are connected or not to private spaces, visible from them, activated by them, colonized by their residents, and so on.

Our actions, then, shape the place networks around us, including the physical structures that delineate them, from the most private to the most public. As we can see, some of these structures are permanent (walls, doors, windows, etc.) and some are quite temporary and even tacit – like the example of the family picnic in a kind of "room" made of baskets, blankets, etc.

Not only can these changes happen at small scales over small intervals of time; these places also can change more significantly over larger spans of time, as users and other agents make larger transformations. One of the authors went back to the same street in London five years later, and observed some significant changes. Not only had some businesses closed and others opened, with some new signage, awnings, and colors; new room-like spaces were also articulated, such as the two terraces above with new fences demarcating them, offering a bit more privacy. The new business demarcated a kind of stoop entry with the two planters at its entrance. Some demarcations also went away. The restaurant that formerly had outdoor seating removed most of it, perhaps because the weather wasn't suitable

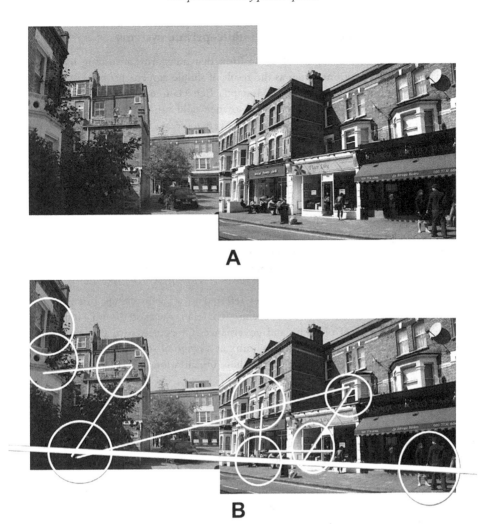

Figure 37.3 An ordinary "high street" in London yet featuring a very complex set of public and private
   spaces – a "place network"

Image credit: Michael W. Mehaffy and Peter Elmlund

at that time, or some other temporal adaptation. In any case, these were all small adaptive
changes that added up to a bigger transformation of the entire streetscape.

We see that small adaptive changes of this kind can occur over five years – but what
kinds of changes can occur from similar processes over a century, or several centuries? The
cadastral plan of the beautiful city of Venice, began quite humbly as a series of straightforward rectangular plots. Over an evolution of several centuries, that simple pattern had transformed into a complex configuration of interlocking spaces, with an even more complex
three-dimensional structure with balconies and bridges. Many small adaptive changes over
several centuries have produced the remarkable urban structure that we all know and love
today.

The evolution of public–private systems

We can see many more examples of urban form that transforms according to such small-scale adaptive transformations, often as the result of simple agreements between buyers and sellers of relatively small plots of land. Mustapha Ben Hamouche (2009) documented the process of transformation of plots of land as owners sold sub-plots (see Figure 37.4a) and then subsequently made agreements with the buyers to create new public streets and lanes, so that they too could sell sub-plots, and/or erect viable commercial businesses on their land (see Figure 37.4b).

Besim Hakim (2010) has also described the way urban form is generated through the iterative application of form-specifying rules, such as those in Islamic building codes. For Hakim, this is an illustration of lessons from the sciences of complexity, which can describe cities as immense "complex adaptive systems" with properties of self-organization. That is, there need not be a central planning agency for many aspects of the city, including its public spaces. Much of the form and character of these spaces can "emerge" from the local actions of agents – the citizens – and form surprisingly well-coordinated frameworks, through a step-wise adaptive process of self-organization.

This form of organization is sometimes referred to as "bottom-up," in that it comes from the decentralized actions of local agents. It is often contrasted with "top-down," or more centralized forms of planning and governance. Of course, in reality there is usually a mix of the two, and moreover, there is a mix at a range of scales from the most bottom-up to the most top-down – and sometimes these organizational forces are overlapping and even conflicting.

For example, a city government may have rules for the use of public spaces, but the users of a park who are playing a game may have a separate set of rules for the movements allowed in the game. Tacit rules (like the one for not invading my picnic blanket) may form yet another overlapping layer of informal governance. Other rules might even be applied nationwide – for example, nation-wide rules for providing access to disabled people, imposed through mandatory standards of construction of ramps and pathways.

This "polycentric governance network" mirrors the polycentric network of public and private spaces in the city, which are "socially produced" (Low 1996). Both are complex, partially emergent, and partly shaped "bottom-up" by the people who occupy public space.

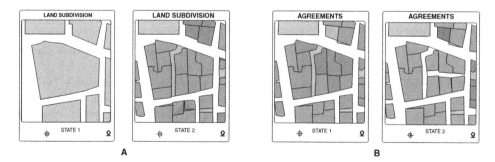

Figure 37.4 Transformation of plots of land as subsequent owners made subdivisions and agreements to create new streets, shown in Ben Hamouche (2009)

Image credit: Michael W. Mehaffy and Peter Elmlund

The death – and life – of public spaces

We may wonder, then, why public space is in decline in so many places, as Habitat III's Joan Clos pointed out. It seems that we have failed to understand how cities actually work, and instead have left ourselves in the grip of a defective model – a point made by Jane Jacobs (1961) in her landmark book *The Death and Life of Great American Cities*. Jacobs argued that our model of cities was based on machine-like segregation and statistical sorting, rather than the "organized complexity" seen in biological systems – and also seen in the most vibrant cities of the past (and seen in the evolving tissue-like structure of place networks). Not understanding this, we have severed the complex tissues of the city's spaces, and its delicate self-organizing relationships between public and private realms.

The automobile is a part of the problem; but as Jacobs pointed out, the problem runs deeper, to the kind of city we build – as it happens in our case, around automobiles. When we enclose ourselves in the capsule of the car, transporting from the capsule of the home to the capsule of the workplace or other destinations, we sever this complex tissue, and change how the city works. We also change how the residents of the city move about and consume resources, in subtle but profound ways.

In this view, cities are networks that help us to connect to one another and to the resources we need. All of us benefit when these networks are richly inter-connected, promoting the interaction of diverse people and ideas. For the same reason, all of us lose something when city networks, or even parts of city networks, are not richly inter-connected.

This was a point made by the physicist-turned-urbanist Luis Bettencourt (2013), who put it this way:

> [This] view of cities in terms of social networks emphasizes the primary role of expanding connectivity per person and of social inclusion in order for cities to real-ize their full socioeconomic potential. In fact, cities that for a variety of reasons (violence, segregation, lack of adequate transportation) remain only incipiently con-nected will typically underperform economically compared to better mixing cities … what these results emphasize is the need for social integration in huge metropolitan areas over their largest scales.
>
> *(3)*

By the same token, we must also be careful to cultivate a healthy balance between the public and the private and avoid letting private spaces and their proprietors over-dominate public spaces – which, at its extreme, is simply a form of destruction of public space. A few sidewalk cafés mixed with pedestrian-filled public spaces is no bad thing, but the closing of large public areas in favor of private activities is a dangerous form of erosion of public space.

In the end, then, as Jacobs argued, a diverse, complex, well-connected public realm is good for everyone's bottom line. As research by colleagues at our Centre has shown, populations with par-ticularly acute needs, like migrant populations, depend upon the opportunities that a fine-grained mix of public and private spaces can offer, particularly along commercial or "main" streets. More fundamentally, the innovation that drives economic expansion thrives on diversity, and on the propinquity, mixing, and serendipitous encounters that happen within public spaces including streets, facilitated by accessible transportation systems. When cities lose this diversity, Jacobs argued, they stagnate economically.

There is a corollary conclusion to be made regarding the urban fabric that we are creating today – an era of rapid urbanization that is unprecedented in human history. It is possible to replace a city-wide system of public spaces with other kinds of networks – data, communications, high-speed transport, and the like – forming a resource-intensive web of homes, offices, and conference rooms. It is possible to operate an innovative and prosperous economy without a robust public space network, as indeed we can see from many sprawling cities with declining public spaces today. These cities show many of the same signs of polycentrism and overlap that we see in the public space networks of great cities.

But something enormously important is lost in the bargain. Not only is this substituted network vastly more resource-intensive – and it may therefore be unsustainable over time – but it excludes many of the people who most need access to the city's public space networks, and the resources to which they connect: the very young, the very old, the poor, and the infirm. This is why the question of public space – and public space systems – could not be more urgent today.

On the other hand, thanks to the research that is now becoming available, we have an opportunity to draw out the lessons from great cities and apply them to a new generation of vibrant, complex, enriching public spaces. What is at stake is not only the economic function of our cities, or the efficiency of their resource use, or their equity and fairness – it is surely all that – but also the renaissance of their public life.

References

Ben Hamouche, M. (2009) Complexity of Urban Fabric in Traditional Muslim Cities: Importing Old Wisdom to Present Cities. *Urban Design International*, 14(1): 22–35.

Bettencourt, L.M.A. (2013) The Kind of Problem a City Is, Working Paper #13-03-008. Santa Fe Institute. https://sfi-edu.s3.amazonaws.com/sfi-edu/production/uploads/sfi-com/dev/uploads/filer/fa/f6/faf61418-fc4f-42d5-8c28-df1197a39018/13-03-008.pdf

Clos, J. (2016) "We Have Lost the Science of Building Cities." *The Guardian*, 18 April. www.theguardian.com/cities/2016/apr/18/lost-science-building-cities-joan-clos-un-habitat

Gutnam, R. (1972) Preface. In: *People and Buildings*, R Gutman. (ed). New York: Basic Books.

Hakim, B.S. (2010) The Generative Nature of Islamic Rules for the Built Environment. *International Journal of Architectural Research: ArchNet-IJAR*, 4(1): 208–212.

Jacobs, J. (1961) *The Death and Life of Great American Cities*. New York: Random House.

Low, S.M. (1996) Spatializing Culture: The Social Production and Social Construction of Public Space in Costa Rica. *American Ethnologist*, 23(4): 861–879.

Madanipour, A. (2003) *Public and Private Spaces of the City*. London: Routledge.

Petrović, E., Vale, B., and Marques, B.. (2015) On the Rise and Apparent Fall of Architectural Psychology in the 1960s, 1970s and Early 1980s. In: *Proceedings of the Society of Architectural Historians, Australia and New Zealand: 32, Architecture, Institutions and Change*, P. Hogben and O'Callaghan, J. (eds). Sydney: SAHANZ, p. 482.

Safdie, M. (1998) *The City after the Automobile: An Architect's Vision*. With W. Kohn. New York: Basic Books.

38

USING BIG DATA TO SUPPORT PUBLIC SPACE RESEARCH

Avigail Vantu and Kristen Day

Introduction

In the last two decades, big data, here defined as real-time, streamed data generated by sensors, have transformed research in the social sciences, health, computer science, and other domains (Batty 2016). This chapter examines how big data may enhance research on public space behavior. For our purposes, research on public space behavior includes empirical studies (both scholarly and practice-oriented) that have a physical environment component. We consider behavior broadly, to include preferences, perceptions, and the use and experience of public spaces.

Research on public space behavior

Research on public space behavior has proliferated since the late 20th century. Researchers examine various place types and "user" groups across numerous geographic locations (c.f. Alsayyad and Guvenc 2015; Cranz 1982; Crossa 2009; Diouf 2003; Hou 2010; Hubbard 2001; Jim and Chen 2006; Low and Smith 2006; Mitchell 2006; Moore 1986; Namaste 1996; Skeggs 1999; Valentine 1996; Whyte 1980). Often, research aims to illuminate diverse experiences of public spaces and to make public spaces safer, more functional, more equitable, and more enjoyable.

Empirical research on public space behavior frequently involves primary data collection including interviews, surveys, archival research, and observations. Researchers also use cross sectional spatial data, including *Geographic Information Systems* (GIS) data and data from *light detection and ranging* (LIDAR) technologies. For example, researchers use GIS data to examine the features of trails and parks that are associated with walking and active play (c.f. Ries, Dunsiger, and Marcus 2009). Public space researchers also use secondary data that are based on population counts from sources like the census and other public or private data sets (Batty 2016).

Big data methods for social science research have expanded in recent years. These approaches include big data generated for research purposes and also secondary use of existing big datasets. Big data offer potential benefits to public space research that include reduced cost; broad representation of both users and spaces; and rich data that are real time, longitudinal, and geographically precise. To explore this potential, we first briefly introduce big data approaches.

What is big data?

A common definition of big data identifies three components known as the three V's: *volume, velocity,* and *variety* (Kitchin and Mcardle 2016). *Volume* implies that big data are enormous in quantity, such as the millions of panoramic images from around the world that comprise *Google Streetviews*. *Velocity* suggests that data are created in real-time, like traffic camera video recordings. Lastly, *variety* acknowledges that the data can be structured, semi-structured, or unstructured. For example, structured data would include pedestrian counts that are organized into columns including date, people per unit, and gender, while unstructured data would include data generated from *Twitter*, which lacks specific organization. Big data are archived digitally, which enables various computations, visualizations, and manipulations (Batty 2016).

Big data sources are often "organic"—meaning that data points are tracked automatically, unlike *a priori* methods of data collection or "designed" data created by users (Florescu et al. 2014; Groves 2011). Because they are automated, big data may be inexpensive to generate in terms of the price per data point, but may require special tools and subject matter expertise to process and analyze, and may have issues such as missing data and noise (Florescu et al. 2014).[1] Big data comprise both quantitative and qualitative information. Rich sources of data include credit card purchases, social media, and quantified human behavior. For example, one study employing big data involves the large-scale measurement of physical activity using mobile data from more than 700,000 people over 68 million days and across 111 countries (Althoff et al. 2017).

Big data's growth and public space research

In recent years, increased collection of data on human activities has multiplied the availability of big datasets (Batty 2013). At the same time, the decentralization of computation and the ability to process data remotely using diverse tasks has democratized data analytics and enabled multidisciplinary collaborations (Batty 2013). The attempt of cities to become "smarter" and more automated has further expanded the prevalence of big data (Batty 2016). More and more services and infrastructural components of the city produce a constant flow of data through sensors, including cameras and audio recorders positioned on vehicles, buildings, and people. These tools facilitate constant, real-time, and temporal understanding of public spaces that can be translated into patterns of use and perceptions. The main techniques described in this chapter derive from *spatial analysis* and *machine learning*, which are two methods within the larger umbrella of data analytics. *Spatial analysis* is a process to study human behavior or systems using geographic data and features (Bailey and Gatrell 1995). *Machine learning* encompasses various models of "smart data analysis," which reduce the number of problems into fewer options to eventually detect and classify behaviors, which allows predictions that can later be applied to new events or data (Smola and Vishwanathan 2008). For example, the development of sophisticated algorithms permits pedestrian detection from surveillance cameras, which enables enriched urban design and transportation planning through accurate counts of numbers of people in specific public places (Wang et al. 2009).

In the following section, we explore specific features of big data for research on public space use and behavior. Key features of big data include high accuracy; convenience (low cost, high speed, and ease of data collection); large scale; the ability to capture fluid behavior; and the ability to definitively locate behavior in space. We tie each feature to potential challenges in current public space research and discuss how big data approaches may help to address these challenges. For the purposes of this chapter, we have highlighted differences between "big data" methods and other primary or more traditional public spaces methods. In reality, this dichotomy is a false one. In many instances, varied methodological approaches can be used in concert. Also, many hybrid research approaches blur boundaries between big data and other methods.

How can big data enhance public space research?

Big data provides (more) objective information about public space behavior

The challenge

Researchers often use surveys and interviews to identify what happens in public space—who uses public spaces, what they do there—and why. These self-reports of behavior can be biased by individuals' moods, recall, and other factors (Hoskin 2012). For naturalistic research that aims to understand behavior as it is understood by individuals themselves, the notion of "bias" is irrelevant. Frequently, though, researchers intend such self-reports as objective accounts of behavior. Concerns include respondent bias (such as providing answers the respondent believes are desired, recollections of behavior that do not match actual behavior, or simply agreeing with the question asked) and also researcher bias (such as stating questions with leading wording). For example, in research on fear of crime in public spaces, respondents' responses may be shaped by a desire to avoid the appearance of racial prejudice (Day 2000). Sampling bias is also a concern—when researchers construct samples of a population that are not representative because, for example, sampling happens at only limited times of the day or year or in limited locations, where users do not reflect the diversity of the research population.

How big data approaches can help address this challenge

Big data are often generated from sensors, such as location, video, or audio recordings, that are produced automatically without human initiation (Lane et al. 2010). Such data represent a constant sample that is triggered by movement or set times, which may limit sampling bias. Data from sensors may also overcome recall bias by accurately reflecting actual behavior as it occurs. Such big data measures could replace self-reports of public space activity in some instances. In the cases of measurement of physical activity in public spaces, for example, smartphone applications demonstrate a high level of accuracy in counting numbers of steps (Case et al. 2015). In comparison, activity diaries recorded directly by study participants report significantly greater activity levels compared to physical activity data that is recorded by sensors (Atienza and King 2005). The accuracy of smartphone measures suggests that they may offer a better measure of step counts, compared to self-reports. Diaries may add another dimension to smartphone-collected data, for example, enhancing understanding of the impacts of individual circumstances such as medical condition, age, or demographics.

Some big data research includes a sufficiently large and representative sample to allow inferences on population behavior. For example, smartphone location data from over 700K individuals in 111 countries produced 68 million days of recorded data on individuals' physical activity. Using this data, researchers identified a correlation between walkability scores and increased physical activity in US cities, including a spike during commute and lunchtime in more "walkable" cities compared to a constant, lower level of activity in less "walkable" cities (Althoff et al. 2017).

Big data are convenient: cheap, fast and easy to collect for many users

The challenge

Primary data collection can be expensive in terms of researchers' time and monetary cost. For example, intercept surveys are often used to assess preferences of open spaces (City of Boston 2002). At a few minutes per survey, the time required of researchers is considerable when calculated over thousands of surveys. The same is true for in-person observations—environmental audits, behavioral mapping, etc.—used to document behaviors in public spaces. For example, observational methods are used to document the numbers and activities of people at locations such as bus stops (Gehl Institute 2017). Applying for human subjects research approval, pilot testing of data collection instruments, training of field researchers, travel time to sites, and entering data for statistical analysis may all extend the time and cost of primary data collection.

How big data approaches can help address this challenge

Big data approaches may reduce the time, cost, and effort needed to acquire data. Often, big data can be obtained and processed rapidly using web tools and remote computing. As with other secondary data sets, data from sensors, social media, census and open data, and other big data sources are not usually collected exclusively for one study and thus can be reused across studies, with careful consideration of context and other factors (Zook et al. 2017).[2] Further, some big data research that uses publicly available data, may not require human subjects research approval (see Metcalf and Crawford 2016), which can reduce the time required before studies begin.

Recent years have seen the simplification of data access and the standardization of research techniques across fields. Accordingly, individuals can now access data and conduct statistical analysis with limited technical expertise and monetary cost. Tools and services like *Application Programming Interfaces (API)*, *Box* and *Amazon Web Services (AWS)* allow researchers to generate "tasks" that automatically retrieve data from websites and store it for no or little cost. These tools can also store, filter, structure, and transfer data from one entity to the other in real-time. For example, researchers captured and analyzed 50 million publicly available *Google Streetview* images from 200 American cities (Gebru et al. 2017). Using this data, the investigators accurately predicted socio-economic status and political orientation based on car make, models, and year in the images. Similar techniques may be used to classify public space types or to predict crime levels. Such research may require some data analysis expertise but is readily scalable, compared to large scale, designed data collection.

One noteworthy tool for low cost, large-scale analysis is *Microsoft Azure*. This personalized computer vision image analysis *API* makes it easy to teach a computer program to detect objects in images. To train the tool, users upload numerous images and tag them by describing their content, for example: tree, bike, bench. With more tagged images, the model becomes better at identifying these objects in new images. Using this Microsoft computer vision tool, a *Twitter* account named *City Describer* was created. *City Describer* automatically describes the content of urban images posted to *Reddit* by users (Boeing 2017). The results appear to be mostly (but not completely) accurate (e.g. "a close up of a busy corner," "a sunset over a city street"). Descriptions even use casual language that sounds "human" (e.g. "a view of a city street filled with lots of traffic.") Using this technique to study the physical environment could save time, cost, and labor by allowing researchers to automatically analyze images to collect data on numerous or remote sites.

While not strictly "big data," new technology tools can also reduce the burden of collecting intercept surveys from large number of people. *Amazon Mechanical Turk* and social media platforms like *Craigslist* or *Facebook* will provide users with an online survey through a *URL*. Using these tools, researchers can distribute surveys applying specific parameters to shape respondents' profiles (e.g. age, geographical location). Those tools can generate large numbers of participants in a short amount of time for no or little cost. Also, such tools may engage younger populations who might not otherwise participate in research (Dalessandro 2018).

Big data enables scaling-up of public space research

The challenge

Time and cost may force researchers to limit primary data collection to a small sample of public places or people. For example, to study how internet provision impacts social interaction in public spaces, Keith Hampton and Neeti Gupta (2008) observed mobile device use in four cafés in Seattle and Boston and interviewed 20 customers in these cafés. A focus on one or a few public spaces may be desirable if researchers aim to understand a single setting in all its richness. Other times, however, the small number of spaces or participants is less a desired feature and more a practical compromise. This limitation becomes problematic when researchers aspire to more general conclusions about public spaces or behavior.

How big data approaches can help address this challenge

Large-scale studies are made possible by the decentralization of sensors and data (Batty 2016) over long periods of time and multiple geographic locations. Images from *Google Streetview* and from social media platforms, offer an increasingly important resource for built environment research (Boeing 2017). Such pictures include geographical coordinates and can be subject to feature extraction as described earlier. Research sites are thus expanded to include all public spaces represented in the image set (Gebru et al. 2017). Research sites can also, to some extent, be extrapolated to others not represented in images using statistical modeling, which becomes powerful if using a large sample. In addition, crowdsourcing of research participation can reach enormous numbers of participants across diverse geographic locations.

The use of machine-learning algorithms can reduce the effort of working with large-scale data, allowing large numbers of individuals to participate in research with minimal effort. Machine learning uses a relatively small amount of data to teach the model specific feature extraction and behaviors. For example, *Place Pulse*, a web interface developed at MIT, asked participants to rate perceptions of specific places by clicking on one of two displayed images, in response to a single question (Dubey et al. 2016; Naik et al. 2014). The participants viewed *Google Streetview* images from various American cities and were asked questions like which looks "livelier," "safer," and "more boring." The researchers used the 3,000 images labeled by participants to identify physical environment features that were linked with perceived safety, based on associations for hundreds of thousands of street images (Naik et al. 2014). This tool and similar others can expand the number and diversity of participants and of environmental features in public space research. The results of such research suggest built environment features that may encourage some perceptions or emotions in humans—here again, findings can be expanded and "deepened" by adding site visits or interviews.

Big data research accommodates changing behavior and ties behavior to geographic location

The challenge

Traditional research methods often generate static data that captures behavior as a "snapshot" at a single point in time and place. Researchers can capture samples of behaviors over short time periods using diaries (in which individuals recall and report their behaviors over a fixed period of time, such as travel diaries) or experience sampling (where individuals complete short questionnaires or provide other information at intervals across a short time period) (Bolger and Laurenceau 2013). Yet longitudinal research on public space behavior remains relatively rare (c.f. Gehl 2010; Mehta 2007; Mehta and Bossen 2018). Additionally, while observational and ethnographic methods do allow behavior to be linked to specific geographical spaces, these methods are hard to scale, as discussed earlier (c. f. Givena and Leckie 2003). *Global Positioning System* (GPS) coordinates are useful in this regard, as demonstrated in research on tourists' spatial behavior (Edwards and Griffin 2012).

How big data approaches can help address this challenge

Mobile phones passively produce highly accurate data to inform on individuals' activities (Jiang et al. 2013). Phones are carried everywhere these days, so data are ubiquitous. Geolocation data can provide information on public space use and behavior, including specific location and time spent in each space. The combination of accelerometer information, which captures physical movement, and *GPS* location estimates, also make it possible to identify modes of travel to public spaces, such as riding a bus, bicycling, or traveling by subway (Lane et al. 2010). For example, researchers developed a system to identify modes of transport for individuals' mobile phones using accelerations from mobile data, as well as frequencies of speed likely to be used for each travel mode. This system can be implemented across a large number of users (Reddy et al. 2010). Audio data is also collected through sensors; these data can produce insights on behavior in streets and parks. For example, researchers have mounted microphones on buildings to track urban noise over time. These data locate noises such as sirens, drilling, and children playing, with high levels of accuracy (Bello, Mydlarz, and Salamon 2018). Such data can answer new questions about noise exposure and activity patterns in specific public spaces.

Social media can help identify users' daily activities in public spaces. When aggregated over time and location, these data can also reveal patterns in behavior and changes over time. In this context, using users' first names, researchers have developed an algorithm to analyze social media check-ins data to map gender segregation in activity spaces in Saudi

Arabia (Alfayez et al. 2017). Changes in short and long-term patterns can also be detected in social media using event-detection systems. Researchers have developed a real-time event detection system to assist media and government in coverage and response by using real-time, geo-tagged social media images posts. By constantly scanning feeds, the system learns to identify "normal" patterns of social media postings in certain geographical areas and flags aberrations in postings to reveal the occurrence and location of hyper-local events like local concerts, weather conditions, or accidents. Results are displayed in real-time on a web-interface visualization (Xia et al. 2014). Similar tools use social media or 311 data as input to detect residents' complaints about public space disturbances in real-time, thus supporting immediate government response.

Big data ethical considerations[3]

As noted earlier, the availability and abundance of big data has created new opportunities for researchers and professionals to better understand people and cities. Yet the same data tools can also produce inaccurate conclusions or perpetuate discrimination against groups or individuals. Computational knowledge and capable data management skills are insufficient to assure responsible use of big data (Donoho 2017). To conduct meaningful and accurate analyses, priorities must include minimizing statistical bias and securing fairness. One example of an ethical concern in public space that is enabled by big data models is China's recent practice of fining and shaming jaywalkers, made possible by the use of facial recognition algorithms (Mitchell and Diamond 2018).

To truly understand data, individuals must engage critically with how models are created. Nate Silver (2012), the founder and editor of the polling website FiveThirtyEight, reminds us that "the numbers have no way of speaking for themselves". To this end, mathematician Kathy O'Neil (2016) identified features in algorithms and models that increase the risk of generating social inequalities. To eliminate those risks, researchers should use data and algorithms that are clear and transparent and that can be updated frequently to account for social changes. In contrast, unclear, opaque, and invisible models can lead to bias and unfairness in the use of data. Also, some models may only be accurate and valid for one specific situation and should not be replicated or scaled-up over other cases (O'Neil 2016).

The need to maintain high standards of data privacy is a specific concern in using big data in the public and urban sphere (Kitchin 2016). For example, consider the unease surrounding the Quayside project—an effort by Google Alphabet's company Sidewalk Labs to transform the Toronto waterfront into a smart city. The fact that a private company will have access to a vast amount of sensitive *personal identifiable information* (PII)[4] has generated significant concern (Fussell 2018). Information about physical location, behavior in space, and online patterns are all considered sensitive and are common urban ethics issues (Kitchin). In the case of a Google-owned company that already possesses a large amount of personal data, these data could potentially be merged to generate detailed and unprecedented understanding about individuals. In an era when data is collected passively from people who interact with public spaces, citizens do not always see a direct gain in exchange for their loss of privacy (Editorial 2018). Also, citizens frequently cannot opt out from having their data recorded (Finch and Tene 2014).

The need is for data to be treated in ways that emphasize human needs and values, and not solely as mathematical and machine-readable form of information (Anderson 2008). Researchers should interrogate their data from all angles with constant skepticism. Researchers must also ask questions and take time to understand data structures, sources of the data, and how well said data fits the research question before making conclusions.

Conclusions

Beyond its utility for research, big data can also support more engaged participation in the planning and design of public spaces. For example, cities, non-profits, and academic researchers can use crowdsourcing and citizen science tools to recruit participants to help detect the features of public spaces (c.f. citizens' contributions to tree counting, NYC Park and Recreation 2016). Big data also offer new tools to support equitable public space distribution by allowing ready comparison of neighborhoods in terms of the quality and accessibility of public spaces. One such example is the *Subway Desert* map, which uses data visualization to display New York City districts that require residents to walk longer than average to reach the closest subway station (Whong 2016).

Big data, including images, audio recordings, and geolocation mobile data, provide a level of detail about people and how they use public spaces, that is unprecedented in its granularity, continuity over time, and extent. Researchers and practitioners now have the flexibility to choose between field work and/or open data and technology tools. This flexibility provides options in how to allocate resources and time to best answer research questions.

Finally, a level of disconnect remains between those who are most likely to engage in public space research (and design), and those with expertise in big data techniques. Researchers in architecture, urban design and planning, sociology, anthropology, history, leisure studies, and similar fields may have limited exposure to big data approaches. These methods are more common in engineering, computer science, and aligned fields, which historically have had limited interest in public space behavior. Going forward, programs that train public space researchers will need to incorporate coding and spatial analysis in their curricula. Researchers should also seek new collaborations between design and planning, humanities, and social science departments and their colleagues across campus.

Notes

1 The term "noisy" data refers to additional information that is delivered with a given dataset which is senseless and not machine readable. Noisy data often derives from errors and from a lack of standards in data collection or entry.
2 Data from social media can be especially useful for longitudinal studies of changes in behavior and public opinions (UK Department of Work and Pensions 2014), for example, research on changing attitudes towards free speech in public space and other contemporary policy issues.
3 This chapter cannot address all the concerns and responsibilities that come with acquiring, maintaining, and creating algorithms to derive meaning from big data. In this discussion, we identify some of the main concerns to be addressed when considering such tools.
4 PII are unique identifiers that can help trace the specific individual to whom data refers. PII include (but are not limited to): pictures, voice, names, emails, addresses, phone numbers, social security numbers, and license plates (Kitchin 2016).

References

Alfayez, A., Awwad, Z., Kerr, C., Alrashed, N., Williams, S. and Al-Wabil, A. (2017) Understanding Gendered Spaces Using Social Media Data. In: *Social Computing and Social Media. Applications and Analytics. SCSM 2017. Lecture Notes in Computer Science, Vol 10283*, Meiselwitz, G. (ed.). Cham: Springer, p. 327.
Alsayyad, N. and Guvenc, M. (2015) Virtual Uprisings: On the Interaction of New Social Media, Traditional Media Coverage and Urban Space during the 'Arab Spring'. *Urban Studies*, 52(11): 2018–2034.

Althoff, T., Sosic, R., Hicks, J.L., King, A.C., Delp, S.L. and Leskovec, J. (2017) Large-Scale Physical Activity Data Reveal Worldwide Activity Inequality. *Nature*, 547: 336–339.

Anderson, C. (2008) "The End of Theory: The Data Deluge Makes the Scientific Method Obsolete." *Wired Magazine*, June 23. Accessed 9 February 2019. www.wired.com/2008/06/pb-theory.

Atienza, A.A. and King, A.C. (2005) Comparing Self-Reported versus Objectively Measured Physical Activity Behavior. *Research Quarterly for Exercise and Sport*, 76(3): 358–362.

Bailey, T.C. and Gatrell, T. (1995) *Interactive Spatial Data Analysis*. New York: Longman Scientific & Technical.

Batty, M. (2013) Big Data, Smart Cities and City Planning. *Dialogues in Human Geography*, 3(3): 274–279.

Batty, M. (2016) Big Data and the City. *Built Environment*, 42(3): 321–337.

Bello, J.P., Mydlarz, C. and Salamon, J. (2018) Sound Analysis in Smart Cities. In: *Computational Analysis of Sound Scenes and Events*, Virtanen, T., Plumbley, M.D. and Ellis, D.P.W. (eds.). Cham: Springer, pp. 373–397.

Boeing, G. (2017) "Describing Cities with Computer Vision." *Geoff Boeing*, 30 September. https://geoff boeing.com/2017/09/describing-cities-with-computer-vision/.

Bolger, N. and Laurenceau, J.P. (2013) *Intensive Longitudinal Methods: An Introduction to Diary and Experience Sampling Research*. New York: Guilford Press.

Case, M.A., Burwick, H.A., Volpp, K.G. and Patel, M.S. (2015) Accuracy of Smartphone Applications and Wearable Devices for Tracking Physical Activity Data. *JAMA*, 313(6): 625–626.

City of Boston. (2002) "City of Boston Open Space Plan 2002–2006: Renewing the Legacy… Fulfilling the Vision." *Boston City Government*, September. www.cityofboston.gov/parks/openspace_doc.asp.

Cranz, G. (1982) *The Politics of Park Design: A History of Urban Parks in America*. Cambridge: MIT Press.

Crossa, V. (2009) Resisting the Entrepreneurial City: Street Vendors' Struggle in Mexico City's Historic Center. *International Journal of Urban and Regional Research*, 33(1): 43–63.

Dalessandro, C. (2018) Recruitment Tools for Reaching Millennials: The Digital Difference. *International Journal of Qualitative Methods*, doi: 10.1177/1609406918774446.

Day, K. (2000) Strangers in the Night? Women's Fear of Sexual Assault on Urban College Campuses. *Journal of Architectural and Planning Research*, 16(4): 289–312.

Diouf, M. (2003) Engaging Postcolonial Cultures: African Youth and Public Space. *African Studies Review*, 46(2): 1–12.

Donoho, D. (2017) 50 Years of Data Science. *Journal of Computational and Graphical Statistics*, 26(4): 745–766.

Dubey, A., Naik, N., Parikh, D., Raskar, R. and Hidalgo, C.A. (2016) Deep Learning the City: Quantifying Urban Perception at a Global Scale. In: *Computer Vision – ECCV, 2016/ / 2016*, Leibe, B., Matas, J., Sebe, N. and Welling, M. (eds.). Cham: Springer, pp. 196–212.

Editorial. (2018) "View on Google and Toronto: Smart City, Dumb Deal." *The Guardian*, 5 February. Accessed 12 February 2019. www.theguardian.com/commentisfree/2018/feb/05/the-guardian-view-on-google-and-toronto-smart-city-dumb-deal.

Edwards, D. and Griffin, T. (2012) Understanding Tourists' Spatial Behavior: GPS Tracking as an Aid to Sustainable Destination Management. *Journal of Sustainable Tourism*, 21(4): 580–595.

Finch, K. and Tene, O. (2014) Welcome to the Metropticon: Protecting Privacy in a Hyperconnected Town. *Fordham Urban Law Journal*, 41(4): 1581–1615.

Florescu, D., Karlberg, M., Fernando, R., Rey Del Castill, P., Skaliotis, M. and Wirthmann, A. (2014) "Will 'Big Data' Transform Official Statistics?" Paper Presented at the European Conference on Quality in Statistics, Vienna, 2–5 June.

Fussell, C. (2018) "The City of the Future Is a Data-Collection Machine." *The Atlantic*, 21 November. Accessed 12 February 2019. www.theatlantic.com/technology/archive/2018/11/google-sidewalk-labs/575551/.

Gehl Institute (2017) *Crossing the Street. Building DC's Inclusive Future through Creative Placemaking*. Washington, DC: District of Columbia Office of Planning.

Gehl, J. (2010) *Cities for People*. Washington, DC: Island Press.

Gerbu, T., Krause, J., Wang, Y., Chen, D., Deng, J., Aiden, E. L. and Fei-Fei, L. (2017) Using Deep Learning and Google Street View to Estimate the Demographic Makeup of Neighborhoods across the United States. *Proceedings of the National Academy of Sciences*, 114(50): 13108–13113.

Givena, L.M. and Leckie, G.J. (2003) "Sweeping" the Library: Mapping the Social Activity Space of the Public Library. *Library & Information Science Research*, 25(4): 365–385.

Groves, R. (2011) "'Designed Data and 'Organic Data'." *The Census Blogs*, 31 May. Accessed 16 October 2018. www.census.gov/newsroom/blogs/director/2011/05/designed-data-and-organic-data.html.

Hampton, K.N. and Gupta, N. (2008) Community and Social Interaction in the Wireless City: Wi-Fi Use in Public and Semi-Public Spaces. *New Media & Society*, 10(6): 831–850.

Hoskin, R. (2012) "The Dangers of Self-Report." *Science for All Brain Waves*, 3 March. Accessed 12 July 2018. www.sciencebrainwaves.com/the-dangers-of-self-report/.

Hou, J. (2010) *Insurgent Public Space: Guerrilla Urbanism and the Remaking of Contemporary Cities*. London: Routledge.

Hubbard, P. (2001) Sex Zones: Intimacy, Citizenship and Public Space. *Sexualities*, 4(1): 51–71.

Jiang, S., Fiore, G.A., Yang, Y., Ferreira, J., Frazzoli, E. and Gonzalez, M.C. (2013) A Review of Urban Computing for Mobile Phone Traces. *Proceedings of the 2nd ACM SIGKDD International Workshop on Urban Computing*, Chicago, IL, USA.

Jim, C.Y. and Chen, W.Y. (2006) Perception and Attitude of Residents toward Urban Green Spaces in Guangzhou (China). *Environmental Management*, 38(3): 338–349.

Kitchin, R. (2016) The Ethics of Smart Cities and Urban Science. *The Royal Society*, 374: 2083.

Kitchin, R. and Mcardle, G. (2016) What Makes Big Data, Big Data? Exploring the Ontological Characteristics of 26 Datasets. *Big Data & Society*, 3(1) January–June: 1–10.

Lane, N.D., Miluzzo, E., Hong, L., Peebles, D., Choudhury, T. and Campbell, A.T. (2010) A Survey of Mobile Phone Sensing. *IEEE Communications Magazine*, 48(9): 140–150.

Low, S. and Smith, N. (eds.). (2006) *The Politics of Public Space*. New York: Routledge.

Mehta, V. (2007) Lively Streets. Determining Environmental Characteristics to Support Social Behavior. *Journal of Planning Education and Research*, 27(2): 165–187.

Mehta, V. and Bossen, J.K. (2018) Revisiting Lively Streets: Social Interactions in Public Space. *Journal of Planning Education and Research*, 27(2): 165–187.

Metcalf, J. and Crawford, K. (2016) Where are Human Subjects in Big Data Research? *The Emerging Ethics Divide*, 3(1), Sage Journals, New York https://doi.org/10.1177/2053951716650211.

Mitchell, A. and Diamond, L. (2018) "China's Surveillance State Should Scare Everyone." *The Atlantic*, 8 February. Accessed 14 February 2019. www.theatlantic.com/international/archive/2018/02/china-surveillance/552203/.

Mitchell, D. (2006) *The Right to the City: Social Justice and the Fight for Public Space*. New York: Guilford Press.

Moore, R. (1986) *Childhood's Domain: Play and Place in Child Development*. London: Croom-Helm.

Naik, N., Philipoom, J., Raskar, R. and Hidalgo, C. (2014) "Streetscore – Predicting the Perceived Safety of One Million Streetscapes." Paper Presented at IEEE Conference on Computer Vision and Pattern Recognition Workshops, Columbus, 23–28 June.

Namaste, K. (1996) Genderbashing: Sexuality, Gender, and the Regulation of Public Space. *Environment and Planning D: Society and Space*, 14(2): 221–240.

NYC Parks and Recreation. (2016) "Treescount! 2015–2016 Street Tree Census." Accessed 16 October 2018. www.nycgovparks.org/trees/treescount.

O'Neil, C. (2016) *Weapons of Math Destruction: How Big Data Increases Inequality and Threatens Democracy*. New York: Broadway Books.

Reddy, S., Mun, M., Burke, J., Estrin, D., Hansen, M. and Srivastava, B.M. (2010) Using Mobile Phones to Determine Transportation Modes. *ACM Transactions on Sensor Networks*, 6(2): 13: 1 – 13:27.

Ries, A.V., Dunsiger, S. and Marcus, B.H. (2009) Physical Activity Interventions and Changes in Perceived Home and Facility Environments. *Preventive Medicine*, 49(6): 515–517.

Silver, N. (2015) *The Signal and the Noise: Why So Many Predictions Fail- But Some Don't*. New York: Penguin.

Skeggs, B. (1999) Matter Out of Place: Visibility and Sexualities in Leisure Spaces. *Leisure Studies*, 18(3): 213–232.

Smola, A. and Vishwanathan, S.V.N. (2008) *Introduction to Machine Learning*. Cambridge: Cambridge University Press.

UK Department of Work and Pensions (2014) *The Use of Social Media for Research and Analysis: A Feasibility Study*. London: Department for Work and Pensions.

Valentine, G. (1996) Children Should Be Seen and Not Heard: The Production and Transgression of Adults' Public Space. *Urban Geography*, 17(3): 205–220.

Wang, X., Ma, X. and Grimson, W.E.L. (2009) Unsupervised Activity Perception in Crowded and Complicated Scenes Using Hierarchical Bayesian Models. *IEEE Transactions on Pattern Analysis and Machine Intelligence*, 31(3): 539–555.

Whong, C. (2016) "Subway Desserts V2." *Webmap*. Accessed 6 September 2018. https://cwhong.carto. com/viz/6dfca01c-47e5-11e6-9fd3-0ee66e2c9693/public_map.

Whyte, W.H. (1980) *The Social Life of Small Urban Spaces*. Washington, DC: The Conservation Foundation.

Xia, C., Schwartz, R., Xie, K., Krebs, A., Langdon, A., Ting, J. and Naaman, M. (2014) CityBeat: Real-Time Social Media Visualization of Hyper-Local City Data. *WWW 2014 Companion*, Seoul, 7–11 April.

Zook, M., Barocas, S., Boyd, D., Crawford, K., Keller, E., Gangadharan, S.P., Goodman, A. et al. (2017) Ten Simple Rules for Responsible Big Data Research. *PLoS Computational Biology*, 13(3): E1005399.

39

DIGITAL TECHNOLOGIES AND PUBLIC SPACE IN CONTEMPORARY CHINA

Tim Jachna

Introduction

The scale and rapidity of the processes of modernization occurring in China in recent years is without precedent in global history, among which are facets of urbanization, the development and implementation of technological infrastructures, and a sweeping metamorphosis of social practices and norms. All of these factors converge on the issue of public life, as new urban forms, new technological affordances, and new social patterns develop in mutually co-defining ways in China's urban public space.

Much has already been written on the emergence of new patterns of spatial practice in the public spaces of Chinese cities in the early years of the 21st century (i.e. Gaubatz 2008; Shannon and Chen 2013; Wessel 2019) and on the particularities of the emergence of the internet as a new public space in China (i.e. Herold and Marolt 2011; Li 2010; Zhou and Lu 2016). Rather than reiterate the many findings presented in this burgeoning literature, this chapter will turn its gaze on the nexus between these two venues of public life to investigate the hybrid practices that unfold in the interplay between public practice in digital and physical public realms in the specific contemporary Chinese urban context.

Digital technologies and social transformation in the early 21st century

Digital venues have been found to serve as rehearsal spaces of sorts for ways of behaving and interacting in society. In some cases, the activities in these venues serve to reaffirm and enforce established norms of (dominant) public behavior. In others, they offer safe haven for identity formation of counterpublics or have served as refuges in which the norms and rules of the official public realm are questioned. Alternative ways of being together are rehearsed, articulated, and fortified, eventually spilling out in revolutionary action in the physical realm as in the case of China's sexual revolution, initiated and sustained through the use of online forums on sexuality in Chinese society and, more pragmatically, to seek potential like-minded sexual partners (People's Daily 2003).

With the arrival of 2G technology and the spread of the mobile phone, short message service (SMS) functionality provided a channel for modes of expression that were potentially awkward in face-to-face or telephone conversations in Chinese culture. Whilst acknowledging the importance of meeting and talking face-to-face, several participants in online Chinese bulletin boards at the time were grateful for the opportunity provided by texting to express oneself in a personal manner that also allows distance and composure to be maintained, especially in romantic matters. This also gave young Chinese a channel of communication in which they could simultaneously transgress and respect their society's disdain for public displays of emotion or intimacy.

China's floating population of workers is a truly mobile demographic, migrating back and forth between their families and villages in the countryside and their jobs in the city, where they have few rights to public amenities or services and are perceived by many "official" urban residents as dirty and unwelcome interlopers in the public space (Zhang 2001). The mobile phone and China's mobile population, who are often far from home with no fixed address and are constantly on the lookout for the next employment opportunity in an informal job market, are well suited to one another. In 2003 it was estimated that 60 to 70% of new mobile phone service subscribers were migrant workers, as these devices served as crucial lifelines to social support networks in their adopted cities as well as to their families back home, connecting them to tenuous publics of their own even as they were alienated from many aspects of urban public life (China Economy 2003).

To give one example, one migrant worker became a minor celebrity as the "cell phone poet" who adopted the SMS as a medium for poetry and used the writing and dissemination of his text message poems as a way to integrate himself into the city where he was a short-term and marginalized resident (CRIonline 2003). Without a physical space in the city to call his own, nor a public platform or venue for self-expression, he was nonetheless able to give a very public airing to his art in a way that suffused the space of the city, becoming both pervasive and anonymous, like the floating population itself in the eyes of the dominant urban public.

Broadening the palette of public venues

At the beginning of the current century there was a broad perception of the internet in China as an alternative social space, where things were possible that were not allowed in the physical space of the city, given a relative lack of government interference and a loosening of social restrictions in the communities of choice that formed around chat rooms, bulletin boards, web logs, and other online social forums. In the words of one Beijing college student, "the atmosphere on the Internet is far freer than the atmosphere in our country in general" (Neumann 2001). This differential in perceived and actual freedom caused people to seek and find sites on the internet for activities banned from the physical space of the city. Anecdotally, a planned memorial service for a female student in Beijing who was murdered just before the eleventh anniversary of the Tiananmen Square massacre was forbidden by authorities (although the proposed memorial had no relation to the massacre), as it was likely feared that the proximity to the historic date would cause the memorial service to turn into a protest. Nonetheless, an online service was organized and attended by over a thousand people, who also used the event to protest for better campus security.

Presently, online microblogs, in particular Weibo, are used as forums in which anger and disappointment with governmental action or inaction are expressed. Such communities of discourse can form around issues of public interest, without the level of risk of drastic repercussions faced by protestors in physical public spaces. Scholars have proposed that these sites serve as forums in which "the identity of the *public* (gongzhong) is gradually replacing the identity of the *people* (renmin)" (Zhou and Lu 2016). In this context, public is understood in a way akin to Habermas' (1989) identification of the bourgeois public sphere that emerged in 19th-century Europe as a counterpoint to the sphere of the state.

Controls on digital technology use of private individuals, however, have tightened in the intervening years. Perhaps in recognition of the role of each internet user as both producer and consumer of content, controls that were previously aimed at reporters and publishers have been explicitly extended to cover individual users as well, as signaled by a 2012 headline in *The People's Daily* newspaper and website, "The Internet is not a Land Outside the Law" (Bandurski 2017). There are contemporaneous indications that the control paradigm that applies to physical public spaces is gradually being applied to digital public forums. A new comprehensive media policy was announced by President Xi Jinping on February 9, 2016, firmly placing the internet in the realm of public media as a channel for the formation of "correct" public opinion. This control of digital public space has also served to quash citizens' practices of using digital technologies as parallel channels that facilitate action in physical public space, as in the detention of citizens who used the social networking site Weibo to spread awareness of the government's violent suppression of protestors in the village of Wukan (Huang 2016).

The physical loci of digital access

Before home computers were widely affordable by common Chinese citizens, the primary points of internet access were internet cafes. The first of these were small and private independent operators, often unlicensed and unregulated. By the early 2000s most of these had been shut down and replaced with chain companies that collaborated with the government. Internet cafes are sites of contact between the public space of the physical city and the online forums of the internet and, as such, a primary focus of state regulation and scrutiny. Even as these businesses were portrayed by the governmental news media as public menaces and lairs of dangerous distractions, they also provided the best points of oversight that the government had on the web use habits of its citizens. Internet cafes were required to keep an online log of all user activity for the past three months, to actively transmit information on customers' web-surfing activities to the police and to install mandatory software that blocks access to hundreds of thousands of sites (China Economic Net 2004). To impose these same strictures on internet users in private homes would have been impractical and more difficult to enforce. Thus, in digitally-mediated interactions, as well as interactions in physical space, the private home was a space of comparative individual freedom for those few who could afford a computer and private internet access.

Fast-forward to 2019 and the physical-spatial context of internet access in China has changed completely. The number of internet cafes fell from 350,000 around 2005 to just over 150,000 a decade later, in part in reaction to the increasing wherewithal of the middle class to afford computers at home, as well as continuing government crackdowns. More recently, though, the growth of online gaming and the desire to play such games in a public place in the presence of other gamers rather than alone at home has reversed this trend and led to a resurgence of internet cafes. More importantly, the preferred means of accessing the internet has shifted drastically from personal computers to smartphones. Between 2008 (the year of introduction of 3G mobile technology in China) and 2017, the portion of Chinese users who accessed the internet via their mobile devices grew from near zero to 97.5% (as compared with 53.0% using desktop computers and 35.8% using laptops), and mobile internet access traffic grew from 400 million to 21.2 billion gigabytes, a fifty-fold increase (CNNIC 2018).

This means that the primary locus of internet access has changed from shared terminals in the semi-public "third spaces" of internet cafes to personal digital devices on the mobile bodies of users as they move through the public spaces of the city, necessitating a shift in strategies of government control. Accordingly, measures have been introduced that facilitate person-centered surveillance of digital devices via the installation of spyware on mobile phones by police and the close collaboration between the government and tech companies in monitoring use of the social platforms that are core to social interaction and economic activity in Chinese society.

Spaces of surveillance

The monitoring of public behavior through personal digital devices is supplemented by tracking equipment, such as digital surveillance cameras in public space. Highly accurate facial recognition technology is then used to analyze the footage of crowds in urban spaces. Facial recognition software has reached a level of sophistication that allows the Chinese government to implement autonomic systems that track millions of individuals' movements through and between large crowded public spaces in cities across the country as a matter of course. One known use of this technology is to identify jaywalkers, who are then shamed by posting their images on web portals. There are speculations that such systems will form a pivotal element in the broader surveillance infrastructure, facilitating President Xi's "social credit" system of rewards for behavior deemed "desirable" or punishment for acts judged to be counter to the interests of the State.

The perceived "value" of the data gathered by surveillance arrays extends beyond regimes of control into the commercial realm. In the words of Qi Lu, chief operating officer of the Chinese internet tech company Baidu, "with far field voice sensors … even at a distance we can hear what we want to hear … We can see your fingers, your eyes … the machine can see what it wants to see" (Lu 2017). These words were not uttered in the context of lauding or decrying the potential of these technologies for surveillance in the service of urban governance, but rather in pointing out the valuable channels of data collection that such capabilities put at the disposal of providers of services and products.

Through the mobilization of the various affordances of digital technologies, public space in China has become a field for the harvesting of data about the Chinese citizenry, contributing to the informational layers accessible to government and commercial actors in the public realm, but it has also augmented the possibilities of grassroots action, as discussed in the following section.

Critical witnesses …

With the advent of text messaging in the early 2000s, news in China could spread extremely quickly and widely, with which government blocking could be evaded. Typically, news is spread throughout a city and the country by individuals using digital communications technologies before being announced by the government (if it ever is). The text message, "a fatal strange flu is spreading in our city," was forwarded 40 million times in Guangzhou on February 8, 2003, establishing a decentralized grassroots network of gathering and dissemination of news that was intentionally suppressed by official news media. The day after it was forwarded 41 million times, and 45 million times the day after that, before the government officially acknowledged the existence of the SARS outbreak (Glaser 2003).

With the rapid development of consumer digital technologies, the range of media accessible to citizen authors of content has expanded from text messages to include sound, images, and videos. In putting the tools of video production within reach of ordinary Chinese people, digital technologies enabled the emergence of citizen journalism that could evade government control and restrictions. Armed with mobile phones, laptops, and digital video cameras, a new generation of citizen journalists – the so-called "dGeneration" (d standing for digital) – began to produce unsanctioned documentaries, turning a critical eye on issues in the public spaces of Chinese cities. Zhao Dayong's 2006 film *Street Life* (Nanjing Lu), for example, documented the daily lives of homeless migrants living on one of Shanghai's most affluent streets. While the agility and mobility afforded by digital devices enabled these filmmakers to evade government efforts to curtail the production of such content, the government has been more effective in controlling channels of content dissemination within the country; as international audiences were afforded an unprecedented vantage into the daily realities of public life in China, the critical perspectives on the problematic aspects and inequities of public life have been withheld from the domestic public.

The government endeavors to consolidate oversight of the country's digital infrastructure to maintain control of the public. The monitoring of public behavior through surveillance is supplemented by the withholding of access to channels of information that would enable citizens to maintain awareness of events in the public realm, blocking the diffusion of perspectives that would encourage critical reflection on these events and patterns. At the same time, as demonstrated in the preceding paragraphs, digital technologies have also been instrumental in enabling citizen journalists, scientists, and activists at times to circumvent these controls, such that the digital realm may be seen as a primary arena in which the battle between these opposing aspirations for the public realm is played out.

... and witnesses of public banalities

The short-circuiting of the feedback loop, by which the critical works of the dGeneration filmmakers could re-enter the public realm that they document, and thus inform public spatial practice, contrasts with the broad, rapid, and unhindered proliferation of another genre of citizen videography: a "micro-cinema" of autobiographical films (Yu 2009). These short clips tend to be personal, exhibitionist, and celebratory of conspicuous consumption rather than overtly critical in their ambitions (Voci 2010). The uploaders of these films constitute a distributed array of sensors, harvesting impressions of Chinese citizens' lives in both nominally public and nominally private spaces and projecting them into the digital public realm.

This network of sentient citizens who consciously author digital content is mirrored by other networks of automatized and machinic "authors" of records of public life, such as video surveillance cameras. The 2017 feature movie *Dragonfly Eyes*, by Xu Bing, is a narrative constructed entirely from publicly available footage from surveillance cameras throughout China (*Dragonfly Eyes* 2017). Xu compared the 170 million cameras (at the time) recording visual data throughout the nation to a dragonfly's thousands of eyes, from which the film draws its title. Though many of the spaces captured by such cameras are private or commercial, the fact that they are all sites of cameras streaming on publicly viewable channels places them all equally under the public gaze. These surveillance cameras, like the cameras that record the personal "micro-cinema" clips, are public eyes in that what they see is nominally available to anyone with internet access, blurring the distinction between public and private spaces and public and private life.

Privatization of the physical/digital nexus

Although in many ways perceived and used as spaces of relative freedom, access to the virtual venues of the internet and digital communications networks is not a public right, but a privilege that must be bought from commercial enterprises, whose rules one must follow in order to participate. This is, in effect, an extension of the privatization of the public realm into virtual space as noted by Sorkin (1992) and others. With the present state of technology, digital communications media are in some ways easier for governments to invisibly monitor than the physical spaces of the city, and the specter of potential surveillance hangs over any exchange that happens via these media.

Digital technologies are also implicated in the colonization of public space by private commercial interests. As with other places in the world, China has seen its share of experiments that seek to use inventive entrepreneurial applications of digital technologies to piggy-back on the provisions of public space for profit, at times to the detriment of public space quality. For instance, a boom of bike sharing providers (systems that rely on digital technologies to facilitate and monitor self-service rental of bicycles) in 2017 led to a sudden and extreme increase in the number of bicycles in urban public spaces. Soon a massive glut of service providers, many of which have since gone out of business, led to scenes common throughout Chinese cities, where tens of thousands of abandoned and broken bicycles were left in mounds, the detritus of failed entrepreneurial experiments in public space at the digital/physical nexus (Taylor 2018).

Conclusion

As has been long established, digital technologies present the potential of enabling democratic ideals and public empowerment, while at the same time providing ever more efficient and pervasive infrastructures of control (Dahl 1971; Lessig 1999). The tension and interaction between these tendencies is currently being played out in societies around the world. This chapter has explored the specific ways in which these dynamics are being manifested in the context of public space in contemporary China.

The geography of digitally mediated public space in China has developed from the archipelagos of internet cafes, which were the primary points of contact between the digital and physical realms, to the current condition in which most internet access is now through smart-phones, bringing new mobility and proximity to the performed overlapping of physical and digital public venues and amenities. This has been mirrored by evolution in the degree and diversity of measures of control, surveillance, and censorship exercised by the government through the application of technologies in the digital realm as well as the physical realm in pursuit of "modernization without democratization" (Ding 2015).

Just as spatial practice in China's cities evinces a state of interdependency between physical and digital places and modes, the information flow that sustains and defines social groups, and through which information is shared, is also dependent on a fluid interplay between digital media and human agents. "Human relays" are important links in the chain of communication, stepping in where digital networks leave off or are impeded, and helping social groups to hurdle both government-imposed and geographic barriers.

In the Special Administrative Region of Hong Kong, in which citizens enjoy exceptional (if increasingly fragile-seeming) freedom to gather and of expression as compared to mainland Chinese citizens, grassroots use of digital technologies has had much freer reign, giving a glimpse into the transformative potential of the forces held at bay by regimes of control elsewhere in the country. Mass political protests in Hong Kong, such as the 2014 "Umbrella Movement" and the 2019 protests against a proposed extradition bill that are in their fifth month as of this writing, provide examples of the use of digital technologies to manifest alternative "framings" of public space and public life. "Mesh network" apps such as "firechat" utilize the Bluetooth functionality of mobile digital devices to form networks, circumventing the need for a phone or internet connection. This functionality has enabled the mobilization and sustaining of massive publics that performatively reconstitute the public spaces of the city, effectively conjuring large public squares in a city that has none, for a time changing the physical fabric and typology of physical public space in the city (Chakroff 2014).

Throughout the first two decades of the 21st century, the evolution of digitally mediated public spatial practice in China has involved testing the limits and opportunities imposed by technological, social, economic, and political factors in an ongoing process of negotiation that is as much situated *in* urban social space as it is the process *from* which urban social space, both physical and virtual, is formed.

References

Bandurski, D. (2017) "The Great Hive of Propaganda" *China Media Project*, 16 September. Accessed 23 August 2019. http://chinamediaproject.org/2017/09/16/the-great-hive-of-propaganda/.

Chakroff, E. (2014) "OccupyCentral Hong Kong Democracy Protests & Public Space." *Archinect Discussion Forum*, 1 October. Accessed 30 August 2019. https://archinect.com/forum/thread/110250556/occupycentral-hong-kong-democracy-protests-public-space.

China Economic Net. (2004) "China Closes 1,600 Internet Cafes." 1 November. Accessed 12 April 2019. http://en.ce.cn/Life/social/200411/01/t20041101_2147294.shtml.

China Economy (China Economic Information Network). (2003) "Unicom Sells Cheaper Pre-paid CDMA Cards." 7 March. Accessed 12 March 2019. http://ce.cei.gov.cn/enew/new_h2/ni00hg46.htm.

CNNIC. (2018) *Statistical Report on Internet Development in China.* China Internet Network Information Center. Beijing.

CRIonline. (2003) "Life of Migrant Workers in the Capital." 13 February, Accessed 14 April 2019. http://english.cri.com.cn/english/2003/Feb/86731.htm.

Dahl, R.A. (1971) *Polyarchy: Participation and Opposition.* New Haven: Yale University Press.

Ding, S. (2015) Modernization without Democratization in the Digital Age: China's Micromanagement of Its Contentious State-Society Relations. *Asian Journal of Political Science*, 23(1): 1–22.

Dragonfly Eyes (Qing ting zhi yang). (2017) Motion Picture. Directed by Xu Bing. Beijing: Xu Bing Studio.

Gaubatz, P. (2008) New Public Space in China: Fewer Walls, More Malls in Beijing, Shanghai, and Xining. *China Perspectives*, 4: 72–83.

Glaser, M. (2003) China's Internet Revolution. *USC Annenberg Online Journalism Review*. www.ojr.org/ojr/world reports/1068766903.php.

Habermas, J. (1989 [1962]) *The Structural Transformation of the Public Sphere: An Inquiry into a Category of Bourgeois Society.* Cambridge: Polity.

Herold, K. and Marolt, P. (eds) (2011) *Online Society in China: Creating, Celebrating and Instrumentalising the Online Carnival.* New York: Routledge.

Huang, Z.P. (2016) "Chinese Citizens are Being Arrested for Sharing News about the Wukan Village Rebellion Online." *Quartz*, 16 September. Accessed 10 August 2019. https://qz.com/783026/china-censorship-chinese-citizens-are-being-arrested-for-sharing-news-about-the-wukan-village-rebellion-online/.

Lessig, L. (1999) *Code and Other Laws of Cyberspace.* New York: Basic Books.

Li, S.B. (2010) The Online Public Space and Popular Ethos in China. *Media, Culture & Society*, 32(1): 63–83.

Lu, Q. (2017) "International Forum on Innovation and Emerging Industries Development. Shanghai Municipal Government and Chinese Academy of Engineering." Shanghai, November 7–9.

Neumann, L.A. (2001) "The Great Firewall." *Committee to Protect Journalists*, January. Accessed 10 April 2019. http://cpj.org/Briefings/2001/China_jan01/China_jan01.html.

People's Daily Online. (2003) "China Undergoing Sexual Revolution." 24 November. Accessed 15 August 2019. http://english.people.com.cn/200306/20/eng20030620_118623.shtml.

Shannon, K. and Chen, Y.Y. (2013) "(Recovering) China's Urban Rivers as Public Space." *Footprint*, January: 27–44.

Sorkin, M. (ed) (1992) *Variations on a Theme Park: The New American City and the End of Public Space.* London: London.

Taylor, A. (2018) "The Bike-share Oversupply in China: Huge Piles of Abandoned and Broken Bicycles." *The Atlantic*, March. Accessed 2 September 2019. www.theatlantic.com/photo/2018/03/bike-share-oversupply-in-china-huge-piles-of-abandoned-and-broken-bicycles/556268/.

Voci, P. (2010) *China on Video: Smaller-screen Realities.* London: Routledge.

Wessel, M. (2019) "How Beijing Is Redefining Public Space with the Temple of Heaven." *The City Fix*, 11 March. Accessed 1 September 2019. https://thecityfix.com/blog/beijing-redefining-public-space-temple-heaven-mark-wessel/.

Yu, H.Q. (2009) 'Just Like Eating Chocolate:' A Reflection on China's DV Culture. *Journal of Chinese Cinemas*, 3(1): 63–67.

Zhang, L. (2001) *Strangers in the City: Reconfiguration of Space, Power and Social Networks within China's Floating Population.* Stanford: Stanford Press.

Zhou, S. and T. Lu. (2016) Social Media and the Public Sphere in China: A Case Study of Political Discussion on Weibo after the Wenzhou High-Speed Rail Derailment Accident. in *Handbook of Research on Citizen Engagement and Public Participation in the Era of New Media (Advances in Public Policy and Administration)*, M. Adria and Y.P. Mao (Eds.). Hershey: Information Science Reference, pp. 410–425.

40

WHAT IF?

Forecasting and composing public spaces

Mark C. Childs

Introduction

A three-story-high holographic ballerina dances amidst the future Los Angeles street packed with pedestrians, display screens, street cafes, and umbrellas (Benatos 2017). This scene from the movie *Blade Runner 2049* updates the earlier *Blade Runner*'s Chinatown with a layer of augmented and interactive reality and provides a palpable vision of how virtual and physical public spaces could merge.

Is it a vision we want? What forces might drive us toward or away from that future? What are the alternatives and how might we compose them? These are questions that the field of forecasting can help urban composers (architects, landscape architects, urban designers, planners, and others who aim to shape built forms) address.

Envisioning futures is a core role of urban composition (see Figure 40.1). Built forms have frequently endured long past their initial purposes, builders, or even civilizations, and our built compositions will be transformed by continuing urbanization, aging populations, migrations, and other consequences of climate change, automation, the continued evolution of virtual worlds, the forces examined in the other chapters of this book, and other emerging and unforeseen trends. Understanding how these forces may (re)shape public spaces is critical to (re)composing spaces that will serve us well during their long lives.

Public spaces can be key components in tempering and adapting to these trends, if our squares, streets, parks, and other public places effectively foster the formation of social capital, encourage physical health, engender a sense of belonging and stability, host peaceful political speech and protest, and provide a sense of conviviality. As the philosopher Karsten Harries (1997) wrote,

> There is a continuing need for the creation of ... places where individuals come together and affirm themselves as members of the community [as they celebrate] those central aspects of our life that maintain and give meaning to existence. The highest function of architecture remains what it has always been: to invite such festivals.
>
> *(365)*

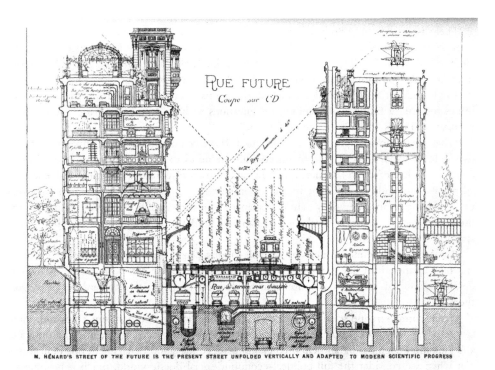

Figure 40.1 Hénard's vision of the future street including a Parisian boulevard, Da Vinci-like flying machines, and the latest conceptions of infrastructure. As a tool of forecasting rather than prediction or advocacy, images such as these can aid in considering resilience and adaptability. Sub-street infrastructure tunnels do ease the repair and installation of utilities. Do the flying machines presage delivery drones?

Image credit: From American City, Jan. 1911, p. 29.

Public spaces, like all built forms, are confluences of cultural practices and physical forms (Childs 2012). The American courthouse square and main street, the Italian piazza and piazzetta, and Tewa pueblo plazas (*bu-ping-geh*), for example, are each civic rooms whose physical pattern developed and continues to develop within its larger cultural context. Third places, such as English pubs and Turkish hookah cafés, are perhaps even more tightly scripted by social practices than civic rooms (Oldenburg 1999). Thus the composition of new public spaces requires understanding, and perhaps shaping, the confluence of social, legal, economic, poetic, and other forces at play.

This chapter borrows the concepts of *framing, drivers of change, causal layered analysis* (CLA), *scenario development,* and *signals of change* from the forecasting literature to provide tools to conceive, consider and compose public spaces that are *pluripotent* – their built-in potentials for change aids resilience and adaptation. Additionally, the practice of *editing the narrative landscape* – that is, curating and critiquing artwork and literature that embodies public space, and publishing informed visions of potential public spaces – can help frame the cogent questions, illuminate drivers of change, provide signals of change, suggest scenarios, and inspire the design of public spaces.

There is an emerging body of literature on planning and foresight (see Fernandez-Guell et al. 2018; Goddard and Tewdwr-Jones 2014; Government Office for Science 2016; Ratcliffe and Krawczyk 2011), and firms such as Arup, Urban Foresight, and IDEO currently offer foresight consulting on various aspects of the built environment. The Lincoln Institute of Land Policy (2019) recently launched the Consortium for Scenario Planning, which brings together a number of planning organizations to help prepare land-use "professionals in urban and rural areas for uncertain futures."

What we do now shapes how events unfold. If we've envisioned and planned for alternative futures, we are better able to purposefully respond to them. Foresight should not be confused with prediction. Rather, foresight can be used to consider and build resilience and adaptability to multiple different futures. Considering, for example, that a town park may need to serve as an emergency camp after an earthquake suggests designing to support this use. Foresight can help us break out of habitual thinking to inspire new possibilities and articulate criteria for success.

With rising temperatures, perhaps we will canopy plazas, and these canopies will serve as artworks, planetarium-like screens, acoustic dampeners, or air filters. Foresight is used to describe preferred and feared futures and to develop tools to steer toward or away from them.

Urban composers use science-based means of prediction to compose our cities and public spaces when they can. Structural calculations, models of available sunlight, environmental and sanitation engineering all help predict and achieve the good behavior of our built forms. Yet when we consider the full political-economic-social-poetic world, much is beyond our ability to reliably predict. "Instead of claiming that anyone can *predict* what is going to happen, we argue that everyone … can see the forces as they are taking shape and not be blindsided…" (Schwartz and Randall 2007, 97).

Framing

Framing is defining the scope and focus of your line of inquiry, with the aim of setting objectives and avoiding "solving the wrong problem" (Hines and Bishop 2015). Are you designing a parking lot or an entry courtyard that provides parking? Are you designing a hotel or a Bourbon Street hotel? Are you gathering insight to design a particular plaza, or investigating changing conceptions of the plaza – that is, are you focused on a particular case, the typology or system, or some combination of both?

Most commissioned work focuses on the singular case at hand. Research, thought experiments, and installations typically examine or construct cases in order to understand and/or change a system. Depending on one's perspective, these future-thinking exercises will have different values. For example, examining the *Blade Runner* ballerina scene from the perspective of an urban composer would yield a different set of questions than considering it from the perspective of someone examining the public perception of augmented reality.

Projects live in multiple contexts (see Childs 2012). Both global climate change and the shade cast by a new neighboring building will shape where ice forms on a plaza. Large trends such as an economic exodus from small towns may be counterbalanced by local factors. The economy of Jackson, Wyoming, for example, is buoyed by its adjacency to Yellowstone Park. Popular books, movies, TV shows, and other stories can cast new meanings over existing places – as the children's book *Make Way for Ducklings* did for Boston's Public Garden, and the TV sitcom *Cheers* did for both a particular bar in Boston and the concept of the bar as a third place. Separating these contexts into large drivers of change, multiple layers of narrative (the stories about the drivers), and local niches (the way a project "sees" a place) can aid in gathering contextual insights and conceiving of futures.

Drivers of change

Drivers of change are forces that change the overall environment around a project. The rise of the automobile in the early 20th century, for instance, was a driver of the redesign and loss of public spaces in the United States. Analyses of current drivers of change are publicly available from multiple sources (for transportation see Rodrigue 2017; for historic cities see Swinney and Thomas 2015). Arup Foresight's set of cards called *Drivers of Change* describes a number of drivers and the questions they raise. It is "designed to facilitate conversation about the trends and issues that are likely to have a significant impact on the built environment and the world at large" (Arup Foresight 2019a). Essentially, Arup has conducted a context analysis at the global scale with specific attention to the built environment. Identifying large-scale drivers of change can help designers both understand a client's motives and concerns and place the project within global and national dynamics.

The history of a site's context can illuminate drivers that helped compose the site. These drivers may still be active. For example, in the summer of 1907, Seattle City Councilman Thomas Revelle proposed that the city create a public marketplace where farmers and consumers could meet directly to sell and buy goods, thereby bypassing wholesalers. In August of 1907 Pike Place Market opened (Pike Place Market 2019). The economic (and perhaps cultural) desire to remove middlemen from between farmers and consumers continues to inform current efforts such as farm business and management assistance programs, farmland preservation efforts, and the five farmers' market sites currently run by a set of Pike Place Market organizations. The history of the market as a locus for artists likewise continues as a driver.

Causal layered analysis

Drivers of change can be examined from multiple different conceptual frames. *Causal layered analysis* (CLA) employs four levels of causes to consider drivers of change: litanies/descriptions, systems/social causes, discourses/worldviews, and myths/metaphors/images (Inayatullah 2004). Examining the rise of illuminated main streets in the early 20th century, these levels might be: electrically-lit main streets were established across the world (description), enabled by the new technologies and changing social patterns (systems), within a celebration of sophisticated modernism and life-style marketing (worldview), using the tagline "the great white way" (image/metaphor) (see Nye 1990).

Designers can use CLA to more deeply understand drivers and to develop proposals. Cities' redesign of their zoning *systems*, for example, led to developers creating privately owned public spaces (Kayden 2000). Designers can develop prototypes of projects based on new *metaphors/images* – in the mid-20th century many U.S. main streets were redesigned as shopping malls. Cynthia Wuellner, FAIA (2011) describes another instance of myths/images driving composition:

> [A] particularly compelling image in America is Small Town U.S.A., mythologized as an ideal way of a life. Drawing on this myth, the New Urbanist movement has transformed thousands of new developments with the purpose of regaining walkable, intimate, mixed use neighborhoods.
>
> *(664)*

Large-scale drivers have significance according to the purposes of the project and local contexts. If we pay attention only to large-scale drivers, we too often get cookie-cutter buildings designed and sited according to generic factors such as traffic counts and neighborhood incomes. More careful investigation can readily uncover specific drivers for each project. National and state highway design standards, for example, are developed in response to large trends and goals. However, when state highways become the main street of a small town, local conditions may need to be given priority (Ewing 2002; United States Genreal Printing Office 2010). Moreover, built forms co-evolve to create mutual benefits – as lawyers' offices and bail-bond offices cluster near courthouses, they may, in turn, support cafés; transforming a parking lot into a plaza can inspire adjacent restaurants to provide sidewalk seating (see Childs 2012). Thus the characteristics of the neighborhood are also drivers.

Scenarios

> Scenarios take us into a possible future. They describe a world to come, making a systematic set of assumptions about the drivers shaping that world. They may be brief and descriptive, or they may include story-like narratives that represent the point of view of personas in the future. They may include a 'history of the future' – how we get from here to there, writes the Institute for the Future (2019).

A range of scenarios provides composers of public spaces conceptual "worlds" to aim toward or avoid. Developing methods for a proposal to adapt to this range of scenarios can build resilience. For example, studying a variety of demographic projections might encourage a designer to make a plaza adaptable to a changing set of activities. Examining scenarios about the rise of driverless automobiles should encourage designers to consider how parking lots may transform into plazas or parks. Considering the rise of telepresence might inspire the design of fountains or artworks that allow people at the site to interact with people at a virtual site.

Four archetypal ways of thinking about the future developed at the University of Hawaii – growth, collapse, constraint, and transformation – are often used to create a set of scenarios (Dator 2009). "Wild card" futures, which attempt to describe "unlikely" changes that would have significant effects, can also be useful to test assumptions. "It is critical to push people's imagination out to the very edges of believability to see the full range of the possible" (Schwartz and Randall 2007, 98).

Comprehensive scenario development can be resource-intensive and may be beyond the time available for the redesign of a plaza or other project. However, "sketch" scenarios may be developed through cogent framing of the questions, consulting with many people holding diverse perspectives, understanding the histories and contexts of the project, and sketching a broad range of futures.

Existing scenarios and scholarly histories of a place or public space type may enrich these "sketch" scenarios. Corporations, cities, and nations have developed sets of scenarios with which to test their decisions (e.g. Royal Dutch Shell 2019, Singapore's Centre for Strategic Futures 2019, Arup Foresight's The Future of Urban Water (ARUP 2019b), Urban Foresight's smart city strategy for The Scottish Urban Foresight 2019). Even outside of the places for which they were developed, these metropolitan and corporate scenarios may provide tools for thought.

Understanding the history of a site and of the project type can help uncover still-active drivers of change, illustrate possibilities, and help us appreciate the complex relationship between intentions and outcomes. Camillo Sitte's 1889 *The Art of Building Cities* used the previous incremental composition of European public spaces to critique then-current practices and advocate for alternative practices. Likewise, the street life of an early 20th-century town may provide a model for a future robust street life in the city. A deep understanding of how the past unfolded can alert us to the complexity of how our actions may play out.

> [P]ortray the past in its own context with all its complexity. … The drama, indeed the tragedy, of history comes from our understanding of the tension that existed between the conscious wills and intentions of the participants in the past and the underlying conditions that constrained their actions and shaped their future.
>
> *(Wood 2009, 11)*

A compelling set of scenarios provides a means to think through how drivers of change and the intentions of a design may interact.

Signals of change

Signals of change, sometimes called weak signals, are current activities that suggest the emergence of significant changes (Hines and Bishop 2015; Institute for the Future 2018). These signals are early indicators that a scenario is emerging. For example, if, in Seattle in the 1980s, you noticed that every morning in the cold February drizzle people were lining up on the sidewalks to buy expensive coffee, you might have seen this as presaging both the designer coffee boom, a reawakening of Seattle's street life, and a revival of street vending.

The professor of communications Everett Rogers developed a model for the diffusion of innovations (diagram at multiple sites see Dans 2016; Rogers 2003). The actions of the "early adopters" are signals of change. Such signals may be ambiguous, unexpected, occurring in unusual places, and difficult to interpret (Lesca 2014). For example, in hindsight, observations of the first users of the safety bicycle were signals of the coming 1890s bicycle craze. The craze, with its development of bicycle manufacturing facilities and techniques, the good roads movement that advocated for paved roads, and the change in street culture created by the presence of (relatively) fast and quiet vehicles, was a signal of changes that paved the way for the rise of the automobile and the transformation of our street culture, cities, and planet (Flink 1990).

Signals of change may be gleaned from multiple sources and are considered stronger when noticed in multiple different arenas. For public space, likely sources include "the street," and art about the street.

Unsurprisingly, "the street," that is, the collection of public spaces, is the prime location to notice emergent patterns of public space. Jane Jacobs' (1961) observations of sidewalks, "the main public places of the city," were central to her urban insights. Likewise, William H. Whyte (1980) and a multitude of other scholars' work is based on observing the street. While much of this work focuses on describing "what is" rather than "what might be emerging" their field methods may aid looking for signals of change.

Fiction, poetry, and art can be a source of signals of change. French writer Charles Baudelaire's observations of Parisian streets informed his 1863 essay *The Painter of Modern Life* and other writings (Baudelairie and Mayne 1964 [1863]). Based in part on these texts, Walter Benjamin elaborated the modern concept of the strolling intellectual observer or *flâneur* (1999). Gordon Matta Clark's art "Fake Estates" in New York City inspired the San Francisco artists of REBAR to create the now widely practiced PARK(ing) Day which, in

Figure 40.2 "Parked Bench" in London, U.K. designed by WMBstudio, 2015. The project illustrates the
rapid spread of the "parklet" but also engages the narrative landscape of London in multiple
ways – perhaps most dramatically by playing with the "London-red" used on buses and many
other public works

Image credit: Ed Butler

turn, is often credited as inspiration for the parklet movement (PARK(ing) day 2019; De
Monchaux 2017, 183; Littke 2016) (Figure 40.2). The use of the courthouse square as
a prime setting in the 1985 movie *Back to the Future* precedes the 1993 founding of the
Congress for New Urbanism and, perhaps, was an early signal of a cultural shift (CNU
2019).[1]

Narrative landscape

The many artists who composed the *Blade Runner* ballerina street scene were acting, intentionally or not, as urban designers, and this scene is part of the narrative landscape of public
space – the mélange of stories of place with which we discuss the meanings and possibilities
of courthouse squares, parklets, and other places.

Fluency with a locale's narrative landscape can help frame better questions, illuminate
drivers of change, alert us to signals of change, and suggest scenarios. Memorials and historic
districts, of course, are essentially tied to their narratives. However, the context for other
types of built forms also includes their narrative landscapes. For example, the collection of
stories that delineate Marfa, Texas as an arts community provide inspiration for, and a frame
by which to evaluate, new built forms in Marfa. After contemplating Donald Judd's minimalist work collected in Marfa, everyday buildings and even the standard squares of the
town's sidewalks seem to take on new meaning (see Figure 40.3).

Figure 40.3 An example of the artistic reframing of everyday built forms in Marfa, Texas; the shop and exhibition space Pure Joy is housed in a reused fuel-storage facility (Pure Joy 2019)

Image credit: Mark C. Childs

The composers of public spaces, however, are not simply observers of the stories of place, but also participants in the dialog. Innumerable artworks have been created about or set in New York's Central Park, Paris's Tuileries, Istanbul's Galata Bridge, and other urban landscapes. Moreover, at least since Palladio, designers have directly contributed to the narrative landscape with published visions and approaches to making built forms (see Habraken and Teicher 2005). For groups such as Archigram and the Paper Architects, the narrative landscape was their primary medium of urban composition as little or none of their work was built (Sadler 2005). However, as Pawley (1998) argues of Archigram's work, it can provide a way of understanding the possibilities of the built environment

> that the world will, in time, come to regard with the same awe as is presently accorded to the prescience of Jules Verne, H.G. Wells, or the Marquis de Sade. Futile to complain (as many do), 'But they never build anything.' Verne never built the Nautilus …
>
> *(428)*

The current practices of pop-ups, temporary urbanism, architectural installations, and New Urbanist pattern books can be seen as a mixture of narrative and built practices.

Figure 40.4 The light filled atrium of the Bradbury Building, Los Angeles, CA. The design was inspired by the 1888 science fiction book *Looking Backwards 2000–1887* (Beaumont 2012). "I was in a vast hall of light received not alone from the windows on all sides, but from the dome, the point of which was a hundred feet above" (Bellamy 1888, 102)

Image credit: Julius Shulman, photographer 1980. Used by permission © J. Paul Getty Trust. Getty Research Institute, Los Angeles (2004.R.10)

Frequently, there is a virtuous cycle of inspiration between built forms and other arts. For example, the 1888 speculative fiction novel *Looking Backward 2000–1887* by Edward Bellemy inspired the formation of a network of Nationalist Clubs in the United States and fueled the rise of the People's Party which, in the 1892 national election, won the votes of six states for their presidential candidate, and elected a governor and two U.S. congressmen. The novel also inspired the design of the Bradbury Building in Los Angeles that, in turn, was used as a primary setting in the original *Blade Runner* (see Figure 40.4) (Beaumont 2012). These virtuous cycles are not limited to specific texts or works. U.S. frontier main streets were lined with "false front" buildings to emulate the image of a "proper" town. These "false front" main streets are now emblematic of the "wild west" (see Heath 1989). The concept of "false fronts" is a fundamental trope of the TV series *Westworld*.

Built form species are confluences of spatial form, building cultures, social practices, and a welter of other factors (Childs 2001). Conceiving of these species existing both as narrative forms (idea, pattern, stories, e.g. "courthouse square") and extant built forms (e.g. Portales, NM's courthouse square), provides a method to think about both built compositions and narrative practices as shaping the built environment (see Figure 40.5).

Urban composers should not only study and contribute to the narrative landscape but also edit that landscape. To shape the dialog about public spaces, urban composers could publish general audience reviews of movies, books, and other works that depict public spaces. Urban composers could curate collections of creative works about Barcelona's Ramblas or other storied public spaces. Urban composers could inform speculative fiction writers and other artists about the histories and theories of urban design and public space to enrich speculative visions (see Childs 2016).

There are dangers to a focus on narrative. The demands of presenting a "good story" may run counter to developing an authentic, messy, open, and vibrant public space. Strongly themed places crowd out a plurality of competing stories. A compelling and popular story may recast and overrun a place as, for a time, the story told in the TV series *Breaking Bad* did to locales in Albuquerque, New Mexico. Nevertheless, the shared narrative landscape is a critical context for specific projects and the conception of species of built forms.

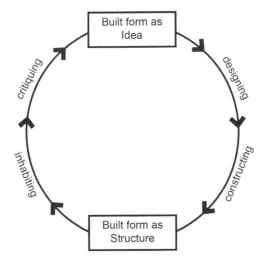

Figure 40.5 Cycle of built forms as ideas and structures
Image credit: Redrawn from Childs, 2001

Pluripotency

Scenarios suggest potential changes in use, disruptions, multiple conceptions of place, adaptations, and reuses that a built form may face. Signals in the built and narrative landscape can alert us to these emerging scenarios.

In 1914 the zoologist Valentin Hæcker coined "pluripotency" to refer to the ability of biological cells to give rise to many developmental possibilities (Hoßfeld, Watts, and Levit 2017). Borrowing the term, pluripotency refers to the ability of a "completed" built form to adapt to changing contexts. A pluripotent design anticipates informal uses, repairs, disruptions, additions, and adaptations. This anticipatory thinking is epitomized by the tale that New College in Oxford managed oak forests for 500 years in order to supply replacement beams for its original dining hall beams that were subject to damage from beetles (Hayer 2014).

Pluripotent design requires testing of proposed designs under each scenario and incorporating capacities for repair, multiple uses, and change. Designing parking lots for weekday traffic, weekend farmers' markets, and storm water detention is a means to provide for multiple futures. Anticipating that potential migrant communities will bring their public space practices with them may suggest plaza designs that can adapt to various uses by changing furnishings, providing for changing utilities, providing a range of sub-spaces, planning for converting adjacent parking spaces into pedestrian spaces or other techniques (see Brand 1994).

Coda

Test runs of sidewalk delivery robots, the virtual souk in *Valerian and the City of a Thousand Planets* (STX Entertainment TVOD 2017), and the artist/group Bored's installation of a Monopoly board on the sidewalks of Chicago (Zimmerman 2012) are signals of potentially emerging futures. The *Fearless Girl* statue by Kristen Visbal that in 2017 faced off *Charging Bull* (also known as the *Wall Street Bull*) may signal the emergence of dialogic or call-and-response artworks. The rise of co-working facilities complete with party and socializing spaces may be the beginning of "fourth spaces" (work-centered third spaces). Composers of public spaces should consider such signals of change, imagine scenarios that elaborate these trends, and help (re)compose pluripotent places that both support desired futures and are resilient in the face of other developments.

The tools of foresight can aid in considering alternative futures and incorporating the capacity to adapt to those futures. Cogent *framing* of the design brief can focus the process of research and design. Identifying multiple *layers* of the *drivers of change* acting on a project and site can help designers understand both how a site came to its present state and forces it may face in the future. *Scenarios* present informed narratives about sets of drivers that provide alternative futures in which design proposals can be evaluated, and *signals of change* alert us to the emergence of these futures. Fluency with a locale's *narrative landscape* can help frame a design brief, illuminate drivers of change, suggest scenarios, and help us design *pluripotent* public spaces.

Note

1 *Back to the Future* emphasized the life of the courthouse square in the future as well as its connection to the past. Signals are often a matter of emphasis and judgment. The stage set for the courthouse square in *Back to the Future* was originally built for the 1948 movie *An Act of Murder* and prior to *Back to the Future* it was used in the 1962 *To Kill a Mockingbird* and thirteen other movies and TV series, but these shows used the set as a historic setting (theStudioTour.com 2019).

References

ARUP Foresight. (2019a) "Drivers of Change." Accessed 3 March 2019. www.driversofchange.com/tools/doc/

ARUP Foresight. (2019b) "The Future of Urban Water." Accessed 19 June 2019. www.arup.com/perspectives/publications/research/section/the-future-of-urban-water?query=scenarios

Baudelaire, C. and Mayne, J. (1964) [1863] *The Painter of Modern Life: And Other Essays*. London: Phaidon.

Beaumont, M. (2012) *The Spectre of Utopia: Utopian and Science Fictions at the Fin De Siècle*. Bern: Peter Lang.

Bellamy, E. (1888) *Looking Backward, 2000–1887*. Boston: Ticknor.

Benatos. (2017) "Blade Runner 2049 – Chinatown Scene." Streamed Video. Accessed 21 March 2019. www.youtube.com/watch?v=opDlMeqRACI

Benjamin, W. (1999) *The Arcades Project*. Cambridge: Harvard University Press.

Brand, S. (1994) *How Buildings Learn: What Happens after They're Built*. New York: Viking.

Centre for Strategic Futures. (2019) "Centre for Strategic Futures." Accessed 19 June 2019. www.csf.gov.sg/

Childs, M.C. (2001) Civic Ecosystems. *Journal of Urban Design*, 6(1): 55–72.

Childs, M.C. (2012) *Urban Composition: Developing Community through Design*. New York: Princeton Architectural Press.

Childs, M.C. (2016) Composing Speculative Cities. *Analog: Science Fiction and Fact*, CXXXVI(4) April 2016: 30–36.

CNU (2019) "The Movement." *Congress for the New Urbanism*. Accessed 19 June 2019. www.cnu.org/who-we-are/movement

Dans, E. (2016) "Airbnb: A Case Study in Diffusion of Innovations." *Medium*, 29 February. Accessed 27 April 2019. https://medium.com/enrique-dans/airbnb-a-case-study-in-diffusion-of-innovations-99b22444f276

Dator, J. (2009) Alternative Futures at the Manoa School. *Journal of Future Studies*, 14(2): 1–18.

De Monchaux, N. (2017) The Death and Life of Gordon Matta-Clark. *AA Files*, 74: 183–199.

Ewing, R. (2002) Impediments to Context-Sensitive Main Street Design. *Transportation Quarterly*, 56(4): 51–64.

Fernandez-Guell, J., Collado-Lara, M., Guzman-Arana, S. and Fernandez-Anez, V. (2018) Incorporating a Systemic and Foresight Approach into Smart City Initiatives: The Case of Spanish Cities. *Journal of Urban Technology*, 23(3): 43–67.

Flink, J. (1990) *The Automobile Age*. Cambridge: MIT Press.

Goddard, J. and Tewdwr-Jones, M. (2014) A Future for Cities? Building New Methodologies and Systems for Urban Foresight. *Town Planning Revie*, 85(6): 773–794.

Government Office for Science (UK) (2016) *Future of Cities: Foresight for Cities*. London: Government Office for Science.

Habraken, N.J. and Teicher, J. (2005) *Palladio's Children: Essays on Everyday Environment and the Architect*. London: Taylor & Francis.

Harries, K. (1997) *The Ethical Function of Architecture*. Cambridge: MIT Press.

Hayer, C. (2014) "Oxford's Oak Beams, and Other Tales of Humans and Trees in Long-Term Partnership." *The Long Now Foundation*, 31 December. Accessed 27 April 2019. http://blog.longnow.org/02014/12/31/humans-and-trees-in-long-term-partnership/

Heath, K. (1989) False-Front Architecture on Montana's Urban Frontier. In *Perspectives in Vernacular Architecture III*, Carter, T and Hermann, B. (Eds.). Columbia: University of Missouri Press, pp. 199–213.

Hines, A. and Bishop, P. (2015) *Thinking about the Future*. Houston: Hinesight.

Hoßfeld, U., Watts, E., and Levit, G. (2017) Valentin Haecker (1864–1927) as a Pioneer of Phenogenetics: Building the Bridge between Genotype and Phenotype. *Epigenetics*, 12(4): 247–253.

Inayatullah, S. (ed) (2004) *The Causal Layered Analysis (CLA) Reader*. Taipei: Tamkang University Press.

Institute for the Future. (2018) "Signals." Accessed 4 April 2019. www.iftf.org/what-we-do/foresight-tools/signals/

Institute for the Future. (2019) "Scenarios." Accessed 23 April 2019. www.iftf.org/what-we-do/foresight-tools/scenarios/

Jacobs, J. (1961) *The Death and Life of Great American Cities*. New York: Random House.

Kayden, J.S. (2000) *Privately Owned Public Space: The New York City Experience.* New York: John Wiley.

Lesca, H. (2014) *Strategic Decisions and Weak Signals.* New York: John Wiley.

Lincoln Institute of Land Policy. (2019) "Consortium for Scenario Planning." Accessed 23 April 2019. www.scenarioplanning.io/

Littke, H. (2016) Revisiting the San Francisco Parklets Problematizing Publicness, Parks, and Transferability. *Urban Forestry & Urban Greening*, 15: 165–173.

Market Pike Place (2019) "History." *Pike Place Market Preservation and Development Authority.* Accessed 23 April 2019. www.pikeplacemarket.org/history

Nye, D.E. (1990) *Electrifying America : Social Meanings of a New Technology, 1880-1940.* Cambridge, Mass: MIT Press.

Oldenburg, R. (1999) *The Great Good Place: Cafés, Coffee Shops, Bookstores, Bars, Hair Salons, and Other Hangouts at the Heart of a Community.* New York: Marlowe.

PARK(ing) day. (2019) "About Us." Accessed 5 May 2019. https://parkingday.org/about-parking-day/

Pawley, M. (1998 [1975]) We Shall Not Bulldoze Westminster Abbey: Archigram and the Retreat from Technology. In *Oppositions Reader*, K.M. Hays (ed). New York: Princeton Architectural Press.

Pure Joy (2019) "Pure Joy." Accessed 19 June 2019. www.purejoymarfa.com/about

Ratcliffe, J. and Krawczyk, E. (2011) Imagineering City Futures: The Use of Prospective through Scenarios in Urban Planning. *Futures*, 43(7): 642–653.

Rodrigue, J. (2017) "Drivers of Change for Future Transportation." *Transportgeography.* Accessed 23 April 2019. https://transportgeography.org/?page_id=1675

Rogers, E.M. (2003) *Diffusion of Innovations.* New York: Free Press.

Royal Dutch Shell. (2019) "Shell Scenarios." Accessed 19 June 2019. www.shell.com/energy-and-innovation/the-energy-future/scenarios.html

Sadler, S. (2005) *Archigram: Architecture without Architecture.* Cambridge: MIT Press.

Schwartz, P. and Randall, D. (2007) Ahead of the Curve. In *Blindside: How to Anticipate Forcing Events and Wild Cards in Global Politics*, F. Fukuyama (Ed.). Washington, DC: Brookings Institution Press, pp. 93–108.

STX Entertainment TVOD (2017) "Valerian Preview." Streaming Video. Accessed 23 April 2019. www.youtube.com/watch?v=Z-yyuqMszPc

Swinney, P. and Thomas, E. (2015) "A Century of Cities." *Centre for Cities*, 4 March. Accessed 23 April 2019. www.centreforcities.org/reader/a-century-of-cities/1-a-century-of-city-performance/#box-2-the-drivers-of-change-in-cities-over-the-20th-century

theStudioTour.com. (2019) "Courthouse Square." Accessed 19 June 2019. www.thestudiotour.com/wp/studios/universal-studios-hollywood/backlot/current-backlot-sets/courthouse-square/

United States General Printing Office (US GPO) (2010) *Using Practical Design and Context Sensitive Solutions in Developing Surface Transportation Projects: Hearing before the Subcommittee on Highways and Transit of the Committee on Transportation and Infrastructure, House of Representatives, One Hundred Eleventh Congress, Second Session, June 10, 2010.* Washington: US GPO.

Urban Foresight. (2019) "Smart City Strategy for Scotland's Seven Cities." Accessed 19 June 2019. https://urbanforesight.org/projects/smart-strategy-for-scotlands-seven-cities/

Whyte, W.H. (1980) *The Social Life of Small Urban Spaces.* Washington, DC: Conservation Foundation.

Wood, G.S. (2009) *The Purpose of the Past: Reflections on the Uses of History.* New York: Penguin Books.

Wuellner, C. (2011) Beyond Economic and Value Wars: Mythic Images of Future Cities. *Futures*, 43(7): 662–672.

Zimmerman, J. (2012) "Street Artist Turns Chicago into a Monopoly Board." *Grist*, 6 July. Accessed 27 April 2019. https://grist.org/cities/street-artist-turns-chicago-into-a-monopoly-board/

41

THE IDEA OF THE URBAN COMMONS

Challenges of enclosure, encroachment, and exclusion[1]

Tridib Banerjee

Introduction

There is a medieval saying, "Stadtluft macht frei" – *city air makes one free* – that referred to the liberation of serfs, as they came to cities and were emancipated by the freedom of ideas, expressions, speech, and civic life. This transformative effect of the urban experience still endures. A recent empirical work in Japan tested the "city air" hypothesis and found increasing levels of independence in choice and self-expression with the size of the urban context (Yamagishi et al. 2012). While not specifically connected to the notion of the public realm that offers freedom of political action, as discussed by Arendt (1958), or the notion of public sphere as a civic buffer between the state and the individual, as argued by Habermas (1989), it is well documented that in liberal democracies the public realm is frequently associated with the idea of freedom: that is, freedom of expression, freedom of speech, and freedom of access and use of the spaces of public life (see Canovan 1985).

In fact, in times of political upheavals, we saw recently protest in the Arab streets and squares – Tahrir Square in Cairo most notably – as well as in the streets of the U.S. and other Western cities. The 2013 uprising in Istanbul's Taksim Square showed that opposition to the government plan to build a shopping mall and rebuild an Ottoman-era military barracks, was not just a protest against possible loss of public space, but an enactment of dissent against growing authoritarian power. The street demonstrations in Hong Kong (ongoing at the time of this writing) underscore why the public realm plays a critical role in staging the insurgency and democracy movements that oppose the specter of autocratic rule. In all these instances, public space – the more palpable experience of the more abstract notion of public sphere – remains a setting for political action, freedom of speech, and democratic engagement.

Political theorist Margaret Kohn (2004) has argued that the idea of public space in the U.S. is fundamentally linked to the First Amendment rights of freedom of speech and expression. The public spaces of the city essentially comprise the arena where the

Lefebvrean notion of the "right to the city" is often invoked; of course, it also serves as the setting for social contact, public life, rest, and recreation. Public health scholars also emphasize access to public space with the imperatives of healthy living (see Jackson and Ferdman 2019).

In the era of globalization, when the new order of a global economy and its neoliberal underpinnings are claiming public space as a good and can be "enclosed" for private benefit, what is the future of public space? This may mean a new form of encroachment (by claim or surveillance) upon the commons (property or resources), which may inexorably result in the exclusion of the "commoner" (that is, the public, the citizenry at large). This specter raises new challenges for urban designers and city planners in designing future cities. This chapter discusses the current trends in the way public space is used, abused, controlled, and managed in the context of rapid urbanization in a globalizing economy and explores ways to redress by design.

The "public" is regularly referenced and discussed in a variety of disciplines. Each imparts a different meaning onto the concept; however, there are clear commonalities. For example, in political theory the term is associated with citizenship, public sphere, democracy, and civil society. In social theory the idea is associated with the notion of conviviality, social life, social contact, sociability, civility, sense of community, and public life. In feminist theory the term stands in contradistinction to the family and the household, and the domestic privacy, especially that of women. In many cultures the public realm is typically dominated by male presence. Women's presence in the public is a form of intrusion, thus requiring women to cover up, as if to be invisible; the private realm, however, presumptively remains a domain of women. In the liberal economic theory, the concept pertains to the functions of the state and market, and in the collective consumption and production of goods and services as public or private process (see Banerjee and Loukaitou-Sideris 2013). Figures 41.1 and 41.2 represent public spaces that convey some of these alternative notions.

Public space as a public good is the most commonly accepted position in Western liberal democracies. What are the implications of this position? When subjected to economic theorizing, public good has been historically justified as a case of market failure, that is, a good that cannot be made available through normal market processes. Contemporary neoliberal economists, however, are challenging that position. They now argue that most goods and services provided by the public sector can be provided with equal or greater efficiency in the marketplace through the private sector (see Richardson and Gordon 1993). More precisely, Webster (2007) has argued, that "planners and designers would be freer to deliver sustainable urban spaces – in which institutional and physical designs are better aligned, [if] urban open spaces were to be depoliticized" (84). He contends that under the condition of overgrazing, it is difficult to meet the non-rival and non-excludability clauses of public good. Of course, it can be argued that the kind of institutional design Webster recommends might be efficient, but will it be just? The idea that public space can be commoditized, produced, and offered in the marketplace is no longer just an abstract academic argument advocated by neoliberal economists, as it is now increasingly pursued by local governments facing a growing deficit in the production of public space.

Accordingly, corporate office sites, shopping malls, and various mixed-use developments are creating ersatz "public" spaces. The problem with these "invented streets and reinvented places" is that they are really not public spaces. One's presence there is not by right, as in the case of a true public space, but only as a privilege granted by their corporate owners

Figure 41.1 Political protest, Naples
Image credit: Tridib Banerjee

(Banerjee 2001; Loukaitou-Sideris and Banerjee 1998). They have every right to exclude the common (the public) – the "undesirables" and "unwashed" – from these spaces according to their whim (cf. Lofland 1973). Users of such spaces may not exercise their right to free speech or engage in public discourse as normally permitted in public spaces. In other words, there is no chance for soapbox oratory or a bake sale for a public cause that would otherwise be permitted in a public space. In the context of shopping malls, the "publicness" notion has been contested and challenged in various state Supreme Courts, which have ruled for or against such causes, reflecting their conservative or liberal orientations. For example, most of these privately owned and operated public spaces (or POOPS, unfortunately) remain under constant surveillance of security guards or cameras and "disruptive" acts, such as systematic photography for study purposes will likely be challenged, even in liberal states like California or Minnesota. This author was escorted out of One Colorado in Pasadena, California in 1998 and The Mall of America in Bloomington, Minnesota in 2010 for taking "too many" pictures. Interestingly, this issue was not addressed by previous well-known studies of the office plazas of New York City by William H. Whyte (1988).

Then the following questions arise: Since the conventional argument about public space elicits the neoliberal preemptive response, does public space have to be seen as a public good? Are there alternative conceptualizations of public space that urban designers may use to advocate increasing the supply of public space and sustaining their social and political imperatives? This chapter explores the idea of the commons, i.e. resources under

Figure 41.2 Millennium Park, Chicago
Image credit: Tridib Banerjee

collective ownership that have remained a central construct in the field of resource management and environmental protection. According to neoliberal economic thinking, the commons becomes the residual realm that may exist once we aggregate and account for all properties under private or corporate ownership after initial entitlements are processed. This view applies particularly to real property. In the monarchic systems, for example, there may still remain significant amount of land under the ownership of the ruler – the "emir's land" or the "king's land" as common in United Arab Emirates or Saudi Arabia. In the heyday of communist regimes People's Republic of China, some 90% of the land belonged to the state, and thus implying a more expansive sense of the commons.

One could argue that public space is a common-pool resource, integral to the rights of citizenship. It should be noted that the notion of urban commons remains uncertain in literature, although there is a growing interest among many sources in addressing the idea. In introducing a recent symposium on urban commons, Dellenbaugh et al. (2015) argue that commons, conceptually, constitute three dimensions: (a) commons as resource; (b) commons as institution; and (c) commons as commoners or communities. While the literature is rich on the arguments concerning resources and institutions, most notably in the works of Hardin (1968) and Ostrom (1990), relatively little has been written about consumers and users (the commoners and the public) of such urban resources. Lefebvre's notion of the "right to the city" and its embellishment by Harvey (2012) have inspired urban scholars to revisit this argument. While much has been written on the former two aspects of the concept, the literature is relatively scant regarding the latter perspective.

There are many compelling arguments for the rhetoric of urban commons. We have briefly mentioned those of Lefebvre and Harvey. As quoted by Kohn (2011), Lessig (2002) adds that "the commons is a resource to which anyone within the relevant community has a right without obtaining the permission of anyone else" (19–20) While conceding these compelling features of the notion of the commons, Kohn remains reluctantly in favor of the term "public" instead. I contend here that the perspective of urban design arguments for public space as a "good" might lose against the neoliberal arguments summarized above. Instead, the notion of public space and realm, as an integral part of the larger construct of urban commons, should support the necessary arguments for sustaining public life and urbanism.

It is important to note that the concept of public space as the commons predated the concept of public good. When conceptualized in this manner, however, public space as part of the larger public sphere still remains somewhat problematic and open to further exploration. The argument that public space can be seen as a common-pool resource is immediately subject to three types of challenges that urban planners and designers may confront. These can be characterized as: (a) Enclosure of the Commons; (b) Obviation of the Commons; and (c) Surveillance of the Commons.

Enclosure of the commons

This refers to the notion of efficiency in the production of services and amenities normally present in the public realm, now made accessible through common interest developments, generally known as gated communities. Gated communities are no longer phenomena of Los Angeles or other U.S. cities, as they are appearing across the prospering Global South, China in particular. This form is also represented in the new urban expansion of many Indian cities; complexes such as South City on Prince Anwar Shah Road in Kolkata, or the many under construction in the Rajarhat expansion of the city, highlight this trend. These are examples of enclosure of the commons, a form of withdrawal from the public city for those who can afford it, leading to private (collective) production and exclusive consumption of what otherwise would have been considered public goods. Some observers have advocated the notion of "voluntary city," with similar aims of fragmenting the urban commons and civil society at large. In his award-winning essay, Webster (2003) argues that the rise of gated communities in China and elsewhere is efficient and a welcome response to the "overgrazing" problems of the urban commons. The problem with this outcome is that it further exacerbates the already overused and overtaxed residual urban commons without any institutional resolution of collective consumption. The tragedy of the commons is simply transferred to the excluded "commoner," who is dependent on streets and public spaces for their daily life and livelihood. Furthermore, the escapism of the gated lifestyle can only be partial. No one can totally escape the quotidian engagements with the city nor the externalities of agglomeration, segregation, and exclusion in the urban sensorium – air quality, for example – that affects all. Thus, the overgrazing of residual urban commons leads to an urbanism of congestion, clutter, and confusion, if not outright public health hazards. Consider, for example, proliferation of homeless encampments on the sidewalks, underpasses, and other residual urban spaces in inner-city Los Angeles. It should be noted, parenthetically, that gated communities are no longer a welcomed outcome in China, as the Chinese government is about to prohibit such new construction. A recent study has explored the possibility of making those gated neighborhoods "permeable" and thus obtaining some welfare gains in the form of walkability and perhaps improved propinquity (Sun, Webster, and Chiaradia 2017) (see Figure 41.3).

Enclosure of the commons also raises the question of whether it involves a form of encroachment into the commons or a form of appropriation. And if this is the case, how can that be justified in the name of collective welfare and wellbeing? Is this outcome Pareto superior, or even Pareto optimal? It is not clear whether such argument, while entertained in the abstract, can be effectively pursued in the spatial context without a full understanding of the property rights regimes in different societies.

Historically, Kolkata has been permissive in allowing the "overgrazing" of urban commons. This has been, arguably, to promote the local informal economy or the conviviality and welfare of the street public, as Figures 41.4 and 41.5 indicate. Cars parked overnight on public sidewalks certainly do not serve any larger welfare functions and suggest lax or corrupt enforcement of city codes and that car ownership has simply outpaced the supply of parking spaces. It also suggests how income inequality manifests in the claims on the public realm, which is supposed to be egalitarian in concept.

The seasonal ephemera and its "urban flux" often lead to episodic encroachment on the commons (Hack 2011). During the annual Durga Puja festival in Kolkata, which lasts almost a week, vehicular traffic is temporarily suspended in many neighborhoods (no doubt causing inevitable parking problems and traffic congestion) and the street is yielded to pedestrians. Open or partially covered arcades are constructed with bamboo framings and decorated with colorful festoons and lighting, lined with temporary food stalls, and other entertainment settings. These streets and tributary alleys house various interpretive images of Goddess Durga, also serving as a venerable tableau of Hindu religious mythology. With the growing prosperity of the Indian economy, these celebratory simulacrascapes in the public realm have become more expensive and exuberant, creating multiple

Figure 41.3 Gated community, Rajarhat, Kolkata
Image credit: Tridib Banerjee

representations throughout the city symptomatic of what Brendan Gill (1991) calls "Disneyitis." Nevertheless, these annual encroachments on the urban commons host a citywide fair, not unlike the Mardi Gras festivals in Rio or New Orleans. Celebratory and imbued with a spirit of convivial public life, despite the hassles of limited parking and pervasive traffic congestion, transformation of the commons in these instances serves to redress obviation of the commons. Ironically, agoraphobic advertisements in newspapers suggest enjoying the Durga Puja by staying home and experiencing events through media coverage.

Figure 41.4 Ephemeral urbanism – encroachment of the urban commons
Image credit: Tridib Banerjee

Obviation of the commons

Stemming from the rise of information and communication technology, and the fundamental changes in social behavior and public life, obviation of the commons leads to decline, not only in our use of public space, but also in our budget of "public time" and social contact that we can afford. It is generally known that average American spends about four and a half hours of daily waking hours watching television. Social critics like Robert Putnam (2000) argue that television watching – thus staying indoors – has decreased social contact in public space, and public life more generally, leading to declining social capital. It can be argued that a fundamental transformation in time budget is underway, as more people are connecting to the internet and using smartphones and other networked devices. The growth of cell phone access and use in India has been impressive; more Indians now have more access to cell phones than toilets. The internet affords opportunities for shopping, socializing, and entertainment, all from the comfort of the air-conditioned spaces of gated communities, enabling the hot, crowded, and polluted residual public realm to be avoided. Aided and abetted by technological innovations, a withdrawal to a world of *cybercivitas* and *cybercommunitas* may presage what Alexander (1967) referred to as "autonomy – withdrawal syndrome" and pervasive agoraphobia. The public realm of cyberspace may continue to obviate the existential realm of the public space from our collective life (Apostol, Antoniades, and Banerjee 2013).

Surveillance of the commons

Surveillance of the commons refers to the emergence of modern nation states and public bureaucracies, and their "seeing" needs (Scott 1998). The modern state derives authority from scientific and technological advances that allow for organization, order, and efficiency in administration. By developing systems to collect and store information (cadastral mapping, census, tax rolls, aerial photos, surveys, land use data, etc.) and requiring traceable social identities (social security cards, driver's license numbers, *aadhar* or biometric ID cards being implemented in India, etc.), the state creates a propensity for and the growth of specialized bureaucratic institutions.

Advances in CCTV and GPS technology have led to a phenomenon of what Cuff (2003) calls "pervasive computing." The public realm, or the urban commons, is under the control of surveillance cameras and other tracking devices that can monitor our individual

Figure 41.5 Informing video surveillance
Image credit: Tridib Banerjee

or collective activities in public settings. This is quite commonplace in the United Kingdom. Used extensively for traffic management, these cameras offer a 24-hour panoptic gaze that can track any activities desired by the state. In the wake of 9/11 such surveillance measures have been introduced at a growing pace in U.S. cities, as well as elsewhere in Europe as the sign at the entrance to a public facility in Copenhagen testifies (see Figure 41.5).

Poverty of the commons and the commons of poverty

Enclosure of the commons involves selling exclusive use rights of part of the commons to individual owners and their gated communities. The proliferation of these spaces in the recently prospering Global South is an example of the expanding enclosure of the commons. Inevitably, this leads to a shrinkage of the residual commons, exacerbating overgrazing and the encroachment of the commons. Examples of this overgrazing are shown in Figure 41.6 by an informal economy in this instance. It is emblematic of the growing inequality between the Global South and the Global North, albeit with different dynamics of change. It is noteworthy also that with new economic growth and the prosperity of the new middle class, car ownership has increased dramatically causing new demand for additional carriageway and parking. As a result the residual commons is farther encroached upon privileging the auto-owning class, depriving the pedestrians even minimal walkable space.

Concluding observations

The images and narratives from the Global North and South attempt to illustrate some of the critical arguments about the future of public space and the associated challenges for urban planning and design. I proposed that the aim of this paper was to initiate a broader discussion of public space, not as a public good, but within a more expansive conception of the public realm and the urban commons, and I have articulated arguments for public space in the context of neoliberal positions, globalization, and their attendant economic order of rising inequality. To conclude, let us address the basic question: What should the planners and urban designers do to confront these challenges? Here are some possibilities:

First, urban designers and planners must engage in a strong advocacy for the urban commons, not as a territorially residual concept, but as a rich assemblage of history, culture, and the urban form of the city. In this view urban commons becomes the principal "protagonist" in urban design.

Second, urban designers must not get co-opted in thinking of public space as a public good. They must consider public space as an essential component of the urban commons, and as said before, the principal protagonist in city design.

Third, urban designers and planners should avoid, and in fact resist, creation of what Trevor Boddy (1992) calls "analogous spaces" – privatized spaces of enclosed commons that exclude the ordinary public.

Fourth, they should connect their design and planning ideas to the concepts of freedom, and liberty – fundamental citizenship rights.

Fifth, they should reclaim the urban commons lost to corporate interests and auto circulation. The final two images from New York City underscore this latter point. Figure 41.7 shows the effort to create new pedestrian and public spaces of social contact and conviviality in Times Square. Figure 41.8 shows public amenities along The High Line, a new linear park created by transforming an abandoned raised railroad track. Kolkata's Millennium Park

Figure 41.6 Competing demands on the urban commons – hawker stalls occupying sidewalk space, Kolkata

Image credit: Tridib Banerjee

Figure 41.7 The new pedestrian Times Square, New York
Image credit: Tridib Banerjee

Figure 41.8 High Line linear park, New York
Image credit: Tridib Banerjee

offers a poignant parallel, where old industrial sites along the Hooghly River were replaced with gardens and amusement rides.

Finally, urban designers should explore ways to afford and facilitate privacy rights in the new surveillance regime of public space.

Note

1 An earlier version of this paper was presented at a conference in Kolkata, India, entitled "Changing Uses of the City" organized by the Centre for Built Environment in October 2012.

References

Alexander, C. (1967) The City as a Mechanism for Sustaining Human Contact. In: *Environment for Man*, Ewald, W. (ed.). Bloomington: Indiana University Press.

Apostol, I., Antoniades, P. and Banerjee, T. (2013) Flânerie between Net and Place: Promises and Possibilities for Participation in Planning. *Journal of Planning Education and Research*, 33(1): 20–33.

Arendt, H. (1958) *The Human Condition*. Chicago: University of Chicago Press.

Banerjee, T. (2001) The Future of Public Space: Beyond Invented Streets and Reinvented Places. *Journal of the American Planning Association*, 67(1): 9–24.

Banerjee, T. and Loukaitou-Sideris, A. (2013) Suspicion, Surveillance, and Safety: A New Imperative for Public Space? In: *Policy, Planning, and People: Promoting Justice in Urban Development*, Carmon, N. and Feinstein, S.S.(eds.). Philadelphia: University of Pennsylvania Press, pp. 337–355.

Boddy, T. (1992) Underground and Overhead: Building the Analogous City. In: *Variations on a Theme Park: The New American City and the End of Public Space*, Sorkin, M. (ed.). New York: Hill and Wang, pp. 123–153.

Canovan, M. (1985) Politics as Culture: Hannah Arendt and the Public Realm. *History of Political Thought*, VI(3) (Winter): 617–642.

Cuff, D. (2003) Immanent Domain: Pervasive Computing and the Public Realm. *Journal of Architectural Education*, 57(1): 43–50.

Dellenbaugh, M., Kip, M., Bieniok, M., Müler, A.K. and Schwegmann, M. (eds.). (2015) *Urban Commons: Moving beyond State and Market*. Basel: Birkhäuser.

Gill, B. (1991) "Disneyitis." *The New Yorker*, April 29.

Habermas, J. (1989) *The Structural Transformation of the Public Sphere: An Inquiry into a Category of Bourgeois Society* (trans. T. Burger and F. Lawrence). Cambridge: MIT Press.

Hack, G. (2011) Urban Flux. In: *Companion to Urban Design*, Banerjee, T. and Loukaitou-Sideris, A. (eds.). New York: Routledge, pp. 432–445.

Hardin, G. (1968) The Tragedy of the Commons. *Science*, 162(3859): 1243–1248.

Harvey, D. (2012) *Rebel Cities: From the Right to the City to Urban Revolution*. London: Verso.

Jackson, R. and Ferdman, R. (2019) Refocusing Planning and Design to Maximize Public Health Benefits. In: *The New Companion to Urban Design*, Banerjee, T. and Loukaitou-Sideris, A. (eds.). New York: Routledge, pp. 425–435.

Kohn, M. (2004) *Brave New Neighborhoods: The Privatization of Public Space*. New York: Routledge.

Kohn, M. (2011) Political Theory and Urban Design. In: *Companion to Urban Design*, Banerjee, T. and Loukaitou-Sideris, A. (eds.). New York: Routledge, pp. 175–185.

Lessig, L. (2002) *The Future of Ideas: The Fate of the Commons in a Connected World*. New York: Vintage.

Lofland, L.H. (1973) *A World of Strangers: Order and Action in Urban Public Space*. New York: Basic Books.

Loukaitou-Sideris, A. and T. Banerjee (1998) *Urban Design Downtown: Poetics and Politics of Form*. Berkeley: University of California Press.

Ostrom, E. (1990) *Governing the Commons: The Evolution of Institutions for Collective Action*. New York: Cambridge University Press.

Putnam, R. (2000) *Bowling Alone: The Collapse and Revival of American Community*. New York: Simon and Schuster.

Richardson, H.W. and Gordon, P. (1993) Market Planning-Oximoron and Common Sense? *Journal of the American Planning Association*, 59(3): 347–349.

Scott, J.C. (1998) *Seeing Like a State: How Certain Schemes to Improve the Human Condition Have Failed.* New Haven, CT: Yale University Press.

Sun, G., Webster, C. and Chiaradia, A. (2017) Ungating the City: A Permeability Perspective. *Urban Studies*, 55(12): 2586–2602.

Webster, C. (2003) The Nature of the Neighborhood. *Urban Studies*, 40(13): 2591–2612.

Webster, C. (2007) Property Rights, Public Space and Urban Design, Urban Planning. *Town Planning Review*, 78(1): 81–101.

Whyte, W.H. (1988) *City: Rediscovering the Center.* New York: Doubleday.

Yamagishi, T., Hashimoto, H., Li, Y. and Schug, J. (2012) Stadtluft Macht Frei [City Air Brings Freedom]. *Journal of Cross-Cultural Psychology*, 43(1): 38–45.

EPILOGUE

Danilo Palazzo and Vikas Mehta

Public space is in continuous evolution. Its occupancy fluctuates from emptiness to crowdedness, sometimes in the arc of the same day. Its publicness is fluid too. Publicness is at risk when reactions to public protests end up with restraints, control, and surveillance. It is at risk when the private sector regulates admission and use of the public realm limiting its functioning and its access—via tickets or through other means that create a perceived sense of exclusion and unbelonging. It is at risk when behavior is monitored through Artificial Intelligence and face-recognition, leading to a slippery slope that can end up in authoritarianism and Orwellian dystopias. Publicness is, in other cases, strengthened by public actions that shift established paradigms.

In the time spent on the production of this *Companion*, we have witnessed rapid changes in the public commons. When we started, the sidewalks of big- and medium-sized cities were invaded by an amorphous and fluid conglomerate of colorful bikes cheaply rented to improve individual urban mobility. A few months later, it was the turn of electric scooters to randomly occupy (and still occupy) the right of way when parked or create a risk for pedestrians when bolting on the sidewalk—a space that is increasingly becoming more and more hybrid. In the following months, we saw cities completely banning scooters or to try to accommodate these new mobility devices and deal with the issue through experimental creative agreements with providers. Today wheeled robots are delivering products, prudently avoiding pedestrians in public space. Autonomous vehicles—based on conceivable scenarios (Meyboom 2018)—might open new opportunities for the production, improvement, or increase of public space due to the anticipated decline of parking spaces and vehicle lanes. Other civic spaces, particularly libraries, are evolving and adding new uses and meanings to their traditional functions, becoming places to meet, spaces for performances, and for making, but also a beacon for those who are marginalized. For public space, a relevant question is: What is the next thing?[1]

In the *Companion to Public Space* we have curated the work of more than 50 authors who originate, live and work in, or investigate, and thus represent a global perspective. We searched for diversity in gender, culture, and education, also looking for variety in geography, scholarship, and experience. We would like to believe that the five parts combine the subjects and interpretations of the authors conceived in the original proposal to the publisher. This is—luckily—only partially true. While the original design and structure of the

Companion has been respected, literature research, personal acquaintances, and third-person connections have enhanced the range of authors and topics we initially envisioned. However, some voices that we invited or topics that we were interested in exploring are not represented here, the result of the over-committed academic or the absence of academic research around some topics.

This *Companion* has taken more than 27 months from the first draft of the proposal to the moment we delivered the 41 chapters to the publisher. In this length of time, we, the editors, decided to commit at least three hours per week in a common physical space, where we have spent our time together at the same table reading, writing, editing, and commenting on the numerous drafts of the chapters that now conjoin this *Companion*. The office that we have occupied always had a door open. Colleagues put their heads in to inquire about academic issues or share mundane topics. These disruptions provided moments of pause and caused intermissions in the flow of reading, writing, and editing, while sometimes suggesting new viewpoints to what we were doing. We also shared a virtual space—somewhere in the "cloud" where we stored the digital files—but we both agreed that personal encounters in a common physical space were the more conducive way to reach the goal of providing a balanced arrangement of voices, interpretations, points of view, angles, and explorations on public space to the readers of this *Companion*.

"Common physical space" and "we" are the key-concepts here. The common space and the interruptions stand for a metaphor of public space where people gather, coalesce to collaborate, to produce, to play, to spend time alone among others, to discuss and take positions, to share ideas and to disagree, to be disrupted, and to be inspired. "We" is another fundamental component in this mix and it's not just due to the undisputed advantage of collaborations and sharing responsibilities versus flying solo. Together we embody three continents, where we have been educated, where we have worked, or where we had and have personal, professional, and academic experiences with public space and its design, theories, and practice. Our backgrounds, the commonalities of readings and references, and the differences of our entry-points and interpretations to the various topics explored by the *Companion*'s authors, have provided the fuel for this long but stimulating endeavor.

Note

1 Unfortunately we are all living the next thing. The COVID-19 pandemic has dramatically and quickly changed the relation between private space and public space, where currently the former is safe while the latter is potentially dangerous and hence to avoid.

We still haven't processed personally, much less scholarly, the impact that this global event has or will have on public space. It will take time to rebuild the world as we know it. Public spaces, common grounds of the universal 'right to the city', might become the loci of the reconstruction.

Reference

Meyboom, A. (2018) *Driverless Urban Futures: A Speculative Atlas for Autonomous Vehicles*. Boca Raton: Routledge.

INDEX

#blacklivesmatter 74, 80–81
#metoo movement 5, 80
#refugeeswelcome 30, 33
9th World Urban Forum (WUF9) 90–91
"20's Plenty for us" campaign 131
30 Rock Building 312
2012 Olympic Games (London) 127–28, 381

access 8, 11–13, 31, 40–41, 67, 80, 106–7, 110, 139–40, 206, 234–35, 237, 239–40, 242, 247, 337–38, 435–36, 442–44, 449–50, 453; access/control 234, 237; access waterways 441; to activities 139–40, 242; differential 449; digital 82, 480; equitable 393, 447; exclusive 439; good 283; handicapped 293; to nature 443; open 269, 441; privileged 434, 439; psychological 308; restricting 422, 453–54; scarce 367; unequal 328; unrestricted 442; and use of public spaces 60
accessible design, visual impairment: ramps 110, 113, 288, 464; staircase navigation 113; tactile paving 110, 113; tactile wayfinding 116
accessible facilities 86, 106–10, 113, 115, 117, 119, 132, 286, 288
accessible spaces 18, 38, 42, 138, 207, 242, 303, 314
accessible transportation systems 465
Accra, Ghana 369–71
ACCRA [dot] ALT 369–70
actions: common 433; governmental 479; space-environment prioritizing 399; spontaneous citizen 335
activating public space 417
activism 29, 76, 322, 341; environmental 64; political 342; pro-democracy 73

activities: common 53; conflicting 460–61; cultural 188, 222, 230, 447; ground-level 185; group 214, 222, 422; large-group 290; lingering 411; nighttime 183; productive 223, 227, 229, 394; small-group 289; sociocultural 378
activity infrastructure 416
Actor-Network Theory 206, 267
actors 36–39, 41, 206, 264, 267, 269, 272–74, 278–80, 333, 339, 409, 411; commercial 481; community-oriented 223; diverse 206–7; intangible 274; invisible 41; local 102; multiple 400; private policy 447; private sector 448; social 43; unexpected 39, 42; unintended 42
acts: autonomous 350; deliberate 329; disruptive 501; extraordinary 301; innovative 361; internal 194; organized 394; political 43, 346; small ordinary 353; transitory 325; unfolding 349; unlawful 416; walk-by 418
adaptations 3, 320, 336, 338, 411, 428, 462, 487, 496
adaptive changes, small 463
admission 162, 240, 513
adopting presentism 359
advertising rights, selling 132
advocacy work 87
Advocates for Privately Owned Public Space (New York City) 211
Afghanistan 88
Africa 81, 86, 90, 159–60, 367, 370; southern 375; sub-Saharan 86; west 370
African Americans 320–21, 327, 329
African and Turkish neighborhoods 137
African cities (generally) 81, 367–68, 371, 375–76
African cultures and spiritualism 155
African languages 159

African origin/descent 155, 159, 164
Africans, enslaved 157
African-Swahili Kikwajuni 157
Afro-Swahili 158, 161
Afro-Trinidadian 158, 160
Age-Friendly City Elements 283
agency: central planning 464; city's parks 454;
    local 223, 409; personal 27; political 394
Agenda 2030 (United Nations) 85, 87, 89
aggravated assault 99–100
aging population 486
agriculture 88, 154–55; community-supported
    290; seasonal 434
Airbnb 21–23
Airfield Berlin Tempelhof 406, 408–9, 411
Aker Brygge neighborhood (Oslo) 224, 227–29
aksorn ซอย (Thai) 194
Alabama Cinema (Cape Town) 371
Albuquerque, New Mexico 326, 495
algorithms 387, 468, 472–73
alienability 439
allocation 291, 447, 451
alternating narratives 358–59, 363
alter-politics 346, 351, 353, 414; spatial
    occupation 353
Amazon Mechanical Turk 471
Amazon Web Services *see* AWS
ambiguities (theory and praxis) 43, 206, 232, 234,
    237, 247
amenities 140–42, 149, 187, 189, 211, 213, 284,
    293, 483, 503; commercial 140, 142–43;
    communal 142–43
American cities (generally) 470, 472
American Civil War 320–21
American Planning Association (APA) 282
Americans with Disabilities Act (ADA) 288
Amour, Salmin 163
Amsterdam, The Netherlands 22–23, 138,
    140–43, 145, 148
analyses of current drivers of change 489
analysis: causal layered 428, 487, 489; content 65;
    historical 213, 441; statistical 470
ancient Greek 401
ancient Greek civilization 191
animating 126, 415–17, 419, 421–22
animation practices 414–16, 422
animation strategies 414–17, 419, 421–22;
    activating 414, 416; aestheticizing 183, 414,
    419; convivializing 414, 420; culinizing 414,
    417; defensive 422; eventifying 414, 418;
    festivalizing 414, 418; gamifying 414, 420;
    naturalizing 414, 416; possibilities and pitfalls
    421; spectaclizing 414, 418; Pop-Up Cinema
    418; whimsicalizing 414, 420
Anpo (US–Japan Security Treaty) 197
Anpo protests 197
anticipated development 329

anti-democratic 438
anti-politics 346, 353
APA *see* American Planning Association
apartheid 80, 322, 372, 374–75; Truth and
    Reconciliation Committee 322
Application Programming Interfaces (API) 470–71
appropriations 35–36, 38–44, 138, 140–41, 197,
    199, 234–35, 254–57, 259, 262–63, 265, 315,
    318, 390, 392; emergent 43; ephemeral 401;
    individual 442; informal 399; insurgent 336;
    linguistic 192; people's 211; temporary 250;
    un-programmed 403; of urban spaces 41
Arab Spring 5, 335–36, 341
arcades 239–40, 242; quasi-public 247
Arche Metropolis (art project) 408
Archigram 399, 493; Living City exhibition 399
architects 7, 37, 196–97, 314, 316, 367, 371, 403,
    410, 486, 493
architectural, characteristics 361
architectural awards 251
architectural issues 458
architectural research 371, 457
architecture 2–3, 56, 91, 164, 364, 371–72,
    376, 429–31, 458, 474, 486; characteristic
    137; colonial 371; conceptualized temporary
    399; hostile 60; memorable 256; object-based
    316; physical 366; schools 458; temporary
    outdoor 403
areas: affluent 129; concealed 287; designated 288;
    heritage 184; historic 394; multifunctional 87;
    outdoor 402; residential 94, 110, 154, 196;
    urbanized 350
arena, cultural 391
Arendt, Hannah 7, 71, 139, 309, 338, 366, 390,
    435, 499
arguments 8, 13, 61, 194, 197, 429–30, 440, 442,
    444, 501, 503–4; controversial 223; political
    443; rights-based 441, 443
Aristotle 191–92, 432–35
Arlt, Peter 277
arrangements: social 271; unequal 13
arrival neighborhoods (immigrants) 334, 378, 384
art-based protest performance 369
Art Deco 171
artificial horizon 54
artificial intelligence 513
artificial landforms 50
artists 269, 271–75, 321, 327, 367, 369, 371, 375,
    419, 489, 492, 495
art projects 408; contemporary 375; temporary 277
arts 2, 255–56, 276, 278, 369–71, 375, 419, 428,
    431, 458–59, 491, 495; ban street 395;
    commissioned 421; digital 375; intersection of
    303, 371; plastic 46; public-facing 334, 366–75;
    small-scale 409
arts community 492
arts festivals 415

arts organizations 280
artworks 254–55, 299, 487–88, 490
Arup Foresight (consulting firm) 488–90
Asia 81, 86, 176, 249, 316, 417
Asociación Madres de Plaza de Mayo 361
asylum seekers 27, 29–32
Athenian agora 76, 80
atmosphere 10, 273, 276–78, 366, 417,
    420–21, 479
atrocities 323
attention: attracting 9; casting 185; inadequate
    126; increasing 267; scholarly 350; shared 41
austerity 72, 121–27, 129, 131, 133, 277
austerity politics 129
austerity urbanism 121
Austria 351, 354
authoritarianism 71, 73, 169, 499, 513; Khmer
    Rouge 179; Nazi atrocities 324; North Korea
    79; Pinochet (Chile) 322–23; state violence 74,
    347; suppression; Uyghurs (China) 179; Wukan
    (China) 480
authoritarian regimes, surveillance and data
    collection 74
authoritarian state-land relationship 79
authorities 9, 167, 179, 187, 196, 326, 333, 339,
    421, 444, 450–51; decision-making 451; legal
    441; monarchical 441; municipal health 103
automobile invasion 174, 176, 178
automobiles 48, 176, 178, 297, 458, 465, 489, 491
autonomous vehicles 513
avenues 65, 76, 195, 297
avoidance 61, 286–88, 292, 395
awnings 255, 270–71, 462
AWS (Amazon Web Services) 470
Aziza binti Salim al-Harthi 155

backstreets 142
Bad Ly (Berlin public art project) 277
Bahasa Seteng ("half-language") 179
Baidu (Chinese technology company) 481
Bandung, Indonesia 171, 177
Bangkok, Thailand 170, 194
Barbadian migrants 159
Barcelona, Spain 334, 338, 381, 386
Barcelona Declaration on Public Space 92
The Barcelona Model 381
Barri Xino (Barcelona) 381
basketball 28, 31, 51
Bastille Day 429
Battersea Park (London) 132
Battery Park (New York City) 450
Baudelaire, Charles 222, 491
Bauman, Zygmunt 364
bazaar 21
beach 60, 74, 239, 269, 272–73, 276–79
Beauvoir Bridge (Paris) 306

behavior 39–40, 64, 71, 167–68, 184, 188, 222,
    227, 232, 400, 406, 467–69, 471–74; anti-social
    182, 184, 187; changing 472; eccentric 232;
    fluid 469; helping 226; managing crowd 189;
    monitoring 1; people's 40, 168; social 222, 506;
    uncivilized/undesirable 186; user 57, 148
behavioral sciences 458
Beijing, China 72, 107–10, 112–13, 115, 119,
    360, 362, 479
Beit al Ajaib (House of Wonders) (Zanzibar) 161
Bellemy, Edward 495
Belmont (Port of Spain) 155, 158–59
Belo Horizonte, Brazil 391
benches 28, 60, 199, 262, 292–93, 299, 313, 471
Beninois 159
Benjamin, Walter 367, 491
Bentway 206, 296, 298–304, 306–8
Bentway Conservancy (Toronto) 296, 302–3
Bentway Project 206, 299, 301–4, 306, 308
Bercy neighborhood (Paris) 306
Berlin, Germany 30, 33, 56, 86, 270–75, 277, 324,
    399–400, 406
Berlin Tempelhof 406
Berlin Wall 269
Bethnal Green (London) 131
Better Finance, Better World (UN report) 88
Better Life Index (OECD) 90
Better Market Street Project (San Francisco) 409
bibliometric networks 94, 103
bicycles 100, 199, 201, 306, 483
BIDs *see* Business Improvement Districts
Big Bang *see* London stock market
big data 428, 467–71, 473–74; approaches 468–72,
    474; datasets 468; diaries 469, 472; employing
    468; feature extraction 471; features 469;
    geolocation 472; machine learning 468, 471;
    measures 469; methods 467, 469; models 473;
    noisy 474; personal identifiable information
    (PII) 473–74; physical activity 468–70; Place
    Pulse (website) 471; prediction 468; research
    428, 470, 472; sampling bias 469; sensors
    467–72, 482; sources 468, 470; Subway Desert
    map 474; techniques 474; use in public space
    research 468–69, 471, 473
Bight of Benin (Guinea) 159
billboards 20, 366
Bing, Xu 482
biometric ID cards 507
BJP *see* Indian People's Party
Black Forest 431
blackness in urban spaces 80
blindness 108, 374, 434
Bloomberg, Michael 404
Bloomsbury, (London) 127
blue-collar workers 12
blurred public boundaries 326
bodies, human 73, 75, 78–79

Bogotá, Colombia 396
Bohigas, Oriol 381, 385
Bois de Boulogne (Paris) 296
Bold St *see* Ropewalks
bonding spaces 378–87
borders 74–75, 157, 162, 264–65
Bored, (artist/group) 496
'Boris Bike' cycle rental (London) 132
Boris Johnson 122–24, 129, 132
Borough Lane Market (London) 129
Boston, United States 63, 98, 470–71, 488
botanical gardens 49
boulevardier 170
Bourbon Street (New Orleans) 488
bourgeois conception 166, 434
Bourgois, Leon 443
Bow River (Calgary) 306
Bradbury Building (Los Angeles) 494–95
Brazil 91, 378, 390–91, 395–96
Breaking Bad (TV series) 495
bridges (infrastructure) 133, 297, 463
Bridges neighborhood (Calgary) 306
bridging and bonding (social) 379, 381, 383, 385, 387
bridging spaces (physical) 381–84, 386
Bristol Urban Beach (Bristol, UK) 273
British colonial regimes 72, 152, 155, 157–58, 161, 163, 171; Caribbean holdings 164
British common law 440–41
Broadway (New York City) 404
Broadway pedestrianization project (New York City 411
Bronx River Park (New York City) 450
Brooke Street Market (Durban) 373
Brooklyn, New York 59
Brooklyn Bridge Park (New York City) 438
Bryant Park (New York City) 404, 449, 452
Bryant Park Restoration Corporation (BPRC) 448–49, 454
Bryggetorget (Harbor Square) (Oslo) 228
Buddhism 168, 179, 185
Buenos Aires, Argentina 361–62, 390, 395–96
buildings: abandoned 100, 418; accessible 237; complex 432; cookie-cutter 490; demolishing 226; demolitions 270, 275; enclosed 459; fabric 316; façades 226, 260; facilities 108; flying 277; high-rise 301; historic 255, 258; private 429; projections 256; residential 110, 210; sustainable 316; unsustainable 317; vernacular 179
building styles 247
building types 310
built environment 7–8, 39–40, 74, 77, 79, 81, 183–84, 378, 380, 419, 471–72, 488–89, 493, 495
Burgemeester de Vlugtlaan (Amsterdam) 143, 146–48

businesses 11, 62, 107, 140, 142, 146–47, 183, 210, 297, 480, 483
Business Improvement Districts (BIDs) 60, 62, 174, 210, 447, 453
Business Leaders Dialogue (UN-Habitat) 88
Busquets, Joan 381
Bute (Scottish Island) 32
Butler, Judith 309
buy-outs 435

cafés 234, 459, 462, 471
Cairns St *see* Ropewalks
Cairo, Egypt 211, 338–39, 499
Calgary, Canada 306
California (United States) 501
Calypso (music) 160
Cambodia 165, 168–69, 178–79
campaign for public space 8, 11
Canada 80, 206, 301–2, 325, 419
Canada Broadcasting Corporation 325
Cañada Real Galiana (Madrid 86
Canary Wharf (London) 121, 316
Canoe Landing Park (Toronto) 303
capacity, productive 8
capacity building 87
Cape Town, South Africa 371, 373
capital 71, 166, 183, 274, 278, 414, 433, 435; historic 304; human 89–90
capitalism 74, 183, 347–50
Caracas, Venezuela 5
Carafa, Giovanni 237–38
Caribbean islands 159
Carlos V, Holy Roman Emperor 335
Carnival celebration 154–55, 158, 160, 163, 186, 222, 225, 434
Carrer de Joaquín Costa (Barcelona) 383, 386
Carrer d'en Robador (Barcelona) 382
Carrer de Sant Pau (Barcelona) 384
Cartesian urban form 310
Casa de Caritat (Barcelona) 382
case studies 17, 20, 72, 103, 205–6, 224, 250, 257, 387, 399–401, 411; comparative 107, 109
Castilian cities' revolt 335
Catholicism 179; Catholic Church 168; pilgrimages 274
Cato Manor *see* uMkhumbane
causal layered analysis (CLA) 428, 487, 489
CBC *see* Canada Broadcasting Corporation
CCTV cameras, private use 250
celebration 67, 138, 155, 158, 161, 225, 359, 380, 396, 489; annual neighborhood events 290; Durga Puja 505; "Pink Dot" (LGBTQ) 78
"celebration street" 185
Central Business District (CBD) 175
centrality 42, 73, 76, 80, 149, 301, 338
Central Park (New York City) 48–49, 77, 157, 163, 296, 301–2, 317, 447–50, 452–53, 493

Central Park Conservancy 301, 448, 450, 452–53
Central Park Experience 301
Centre for the Future of Places (Stockholm) 458, 465, 511
Centre for Urban and Community Studies (University of Toronto) 304
Chain of Lakes (Minneapolis) 307
"Chair-bombing" 420
Chale Wote Street Art Festival 369–71, 375–76
Chaoyang (Beijing) 109
Chapultepec Park (Mexico City) 296
characteristics: environmental 338; locational 223; social 141–42, 149, 294; socio-cultural 294; street-pattern 100; structural 457
Charging Bull (Wall Street Bull) 496
Charleston, United States 320
Cheesegrater *see* Leadenhall Street Tower
Chelsea Barracks (London) 125
Chemnitz, Germany 324
Chicago, United States 496, 502
children 13, 67, 88–89, 99, 147, 196, 225, 273, 297, 304, 367
Chile 322
China 9, 77, 79–80, 107–9, 119, 126, 168, 362, 473, 478–84, 503–4; population mobility 479; public space 478; urban 478
China Association for the Blind 108, 119
China Disabled Persons' Federation 119
Chinatown (Singapore) 178, 184, 381
Chindos (Chinese-Indonesians) 179
Chinese: cities 428, 478, 482–83; citizenry 480–83; culture and ethnicity 66, 109, 179, 184, 478, 481; dynasty 179; government 108, 362, 481, 504; internet use 480, 482; workers 342; youth 479
Chinese New Year celebrations 290
Christian: churches 381; missionaries 179, 194
Christianity 169
churches (generally) 79, 237, 239–40, 242, 247, 280, 288
Churchill, Winston 121
Ciclovia (Bogotá) 419
cinema 371–72
cities: 21st-century 6, 237; adopted 479; authoritarian 77; beautiful 463; best 460; better mixing 410, 465; bomb-ravaged post-war 121; brand 137; caring 64; central 76, 242, 284, 447–48; colonial 163; constructing 90; contemporary 1, 49, 54, 182–84, 189, 234, 318, 333, 346, 396; crowded 25; deindustrialized 9; dense 109, 306, 381; digital 82; diverse 379; equitable 375; ever-heterogenizing 6; expanding 49; feudal 196; flourishing 250; from fragmented to integrated 7; global 62, 80, 129; good 64; great 306, 308, 466; growing 11; historic 489; human-scaled 46; inclusive 72, 87, 137–38, 447; industrial 12, 175; informational

75; inner 149, 174–75, 373; just 59; large 12, 23, 103, 267; legitimate 86; lived 192, 350; low-density 125; major 121; making 10; material 387; mature 191; medium-sized 513; modern 53–55, 138–39; neo-liberal 16; new 27, 31; nocturnal 183; permanent 399–400; physical 480; planned 306; post-colonial 158; post-industrial 205; public 504; safest 185; science of 458; seaside 429; service 12; single 133; small 164; smart 74, 473; southern 368; sprawling 47, 466; sustainable 184; tent 65; traditional 165, 192; unequal 366, 371; united 303; vibrant 465; voluntary 504; walkable 470; wealthy 89
citizen authors 482
citizen engagement 400
citizen initiatives 272
citizen interaction 36
citizen journalism 482
citizenry 2, 5, 500
citizens 1–2, 88, 170, 195–97, 225–26, 265, 333–34, 359, 399, 409, 438–40, 443–44, 473–74, 480, 482–83; active 317; civic 443; ordinary 182–83, 196; quashed 480; sentient 482
citizenship 165, 167, 359, 367, 500, 503
"citizens" plazas *see* shimin hiroba
city authorities 137
city beaches 267–71, 273–80; artificiality 206; concept 267, 269–70, 272–73, 277; Costa Hamburgo (Hamburg) 269; décor 276; environment 270–71, 273, 276; nature 267; palm trees 267, 270–71; sand 267, 270–71, 274–75, 302; thatched huts 267, 270–71; Elbiza (Dresden) 269; event 277; free wireless internet 276; Galeria Kaufhof (Stuttgart) 270; Minimumverzehr ("minimum purchase") 277; private-sector 272–73; programming 275; public space 278; relationships 272; Skybeach (Stuttgart) 269, 273; Skybeach model 274; Strandbar Mitte (Berlin) 275; Strandleben (Vaihingen) 280; urban brownfields 277
city building 128
city centers 11, 19, 29, 42, 152, 171, 175, 304, 417; historic 137, 333
city councils 318, 327, 439, 449, 454
City Describer (Twitter account) 471
city design 13, 508
city designers 37
city environments 102
city governments 212, 459, 464
city life 222, 290, 369, 375, 415; contemporary 369; inclusive 18
city management 394
city networks 465
City of Sanctuary (UK) 32
city parks 31, 296, 444, 449
City Place (Toronto) 303

city planning *see* urban planning
City Prosperity Index (CPI) 90
City Space Architecture 87
city squares 17
city streets 326, 471
Ciutat Vella (Barcelona) 378, 382
civic activities by citizens 225
civic boosterism 183
civic buffer 499
civic education groupings 163
civic generosity 301
civic identity 71, 86
civic inculcation 77
civic innovators 410
civic interaction 7, 13, 49
civic life 368, 499
civic muscle 299
civic participation 91, 379, 409
civic rooms 487; American Courthouse Square
    487; Italian piazza and piazzetta 487; Main
    Street 487; Tewa pueblo plazas 487
civic space 5–7, 71, 77, 86, 88–89, 91, 214,
    224–25, 299, 301–2, 358–59, 409–10, 443–44,
    487, 499
civic spaces 53, 336, 513; great 301; planned 81
civil actions 363
civil inattention 40
civility 38, 40, 62, 71, 139–41, 149, 192, 254, 500
Civil Rights Movement (United States) 327, 336
civil society 71, 89–90, 296, 308, 390–91, 500,
    504; democratic 433; engaged 1
civil society activists 12
civil society groups 340
civitas 192
CLA *see* causal layered analysis
Clark, Gordon Matta 491
class conflict 28–29, 39, 42, 80, 167, 177, 179,
    367, 434
Classical Antiquity 169
classification 3, 205, 209, 212, 221, 223–25,
    228–29, 231; applied 205–6, 218
classification system 212–14, 221, 223–24, 228–32
Clean and Green Cities Programme
    (Afghanistan) 88
Cleveland, Horace 307
clienteles 149, 272–73, 275
Clos, Joan 465
Closed-circuit television (CCTV) 97, 256,
    262, 507
CNU *see* Congress for New Urbanism (CNU)
Coal Line Urban Park (London) 131
co-design 92
coding spaces for function 214
co-ethnic workforce 384
collaborations 33, 88, 371, 375, 395, 399, 404,
    408–9, 457–58, 514
collective actions 38

collective approach 72, 121
collective enjoyment of public 444
collective spaces 36, 417
collectives 42–43, 179, 336, 353, 391
Cologne, Germany 277, 327
colonial plans 158
colonial power 391
colonial protectorate (Zanzibar) 157, 161
Colorado (United States) 501
color contrast 113, 293
co-management 308
comfort 40, 62, 270, 283, 285, 292, 418, 506;
    psychological 292
comfortable movement 293
comfortable seating 292
commercialization 8, 16, 211, 418, 438, 450
commodification 137–38, 394, 427, 443
commodified spaces 137
commodity 137, 185, 199, 432–33, 443
common causal processes 102
"common good" 396
common ground 46, 57, 297, 299, 301, 303–4,
    307–8
the commons 77, 188, 206, 248, 296, 308, 366,
    396, 440–41, 499–501, 503–8
commons: enclosed 508; residual 508; shared 308;
    slippery 247
common spaces 8, 170, 433, 442, 514; creating 11;
    green 284; metatopical 433
Communist Party of China (CPC) 362, 448, 450,
    453–54, 503
communist regimes 168–69
communities: adjoining 199; cohesive 379;
    cultural 302; imagined 323, 358; marginalized
    449; neighboring 403; new 387; small 142;
    suburban 387; tight knit and homogeneous 379
community action 272, 278, 326–27
community building 32
community centers 116, 210, 288, 303, 383–84,
    417; neighborhood-based 385; senior services
    284, 288; youth drop-in 422
community empowerment 131
community engagement workshops 406
community gardens 210, 383, 416
community groups 131, 163, 320, 327, 385, 418;
    local 272; small not-for-profit 249
Community Re-Forestation Project (Fondes
    Amandes) 160
community services 108
complexity, contemporary 85
complex public processes 323
composing public spaces 486
comprehensive scenario development 490
computational knowledge 473
concept: contested 444; crucial 35; social 107;
    utopian 311

concert halls 137
concerts 225, 392, 418, 421, 451, 469, 473
Concert Square (London) 262
conditions: contextual 278; contingent 390;
  controlled 43; cultural 333; economic 71, 269;
  ephemeral 401; geographic 71; inadequate 89;
  inevitable new urban 312; insufficient 40; living
  348; multi-layered 307; natural 433; new
  environmental 199; new social 195; physical
  196; political 165, 391; social urban 401;
  topographic 50
conflicts 16–17, 36, 38, 40, 65–66, 86, 170–71,
  267, 269, 321, 333, 339–40, 460–61; creative
  36; inter-group 167; potential 188, 460–61;
  unresolvable 43; violent 169, 182
confluences 487, 495
Confucianism 179; nei-wai 168
Confucian tradition 168
Congolese diaspora 80
Congress for New Urbanism (CNU) 492
connectivity 54, 87, 90, 122, 139, 194, 460–61
consensus 85, 87, 168, 191, 320, 435
consequences 36–37, 39, 192, 250, 338, 392,
  486; environmental 11; unintended 35, 39,
  42–43, 338
conservation 184, 202
Consortium for Scenario Planning (Lincoln
  Institute of Land Institute) 488
consumerism 10, 179; contemporary 19
consumerist values 178
consumers 197, 448, 480, 482, 489, 503
consumption 9–10, 12–13, 178, 182–83, 185, 188,
  211, 214, 226, 232, 236; spaces 138, 214, 448
contemporary homogenization of public space 173
contemporary neoliberal economic thought 500
contemporary public space 1, 76, 81, 206, 210,
  230, 236, 390
context: critical 495; diverse 73; historic 269; local
  75, 490; modern 106; multiple 488; non-
  Western 77; physical-spatial 480; policy
  122–23; political 336; socio-historic 350;
  southern 369
control 154–55, 187–88, 199, 206, 211, 234–37,
  239, 247–48, 254, 277, 285–86, 349–50,
  450–51, 459, 461–62, 482–83; citizen 202;
  corporate 59; crowd 342, 453; democratic 441;
  direct 304; excessive 210; legal 267; military
  347; political 335; regimes of 40, 236, 481, 483;
  religious 166; technological 364
controlled access 53
controlled use 62
control paradigm 480
control privacy 460
Convention on the Rights of Persons with
  Disabilities (CRPD) 108
co-production 92, 271, 280
Corman, Roger 430

corporate plazas/parks 239, 500
corporate sponsorship 132, 277, 298; Barclays and
  Santander 132
Costa Rica 59–60
counter-publics 166, 347, 478
counter-spaces 395
counter-spectacle 418
CPC *see* Communist Party of China
creating diverse spaces 334, 379
creating diverse spaces of encounter 334, 379
creative energy 273–74
creative interventions 375
Creek Road (Zanzibar) 163
crime 13, 38, 60, 86, 94–100, 102–3, 177, 284,
  287, 381, 434
crime attractors 99
crime concentrations 95, 98–99
crime concentrations in public spaces 98
crime levels 94, 470
crisis 8, 121, 134, 297, 339, 349; fiscal 448, 450;
  good 121
critical designers 353
critical witnesses 481
critique 5, 7, 13, 61, 211, 235–37, 240, 316, 448,
  450–51, 453; normative 240, 247; orthodox
  Marxist 434; public space 9, 11, 13
Cuba 79
cultural associations 382
cultural backgrounds 106, 119, 137
cultural clashes 179
cultural contexts 109, 231
cultural continuities 160
cultural experiences 50, 183
cultural expressions 87, 122, 140
cultural factors 71
cultural features 137, 149
cultural heritage 353, 379
cultural nightlife experience, unique 182
cultural performances 186
cultural practices 32, 159, 175, 379, 393, 487
cultural preferences 11
cultural prejudice 199
cultural processes 3
cultural quarters 10
cultural refinement 193
cultural shift 492
cultural-spatial delineation 24, 459, 462, 492
cultural status 419
cultural systems 358
culture: café 122; civic 166; collectivistic 2;
  consumerist 368; contemporary 368; creative
  155; diverse 402; face-and-shame 167; high
  402; immigrant 138; improvised 228; local 165,
  171, 178; mainland 157; material 185; political
  342; post-colonial 163; tea 196; transnational
  33; world-dominating 194
Curitiba, Brazil 395

customs, cultural 379
Cuzco, Peru 395
cyber-commons 77
cyber-extension 77
cyberspace 5, 75, 77, 80, 82, 506
cycling 22, 25, 86, 141, 288, 297, 304, 307

dala's CityWalk (Durban) 369, 372, 374–75
Damon, Wilfred 372
Dandora, Nairobi 87
Dappermarket (Amsterdam) 146
data collection 67, 74, 119, 141, 467–71, 474, 481
Davis, Jefferson 327–29
Dayong, Zhao 482
Death and Life of Great American Cities 209, 465
de Blasio administration (New York) 454
Debord, Guy 236, 429, 433–34
de-centering 80
decentralization 468, 471
de Certeau, Michel 194, 341, 367
deck chair 274
Demnig, Gunter 324, 327
democracy 5–6, 71, 74, 77, 80, 169–70, 192, 195,
    334–35, 337, 339, 341–42, 346, 349, 362–63;
    contemporary 322, 337; radical 337, 352
democratic resilience 335, 337, 339, 341
demolition 99, 235, 251, 371–72, 375
densification challenges 125
design: permanent 334, 400; physical 214, 218,
    458, 500; planning and urban 184, 282, 349;
    pluripotent 496; site 265
design and planning 474, 508
design disciplines 2, 458
design elements 288, 293
designer coffee boom 491
designers 41, 43, 106–7, 116–19, 147, 149, 250,
    264–65, 294, 312–13, 378–80, 387, 408–9,
    411, 489–90
design exclusion 107, 109
design features 211–12, 214, 293
Design for London 122
design guidelines 206, 282, 284, 287, 294
designing for the publicness of spaces 41
designing parks for older adults 283, 285, 287,
    289, 293
design process 35, 37, 43, 116
design qualities 253, 263
design standard 125
design tools 265
Detroit, Michigan 81
developers 37, 90, 121, 124, 127, 130, 137, 146,
    309–10, 312, 315–16
development 7–8, 10, 12, 36, 38, 87–88, 107,
    109, 119, 121–22, 124–26, 273–74, 380–81,
    406, 408–9; sustainable 11, 85, 87, 90, 138, 369
development process 13, 36–37, 126
developments, skyscrapers 312, 315

development zones 406, 408
deviance 187, 232
Dhaka, Bangladesh 381
dialectical approach 35, 39, 41–42
dialogue, engaging in 2, 6, 39, 87–88, 92, 231,
    236, 337, 341, 367, 375–76
Die Vlakte (Stellenbosch) 372
digital communication 230
digital forums 2
digitally connected world 76, 387
digital networks 73, 483
digital public space, surveillance 481
digital technologies and social transformation 230,
    428, 478, 480–83
digital venues 478
dimensions, political 3, 187, 334, 390
discursive walking 22–23
Disneyitis 505
Disneyland 249
distributive justice 63, 66, 91
Divine Universal Scheme 434
Docklands (Melbourne) 240, 242
do-it-yourself landscape 271
do-it-yourself urbanism 212, 399, 411
Dongcheng (Beijing) 109
Dresden, Germany 269
drivers of change 428, 487, 489–92, 496
Dubai, United Arab Emirates 316
Durban, South Africa 369, 372–74
Düsseldorf, Germany 277

Early Morning Market (Durban) 373
East Berlin Park (Berlin) 275
East Dry River, Port of Spain 158
Eastern contexts 165
Eastern cultures 72, 165–66
East Side Gallery (Berlin) 269
East Village (Calgary) 306
eateries 145–46, 172, 185
ecological requirements 316
economic development 10, 12, 186, 368,
    418, 428
economic space 38
edges 28, 42, 51, 53–54, 65, 129, 255–56, 406,
    408, 444, 448
Egypt 342
Eisenman, Peter 324
Elbe River (Germany) 269
elements: natural 284–85, 291, 416; physical
    254–56, 409; vertical 255
el Raval: Barcelona 334, 378, 380–87; long-time
    residents 384; Pakistani women 386; Pakistani
    workers 384; public space and immigrant
    integration 378
emancipation 159, 327, 334, 347–50, 352–54;
    active 354; capacity of planning for 352;

political 347–49; social 347–49, 353; struggles in 347–49, 352, 355
emancipatory public spaces 74
Emanuel African Methodist Episcopal Church (Charleston) 320
Emerald Necklace (Boston) 307
Emperor Valley Zoo (Trinidad) 158
empirical evidence 140
enclosure 60, 74, 188, 253, 255, 258, 260, 431, 433, 503–4, 508; jersey barricades 60
enclosure of the commons 188, 503–4, 508
entertainment 166, 183, 188, 194, 224, 230, 263, 418, 506
entrances 65–66, 110, 113, 142, 262, 286, 462, 508
entrepreneurs 18, 86, 273, 275–77, 419
environment: external 22; natural 82, 159, 416, 427, 438; social 21, 94, 97, 133
environmental, water 30, 52, 291–92, 367, 416, 432
environmental conditions: breezes 291–92; glare 292–93; nature 30, 48, 99, 152, 163, 284–85, 291, 416, 427, 431, 442–44; shade 287, 290–92; sunlight 293, 443; weather 76, 271, 273–74, 278, 463
environmental sun 274–75, 278, 283, 287, 293, 302, 504
equality 29, 33, 41, 64, 165, 310, 321, 340, 348, 350, 427; political 347, 349; in public space 350
Equal Justice Initiative 329
equity 6, 41, 59, 63, 86, 90, 166, 170, 304, 348, 453–54
Ericsson (technology company) 89
The Esplanade (Fremantle) 54–55
ethical engagement 352
ethic of care 59, 64, 67, 193
European grid pattern 171
evaluating public space 6, 59, 61, 63, 65, 67
eventifying 414, 418–19, 447–48, 450–51, 453
eventscapes 447, 449
everyday 41, 128, 130, 248–49, 350, 352, 354, 367–68, 392, 395, 415, 417, 419, 433, 436
everyday acts 341
everyday experiences 20–21, 40, 349, 378, 392, 396
everyday life 57, 60–61, 137, 139, 187, 191, 196, 230, 237, 348–50, 352–55, 390–91, 393; non-movements 341
everyday micro-relations 390–91
everyday practices 35, 320, 375, 396; conflicts 396
everyday resistance 335, 341
everyday spaces 125, 367
everyday stories 372
everyday urbanism 41, 211
exclusion 12–13, 106–7, 109, 113, 115, 117, 119, 182, 184, 348–49, 393, 396, 499–500

exercise 192, 283, 290, 308, 312, 347, 395, 437, 443, 450, 501
exercise incentives 290, 361
experts 21, 346, 350
externalities 309, 311, 313, 504
eye level 260

Fabian, Errol 163
façades 260
fair allocation 66, 443
fake news 166, 169
"false front" main streets 495
farmers' markets 415, 417, 496
Fatahillah Square (Jakarta) 171
The Fearless Girl 496
features, natural 308, 416
federal land 439, 442, 444
Federation Square (Melbourne) 52
Fenchurch Street Tower 124, 309, 312–14
Ferrarotti, Franco 202
Festa Major del Raval 382
festivalizing 414, 418–19; staging 225, 499
Festival of Neighborhoods (Kitchener, Ontario) 419
festivals 370; Durga Puja 504; Mardi Gras 505
Filmoteca de Catalunya (Barcelona) 382
financers 37
Finland 283, 352
flâneur 170, 222, 307, 367, 491
fluidity of use 29, 32–33
Fondes Amandes (Port of Spain) 160
food 31–32, 65–66, 145, 147, 225, 271, 277, 293, 417
food banks 290
forecasting 428, 486–87
former British colonies 152, 184
Forodhani Gardens (Zanzibar) 161
Fort York Boulevard (Toronto) 299, 302
"found space" 211, 306
framework, public space policy 91
France 296
Francoism (Spain) 381
Freedom Hill (Tokyo) 195
freedom of speech 499
Freedom Park (Pretoria) 322
Freetown (Port of Spain) 159
Fundació Tot Raval (Barcelona) 385
futures: alternative 488, 496; of public space 3, 315, 427–37, 488–90, 496
The Future We Want (UN Resolution) 90

Gaiety Bioscope (Stellenbosch, ZA) 372
Galata Bridge (Istanbul) 493
games 290, 368, 417, 420, 436, 464, 480
Gandhi, Mahatma 336
Garden Bridge Trust (London) 124
gardening 31, 211, 226, 282, 287, 290

Gardiner Expressway (Toronto) 299,
301–3
Gasworks Park (Seattle) 50, 53
gated communities, "four-pack" dream 435
gated community: Rajarhat (Kolkata) 504–5;
South City (Kolkata) 503
"Gay Pride" event (Singapore) 78
Gehl, Jan 16, 18, 28, 40, 125, 209, 222–24,
227–28, 247, 249, 252, 400, 404, 406, 410–11,
470, 472
gender 28, 30, 32, 42, 137, 141–42, 167, 284,
348, 367, 468
Geographic Information Systems (GIS) 467
George C. King Bridge (Calgary) 306
German towns (generally) 324
Germany 324
Geuzenveld-Slotermeer neighborhood
(Amsterdam) 146
Ghana 371
GIS (Geographic Information Systems) 467
globalization 8–10, 12, 24, 72, 79, 166, 178, 230,
236, 500, 508; expressive architecture 10; high-
quality public spaces 10, 125; investment 9–10,
88, 134, 178, 296; small towns 488–90; tall
buildings 10
Global North 89, 92, 508
Global Platform for the Right to the City
Initiative 184
Global Positioning System (GPS) 132, 472
Global Public Space Programme (UN-Habitat) 85,
87–89
Global Public Space Toolkit (UN-Habitat) 89
Global South 80, 82, 85–86, 89, 103, 503, 508
Goddess of Democracy (Tiananmen) 362–63
Google Alphabet 473
Google Earth 383
Google Streetview 468, 470–72
Government House (Zanzibar) 152, 154
government officials 428
Governor's Island (New York City) 449
GPS (Global Positioning System) technology 132,
472, 507
Granby4Streets 251–52, 258–62, 264; community
land trust 252
Grand Phnom Penh International City,
Cambodia 178
Grand Rounds (Cleveland) 307
"Grands Boulevards" (Paris) 302
Grand Staircase-Escalante National Monument
(Utah) 438
Gray Area Foundation 409
Great Chain of Being 434
Greater BH Youth Forum (Belo Horizonte) 391
Green agora 71, 74, 76–77, 80, 170, 196, 433
Greenbelt Mall (Manila) 178

greenspace 10, 32, 40, 99, 283, 416; parkland 288,
438, 447, 451–52; trees 30, 158, 161–62,
255–56, 258, 260–61, 290–91, 293, 313
greenspace management sector 30–31
Greenwich Village (New York City) 381
Gropius, Walter 458
Guangzhou, China 481
guerilla urbanism 211, 256
Gunduz, Erdem 347

Habermas, Jürgen 7, 16, 73, 82, 165–67, 309, 366,
480, 499
Habitat II (United Nations) 91
Habitat III (United Nations) 71, 92, 459, 465
Hæcker, Valentin 496
The Hague Market (Netherlands) 18
Hall, Tony 163
Halprin, Lawrence 322
Hamburg, Germany 269
Hanoi, Vietnam 171
Happold, Buro 324
HBD flats (Singapore public housing) 79
health outcomes 94, 98–99, 102
Healthy Streets for London 123
Healthy Streets Indicators (London) 123
Heidegger, Martin 431–32
Helsinki, Finland 283, 352
Herald Square (New York City) 404
Heroes Monument Field (Surabaya) 173
Hester Street Playground 66
High Court (Zanzibar) 157
The High Line (New York) 298, 301, 306, 447,
449–50, 454, 508, 510
The Highline Network 298, 304
Hillier, Bill 309
Hinduism 167, 179, 185, 504; ethos 168; festival
(Deepavali) 185; Goddess Durga 504; minorities
168; temples 381
Hiroba 191, 196–97, 199, 202
Hirshberg, Peter 409
historical narration 358–60
history, political 155, 321, 371–72, 375–76
Hofburg (Vienna) 353
Holmens Street 229
homes 19, 22, 79, 82, 108, 115, 160, 163, 166,
230, 432, 435, 460–61, 465–66, 479–80; private
461, 480
Honduras 74
Hong Kong 5, 72–74, 76–77, 107, 109–10, 113,
115, 119, 346, 353, 483–84
Hong Lim Park (Singapore) 78–79
Hood, Raymond 312
Hoog Catharijne (Utrecht) 17, 20
Hooghly River (India) 511
Houston, Texas 74

Houston Street (New York City) 310
Hudson River (New York) 315
Hudson Street (New York City) 209
Hudson Yards 315
Hulchanski, David 304
Human Development Index (United Nations) 90
Human Rights Act (1998) 129
human settlements 2
Humber River (Toronto) 306
Hunan Province, China 179
Hungary 378
hutong (narrow alley) 110

IDEO 488
IDS Center (Minneapolis) 210
"I'm a City Changer" (UN-Habitat) 90
imageability 253, 255, 258
images 127, 137, 139–40, 147–48, 358–59,
    362–63, 462, 470–72, 474, 481–82, 487,
    489, 508
imaginaries 19, 375
imaginary of contemporary consumerism 19
immigrant: integration support 378–79, 387;
    reception 334, 378–80, 384, 387
immigrant amenities 138–42, 148–49; spatial
    characteristics of 141; visibility of 140–41
immigrant community 141; Asian, African, and
    Turkish neighborhoods 137; experience 378,
    380; groups 140, 146, 387; integration; el Raval
    334, 378–80, 387; positive contexts of reception
    in el Raval 378
immigrant neighborhoods, major western
    European cities 137
immigrants: Filipino; salons 383; teens 384;
    Pakistani 382–83; Pakistani shops 383–84;
    transforming neighborhoods 137, 139–41;
    Turkish 140, 145, 148–49; communal amenities
    138, 140–44, 146–49; shops & restaurants
    142–43, 145–49, 383; women 146
immigration 60, 62, 140–42, 149, 265, 334,
    378–81, 384–87; assimilationist 379; illegal 60;
    Immigration and Customs Enforcement (ICE)
    60, 299; multiculturalism 379
implementation (policy/strategy) 37, 62, 85, 88,
    90–91, 101, 113, 116–17, 119, 401, 411
implementation processes 37, 117, 119
inclusive design 107, 109, 119, 316; facilities 110,
    113, 116; public memorials 330; public spaces
    138, 148–49, 298, 304, 321
inclusiveness 1, 40, 62, 72, 86, 106–7, 109, 119,
    134, 138, 142
Independence Day 173
India 86; cities (generally) 503
Indian (ethnicity) 179, 185, 506
Indian People's Party (BJP) 168
indicators 87–88, 90, 139
Indische Buurt neighborhood (Amsterdam) 145

individuals experiencing homelessness 60, 65–67
Indochina peninsula 169
Indonesia 165, 167–69, 178–79
inequality 12–13, 61, 76, 86–87, 329, 348–49,
    367, 393, 396, 508
informal settlements 12, 86, 175–76, 372;
    uMkhumbane (Cato Manor) 372, 374–75
infrastructure 35, 43, 86, 88, 94, 127–28, 226,
    408, 417, 443, 487
innovation 10, 121–33, 272, 347, 421, 436, 465,
    491
Institutes of Justinian 440–42
"insurgent urbanism" 211, 347, 353
intention–outcome gap 36
interaction (social and physical) 17–18, 21, 46,
    289, 324, 379, 418–19, 421, 442–43, 473,
    490–91
internet 77, 82, 107, 128, 166, 169, 178, 225,
    478–80, 483, 506; access 74, 480–83; mesh
    networks 484
investment in public space 9–10, 85, 88, 128, 296,
    410, 448
Iran 27, 77
Islam 167–68, 179
Islamic: building codes 464; property inheritance
    155
Islamic Republic of Afghanistan 88
Israel 338, 342, 361
Istanbul, Turkey 5, 333, 338
Italy 296, 378, 442
Ithaca, New York 419

Jackson, Wyoming 488
Jacobs, Jane 39, 48, 94, 125, 209, 222, 231, 247,
    309, 368, 415, 465, 491
jahangeer, doung 374
Jakarta, Indonesia 169, 171, 338–39
Jamaica 27
James Town (Accra) 370–71
Japan 193–94, 196–97, 199, 201, 499; feudal
    period 193–94, 201
Japanese Americans 321
Japanese Belle Époque 195
Japanese politics of space 197
Japanese roji (garden) 194
Japanese ruling class 197
Japanese society 195
Japanese urban culture 201
Japanese urbanity 194
Javastraat (Amsterdam) 143, 145–46, 148
Jersey City, New Jersey 59
Jerusalem, Israel 74
Jiyūgaoka (Tokyo) 195–96, 199, 202
Jiyūgaoka Hiroba (Tokyo) 196–99
Johannesburg, South Africa 368
Johnson, Boris 122–24, 129, 132
Judd, Donald 492

Jullien, François 192
justice 1, 61, 63–64, 323, 329–30, 340, 348, 435;
    interactional 63–64, 66
justification, public space 10, 221, 439
Justinian code 440

Kafka, Franz 431–32
kanji 路地 (Japanese) 194
katakana 191, 201
Keynesian economics 8
Khan, Sadiq 123
King, Martin Luther, Jr. 321, 327
Kings Cross redevelopment (London) 126
King Street pilot project (Toronto) 404
Kishi, Nobusuke 197
Kisiwandui (Smallpox Island) (Zanzibar) 163
Kōichi 196
Koinonia 192
Kolkata, India 503–4, 511
KTH Royal Institute of Technology (Stockholm)
    458
Kuala Lumpur, Malaysia 171, 179
Kuhonbutsugawa Street (Tokyo) 199
Kulturstrand (Munich) 277, 280

La Bohème (movie) 372
Lagos, Nigeria 74
Lahouaiej-Bouhlel, Mohamed 429
land 10, 79, 251, 299, 301, 306, 308, 310, 312,
    367, 406, 438–43, 464, 503
landform 46–48, 50, 52–54
landowners 37, 394
landscape 22, 27, 43, 46, 53–54, 56, 154, 157–58,
    163, 302, 308, 416, 421; narrative 487, 492–93,
    495–96
landscape features 255
landscaping 256, 258, 269, 291, 293, 459
lanes, pedestrian alleyways 187, 194, 196
Laos 165, 169
Las Vegas, Nevada 430
Latin America 59, 390–91, 393–97
Latin American public space and cities 334, 391,
    393, 395–96
Latinos 66
Latour, Bruno 206, 267
Laventille (Port of Spain) 152, 155, 158–59
Leadenhall Street Tower (London) 309, 312–14
Lefebvre, Henri 28, 36, 39, 61, 72–73, 75–76, 80,
    187–88, 192, 194, 197, 199, 235, 263, 310,
    315, 318, 337, 347–49, 352, 367, 390, 395,
    422, 435, 503
legibility 40, 50, 54–55, 140, 142, 144, 149,
    253–54, 256, 286, 316
Leipzig, Germany 57
leisure 2, 6, 10, 38, 110, 122, 128–29, 187, 403,
    432–33, 436; mass 184, 188; nighttime 186, 189
Les Rambles (Barcelona) 382, 384, 495

Levitas, Jake 409
LGBTQ+ 74, 77–78, 80, 421
liberal democracies 337, 349, 499
libraries 76, 210, 384
Light Detection and Ranging (LIDAR)
    technology 467
Lincoln, Abraham 321
Line art walk (London) 131
linkages 90, 102, 253–54, 256, 259, 265, 297,
    346–47
Linnaeus, Carl 3
Little India (Singapore) 72, 182, 184–88
Little India Shopkeepers & Heritage Association
    (LISHA) 185
Liverpool ONE (Liverpool) 251–52, 258–60,
    262–63
Livingstone, Ken 122–23
London, United Kingdom 30–31, 33, 38,
    72, 74, 79–81, 121–34, 194, 207,
    213, 236, 248, 304, 309, 312, 314–16,
    462–63, 492
London Development Agency 122
London Olympics (2012) 122
London Plan 125; revised 127
London Stock Market 79
loose space 211
Lordstreet Theatre Company 163
Lorenz, Susanne 277
Los Angeles: California 62, 81, 206, 212, 284,
    494–95, 503–4; future 486
Lower East Side (New York City) 65
Lower Manhattan (New York City) 316
low-income residents 60
Lu, Qi 481
Luanda, Angola 74
Luang Prabang, Laos 171
Lynch, Kevin 39, 50–51, 235–36, 247, 286

Macapagal-Arroyo, Gloria 179
MACBA *see* Museu d'Art Contemporani de
    Barcelona
Madrid, Spain 5, 47, 86, 339, 342
main street 487, 490
maintenance 102
Makers on Market (San Francisco) 411
Malaysia 165
Malcom X 327
Mall of America (Bloomington, Minnesota) 501
management 8–9, 12–13, 38, 41, 62–63, 117, 130,
    132–33, 209–10, 213–14, 236, 254, 264, 267,
    272–73
management teams 117–19
Manhattan (New York City) 59, 65, 211, 309–10,
    312, 315, 435, 451
Manila, Philippines 177–78
Manitoba Province, Canada 325
mapping 94, 206, 234–37, 239–40, 247–48

maps 2, 21, 141, 235–37, 240, 242, 247, 269, 279, 286, 367
Marfa, Texas 492–93
marketing 140, 221, 225; "Iamsterdam" 140; place branding 371, 375, 403, 411; real-estate 140
market-oriented growth models 137
the marketplace 18, 201, 500
markets 8–10, 74, 146, 172, 205, 225, 236, 273, 358, 439, 443
Market Street (San Francisco) 409–11
Market Street Prototyping Festival (San Francisco) 409, 411
Marlene Dietrich Platz (Berlin) 56
Marquis de Sade 493
Marx, Karl 76, 348–49
materiality of public space 41, 54–55, 368, 375, 396
Matthews, Judy 299, 301
Matthews, Wil 299, 301
Maunula Democracy Project (Helsinki) 352
May Pyramid *see* memorialization, Mothers of the Plaza De Mayo
McEwen, David 304
Mecca, Saudi Arabia 74
media 41, 169, 187, 226, 273, 277, 315, 337, 339, 473, 482–83
mediating structures 460–62
Meguro, Japan 197, 199
Melbourne, Australia 52, 206, 234, 240, 242, 247, 296
memorialization 321, 363; informal, ghost bikes 325–26; Mothers of the Plaza De Mayo 361; official sanction (New York City) 326; Rabin's assassination 361; state sanctioned, Street Memorial Project (New York City) 326
memorials 205, 210, 320–27, 329–30, 479; grassroots 325; informal; Missing and Murdered Indigenous Women and Girls 325, 327; Selkirk Bridge (Canada) 325; public 320–21, 323, 326–27; state sanctioned; The Memorial to the Murdered Jews of Europe 324–25; The National Mall and Memorial Parks in Washington 321–22; "Peace Park" (Villa Grimaldi) 323; Sharpeville Massacre 329; Stumbling Blocks 324
Merdeka Square (Kuala Lumpur) 171
Mexico 86
Mexico City, Mexico 296, 394–95
Michael Braum & Partners 406–7
micro-actions 38, 40–41
micro public space 409
Microsoft (technology company) 89, 471
Mid-market neighborhood (San Francisco) 409
Midtown Manhattan (New York City) 448
Migombani Botanical Garden (Zanzibar) 155–57, 162–63
migrants 27, 29, 166, 380; new 28; newcomers 6, 28, 31–33, 138, 379–80, 386

migrant worker death, Indian 182
migrant workers 186–87, 479; South Asian 155, 157, 182, 185–86, 188; transient South Asian 72
migration, forced 28, 30
Miguilim (Belo Horizonte) 391
Millennium Park (Chicago) 296, 502
Millennium Park (Kolkata) 508
Minassian, Alek 429
Minecraft (video game) 89
Minneapolis, Minnesota 210, 307
Mnazimmoja (One Coconut Tree) Park (Tanzania) 155–58, 161–63
mobile phones 17, 472, 478–79, 481–82
mobility (individual) 22, 88, 100, 373, 395, 482
modernism 47–48, 183, 195, 236, 349, 429, 443, 489
monitoring, public behavior 97, 395, 404, 481–82
Montjuic (Barcelona) 381
monuments 76, 320–28, 330, 358, 362, 369, 432; American presidents 321–22, 327, 444, 480–81; black South African 322; capital cities 321; Civil Rights Movement 327; commemorative 321–22, 330; Confederate 321, 327; removal 327–28; counter-memorials 323, 330; George Washington 321; installation 327; Korean War 321; national 173, 322–23, 438, 444; National Memorial for Peace and Justice (Montgomery) 328; public 207, 320, 326; Santiago General National Cemetery 323; vernacular 323–24; Vietnam War 169, 321; Vimean Ekareach (Phnom Pehn) 173
Moore Street Market (New York City) 63
Morsi, Mohamed 342
Moses, Robert 65
movement activists 339, 342
Muang Thong Thani (Nonthaburi) 177
multi-dimensionality 35, 38
multi-scale 36
Mumford, Lewis 429
mundane 22, 32–33, 189, 228, 230, 458
Municipal Arts Society (New York City) 211
Museu d'Art Contemporani de Barcelona (MACBA) 382
Museums Quarter (Vienna) 401–3, 411
Muslims, Rohingya 179
Muslims, Thai-Malay 179
Mustafa Centre (Little India) 185
Muthi Market (Durban) 373
Mwaka Kogwa (Persian New Year) celebration (Tanzania) 160
Myanmar 165, 168–69, 179

Nagasaki, Japan 47
Nairobi, Kenya 47, 87
Naples, Italy 237–38, 501
narrative landscape 492, 496
National Park Service (NPS) 321

Neely, Sionne 369
neglect 86, 175–76, 178, 296, 416, 450–51
neighborhood level 141–42, 149, 342
neighborhood parks 48, 206, 282–83, 287, 294
neighborhoods 17–19, 21–22, 99–100, 108–10,
    116, 118, 141–42, 145–47, 158, 282–84,
    301–2, 371–72, 380–82, 384–86, 419–20
neoliberalism 6, 8, 12–13, 36–37, 57, 59–60, 62,
    71, 73, 240, 242, 394, 500–501, 503
neoliberal planning agenda 242
neoliberal public–private partnerships 91
neoliberal urbanism 59
The Netherlands 17–19, 22, 296
networks 77, 237, 242, 247, 304, 308, 374, 380,
    459, 461–62, 465–66, 482; actor 24; place
    461–65
network society 74, 80
New Humanist perspective 91
New Orleans 327–29, 505
New Urban Agenda (NUA) (United Nations) 71,
    85, 87–88, 90, 138, 221, 458
New Urbanist Movement 489
New York City 19, 59–60, 62–65, 207, 209,
    211–13, 296, 298, 301, 338–39, 404, 447–51,
    453–54; Community Parks Initiative 454;
    Department of City Planning 211–13;
    Department of Parks and Recreation 64, 67,
    449–54; Department of Transportation 404;
    Public Plaza Program 212, 404, 411; New York
    City Council, Committee on Parks and
    Recreation 450; public spaces 213–14, 217
New York City. Sanitation Department 326
New Yorkers for Parks 450, 453
Nezu (Tokyo) 193
Ng'ambo (Zanzibar) 155, 157, 161, 163
NGOs (non-governmental organizations) 115–16,
    118, 374
Nicaragua 390
Nice, France 429–30
Nicklin, Adam 301
nightlife 146, 148, 175, 184–85, 188–89; bars 102,
    210, 262, 488; entertainment 183, 188
nighttime economies 182–83
Nolli, Giambattista 237–38
Nolli map 237–38
non-governmental organizations *see* NGOs
non-rival and non-excludable 427, 430, 435, 500
non-spatial public space 12
North American cities 27, 74, 298, 417, 419, 421
North York (Toronto) 430
Norway 224, 431
NPS *see* National Park Service
NUA *see* New Urban Agenda

obviation of the commons 503, 505–6
occupancy levels 213–14, 218, 342
occupants 211, 317–18

Occupy London Stock Exchange 79
Occupy Movement 338–39
Occupy Wall Street 5, 19, 61, 64, 335–36, 339,
    342
Oceania 165
Ocean View (Cape Town) 372
OECD *see* Organization for Economic Cooper-
    ation and Development
Öffentlichkeit ("public sphere") 166
older adults 100–101, 206, 282–94; American
    parks 282; benefits of parks for 282–83; fall
    incidents 94, 101–2; open space preferences 30,
    282, 470; outdoor gardens 283; parks design,
    handrails 288; parks design 291; park use 30,
    282; pedestrians 100–101; reduced mobility 13;
    traffic injuries 100–102
Olmstead, Frederick Law 307, 443
Omani overlords (Zanzibar) 155
Omani Sultan's Palace (Zanzibar) 161
One Colorado (Pasadena, California) 501
online bulletin boards 479
Ontario, Canada 419
"open-private" space 237, 239–40, 242, 247
"open-public" space, internal 239, 313
open spaces 188, 197, 274, 278, 282–84, 296–97,
    301, 306, 311–12, 406, 408; flat 51; inclusive
    72, 107, 116
Open Streets Initiative 419
ordinary people, public space use 72, 91, 161, 163,
    197, 323, 341
Organization for Economic Cooperation and
    Development (OECD) 9, 12
organizations: grassroots 49; local 87, 212, 326
orientation, sense of 256, 285–86
Orwellian dystopias 513
Oslo 206, 224, 227–29, 231, 360
Oslo Accords 360
outdoor public furniture: deck chairs 267–68, 271,
    275; seating 107, 110, 112–13, 115, 211,
    214–15, 289, 291–93, 312, 420, 459; table 212,
    262, 264, 293; table and chairs 212, 232, 420
over-managed space 38, 173–74, 176
Overmeyer, Klaus 406

Pakistan 86
Paper Architects 493
Paradise St. (London) 257–60
Parc des Buttes-Chaumont (Paris) 48
Paris, France 5, 31, 48, 76, 86, 197, 235,
    302, 306, 338
Parisian banlieues 379
Parisian boulevard 487
Paris Plage 267, 269, 275, 302
park conservancies 447–51, 453
parking (garage and street) 146, 212, 240, 242,
    269–70, 276, 416, 420, 488, 490, 504, 508, 513
parking design 496

parklets 212, 409, 420, 492; PARK(ing) Day 212, 409, 416, 491–92; *Parked Bench* 492
Park Rabet (Leipzig) 57
parks: colonial 155, 157, 159, 161, 163; eventization of 450, 452, 454; municipal 161, 430; neighborhood 31; new 294, 306; pocket 59, 65, 212, 251, 262; private 239–40, 247
parks design 282
parks equity 448–50
park visitors 53, 65–67, 285, 287, 291–92
Parliament Square (London) 122
participatory engagement 21, 280, 303, 352, 419
partnerships, public–private 60, 212, 239, 250–51, 262, 296, 298, 304, 449
Pasadena, California 501
Paternoster Square (London) 79, 236
Patuxai Park (Vientiane) 173–74
pavements 183, 188, 212, 256, 258, 324, 330, 334, 366–76, 420
Pavement to Parks Program (San Francisco) 212
Peace Gardens (Sheffield) 29
Peace Memorial (Zanzibar) 161–62
Peace Memorial Museum (Zanzibar) 157
Peckham (London) 131
pedestrianism 100–101, 125, 128, 256, 258, 260, 404, 418, 421, 430, 504, 508, 513
pedestrian shopping street 60, 269
pedestrian spaces 239, 496
Pelourinho, Salvador 396
Penang, Malaysia 171
people-centric solutions 189
People's Republic of China (PRC) 362
periphery 48, 73, 76, 309, 312, 315–16, 338
permeability 54, 57, 140, 436
personal mobility: cars 11, 89, 94, 174, 236, 239, 262, 373, 416–17, 419, 429; wheelchairs 13, 288–89, 292–93, 322
Perth, Australia 47
Peru 395
Petticoat Lane Market (London) 129
Philadelphia, Pennsylvania 59
philanthropy 88, 301, 304, 308
Philippines 42, 165, 168–69
Phnom Penh, Cambodia 173
physical accessibility 37, 42, 139–40, 288
physical activity 94, 98–99, 283–85, 287, 290, 416–17, 468–70; initiatives; Playborhood 417; Play Street Movement 417
physical boundaries 207, 312, 401, 411
physical spaces/places 3, 5, 19, 36, 40–43, 165–66, 338–39, 403–4, 439, 442, 479–80, 483
Piazza del Campo (Italy) 52, 249
piazzas 195–97, 199, 201, 213–14, 249
Pike Place Market (Seattle) 489
PIM (policy-implementation-management) model 117–18

"Pink Dot" (Singapore) *see* LGBTQ+ community
pixo 392–94
Plaça de Castella (Barcelona) 384
Plaça del Pedro (Barcelona) 383
Plaça dels Àngels (Barcelona) 382, 384–85
Plaça de Pere Coromines (Barcelona) 384
Plaça de Salvador Segui (Barcelona) 382
Plaça de Terenci Moix (Barcelona) 384
place-based approaches 16–17, 21, 99
Place de la Bataille de Stalingrad (Paris) 31
placemaking 133, 182, 184, 278, 414
planetary public spaces 74, 79, 81–82
planned community, Bumi Serpong Damai (Tangerang) 178
planners 2, 37, 41, 61, 67, 137, 206, 250, 352, 448–49, 508
planning (act of) 35, 37, 43–44, 138–39, 183–84, 188, 250, 263, 267, 348–49, 454
planning, ecological 48
planning authorities/departments 183, 358, 399, 409
planning for individuals with disabilities 88–89, 108–10, 116, 119, 322
Plataforma de Afectados por la Hipoteca (Madrid) 353
Plato 192, 434
Plaza Catalunya (Barcelona) 338
Plaza del Sol (Madrid) 347
Plaza de Mayo (Buenos Aires) 361–62
plazas 59–61, 63, 65, 74, 80, 195–96, 205, 209, 333, 335, 490; corporate (private) 174, 237, 240
Pokémon GO (video game) 420
Poland 21, 327
policy-implementation-management (PIM) 117
policy makers 61, 72, 85, 90, 116, 118–19, 444
Policy Reform and Social Responsibility Act (2011) 129
polis 191–92
Polish Bazaar (Slubice) 21
political: actions 42; arrests 323; occupation 376; resistance 336–37, 390; system 172, 337
political demonstrations 129, 225, 228, 299, 337–38, 353, 362
political economies 37
politics 13, 43, 64, 71, 73, 77, 80–81, 192, 350, 369, 376; democratic 338, 353, 435–36; emancipatory 350
politics of public space 138
politics of public story 320
political demonstrations, Latin America 390
POOPS *see* Privately Owned and Operated Public Spaces (POOPS)
Porchfest (Ithaca) 419
Porto Alegre, Brazil 91
Port of Spain, Trinidad 152, 154–55, 158–60
ports 239, 269, 442

post-9/11 (New York City) 60
post-Fordist 271, 278
post-politics 346
potentials of public space 29, 35, 42, 342, 487
Potomac River (Washington DC) 321
Potsdamer Platz (Berlin) 56
PPAG Architects 403
PPPs (public–private Partnerships) 91, 212, 262, 265, 296, 298, 428, 447–51, 453–54
PPPs: governance 447, 449–51; paradigm 450, 454; political arrangements 447–49, 451, 454
PPS *see* Project for Public Spaces
practitioners 2, 35, 316, 342, 386, 393–94, 417, 419, 428, 474
Pretoria, South Africa 322
Prince Anwar Shah Road (Kolkata) 503
privacy 9, 193, 239, 285, 287, 290–91, 427–28, 461–62, 473
privacy in public spaces 290–91
private actors 251, 428, 439, 447
private agencies 8–9
private developers 9–10, 126, 178, 264–65
private funds 450–51, 454
Privately Owned and Operated Public Spaces (POOPS) 501
Privately Owned Public Spaces (POPS) 60, 207, 211, 309; "bonus spaces" 211, 213
private partners 178, 428, 448–50, 453–54
private spaces 37, 79, 234, 236–37, 247, 364, 368, 434, 442, 461–62, 465
private spheres 8–9, 12, 168, 391
privatization 7, 9, 16, 174, 178, 211, 234–36, 247, 250, 438, 441, 444, 483; gated communities 61, 63, 174, 176, 209, 236–37, 239–40, 247, 296, 503–6, 508; gated parks 8, 53, 408, 439; New Public Management 9
privatized public space 37, 63
procedural justice 63, 66
processes 6–8, 16–17, 35–38, 41–43, 56–57, 66, 90–92, 118, 121, 206–7, 272, 335–37, 341–42, 346–47, 349, 364, 403, 407–9, 418–19, 421–22; constructive 56; political 161, 339, 342
process-oriented approaches 16–17, 24–25
process-oriented design and temporary spaces 399
process-oriented investigations 17
Programa Rescate (Mexico City) 394
Programme, The *see* Global Public Space Programme
programming 275, 278, 282, 284–85, 289, 298–99, 304, 403, 406, 410–11, 414–16, 447, 449
Project for Public Spaces (PPS) 223
Promenade des Anglais (Nice) 429
promenades 51, 199, 223, 307, 312, 333, 429

property 10, 199, 315, 318, 439–42, 444, 452, 461, 464, 500, 503; non-exclusive (res nullius) 440; public 197; vacant 256, 262
property market 10, 86
property rights 79, 439–40, 504
Prospect Park (New York City) 449–50
Prospect Park Alliance 450–51
prostitution 225, 381–82
prosumers 279–80
protestors 74, 187, 340, 346–47, 358, 360–61, 364, 479–80
protests 76, 78, 129, 339, 342, 346–47, 358, 360, 362–63, 392–93, 479, 484, 486, 499; military crackdown 362–63; Paternoster Square 79; political 179, 484, 501; Sana'a 340; Sharpeville 329; Tahrir Square 340; University of Tokyo 197; West Shinjuku Station (Tokyo) 197
PSI (Public Space Index) 62–63
public access 206, 237, 239–40, 242, 312, 427–28, 438, 444, 451
public accessibility 315, 408
public actions 321, 327, 376, 513
public aid activities 226, 228, 232
public amenities 312, 448, 451–52, 479, 508
public and private spaces 38, 79, 234, 240, 247, 249, 435, 459, 464–65, 482
public areas 90, 207
public art 214, 366, 368–69, 419; *Fake Estates* (New York City) 491; *Monopoly Board* (Chicago) 496; TIME FRAME 2010 375
public authorities 8–10, 12–13, 103, 126
public behavior 478, 481–82
public benefits 206, 240
public budgets 12–13, 448
public buildings 132, 320
public bureaucracies 272, 507
public character 139, 231, 442
public control 206, 237, 239–40
public conversations 187, 372, 454
public culture 40, 371
public defamations 166, 169
public dimensions 79, 367
public discourse 82, 168, 309, 339, 371, 501
public domain 36, 86, 262, 311
public encounter 368
public expenditure 127
public facilities 97, 108, 115, 117–19, 304
public functions 38, 42
public funding 273, 298, 449
Public Garden (Boston) 488
public gatherings 38, 78, 119, 366, 418
public goods 66, 126, 310, 427, 435, 438–39, 442, 504; basic needs 438–39, 442; market failures 438–39, 500
public health 94–99, 103, 221, 282, 304; place-oriented perspective 94, 98, 102–3; tobacco drugs and smoking 96–97, 384

public health opportunities 102
public housing 74, 80
public infrastructure 10, 59, 205
public institutions 2, 9, 88, 169
public interest 37, 125, 312, 315, 391, 444, 479
public investments 10, 414
public issues 166, 364
public lands 240, 444, 450–51, 454
public life 1–2, 138, 166–68, 236–37, 240,
    309–10, 366, 368, 427–28, 478, 482, 484,
    499–500, 506
publicly accessible spaces 237
public management approaches 130
publicness 5–6, 35–44, 62, 81–82, 169–70, 205–6,
    234–37, 239–40, 247, 249–50, 254, 310, 312,
    367–68, 513; degree of 62; design of 6, 35, 37,
    39, 41–44, 237; levels of 36–38, 42, 263;
    perceived 39; producing 35–36, 43;
    transformation of 39–40, 43
publicness of public space 6, 79, 206, 234–35, 237,
    239, 247, 427
publicness of space 35–37, 42–43, 62, 249
publicness of urban spaces 35, 237
public officials 325, 402, 408, 447, 454
public open spaces 11, 32, 72, 106–7, 109–10,
    113–15, 119, 297, 367, 420
public order 172, 187–88
public ownership 42, 133, 210, 444
public parks 12, 48, 110, 121, 152, 428–29, 443,
    447, 450, 459
public participation 166
public physical activity, low-impact exercise
    equipment 290
public places, hate crimes 28
public plazas 327, 346
public policy 106, 393, 450
public–private binaries 235, 237, 247, 315,
    367, 375
public–private Partnerships *see* PPPs
public–private–People Partnerships (PPPP) 91
public–private relations 13, 247
public project 206, 250
public property 421, 440–41
public purposes 197, 229, 439
public purse 125, 128
public realms 36, 91, 142, 457
publics 36–38, 40–43, 76–77, 80–81, 106, 191,
    193, 196–97, 273, 310–13, 315–17, 323, 346,
    350–53, 366–75; multiple 184, 189, 338, 353,
    375–76
public safety 99, 189
public sector 2, 133, 182–83, 251, 264, 278,
    450–51, 454, 500
public services 8–9, 107, 131
public space: amenities; chess 66–67, 382;
    playgrounds 64–66, 163, 205, 210–11, 271,
    276, 367–68, 417; pool 163; pool table 368;

appropriation; transformative power 39, 41; to
    transform publicness 41–43; bio-diversity of
    416; black bodies in 74, 80; changes in 3;
    contemporary challenges with 428; democratic
    approach to 439; democratic structure of 74;
    distinct 50, 269; dynamic 363; encounters
    and negotiations in 339; exclusive categories
    of 213–14; fencing 64–66, 453; financial
    exclusion of 174; fine-grained categorization
    of 209; interactions in 415; materiality of
    73, 76, 78; mental health 86, 283, 297,
    304, 306; Modernist 121; modification
    235, 363, 404; naturalizing; planters
    255–56, 262, 404, 420, 462; plants 152,
    158–60, 260, 293, 416; overlapping
    categories 213, 240; physical exclusion 174;
    pilot projects 87, 404, 406; psychological
    exclusion 174; under-managed, uncollected
    garbage 174; undesirables 60, 174, 501; user
    benefit 118, 211; Western and Southeast
    Asian contexts 166, 168, 172
public space and democracy 337
public space and justice 63
public space and public goods 438
public space and recognition 64
public space animation, formal tactics 414
public space behavior 467, 469, 472, 474
public space challenges in Latin America 391, 393,
    395, 397
public space changes 59, 205
public space classifications 221, 223
public space culture 90–91
Public Space Database Project (Stockholm) 458
public space design 46, 57, 62, 88–89, 107, 235,
    414, 487
public space dynamics 337, 391
public space for defiant denunciation and
    resistance 392
public space functions 333, 335, 337
public space geographies 74
public space improvements 12, 264
Public Space Index (PSI) 62–63
public space investment 123, 133
public space issues 2, 94, 126, 166, 174, 223
public space literature 2, 336
public space narration 334, 358
public space neglect 174–75
public space networks 242, 247, 302, 458, 466
public space opportunities 419
public space policing 60, 67, 97–99, 103, 171,
    182, 256, 262, 393, 395, 480–81
public space production 3, 178, 249, 264–65,
    391, 500
public space projects 123, 178, 298, 381
public space provision 12, 88, 264
public space research 16, 24, 64, 205, 209–11,
    347, 354, 428, 467–69, 471–74

Public Space Research Group (PSRG) 59, 63
public spaces 1–3, 5–13, 16–25, 35–44, 53–67,
71–77, 79–82, 85–92, 97–103, 121–34, 137–41,
165–79, 182–89, 209–14, 306–18, 333–42,
390–96, 414–23, 427–30, 457–74; access to 13,
40–41, 73–74, 80, 237, 304, 436, 442, 500;
adjoining 461; analysis of 21, 364; animate 3,
334, 414, 416, 422; appropriation of 5, 37, 39,
41–43, 140, 403, 419; boundaries of 1, 427;
central 80, 158, 361; character of 9, 11, 316,
422; classifications of 205–6, 218; collective
approach to 72, 121; complexity of 106, 461;
composers of 490, 493, 496; construction of
401, 406; context of 36, 250, 415–16, 483, 503;
creating 88, 212, 221, 303, 367, 385, 404, 478,
487; culture in 303; defining 85, 350;
democratic 74, 133, 308; democratization of
235; diverse 22, 24; drawing people to 415–18;
evaluation of 61, 64, 206, 250; experience of 6,
22, 236, 467; exploration of 2, 353; focus on
123, 301, 354; function of 36, 339; global 78,
87; hybrid 78, 301; ideal 10, 39, 60, 87, 134,
170, 317, 427; importance of 72, 76–77, 87–88,
306, 335, 337, 457; inaccessible 239, 242–43;
inclusive features of 137, 149; inclusiveness of
138, 142; inquiry of 2, 5; insertion of 313,
316–17; interpretations of 2–3; investment in
10, 85, 88; issues of safety in 94, 98; key feature
of 13, 139; legacy of work on 123; local 12, 19;
maintenance of 94, 174, 249; management 2, 8,
60, 72, 87, 121–22, 126, 130–32, 178, 205,
223; meanings of 1, 3, 6, 71, 395, 427; multiple
dimensions of 333, 335; nature of 72–73,
165–66, 430; occupation of 141; open 110,
173, 221, 231, 243; outdoor 99–100, 186, 210,
224; performance of 5–6; physical 5, 165–66,
310, 479–80, 484, 486; power of narration of
358, 360; privately-owned 42, 218, 309, 315;
privatization of 9, 126, 234–36, 240, 247, 394;
processual understanding of 17, 21, 24–25;
producing 44, 428; provision and maintenance
of 12, 49; quality of 62, 90, 100, 102, 122, 228,
296, 483; relational understanding of 17, 20;
roles and significance of 1, 3, 16, 71, 205,
333–34, 337, 347, 366, 427; safety 95–96,
98–99, 101, 103; security in 187; shared 304,
308; small-scale 317, 411; socio-political
dimensions of 391, 396; temporary 334, 400;
traditional 121, 210, 250, 263; understanding of
5, 24, 205; value of 5, 366; vibrant 51, 72, 191,
366; virtual 5, 19, 77, 80, 479–81, 483
public spaces and public life 375, 484
public spaces of Chinese cities 478, 482
public space typologies 49, 206, 234, 242, 409,
470, 490
public space use 3, 29, 60, 71–72, 140, 182,
188–89, 221–31, 464, 469, 472; classifying

222–23; list of examples 225–27; uncivil
acts 227
public sphere 1–2, 5, 9–10, 13, 82, 86, 165–69,
359, 364, 366, 433, 499–500; virtual 12, 166
public squares 78, 211, 360, 382, 417
public stories 320, 326
public territories 39, 41–42
public transportation 90, 125, 127, 225, 234, 240,
372–73; buses 373, 492; minibus taxis 373; rail,
Tokyu Line (Tokyo) 195; taxi 225; trains;
routes 373; station 195–96, 234; stations 17, 99,
234, 327, 351
public trust 439–41, 444; trustee 441, 444; trustor
441, 444
public trust doctrine 427, 438–41, 443–45
public visibility 31, 393
Public Work (architectural firm) 301–2
public zones 459, 461
PubMed 95–96
Puerta del Sol (Madrid) 339, 342
pumflet 369, 371–73, 375–76; "art,
architecture, and stuff" 371; "Daar gaan
die Alabama" 371; "Gaiety" 372; "Gladiolus"
372; "Rondhuis" 372
Putrajaya, Malaysia 179
Pwani Ndogo (Zanzibar) 157, 163

qualitative assessment 256, 260
quantitative assessment 255, 257–58
quasi-public space 236, 240, 242–43, 247
Quayside project (Toronto) 473
Queen Elizabeth Olympic Park (London)
128, 131
Queen's Park Savannah (Port of Spain) 152,
154–55, 157–60, 163
Querétaro, Mexico 395

Rabin, Yitzhak 360
Rabin Square (Tel Aviv) 360
radical democracy 337, 351–52
Ramadhan 157, 161
Rambla del Raval (Barcelona) 382, 384
Randall's Island (New York City) 449–50, 452
Rangoon, Myanmar 171
Ratchadamnoen Klang Avenue (Bangkok) 171
Raumlabor 406, 408
real estate 140, 310–11, 438
Rebar 212, 491
recognition of urban identities 28, 59, 61–62, 64,
66–67, 138, 140, 325, 352–53, 420, 480
recreation activities: active 287, 290; individual
224–25, 228, 232
Reddit (website) 471
Red Square (Moscow) 74
Reference Center for Youth (Belo Horizonte) 391
refugee, SLAM (London) 31

refugee camps, Middle East 86
refugees: asylum seekers 29–32, 80, 88, 351;
    Plymouth (UK) 31; Syrian 32; tensions with
    host communities 88, 379
Regents Street (London) 126
regimes, colonial 155, 157–58, 160, 163
religion 28, 166–68, 172, 179, 358, 392
remembrances 324, 326–27
Renaissance Europe 46
renaming places 363
representations 28, 73–74, 76–77, 80, 320, 334,
    337, 341, 352, 358–59, 361
research: empirical 138, 140–41, 282; public
    space behavior 467; ethnographic 6, 32, 59–61;
    urban 347
residential clustering 96, 141, 333, 380, 383–84
residential segregation 179
residents 19, 62, 64–66, 100, 102–3, 138, 140–41,
    251, 260, 262, 379–81, 386, 410–11, 460,
    473–74; local 174, 209, 260, 264, 273, 277,
    338; long-time 381, 385, 387
resilience 333–35, 337, 487–88, 490
resistance 76, 160, 322, 327, 334–38, 341, 346–47,
    353, 376, 392, 394–95; neighborhood 385
restaurants 11, 79, 102, 137, 142–43, 146, 148–49,
    185, 239–40, 314–15, 462
retail streets, traditional 129, 211
revenue 173, 440, 448, 450–54
rhetoric of space 5–7, 9, 11–13, 503
"the right to the city" 61, 92, 182, 184, 189, 192,
    234–35, 237, 240, 247–48, 316, 318, 435–36,
    500, 503
Rio de Janeiro, Brazil 47, 396, 505
riot 13, 129, 182, 186–87, 251
roads 126, 157–58, 163, 196, 199, 226, 307, 373,
    421, 439, 442
road traffic 196, 373, 464
Rockefeller Center (New York City) 309, 312
Rocky Park community garden (London) 131
Rogers, Betsy Barlow 301
Rogers, Richard 125
Roger Stirk Harbour + Partners 312–13
Roma (ethnicity) 29, 327
Roman concept: public 441; Stoic moral
    principle 441
Roman early modern legal literature 440, 442
Roman Empire 46, 55, 440, 444
Roman forum 170
Roman jurists 440
Rome, Italy 86, 237–38
Roosevelt, Franklin D. 321–22
Ropewalks 251, 253, 258–62
Rotterdam, Germany 62
rough sleepers (unhoused individuals) 129–30
Royal Botanic Garden (Port of Spain) 47, 152,
    154–55, 157–58, 160, 163

rules 42, 60, 62, 74, 168, 224, 313–14, 393, 464,
    478, 483; and regulations 108–9, 166, 170, 172,
    174, 183–84, 310, 313–14, 349, 391, 395; tacit
    457, 459, 464
rural-to-urban transect 48
Russian Hill District (San Francisco) 47
Ryan, Marc 301
Ryokudō, Japan 199

Sabolai Radio 370
Safdie, Moshe 458
safe cities 100
safe/secure environment 285
Saint Barnabas Senior Services (Los Angeles) 284
sakariba ("flourishing place") 201
Samba Nights (Belo Horizonte) 392
Sana'a (Yemen) 340
San Diego, California 47
San Francisco, California 47, 50, 62, 210, 212–13,
    399, 404, 409, 411, 491
San Francisco, Planning Department 212
San José, Costa Rica 59–60
San Juan (Port of Spain) 158
Santa Tereza Overpass (Belo Horizonte) 391–92
Sant Gervasi (Barcelona) 386
Santiago (Chile) 323
San Victorino, Bogota 396
São Paulo, Brazil 47, 396
SARS outbreak 481
Saudi Arabia 472, 503
scenario planning 403, 487, 490–92, 496
Scopus (database) 95–96
Sea Lots (Port of Spain) 158
Seattle, Washington 50, 336, 471, 491
security 32, 66, 86, 90, 124, 133, 187, 284–85,
    287, 430, 432
security cameras 8, 174, 187, 262, 393, 468, 482,
    501, 508
semi-public spaces 185, 234, 309, 312
Sennett, Richard 16, 18, 48, 55, 106, 139–40,
    192, 249, 264, 364, 390
sensescapes 22
sensory stimulation 28, 287, 293, 417; amenity
    placement 291, 293
Sert, Josep Lluís 458
Setagaya, Japan 199
Shanghai, China 362, 482
Sheffield (London) 27–31
shelter from elements 50, 172, 313, 431
shimin hiroba 196
Shogun Tokugawa Iemitsu 194
shop owners 142, 175, 374
shopping malls 17, 20, 115, 174, 177, 210–11,
    236–37, 239–40, 489, 499–501
shopping street 145–46
shops 137, 142, 145–46, 179, 182, 185–86, 234,
    239–40, 242, 247, 383–84; immigrant 145–47

short message service (SMS) (texting) 478–79
Sidewalk Labs *see* Google Alphabet
sidewalks 16, 19, 59–60, 87, 89, 210–11, 262, 264,
    288, 368, 417, 419, 421, 491, 513; widened 146
Siena, Italy 52
Sierra Leone 28
sight lines 255, 258, 260; long 257
sightseeing tours 225
Sikh gurdwara 381
Silent Revolution of Public Spaces in Afghanistan
    (UN-Habitat report) 88
Silicon Valley (California) 435
Simone de Beauvoir Bridge (Paris) 306
Singapore 72, 77–79, 165, 169, 178–79, 182,
    184–88; citizens 186–87; government 78–79,
    184; society 186; Urban Redevelopment
    Authority 184
Singapore Civil Defence Force 186
Singapore Police Force 186
Sitte, Camillo 11, 491
situationist 235, 399
"sketch" scenarios 490
"sketch" scenarios: Centre for Strategic Futures
    (Singapore) 490; The Future of Urban Water
    490; The Scottish Cities Alliance 490
skhōle (ancient Greek) 432
skyscraper, architectural type 310, 312
skyscrapers 124, 207, 309–18; American
    projection of society 310; bigness (scale)
    196, 317–18; contemporary 207, 309;
    context of 311, 317; sky-courts 317; Sky
    Garden (Walk Talkie tower) 124, 313–15;
    vertical urbanism 317
Slubice, Poland 21
"Smart City" 74, 473, 490
Snohetta (architecture firm) 404
sociability 6, 11, 139–41, 206, 250, 415, 420, 447,
    449, 500
social access 139–40
social construction of public space 2
social contacts 142, 250, 500, 506, 508
social diversification 12–13
social encounters 40, 42, 141, 149, 366
Social Farms and Gardens (UK) 32
social fragmentation 7, 11, 30, 166–67
social groups 12, 61, 86, 137–39, 186, 188, 358,
    363, 483
social hierarchies 167–68, 176, 179, 361
social imaginations 369, 371
social inequality 12, 347, 349, 352
social interaction 18, 59, 66, 87, 89, 273, 283–85,
    289, 291, 415, 418–20
socialization 12–13, 141, 175, 368
social justice 5–6, 59, 61–67, 168, 353
social media 5, 19, 73, 77, 168, 275, 299, 338,
    435, 468, 470–74; Facebook 77, 471; fostering
    protest movement 19

social movements 13, 80, 325, 327, 335–36,
    338–40, 342, 350, 362, 391, 393
social narration 358–60, 363
social orders 346, 348, 350, 364
social power 348–49, 358
social relations 36, 43, 66, 75, 186, 348, 420, 433
social space 76, 400, 420
social support 285, 289–90
social theory 443, 500
society 1–2, 8, 11, 13, 166–67, 177–79, 321, 323,
    337, 339, 341–42, 347–48, 364, 367, 433–34;
    contemporary 71, 416; modern 443
socio-political dimensions 390–91, 396
Socrates 192
solidarism 427, 442–44
solidarism and public space 443
solidarity 29, 61, 86, 339, 395; differentiated 380
Sorkin, Michael 16, 60, 210, 249, 264, 336,
    448, 483
South Africa 322, 329, 334, 372–73
South Africans (black) 322, 329
South African squatter settlements 373
Southeast Asia 42, 72, 165, 167–74, 176–79
South Etobicoke (Toronto) 306
South John St. (London) 258–60
space of appearance 338
space of visibility and meanings 338–39
spaces: accessing central 80, 202; activity 197, 472;
    appropriation of 1, 40, 42; architectural 201,
    311; colonial 152, 154–55, 341; communal 79,
    170; consumer-oriented 177, 211; controlled
    155, 239–40; democratic 5, 129; embodied 352;
    enclosed 54, 312, 314; hybrid 79, 249, 263;
    inclusive 321, 323, 325, 327, 329; indoor 177,
    210; interstitial 98, 368, 375, 449; invitation
    239–40, 242–43, 247; leftover 85, 236, 265;
    lived 38, 346, 349–55; local 350, 459; making
    5, 38; marginal 213, 381; outdoor 206, 210,
    221, 223, 283, 403; political 80, 358; privatized
    1, 242, 250, 427, 508; residential 10, 176; safe
    27, 31, 66, 419; shared 248; under-managed
    173–75; vehicle 247; virtual 3, 74–77, 165, 483,
    514; visible 187, 335
spaces of consumption 13, 183, 185, 236;
    carnivalesque 183, 185–86; privatized 1, 37, 82,
    86, 174, 183, 242, 250, 254, 263, 265
spaces of resistance 335, 337, 339; Azadi Square
    (Tehran) 333; Gezi Park (Istanbul) 76, 333, 338;
    Hong Kong International Airport 333
spaces of social assembly 1, 73, 79, 333, 335, 338,
    342, 361, 427
Spadina Avenue (Toronto) 299, 303
Spain 152, 154–55, 158–60, 353, 381
spatial analysis 87, 351, 428, 468, 474
spatial dimensions 352, 355
spatiality 38, 41–42
spatial production 6

spatial sanitizing: Belo Horizonte, Overpass Upgrade Project 391; Belo Horizonte, physical renovations 391
spatial segregation 12, 86, 179, 394, 460, 465, 504
Speakers' Corner *see* Hong Lim Park (Singapore)
spectacle 10, 124–25, 182–86, 188, 236
specticalizing: buskers 230, 234, 254, 262, 418; graffiti 41, 174, 256, 258, 262, 276, 368–69, 393, 419; street artist 262
spheres: domestic 13, 74; public and private 8–9, 13, 73, 108, 168, 179, 391, 480
sport and activity 273, 276, 408, 416–17, 451
squares 21, 29, 32, 51, 74, 76–77, 79, 210, 214, 336, 340; central 76, 353; courthouse 492, 495–96; garden 125, 213–14; station 17, 20, 196
Standing Man see Erdem Gunduz
St. Ann (Port of Spain) 152, 160
state (government) 8–9, 74–75, 79, 161–62, 166, 337, 341, 348, 390–91, 431–32, 438–41, 464, 480–81, 499–500, 507–8
State Senator Squadron (New York City) 454
Statue of Liberty 362
statues *see* monuments
Stellenbosch, South Africa 372
step-wise adaptive process 464
stewardship models 301, 304
St. Louis, Missouri 5, 325
Stockholm, Sweden 101, 458
Stone Town (Zanzibar) 157
St. Paul's Cathedral (London) 76, 79, 313
St. Quentin, France 267
Strachan Avenue (Toronto) 299
Stratford and Nine Elms (London) 125, 128
street amenities 146, 148
street conversion 404
street corners 74, 98, 102, 195, 326
street culture 491
street dwellers 391, 395
street furniture 115, 255–58, 262, 404, 420
street grid 47, 242, 306, 310–11
streetlamps 183, 187
street level 56, 140, 142, 144, 257
street life 148, 491; in Seattle 491
Street Life Zone (San Francisco) 410
street markets 18, 139, 415
street networks 90, 140, 237, 294
street noise 291
street performances 225, 228, 256, 418
streets 89, 101, 141–43, 146–49, 172–76, 182–88, 195–97, 209–11, 225–26, 255–56, 258, 260, 287–88, 368–70, 381–83, 394–95, 417–19, 461–62, 489–91, 504; commercial 143, 149; high 122, 462–63; invented 500; pedestrianized 383; selected 141–42
Streets as Public Spaces and Drivers of Urban Prosperity (UN-Habitat Report) 89

streetscape 148, 419, 459, 463; weekend closure 404
streetscapes, claiming 404
street segments 99–102
street vending 60, 171, 174, 228, 368, 373–74, 394–95, 491; blankets 395, 462; informal selling 225; *toreo* 395
street vendors 187, 391, 394–95; informal 373, 396; workspace 395
street vitality 182
Strolling Goats 370
suburbs 11–13, 76, 155, 211, 372, 378–79, 381, 430, 448; Holmwood estates (Zanzibar) 155; middle-class 372–73
Success Laventille Networking Committee (Zanzibar) 159
Suleiman Maisara (Zanzibar) 161
Sunday Social Yoga 305
Sunday Streets 404
Sunflower Movement (Taipei) 340
supra-authorities 171
surveillance 60, 62, 73–74, 182, 184, 186–89, 250, 481–83, 500, 503, 507, 513; internet activities 480; panoptic state 1
surveillance cameras 60, 287, 468, 482, 507
surveillance of the commons 503, 507
sustainability 90, 96, 98, 122, 316, 408
Sustainable Development Goals (SDGs) (United Nations) 87–88, 91, 310
Swahili speaking indigenous communities (Zanzibar) 157, 159–60, 163–64
Sweden 100–101, 378
Sydney Harbor 47
Syntagma Square (Athens) 311, 338–40

Tahrir Square (Cairo) 76, 338, 340, 499
Taipei, Taiwan 5, 72, 107, 109–10, 113, 115, 119, 340, 346
Taisho Era (Japan) 195
Taisho Era, cultural imports 195, 199
Taiwan 179, 346
Takamasa 196
Taksim Square (Istanbul) 339, 499
Talk Party Series 370
Tampa, Florida 63
taxonomy 75, 223–24
technology 427, 433, 458, 478, 483; facial recognition 481; mixed-reality 89; smart 409
Tehran, Iran 333
Tel Aviv, Israel 338, 360
Telegraph Hill District (San Francisco) 47
Tempelhofer Feld (Berlin) 406, 408
temporality 21, 32, 41, 358, 400–401, 403, 436; experimental 400, 408
temporary art installations 419
temporary city 212, 334, 399–400; contemporary 399

temporary spaces 399, 459

temporary strategies 399, 402, 404, 406, 408–9, 411, 414

temporary urbanism 126, 211, 336, 400–401, 493; short-term action 401

temporary use 211, 276

tensions 65, 72, 88, 92, 337, 340, 349–50, 392–93, 396, 483, 491

territories 2, 13, 37, 40–42, 74–75, 79–81, 196, 392–95, 458–59

terrorism 179, 227, 329, 427, 429, 431, 434, 436; Holocaust atrocities 324, 327; IRA attacks 133; London (2017) 132; radical Islamist 133

TESS (Toolkit for the Ethnographic Study of Space) 65

Thailand 165, 168–69, 178–79

Thames River (London) 124, 133

Thatcherite politics 250

theatre space and outdoor classrooms 160

theory: political 9, 348, 443, 500; urban design 74–75, 82, 149, 173, 427, 442; urban planning 74–75, 82, 149, 173, 427, 442

Thiel, Peter 435

third places 487

third spaces 174, 283–84, 481, 496; bars 102, 210, 262, 488; English pubs 487; internet cafes 480–81, 483; mosques 79, 142–43, 148, 381; personal digital devices 481; pubs 146; teahouses 142, 146; thresholds 42, 54–55, 367, 434–36; Turkish hookah cafés 487

Third-Way approach (Tony Blair) 250

Thomas, Gladys 372

"throwntogether" places 267, 280

Tiananmen Square (Beijing) 74, 338, 360, 362, 479

Tiananmen Square protests 342

ticketed space 124, 225, 240, 242–43, 247, 453, 513

Times Square (New York City) 404, 406, 508

Times Square: new pedestrian spaces 510; traffic-jammed 406

Tjuvholmen neighborhood (Oslo) 224, 227–28

Tokugawa periods (feudal Japan) 194, 196, 201

Tokyo, Japan 72, 191, 193–97, 199–202

Tokyo District Court 193

Tolbiac neighborhood (Paris) 306

Tompkins Square Park (New York City) 59, 63–65, 67

Toolkit for the Ethnographic Study of Space (TESS) 65

topographic gradients 50

topographic horizon 54

topographic motifs for public space 48, 50, 53

topography 5–6, 46–48, 50–52, 54, 58, 294, 308, 364; designed 50; Mesopotamia 46; micro; level changes 56; textured surface 50, 55–56; uneven surface 101

topography of place 50

topography of public space 46

topography and urban theory 47

Toronto, Canada 206, 296, 299, 301–4, 306, 404, 430, 473

Toronto, Planning and Development Department 301

Torrons Licors (Barcelona) 383

Tory, John 299

touch (sense) 32, 50, 54–55, 433

Tower of London 126

towers *see* skyscrapers

town center 11, 20–21; historic 394, 396

traffic signals 115, 229, 421, 442, 491, 496

Transport for London (TfL) 125

Trinidad 152, 154–55, 158–59, 163; Director of Agriculture 154–55

Trinidadian African community 159

Trinidadian culture 154, 160, 163

Trump, Donald 28, 438, 444

Truth and Reconciliation Committee (South Africa) 322

Tuileries (Paris) 493

Tunisia 77

Turkey 19, 347

Twitter 77, 468, 471

typing (as an analytic tool) 205, 209, 213–14, 218

typologies 48, 54, 205, 223–24, 234, 236–37, 240, 247, 400–401, 406, 484, 488

typology, Kevin Lynch 235

Ujamaa Newsletter ("Family-ness") (Port of Spain) 159

Umbrella Movement 335, 484

UNDP *see* United Nations Development Programme

Unguja (Zanzibar) 157

United Arab Emirates 503

United Cities and Local Governments (UCLG) 91

United Kingdom 251, 508

United Nations 85, 89–92, 108

United Nations Development Programme (UNDP) 88

United Nations Human Settlements Program (UN-Habitat) 85, 87–90, 458

United States 74

United States Capitol (Washington DC) 321

Universal Declaration of Human Rights (United Nations) 90

Universitat de Barcelona 384

the urban commons 89, 248, 428, 503–4, 506–9

urban design 2–3, 6, 16, 36–38, 41–44, 46, 57, 62, 221, 229, 428; assessment 249, 253, 258; knowledge 35, 43; models 59, 62; parks 290, 293; practice 35–39, 267; principles 247; process 35, 37; projects 35–37, 179, 222; quality

250, 252, 258; tool of neoliberalism 37; traditional 48; transform publicness 41–43
urban design aspects 253, 258–59; coherence 253; complexity 253, 255, 258, 260; enclosure 253, 255, 258, 260; human scale 253, 255, 257–58, 260; imageability 253; legibility 253–54, 256; linkage 253–54, 256, 259, 265; tidiness 253–54, 256, 258–59, 262; transparency 253, 255, 258, 260
urban designers 7, 37, 39–40, 42–43, 206, 264–65, 486, 492, 500–501, 508, 511
urban development 6–7, 11–13, 44, 65, 97, 182, 186, 346, 354, 399, 406; processes 7, 36; projects 8
urban emancipation 334, 346–47, 349, 352
urban environments 10, 13, 27, 50, 79–80, 86, 89, 98–99, 152, 416–18
urban fabric 46, 56, 302–3, 306, 314, 350, 367, 381–82, 465
urban farming 226
Urban Foresight 488
urban form 37, 46, 82, 90, 195–96, 282, 286, 304, 358, 464, 508
urban gardening: Qigong 408; Shiatsu 408
urban greenspace: flowers 28, 131, 290–91, 293, 416, 421; gardens 32, 46, 48, 65, 72, 126, 131, 152, 158, 161, 163; vegetation 287, 291, 293, 302
urbanism 7, 29, 81, 155, 165, 193, 221, 303, 317–18, 460, 503–4
urbanity 176
urbanization 86
urbanization processes 92, 347, 354
urban land 100, 443
urban life 7, 12, 183, 189, 192, 230, 235, 240, 310, 317, 414, 419
urban morphology 50, 237, 318, 401
urban parks 29, 32–33, 48–49, 53–54, 61, 63–64, 98, 107, 448, 453; central 72, 152; commercialization 438, 442, 448, 450; eventization of 453–54; evolution 209, 213; Pleasure Ground (1850-1900) 213; Reform Park (1900-1930) 213; natural spaces 292, 427; networked 27
urban pioneering projects 269
urban pioneers 399, 408
urban planners 24, 103, 149, 367, 503
urban planning 2, 86, 89, 99, 149, 211–13, 229, 400, 458, 508; design 89, 229, 400; design challenges 508; core issues 231
urban policy 97, 448–49, 458
urban populations 11–12, 75, 86, 173, 206
Urban Prototyping Events 409
Urban Prototyping Festival 411
urban public cultures 229, 366
urban public spaces 1, 29, 71, 152, 221, 296; physicality 46, 102, 137, 483; potentials 7, 18,

75, 81, 102, 137, 152, 221, 309; social 18, 29, 74–75, 77, 80, 478
urban redevelopment 80, 224, 252–54, 269, 278, 394
urban rejuvenation schemes 183
urban resistance 334–35, 346–49
urban shorelines 439–42
urban social life 139
urban social space 484
urban societies 152
urban space: regulatory instruments 60; tactics 41, 237, 334, 341, 353, 391–93, 395, 415, 421, 436
urban space "composers" 486, 488, 495
urban space pioneer fields 408
urban spaces 13
urban theory 47, 73, 82, 192, 348
urban tourism 10, 138, 161, 183
urban water 490
users of public space 106, 118, 149, 211, 224, 236, 254, 417–18, 420–21
Utah (United States) 438
Utrecht, The Netherlands 17, 20, 22

vacant urban land 94, 98, 100, 186, 211, 256, 262, 269, 274, 400
Vatican 74
Venice, Italy 463
Venice Architecture Biennale (exhibit) 372
Verne, Jules 493
the Vessel *see* Hudson Yards
Victoria Garden (Zanzibar) 161
Victoria Harbour (Hong Kong) 242, 247
Victoria Street Market (Durban) 373
Vienna, Austria 351, 353–54, 399, 401, 411
Vienna New Main Station 351
Vientiane, Laos 173–74, 177
Vietnam 165, 168–69, 179
villages 76, 170, 184, 318, 479–80
Villa Grimaldi (Santiago) 323
Viñoly, Rafael 313
violence 86, 99–100, 103, 155, 179, 227, 361, 392, 427, 436; western ghettos 176
VIPs *see* Visually Impaired Persons
Visbal, Kristen 496
visibility (inclusivity) 28, 33, 38, 41–42, 138–42, 149, 333, 335, 338–39, 342
visibility: distinctive urban groups 72, 138, 149; mass gatherings 335, 338–39
vision (futuring) 90, 158, 191, 317, 323, 326, 352, 433, 486, 493
Visually Impaired Persons (VIPs) 106–10, 113, 115–19; challenges 108; inclusion and exclusion from public space 107, 109, 113, 115, 117, 119; mobility obstacles 110, 113, 115
vitality 141, 143, 145–46, 148, 185, 235–37, 415–16
voices 166

Volkspark (Potsdam) 51
volunteers 67, 115–16, 131, 351, 449–51
von Erlach, Fisher 401
VOSviewer 94–95
Vuga (Zanzibar) 161

Wa Lehulere, Kemang 371, 373
Walkie-Talkie Tower (London) *see* Fenchurch
  Street Tower
walking paths 57, 113, 287–88, 290, 293, 322, 373
walkways, elevated 258–59
walls (structure) 8, 41, 74, 194, 255, 260, 269,
  302, 392–93, 418–19, 460–62
Wall Street 19
Wall Street Bull (New York City) 496
waqf *see* Islamic property inheritance
Warwick Triangle (Durban) 372–73, 375
Washington, DC 322, 359
Waterfront (Toronto) 473
Waterfront City (Melbourne Docklands) 242, 247
wayfinding: braille 286; cues 55; devices 50, 286
Weber, Max 347
weekend enclaves 187
Weibo 479–80
Wells, H.G. 493
West African practices 159–60
western-based contexts and traditions 92, 166,
  168–69
western-based frameworks 71, 91, 165–74, 176,
  178, 195, 201, 231; Anglophone and Northern-
  centric perspectives 87, 369; Australia 165, 167,
  206; Europe 48, 82, 85–86, 103, 167; events
  274; preconceptions 31, 271, 391; public spaces
  157, 491; urban contexts 29; North America
  82, 96, 103, 165, 167; public space 85, 133,
  169–70, 174, 337, 431
Western Beaches (Toronto) 306
Western Cape (South Africa) 369
Westminster (London) 81
Westworld (TV series) 495
WhatsApp 19
WHO *see* World Health Organization
Whyte, William H. 16, 24, 28, 39, 209–11,
  222–23, 420, 467, 491, 501
Williams, Robert Orchard 154
windows 142, 145, 255–56, 262, 461–62, 494;
  active 260; boarded 251, 262

Windrush Square (London) 123
Wittgenstein, Ludwig 431
WOHA Architects 316
Wolff, Ilze 371, 373
women 13, 64, 67, 74, 80, 82, 86, 88, 158, 160,
  166–67, 177, 179, 211, 500; neighborhoods 158
workers 42, 186–88, 338, 432, 434, 479; creative
  276; domestic 42; foreign 182; skilled 145;
  underpaid 12
workplaces 342, 386, 465
World Bank 90
World Cup 274–76, 278–79, 375
World Health Organization (WHO) 89, 108,
  282–83, 285
World War I 322
World War II 8, 169, 195–96, 321–22
WUF 9 *see* 9th World Urban Forum

Xi, Jinping 480
Xicheng (Beijing) 109

Yellowstone National Park 488
Yerba Buena Center for the Arts
  (San Francisco) 409
"Yes to Peace, No to Violence" 360
yoga 273, 287–88, 290
Yoruba *esusu* 159
Yoruba village (Trinidad) 159
Young Artists' Project (Cape Town) 375
youth 86, 89, 280, 290, 302, 371, 376, 391

Zanzibar 152, 154–55, 157–58, 160–63;
  Agriculture Department 157; Parks and Gardens
  Department 155; Town Council 157
Zanzibari revolutionary regime 163
Zanzibari society 72
Zhongzheng (China) 109
Ziwani Police Barracks (Tanzania) 155
zones: functional 235; liquor control 187;
  no-go 373
zoning 459; codes 400; ordinance 212; plans 229;
  systems 489
Zuccotti Park (New York City) 19, 61, 76,
  333, 336
Zwischennutzung 276